Deutsches Institut für Ernährungsforschung
Potsdam-Rehbrücke
Arthur-Scheunert-Allee 114-116
14558 Nuthetal
Dr. Reinhart Kluge

The COST Manual of Laboratory Animal Care and Use
Refinement, Reduction, and Research

The COST Manual of Laboratory Animal Care and Use

Refinement, Reduction, and Research

Edited by
Bryan Howard
Timo Nevalainen
Gemma Perretta

COST is supported by the EU RTD Framework programme.
ESF provides the COST Office through an EC contract.

CRC Press
Taylor & Francis Group
Boca Raton London New York

CRC Press is an imprint of the
Taylor & Francis Group, an **informa** business

EUROPEAN COOPERATION
IN SCIENCE AND TECHNOLOGY

Neither the COST Office nor any person acting on its behalf is responsible for the use which might be made of the information contained in this publication. The COST Office is not responsible for the external websites referred to in this publication.

CRC Press
Taylor & Francis Group
6000 Broken Sound Parkway NW, Suite 300
Boca Raton, FL 33487-2742

© 2011 by Taylor and Francis Group, LLC
CRC Press is an imprint of Taylor & Francis Group, an Informa business

No claim to original U.S. Government works

Printed in the United States of America on acid-free paper
10 9 8 7 6 5 4 3 2 1

International Standard Book Number: 978-1-4398-2492-4 (Hardback)

This book contains information obtained from authentic and highly regarded sources. Reasonable efforts have been made to publish reliable data and information, but the author and publisher cannot assume responsibility for the validity of all materials or the consequences of their use. The authors and publishers have attempted to trace the copyright holders of all material reproduced in this publication and apologize to copyright holders if permission to publish in this form has not been obtained. If any copyright material has not been acknowledged please write and let us know so we may rectify in any future reprint.

Except as permitted under U.S. Copyright Law, no part of this book may be reprinted, reproduced, transmitted, or utilized in any form by any electronic, mechanical, or other means, now known or hereafter invented, including photocopying, microfilming, and recording, or in any information storage or retrieval system, without written permission from the publishers.

For permission to photocopy or use material electronically from this work, please access www.copyright.com (http://www.copyright.com/) or contact the Copyright Clearance Center, Inc. (CCC), 222 Rosewood Drive, Danvers, MA 01923, 978-750-8400. CCC is a not-for-profit organization that provides licenses and registration for a variety of users. For organizations that have been granted a photocopy license by the CCC, a separate system of payment has been arranged.

Trademark Notice: Product or corporate names may be trademarks or registered trademarks, and are used only for identification and explanation without intent to infringe.

Visit the Taylor & Francis Web site at
http://www.taylorandfrancis.com

and the CRC Press Web site at
http://www.crcpress.com

Contents

COST and COST Action B-24 vii
Preface ix
Editors xi
Contributors xiii
Editorial Board of the Manual xvii

Chapter 1   Introduction 1

                     Bryan Howard, United Kingdom

Chapter 2   Design and Oversight of Laboratory Animal Facilities 7

                     Dag Sørensen, Norway; Heinz Brandstetter, Germany;
                     Nikolaos Kostomitsopoulos, Greece; and Richard Fosse, France

Chapter 3   Housing and Care of Laboratory Animals 29

                     Hanna-Marja Voipio, Finland; Ping-Ping Tsai, Germany; Heinz Brandstetter,
                     Germany; Marcel Gyger, Switzerland; Hansjoachim Hackbarth, Germany;
                     Axel Kornerup Hansen, Denmark; and Thomas Krohn, Denmark

Chapter 4   Animal Needs and Environmental Refinement 75

                     Vera Baumans, The Netherlands; Hanna Augustsson, Sweden;
                     and Gemma Perretta, Italy

Chapter 5   Ethical Evaluation of Scientific Procedures: Recommendations for Ethics
                     Committees 101

                     Rony Kalman, Israel; I. Anna S. Olsson, Portugal; Claudio Bernardi,
                     Italy; Frank van den Broek, The Netherlands; Aurora Brønstad, Norway;
                     Istvan Gyertyan, Hungary; Aavo Lang, Estonia; Katerina Marinou,
                     Greece; and Walter Zeller, Switzerland

Chapter 6   Reduction by Careful Design and Statistical Analysis 131

                     Michael Festing, United Kingdom

Chapter 7   Animal Models: Selecting and Preparing Animals for a Study 151

                     Patrick Hardy, France and Sarah Wolfensohn, United Kingdom

Chapter 8   Creation of Genetically Modified Animals 179

                     Belen Pintado, Spain and Marian van Roon, The Netherlands

Chapter 9 Management of Genetically Modified Rodents 205

   Jan-Bas Prins, The Netherlands

Chapter 10 Impact of Handling, Radiotelemetry, and Food Restriction 227

   *Timo Nevalainen, Finland: Marlies Leenaars, The Netherlands;
   Vladiana Crljen, Croatia; Lars Friis Mikkelsen, Denmark; Ismene Dontas,
   Greece; Bart Savenije, The Netherlands; Carlijn Hooijmans, The Netherlands;
   and Merel Ritskes-Hoitinga, The Netherlands*

Chapter 11 Basic Procedures: Dosing, Sampling and Immunisation 257

   *Ismene Dontas, Greece; Jann Hau, Denmark; Katerina Marinou, Greece;
   and Timo Nevalainen, Finland*

Chapter 12 Imaging Techniques 287

   Aurora Brønstad, Norway and Ismene Dontas, Greece

Chapter 13 Anaesthesia and Analgesia 313

   Patricia Hedenqvist, Sweden and Paul Flecknell, United Kingdom

Chapter 14 Use of Humane Endpoints to Minimise Suffering 333

   *Coenraad Hendriksen, The Netherlands; David Morton, United Kingdom;
   and Klaus Cussler, Germany*

Chapter 15 Euthanasia 355

   Luis Antunes, Portugal

Chapter 16 Education, Training, and Competence 369

   *Bryan Howard, United Kingdom; Katey Howard, Spain; and Peter Sandøe,
   Denmark*

Chapter 17 Animal Experimentation and Open Communication 391

   *Ann-Christine Eklöf, Sweden; Anne-Grethe Berg, Norway;
   and Jon Richmond, United Kingdom*

Index 405

COST and COST Action B-24

COST is an intergovernmental framework for European Cooperation in Science and Technology, which promotes and coordinates nationally funded research in Europe. COST provides funds for research networks (Actions) and in this way helps consolidate European research investment and opens the European research area to cooperation worldwide.

COST is solely funded from a specific part of the EU RTD Framework Programmes. The financial support that COST provides to Actions is used for coordination and networking activities, while the research itself is funded at the national/EU level. In this way, COST reaches out to over 30,000 researchers across Europe and levers approximately EUR 2 billion of research funding, although its direct support is less than 1% of this sum.

One of the key scientific domains that COST covers (out of a total of nine) is the Domain of Biomedicine and Molecular Biosciences (BMBS), which at the time of writing (June 2010) is funding a total of 29 Actions.

This manual summarises the output, over a period of four years, of BMBS COST Action B-24 on "Laboratory Animal Science and Welfare". The remit of Action B-24 was to increase knowledge and awareness of the scientific uses of laboratory animals within the context of the Three Rs, and to promote the conduct of high-quality research. The Action began its activities in March 2004 and came to an end in April 2009. However, many participants have links with national associations and with the Federation of European Laboratory Animal Science Associations (FELASA) with which the Action has collaborated closely throughout. It is anticipated that FELASA will continue to serve as a coordinating body for the progress of laboratory animal science and the Three Rs within Europe. Twenty-four countries participated in the Action, as shown in Table 0.1.

Participants of Action B-24 have published many papers in specialist journals and offer this manual, as a collated source of up-to-date information about ethically sound approaches to the care and scientific use of animals. It draws heavily from the widely accepted principles of Russell and Burch, usually known as the Three Rs. We hope that this volume will be of value to those who wish to benchmark or improve practices relating to the scientific use of animals, and will serve as a convenient source of evidence-based information about good practices and laboratory animal science and welfare to all those engaged in this field, including educators and regulatory/oversight authorities.

For more information: http://www.cost.eu

TABLE 0.1
Nations Participating in Cost Action B-24

Austria	Germany	The Netherlands
Belgium	Greece	Norway
Croatia	Hungary	Portugal
Czech Republic	Ireland	Spain
Denmark	Israel	Sweden
Estonia	Italy	Switzerland
Finland	Lithuania	Turkey
France	Malta	United Kingdom

Dr. Kalliopi Kostelidou
Cluster Leader, Cluster of Life Sciences
Science Officer, BMBS Domain

Preface

A revised Appendix A to the European Convention ETS 123 came into force in 2007. This represented a significant change to baseline standards for the care of laboratory animals within Europe. Subsequent adoption of these guidelines by the European Commission and development of a new Directive to replace 86/609, marks greater emphasis on the Three Rs in laboratory animal science.

In 2004, COST (European Cooperation in Science and Technology) established Action B-24 (Laboratory Animal Science and Welfare) as an initiative to increase knowledge necessary for the ethically and scientifically sustainable use of laboratory animals in research. The Action attracted 24 Member States, each represented by one or two members, who contributed to on-going dialogue and research about the most humane and effective way of using laboratory animals for scientific purposes. One result of this initiative has been the preparation of this text—*The COST Manual of Laboratory Animal Care and Use: Refinement, Reduction, and Research*. The manual is a joint effort involving many authors and it represents a truly international perspective on best practice, as currently perceived. Although 15 different COST countries have each contributed one or more authors in preparation of the text, the information contained includes a great deal of less formal input by other members of the COST Action. Although it is not possible to recognise everybody who has participated in preparation of the manual, the editors are extremely grateful for the support received from them and also from a number of colleagues who, though not directly related to the Action, nonetheless contributed material where specific expertise was not immediately available. The manual can be viewed as an international venture, made possible by the funding and support of COST.

The reason for not addressing "Replacement" in this manual is that the underlying technologies involve different scientific disciplines and often require specialisms that are beyond the reach of laboratory animal scientists. Omission in no way reflects a view of the editors that Replacement is in any way less of a priority in laboratory animal science than Reduction and Refinement. Indeed, we firmly believe that Replacement is the ultimate objective for laboratory animal science, although we are of the opinion that science should not be sidestepped simply because appropriate replacement techniques are not yet currently available. In circumstances where no replacement strategy is capable of generating the scientific information necessary to test the hypothesis being addressed and where the work proposed can be shown to have important applications, then every measure needs to be taken to minimise the number of animals used and the impact of procedures on each of those animals.

Only the more commonly used laboratory animal species are considered: rodents and rabbits. Our intention is to offer a one-stop source of best practice, which will be of value to personnel responsible for the care and welfare of animals and scientists conducting activities related to the use of animals for scientific purposes. It is also addressed to those with management responsibilities including facility engineers, architects and lay persons involved with ethical review, and the interested general reader.

The manual presents perceived best practices and, as far as possible, the contents are evidence-based. In the many cases where the evidence is poor, the authors have endeavoured to demonstrate the rationale underlying their contentions by including relevant references. The emphasis is not on describing current practices, but rather what is seen as good practice; moreover it is not our intention to tell the reader what to do, but rather to offer options and provide advice on what not to do.

Each chapter follows a similar structure, comprising six sections, the first of which, "Objectives", outlines the breadth of what the chapter is addressing. This is followed by "Key Factors", which summarises the central issues that underpin good practice. Each chapter ends with a short section, "Questions Unresolved", which identifies areas of uncertainty and may serve as a prompt to those wishing to undertake investigations directed at progressing knowledge and application within particular fields.

COST offers the manual as a benchmark against which practices can be judged, and it is hoped in this way to provide a mechanism for the advancement of laboratory animal science at a time when the regulatory paradigm is changing in the same direction.

Editors

Bryan Howard retired from the post of Director of Animal Welfare at the University of Sheffield in 2005 but retains an active interest in promoting the Three Rs in relation to the scientific use of animals. He qualified as a veterinary surgeon from the University of Glasgow in 1962 and completed a Doctorate in neurophysiology at the University of Edinburgh in 1966. After a period working on the welfare of poultry at slaughter, he took up posts as visiting professor of Veterinary Physiology at several universities in the Middle East.

Since returning to the UK in 1986, he completed the Royal College of Veterinary Surgeons Certificate in Laboratory Animal Science followed by a Master of Science (with distinction) in Laboratory Animal Science at the University of London, and M Ed in Teaching and Learning for University Lecturers; he remains a Fellow of the Higher Education Academy of UK. Apart from pursuing research into best practice for the care of laboratory species, he has been President of the Laboratory Animal Science Association, a Board member of the UK National Centre for the Three Rs (NC3Rs) and Chair of the Trustees and the Council of the Universities Federation for Animal Welfare (UFAW). He was also a member of the Governing Board of COST Action B-24. He is currently a member of the European Board of the Association for Assessment and Accreditation of Laboratory Animal Care (AAALAC International), the Board for Accreditation of Training established by the Federation of European Laboratory Animal Science Associations (FELASA) and of the Reduction Committee of the Fund for the Replacement of Animals in Medical Experiments (FRAME). In addition Howard has contributed internationally to a large number of training courses designed to advance laboratory animal science and welfare and has participated in the development of printed and web-based guidance for best practice in these areas.

Timo Nevalainen earned his Doctor of Veterinary Medicine in 1972 in Helsinki, graduated from the MS program in Laboratory Animal Medicine at the Pennsylvania State University in 1976, and defended his PhD thesis in physiology (1984) in Kuopio. His professional career started as a physiology teacher, from where he progressed to being named director of the National Laboratory Animal Center, both at the University of Kuopio, Finland. Since 1990 he has served as the Professor in Laboratory Animal Science and Welfare both at the University of Kuopio (now University of Eastern Finland) and at the University of Helsinki. While on sabbatical leaves, he has worked as a visiting professor at the University of Cincinnati and at the Pennsylvania State University.

He has had an active role in scientific and professional associations; he is a past president of both Scand-LAS and FELASA. He has chaired an EU COST Action B-24 and is involved with various other EU activities, such as ethics review of the Seventh Framework Program. In 2009 he joined the ILAR Council. He has a keen interest in training the future generation of scientists; in addition to being instrumental in establishing competence education of all categories in his country of origin, he has organised six competence courses for scientists in the Baltic countries and Russia starting immediately after the collapse of the Soviet Union. Since then, he widened this activity to teaching such courses in Thailand. He has had a key role in establishing FELASA education guidelines of all four categories, and FELASA Accreditation of training; most recently he was invited to join the FELASA Accreditation Board.

The common theme of his research emphasises the Refinement and Reduction alternatives, and their interplay in the use of laboratory animals. More recently, he has become interested in a third dimension; that is, to examine how those two Rs alternatives can also achieve better science. Currently his group is actively pursuing various aspects of bedding, evaluation of variants of the most common procedures (handling, IG-gavage, blood sampling, IP-injection, identification), comparing efficacy of cage furniture, impact of group size and a novel solution to a "poorly controlled variable" called *ad libitum* feeding suitable for use in group-housed rats.

Gemma Perretta is the head of the Division of Special Zoology and Animal Models of the Italian National Research Council in Rome. She graduated in veterinary medicine and specialised in laboratory animal science and medicine at the University of Milan for which she also served as professor from 1993 through 2006.

Over the past 25 years, Perretta has developed relevant competence in the field of laboratory animal welfare and science and gained wide knowledge on international guidelines and regulations on the protection of animals. She has been an invited expert in several commissions and working groups and has provided technical and scientific support to public regulatory authorities at national and European level. As a participant in national and international commissions, she has been involved in the promotion and implementation of high-quality standards in laboratory animal care with particular focus on education and training requirements and recommendations for staff working with animals.

She is a former president of the Federation of European Laboratory Animal Science Associations (FELASA) and the Italian Association for Laboratory Animal Science (AISAL) and serves as special consultant for the International Council for Laboratory Animal Science (ICLAS).

Perretta has been vice-chair of the Management Committee in the EU COST Action B-24 (Laboratory Animal Science and Welfare).

Contributors

Luis Antunes*
Veterinary Sciences Department
Universidade de Trás-os-Montes e
Alto Douro, Vila Real, Portugal
lantunes@utad.pt

Hanna Augustsson
Sweden

Vera Baumans*
Department of Animals in Science
  and Society
Division of Animal Welfare and Laboratory
  Animal Science Utrecht University
The Netherlands
v.baumans@uu.nl

Anne-Grethe Trønsdal Berg
Nord-Trøndelag University College
Norway

Claudio Bernardi
Italy

Heinz Brandstetter
Germany

Aurora Brønstad*
Vivarium, Haukeland sykehus
Faculty of Medicine and Dentistry
University of Bergen
Bergen, Norway
aurora.bronstad@ffhs.uib.no

Vladiana Crljen
Croatia

Klaus Cussler
Germany

Ismene Dontas*
School of Medicine
University of Athens
Athens, Greece
idontas@med.uoa.gr

Ann-Christine Eklöf*
Researchlab Q2:09
Karolinska Institutet
Stockholm, Sweden
Ann-Christine.Eklof@ki.se

Michael Festing*
c/o Understanding Animal Research
London
United Kingdom
michaelfesting@aol.com

Paul Flecknell
United Kingdom

Richard Fosse
France

Istran Gyertyan
Hungary

Marcel Gyger
Switzerland

Hansjoachim Hackbarth
Germany

Axel Kornerup Hansen
Denmark

Patrick Hardy*
AFSTAL (Association Française des Sciences
  et Techniques de l'Animal de Laboratoire)
Paris, France
afstal@free.fr

Jann Hau
Denmark

---
* Senior author.

Patricia Hedenquist*
Unit for Comparative Physiology and Medicine
Department of Large Animal Clinical Sciences
Swedish University of Agricultural Sciences (SLU)
Uppsala, Sweden
Patricia.Hedenqvist@kv.slu.se

Coenraad Hendriksen*
Netherlands Vaccine Institute (NVI)
Bilthoven
The Netherlands
Coenraad.Hendriksen@nvi-vaccin.nl

and

Department of Animals in Science and Society
Utrecht University
Utrecht, The Netherlands

Carlijn Hooijmans
The Netherlands

Bryan Howard*
Formerly of University of Sheffield
United Kingdom
B.R.Howard@sheffield.ac.uk

Katey Howard
Spain

Rony Kalman*
Authority for Animal Facilities
Hebrew University
Israel
ronyk@huji.ac.il

Nikolaos Kostomitsopoulos
Greece

Thomas Krohn
Denmark

Aavo Lang
Estonia

Marlies Leenaars
The Netherlands

Katerina Marinou
Greece

Lars Friis Mikkelsen
Denmark

David Morton*
Formerly University of Birmingham
United Kingdom
d.b.morton@bham.ac.uk

Timo Nevalainen*
Laboratory Animal Center
University of Eastern Finland
Kuopio, Finland
Timo.Nevalainen@uef.fi

I. Anna S. Olsson*
Institute for Molecular and Cell Biology—IBMC
Porto, Portugal
olsson@ibmc.up.pt

Gemma Perretta
Institute of Neurobiology and Molecular Medicine
National Research Council
Rome, Italy
ims@casaccia.enea.it

Belen Pintado*
CSIC
Centro Nacional de Biotecnología
Madrid, Spain
bpintado@cnb.csic.es

Jan-Bas Prins*
Leiden University Medical Centre
Leiden, The Netherlands
j.b.prins@lumc.nl

Jon Richmond
United Kingdom

Merel Ritskes-Hoitinga
The Netherlands

Peter Sandøe
Denmark

Contributors

Bart Savenije
The Netherlands

Dag R. Sørensen*
Centre for Comparative Medicine
Rikshospitalet
Oslo University Hospital
Oslo, Norway
dagrs@rr-research.no

Ping-Ping Tsai
Germany

Frank van den Brook
The Netherlands

Marian van Roon
The Netherlands

Hanna-Marja Voipio*
Laboratory Animal Centre
University of Oulu
Oulu, Finland
Hanna-Marja.Voipio@oulu.fi

Sarah Wolfensohn
United Kingdom

Walter Zeller
Switzerland

Editorial Board of the Manual

Heinz Brandstetter
Bryan Howard (Chair)
Nikolaos Kostomitsopoulos
Aavo Lang
Timo Nevalainen
Anna Olsson
Gemma Perretta
Jan-Bas Prins
Annie Reber
Merel Ritskes-Hoitinga
Hanna-Marja Voipio
Johannes Wilbertz

1 Introduction

Bryan Howard, United Kingdom

CONTENTS

1.1 Public Concerns 1
1.2 Replacement 2
1.3 New Technical Advances 3
1.4 The Culture of Care 4
References 5

1.1 PUBLIC CONCERNS

The use of animals in scientific experimental and testing laboratories within Europe is the subject of considerable public concern. Although a small proportion of the population believes that animal testing can never be morally justified, a similarly small proportion believes that animal testing presents no ethical difficulties. The great majority of people recognise that some animal experimentation may be necessary, but dislike it. The outcomes of surveys of public attitude are strongly influenced by the nature of the questions and the way they are asked. For example, a poll conducted in UK in 1999 (Ipsos Mori 1999) found that of 1014 people questioned, just over a quarter stated that animal experimentation was acceptable, providing there was no alternative; the number rose to two-thirds if the research was for medical purposes. In the same survey, 39% of respondents to another question said they did not support the use of animals for any experiments.

These findings were broadly confirmed by a similar survey of 1,125 people in 2002. Both of these surveys were commissioned by bodies that supported biomedical research. In contrast a poll conducted in 2003 by the British Union for the Abolition of Vivisection found that three-quarters of respondents believed that the British Government should, as a matter of principle, prohibit experiments on live animals (via The Nuffield Council on Bioethics 2005).

In 2005 a poll carried out on behalf of a major UK newspaper found that only about 10% of the UK population would support painful experiments on mice for testing cosmetics, but that over 60% would support such experiments to develop a treatment for leukaemia (Daily Telegraph 2006). Providing experiments did not cause pain, acceptance for using mice rose to over 80%. Many of the differences between the findings of these polls in UK are influenced by the precise way in which the question is asked and the caveats that respondents place on their answers (Ipsos Mori 2006). For the public to become convinced that animal experimentation is ethically justified, not only must there be a clear beneficial outcome, but there should be no means of acquiring the information without using live animals, experiments should involve the smallest possible number of animals and these must be cared for and used in the most humane way possible. These issues translate to the three Rs of Russell and Burch (1959).

Similar findings have arisen from polls carried out in several other European countries, but perhaps the most rigorous examination of public opinion has taken place in Switzerland. Under the Swiss constitution, the Government is obliged to call a national referendum if it receives a petition containing at least 100,000 citizens' signatures. The collection of signatures and subsequent referendum encourage considerable debate about the key issues. Referenda relating to the use of laboratory animals in Switzerland have been conducted in 1985 and 1992; the first, seeking an outright ban, was rejected by almost three-to-one; the second, which proposed restrictions on the ability of researchers to obtain licences, was rejected by a vote of 56% (von Roten 2008).

The process of revising European Directive 86/609/EEC on the protection of animals used for experimental and other scientific purposes (European Commission 1986) involved consulting European citizens about their attitudes towards the use of animals for experimental purposes. The European Commission conducted an internet consultation between June and August 2006. Although the findings are undoubtedly influenced by the same factors discussed previously, and the solicitation of responses by open invitation rather than statistically selecting individuals may have resulted in additional bias, interesting conclusions can be drawn. Most respondents (75%) felt that the level of welfare and protection of experimental animals within the European Union was either poor or very poor although rather fewer (68%) felt that the situation was as poor in their own country. Ethical acceptability of research depended on the purpose and scientific objectives (52%); the amount of pain, suffering, and distress caused (76%); the absence of alternative methods (70%); and standards of animal care and husbandry (51%), again reflecting the three Rs of Russell and Burch; they also argued for harm-benefit assessment. Only 40% of respondents considered that the use of animals to develop treatments for disease, to develop medicines, and test their safety was acceptable. The European Directive provides a framework within which these issues can be addressed at a Europe-wide level.

Against this background of public concern, the European Commission introduced a number of measures into the draft proposal for a new directive, which were intended to raise public confidence, either by specific prohibitions or by requiring member states to establish initiatives to deal specifically with some of the topics raised. These developments are taking place in a climate of heightened concern about the environment and call for more rigorous safety standards of consumer products.

In 2005, statistics collected by the European Union across the 25 member states (European Commission 2007) indicated that 12.1 million animals were used for scientific procedures. Mice accounted for 53% of these, rats 19% and rabbits 2.6%—three-quarters of all animals used. This manual deals only with these three species—it is difficult to envisage a text that could deal adequately with the different requirements and characteristics of all laboratory animals, although many of the principles described here apply also to other species. Almost two-thirds of the animals were used in developing human and veterinary therapies, dentistry and fundamental biological studies. About 12% were used for production and quality control of human medical and dentistry products and 8% for toxicological and other safety evaluations; the number of animals being used for toxicological and other safety evaluations appear to be declining over time. Considerable efforts are being made to harmonise the requirements for such testing and it is anticipated the number of animals used in this way will decline further.

1.2 REPLACEMENT

Replacement is the ultimate objective of laboratory animal science. Laboratory animals are not perfect models of the human; moreover, they are expensive, their care and use demand great technical skill and precision, and their use is the object of considerable ethical concern. Russell and Burch had identified a number of strategies for the replacement of laboratory animals, including both *in vitro* and *in silico* approaches; there has been progress in the former—models of human skin and vascular beds, for example—which are intended to replicate human tissues. Computational capabilities, using computer-generated models to replace animals, have become much more powerful with improvements in information technology (IT), but many of them still require validation, with the notable exception of toxicity prediction based on quantitative structure-activity relationships. Replacement technologies have also been introduced into university teaching, including interactive DVD educational software intended to substitute for the use of animals in live practical classes. There is evidence that this approach is a useful adjunct to traditional teaching. However, it remains important for laboratory animal scientists to collaborate with others involved in replacement disciplines to conduct parallel studies so as to rigorously compare the results obtained and to assist the validation process of new replacement methods.

Obviously, where it is possible to identify replacement strategies that make the use of animals unessential for part of all of an investigation, they should always be adopted. Unfortunately, progress in developing replacement strategies has been slow; in part, this is because approaches such as the use of human volunteers and *in vitro* or *in silico* strategies require reformulation of research objectives, involve technologies with which traditional laboratory animal scientists are unfamiliar, and require deployment of equipment and research strategies that are often of a highly specialised nature. Although slow, there is no doubt that progress has been made, particularly in the field of toxicity testing (see, for example, European Commission 2009); advances in other fields are often more difficult to identify and the sources of funding available for novel experimental approaches are disappointingly small. For these reasons, this manual does not address the first R of Russell and Burch—Replacement. There are texts, Web sites and organisations that deal specifically with the development and deployment of replacement technologies, and the reader is strongly encouraged to make use of these.*

1.3 NEW TECHNICAL ADVANCES

As with all scientific disciplines, laboratory animal science continues to progress by innovation, technical developments, raised awareness and application of good practises. These constitute the subject matter for this manual. Over the last few years, much emphasis has been placed on enhancing animals' immediate surroundings; including the provision of palatable, high-quality diets; provision of safe, flexible and effective caging systems; and refinement to the animals' environment by group housing and attention to the quality as well as the amount of living space. Emphasis now appears to be moving towards creating more uniform environments by attention to patterns of airflow, monitoring cage conditions and, above all, improvements in animal health status to ensure freedom from clinical and sub-clinical infection. Most suppliers now deliver specified pathogen free (SPF) animals, which are maintained under rigorous biocontainment conditions such as strict barrier entry and exit routines or by housing within a permanent microbiological barrier such as a flexible film isolator, ventilated rack, cabinet biospheres and so on.

Considerable advances have been made in our ability to monitor animals' physiology and anatomy using non-invasive techniques, such as magnetic resonance imaging (MRI), computed tomography (CT) scanning, diagnostic sonography and ultrasonography, movement detection and analysis equipment and so on. Technological improvements have also been made in the ability to equip animals with sensitive dosing and measuring equipment, which enable substances to be infused parenterally over prolonged periods, for samples to be withdrawn without the need to restrain animals on each occasion, or biological information to be collected by radiotelemetry. Developments such as these make it possible to conduct longitudinal studies, in which each animal serves as its own control, thereby considerably reducing the number of animals needed for a study and increasing scientific precision; however, the 'lifetime impact' of these procedures on animals' well-being may sometimes be increased. This has been partly ameliorated by the availability and uptake of more effective anaesthetic and analgesic techniques.

There is continuing ethical debate about whether it is better to subject a relatively small number of animals to remote monitoring and dosing techniques that are non-invasive but may involve previous surgical preparation, or to use substantially greater numbers of animals but that does not require

---

* See, for example, http://www.frame.org.uk, http://altweb.jhsph.edu/; http://oslovet.veths.no/fag.aspx?fag=57&mnu=databases_1; and http://caat.jhsph.edu/ All accessed 16 December 2009.

this intervention. Another unresolved question is whether animals already prepared should be used for only one investigation, or whether, providing they are healthy and have recovered fully, it is better to use them again or to surgically prepare new animals.

During the last decade, the number of genetically altered animals (principally mice) used in scientific studies have been increasing steadily; this is a consequence of the ability to manipulate the genome to simulate diseases that occur in the human, or to suppress or enhance gene function in order to better understand the way in which genotype influences biology. Greater impetus was added to such studies when the mouse genome was decoded, although the complexities of gene interaction considerably complicate straightforward analysis of gene function in this way (Mouse Genome Sequencing Consortium et al. 2002). It seems likely that the use of genetically altered animals will increase during the coming years, and will considerably increase our understanding of gene biology and the consequences of gene malfunction.

Coupled with enhanced understanding of gene function, there have been developments in somatic and stem cell cloning, wider application of reproductive manipulation techniques, including assisted fertilisation (such as intracytoplasmic sperm injection), gamete analysis and manipulation, greater availability of methods for transporting and 'archiving' animals as gametes or germ cells and their subsequent recovery, and implantation techniques; these procedures ensure that unique genetic lines of animals can be retained within freezer cabinets, rather than retaining animals in cages with attendant welfare compromises.

It is the purpose of this manual to explore the ways in which these new technologies can be applied in further reducing and refining the use of animals for scientific purposes, recognising that until such time as satisfactory replacement strategies make further animal experiments unnecessary, we have a duty to minimise our impact on those animals we do use, and to ensure that only high-quality science is conducted.

1.4 THE CULTURE OF CARE

All experimentation including that involving living animals is only as valuable as the care that is taken with its design, performance and analysis. Those engaged in the scientific process require competencies not only to ensure application of the three Rs, but also to assure the quality of the scientific process, and the eventual publication and dissemination of reliable, reproducible and accurate results. Good science requires collaboration between a team of persons skilled in their respective fields. At the very least, these involve the scientist, those responsible for caring for the animals (including the facility director), the regulatory authorities, and the veterinarian. Before commencing studies, it is essential that availability of the necessary facilities is assured (including equipment and funding) for the entire study and that all necessary regulatory and legal requirements have been addressed. The process of obtaining authorisation may well involve ethical review by peers who must be presented with a clear scientific justification for the proposed use of animals, assurance that each of the three Rs has been addressed, that the work proposed would generate findings of value, and that contingencies are in place to ensure that animal welfare is safeguarded, even when issues arise unexpectedly that could impact negatively on an animal's well-being.

Scientific and welfare integrity depends on participation of everyone involved in the study, operating within an open climate of mutual support, concern for scientific rigour, and, above all, commitment to animal welfare through implementation of the three Rs. In such an environment unauthorised procedures do not take place, deviations from good practice are instantly corrected and unnecessary animal suffering does not occur. This condition of mutually assured cooperation is commonly known as a culture of care, and should underpin the working environment in all establishments where animals are used for experimental purposes.

REFERENCES

Daily Telegraph. May 29, 2006. Animal rights wrongs. [Cited July 30, 2009]. Available from http://www.telegraph.co.uk/comment/telegraph-view/3625317/Animal-rights-wrongs.html

European Commission. 1986. Council directive 86/609/EEC of 24 November 1986 on the approximation of laws, regulations and administrative provisions of the member states regarding the protection of animals used for experimental and other scientific purposes. *Official Journal of the European Union* L358:1–29.

European Commission. 2007. *Report from the Commission to the Council and the European Parliament; fifth report on the statistics on the number of animals used for experimental and other scientific purposes in the member states of the European Union.* Brussels: European Council, COM(2007)675 final.

European Commission. 2009. *Alternative testing strategies—progress report 2009; replacing, reducing and refining use of animals in research—genomics & biotechnology for health.* Luxenbourg: Office for Official Publications of the European Communities, EUR 23886.

Ipsos Mori. 1999. Attitudes towards experimentation on live animals. [Cited July 30, 2009]. Available from http://www.ipsos-mori.com/researchpublications/researcharchive/poll.aspx?oItemId=1883

Ipsos Mori. 2006. Views on animal experimentation. [Cited July 30, 2009]. Available from http://www.ipsos-mori.com/Assets/Docs/Archive/Polls/dti.pdf

Mouse Genome Sequencing Consortium, R. H. Waterston, K. Lindblad-Toh, E. Birney, J. Rogers, J. F. Abril, P. Agarwal, et al. 2002. Initial sequencing and comparative analysis of the mouse genome. *Nature* 420:520–62.

Nuffield Council on Bioethics, the. 2005. *The ethics of research involving animals.* The Nuffield Council on Bioethics, London.

Russell, W. M. S. and R. L. Burch. 1959. *The principles of humane experimental technique.* Potters Bar, England: Special edition, Universities Federation for Animal Welfare.

von Roten, F. C. 2008. Mapping perceptions of animal experimentation: Trend and explanatory factors. *Social Science Quarterly* 89:537–49.

2 Design and Oversight of Laboratory Animal Facilities

*Dag Sørensen, Norway; Heinz Brandstetter, Germany;
Nikolaos Kostomitsopoulos, Greece; and Richard Fosse, France*

CONTENTS

Objectives 8
Key Factors 8
2.1 Facilities 9
    2.1.1 Facility Planning 9
        2.1.1.1 The Design Team 9
        2.1.1.2 Defining Activities 9
        2.1.1.3 Capacity 10
        2.1.1.4 Strategic Limits and Questions 10
2.2 Facility Oversight 10
    2.2.1 Qualifications of the Animal House Director 10
    2.2.2 Duties of the Animal House Director 11
    2.2.3 Qualifications of the Veterinarian Responsible for the Veterinary Care in
        Laboratory Animal Units 11
    2.2.4 Qualifications of the Supervisor of the Technical Staff 12
    2.2.5 Qualifications of Persons Taking Care of Laboratory Animals 12
2.3 Key Components in Design 12
    2.3.1 Barriers, Barrier Elements 13
    2.3.2 Corridors: Flow Cycles for Animals, Scientists and Equipment 14
        2.3.2.1 Dual or Single Corridor Systems 14
    2.3.3 Reception Areas 14
    2.3.4 Animal Holding Rooms 14
    2.3.5 Ventilation of Secondary Enclosures 15
    2.3.6 Procedure Rooms 16
        2.3.6.1 General Requirements for Furniture and Equipment 16
        2.3.6.2 Laboratories 16
        2.3.6.3 Surgical Suites 16
        2.3.6.4 Necropsy 17
    2.3.7 Transgenic Core Facilities 17
    2.3.8 Teaching and Training Rooms 18
    2.3.9 Architectural Finishes and Materials 19
2.4 Distribution Systems 19
    2.4.1 Drains 19
    2.4.2 Animal Watering Systems: Bottle or Line Delivery 19
2.5 Building Management Systems (Environmental Control) and Monitoring 20
    2.5.1 Maintenance Programmes of the Facility and Equipment 20

2.6 Other Functional Areas 21
    2.6.1 Washing Area 21
        2.6.1.1 Robot Systems...21
    2.6.2 Storage (Clean and Dirty Area) 21
    2.6.3 Cold Room 22
    2.6.4 Waste Area: Waste Disposal 22
    2.6.5 Security: Alarm Systems—Contingency Plans 22
2.7 Communications 22
2.8 Barriers: Levels 23
    2.8.1 Bioexclusion (Microbiological Standardisation) 23
    2.8.2 Health Monitoring 23
    2.8.3 Immunocompetent Animals 24
    2.8.4 Immunocompromised Animals 24
    2.8.5 Experimentally Infected Animals 24
2.9 Biocontainment 24
    2.9.1 Health and Safety S 0–4 24
2.10 Quarantine 25
2.11 Conclusions 25
2.12 Questions Unresolved 26
References 26

OBJECTIVES

The building and its associated structures represent the life-long environment of laboratory animals that are housed and used in it. Therefore it is essential that this environment is designed and constructed so as to provide for all of the animals' needs throughout the time they spend in the facility. This is a total concept that includes the quality of the building and the competence of the staff. When animals are kept and used for scientific purposes, this total concept should always be applied under the best possible conditions to ensure that the greatest possible amount of valid and precise data are obtained.

The planning and building of a laboratory animal facility is a complex process involving the contribution of expertise in several fields. Experts in a number of different fields must cooperate at all stages of the design, construction and commissioning, if the facility is to meet fully all requirements for the intended research activities and to be sufficiently flexible for future developments.

Personnel training is a prerequisite for working with laboratory animals, and central animal facilities should provide laboratories and other facilities for delivering training programmes for categories A, B, and C in accordance with the FELASA recommendations (see Chapter 16).

Particular attention should be given to the overall management of the facility. If all factors in the following sections are carefully addressed, the achievement of reduction and refinement will be greatly simplified.

KEY FACTORS

The following key components should be taken into account at the planning, design and construction stages:

- Types of activities proposed and degree of flexibility required
- Species to be housed and the maximum number of each that must be accommodated at anytime
- Extent of environmental consistency desired

Design and Oversight of Laboratory Animal Facilities

- Barriers and biosafety levels
- Maintenance programmes for the facility and its equipment
- Facility oversight, including qualification of key personnel
- Organisation and structure, including access of staff, animals and equipment as well as disposal routes

2.1 FACILITIES

2.1.1 Facility Planning

2.1.1.1 The Design Team

The planning and building of a laboratory animal facility is a complex process involving the participation of expertise drawn from several different areas. The first step is to set up a design team or committee consisting of the following persons:

Representatives of the client

- Head of the facility being built (if appointed) or senior facility manager
- Experienced laboratory animal scientist
- Veterinarian
- User representatives (scientists using the animals)
- Animal technicians
- Administration representatives
- Staff responsible for future maintenance of the building
- Animal welfare officer

Professional team

- Architects
- Engineers
- Consultants

It is the responsibility of the client and professional teams to involve further specialists if appropriate, external experts, and relevant authorities in advance. This has to be done before crucial decisions are made.

2.1.1.2 Defining Activities

The following should be determined before attempting to estimate the capacity within the given budget:

- Types of research activities: Development (toxicology, product testing, etc.) or discovery (basic research, applied research, etc.)
  - Special equipment (e.g., for imaging)
- Regulatory control: Good Laboratory Practise (GLP) or non-GLP
- Species: Identify the variety of species expected to be accommodated over time:
  - Mammals: Rodents, lagomorphs, large animals (dogs, pigs), primates
  - Aquatic species: Marine, fresh water
  - Other vertebrates
  - Insects
- Containment: Risk of contamination to researchers and/or the outside environment
  - Infection
  - Chemical
  - Radioisotope

- Genetically modified organisms (transgenic animals)
  - Virus vector
  - Non-virus vector
- In-house breeding
- Sources of animals
  - Commercial sources (SPF quality only)
  - Non-commercial (academic institutions, wild caught animals)
- Education and training

2.1.1.3 Capacity

Determining the capacity of a large laboratory animal facility somewhat resembles the process for a hotel. Plans must be based on the average number of housing units per day, and that is balanced against the total number of animals that can be housed in the course of any given research period. This can be exemplified by a project that requires the use of 1000 mice. The "hotel" needs to know whether all 1000 arrive on one day and leave at the end of one week or whether they will arrive weekly, for example in groups of 100 over a period of 10 weeks? This is not easy to estimate but is essential. The requirement must be evaluated species by species, including considerations of isolation, number of animals per room, and number of investigators and the number, nature and type of projects expected to take place in different parts of the facility.

2.1.1.4 Strategic Limits and Questions

It is also important to estimate long-term trends in likely demand for access to the facility. An example is whether it will be necessary to provide facilities for large and small species in the future. Exclusion of large species is a major strategic decision that strongly influences the planning process.

The animal facility should be designed with as much flexibility and allowance for future expansion as possible.

Questions to ask:

- Small and/or large animals?
- Is there a requirement for species that need specialised environments (e.g., nude mice or nude rats that require a room temperature 2–3°C higher than normal small rodents)?
- Is there a requirement for primates?
- May there be a need for work with human or animal pathogens?
- Is the facility to be a separate building or integrated into a research building?
- Will all *in vivo* laboratories be included within the facility (behind the barrier)?
- Is there a requirement for in-house breeding (stringent barrier requirements)?

2.2 FACILITY OVERSIGHT

To guarantee effective overall management of the facility, ensuring high standards of animal welfare as well as good scientific practice, the head of the animal facility needs a high level of competence and the support of a team of qualified people, specialised in their respective roles.

2.2.1 QUALIFICATIONS OF THE ANIMAL HOUSE DIRECTOR

In an ideal situation, qualifications should start with an academic degree in either veterinary medicine, biology or agricultural science, followed by postgraduate education of at least 3–5 years, and a qualification in laboratory animal science recognised by national or international authorities (e.g., GV-SOLAS, FELASA Category D, ECLAM) or equivalent experience, appropriately documented.

Design and Oversight of Laboratory Animal Facilities

The facility director needs a comprehensive and broad knowledge of physiology, reproduction, genetics, and technologies for genetic modification, microbiology, hygiene, anaesthesiology, surgery, pathology, behaviour and animal welfare. Depending on the research activities proposed, this competence should extend to the commonest species of laboratory animals (mouse, rat, Syrian hamster, guinea pig, rabbit, cat, dog, sheep/goat, swine, primates, poultry species, amphibians/fish, etc.).

In addition, an animal house director must be familiar with ethics, national and international legislation, and financial management and be skilled in general and personnel management including specification, construction and operation of animal facilities.

Naturally, this wide specification of knowledge and experience is unlikely to be available in one individual, so the director should at least possess competence for the species that will be his or her responsibility; depending on the size, diversity and research programmes taking place in the facility, additional specialists such as administrative managers, biologists, surgeons, pathologists and so on might be required to work under the supervision of and in close collaboration with the director.

2.2.2 Duties of the Animal House Director

The head of a laboratory animal facility and his/her management team are responsible for all aspects of running the entire facility. Some duties might be the responsibility of other professionals and may be regulated differently in different countries. The principal responsibilities are

- Organisation of work flow
- Personnel management
- Maintaining hygiene standards and pest control
- Monitoring the health of the animals
- Veterinary care
- Activities related to regulatory and administrative compliance
- Support for scientists
- Animal transport
- Animal welfare
- Occupational health and safety
- Protection of the environment
- Ethical evaluation of animal care and use
- Designing animal facilities (as a partner together with architects, engineers, user representatives, etc.)
- Research in laboratory animal science
- Education of non-academic and academic people working with animals (caretakers, students, diplomats, and postgraduates—performing training courses for categories A, B, C, and D according to the FELASA recommendations)
- Public relations and liaison with regulatory authorities

In some countries there is a special position for someone to oversee issues related animal welfare (called e.g., animal welfare officer or named animal care and welfare officer).

2.2.3 Qualifications of the Veterinarian Responsible for the Veterinary Care in Laboratory Animal Units

If the animal house director is not a veterinarian, one must be appointed to take responsibility for veterinary care. In addition to basic veterinary skills relating to farm and domestic animals, this person should have a profound knowledge of the common laboratory animal species, particularly those kept and used in that animal facility. In addition to competence in diagnosis and medical treatment, the veterinarian in charge should possess specific knowledge of the following:

- Health management, health monitoring, and prevention of diseases
- Husbandry
- Anaesthesia, analgesia and euthanasia
- Surgery
- Humane endpoints
- Animal transport
- Animal welfare
- Relevant legislation

Such qualifications can be demonstrated by completing an examination in laboratory animal science recognised by national or international organisations (e.g., GV-SOLAS,* FELASA Category D; Nevalainen et al. 1999), ECLAM,† or equivalent; alternatively, competence can be obtained by experience over at least 3–5 years. The responsibilities of the veterinarian are also defined in the European Convention (ETS 123), Council Directive 86/609/EEC, and in national laws of many European countries.

Depending on the research projects performed at the establishment, it might also be necessary for the veterinarian to possess specialist knowledge, for example, relating to specific studies with infectious agents at the Biosafety Level (BSL) 1, 2, 3, or 4; to safety and toxicity studies under GLP; or to principles in the field of quality assurance.

2.2.4 Qualifications of the Supervisor of the Technical Staff

A supervisor of technical staff would normally be a qualified senior laboratory animal technologist with at least 3–5 years of relevant experience working with laboratory animals and an additional professional qualification in supervisory and managerial skills. This expertise can be demonstrated in a variety of ways, for example, possessing FELASA Category Level A2, Fellow of the Institute of Animal Technology (FIAT) in the United Kingdom, or the "Master for Laboratory Animal Technicians" recognised in Germany.

2.2.5 Qualifications of Persons Taking Care of Laboratory Animals

Persons taking care of laboratory animals (laboratory animal technicians) should have a basic knowledge of mathematics, biology, general science, computer use and oral and written communication. Their minimum qualification is in accordance with the FELASA recommendations for the education and training of persons working with laboratory animals—Category Level A1-or with nationally recognised standards such as Membership of the Institute of Animal Technology (MIAT; United Kingdom), or have demonstrated professional competence as a laboratory animal technician in countries such as Austria, Finland, Germany or Switzerland.

2.3 KEY COMPONENTS IN DESIGN

The layout of an animal facility is determined by the activities that take place within it (see Figure 2.1).

---

* http://www.gv-solas.de/auss/fac/richtlinien.html
† http://eclam.org/

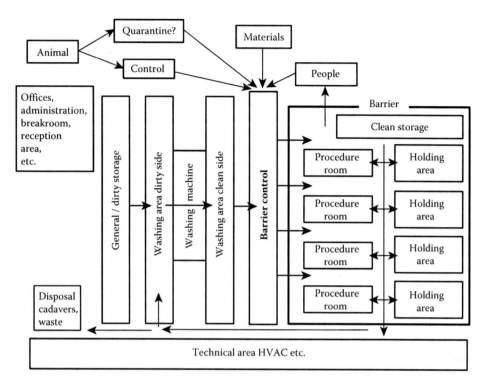

FIGURE 2.1 Flow chart of activities in a laboratory animal facility.

2.3.1 BARRIERS, BARRIER ELEMENTS

There is ample evidence demonstrating that intercurrent disease has a negative impact on biomedical research so that research in completely open, uncontrolled systems is no longer acceptable (Morris 1999).

Modern animal facilities should be planned with appropriate levels of barrier confinement. The barrier concept is designed to provide a stable and standardised environment that ensures reproducible research. The barrier is also intended to exclude from the facility undesirable infectious agents that may be present in the outside environment (bioexclusion) and also to reduce exposure of personnel to allergens and chemical hazards that may be used within the facility.

Isolation is defined as "keeping out" or providing a barrier whereas containment means "keeping in" (biocontainment) or providing an envelope around an experiment or animal room. These measures are provided in accordance with specific requirements, and may have to be met simultaneously. Because there is no such thing as a perfect barrier, the design and planning of a laboratory animal facility involves sorting out priorities and establishing compromises (Ruys 1991).

Important questions to be considered are:

- How much control?
- What level of barrier?
- Access control?
- Microbiological control?
- Environmental control?

2.3.2 CORRIDORS: FLOW CYCLES FOR ANIMALS, SCIENTISTS AND EQUIPMENT

2.3.2.1 Dual or Single Corridor Systems

The general availability of high-quality, disease-free animals has led to the dual corridor layout becoming largely redundant. A well-managed, single corridor system that preserves a clean to dirty work flow can be just as efficient and saves space.

Corridors should be wide enough to allow standard width racks and transport cages to pass freely in both directions. However, the width required depends on the type of activity in the facility (species, etc.). In general corridors should not be used for daily storage.

2.3.3 RECEPTION AREAS

Reception areas are public zones outside barriers; they include entrances, public corridors and elevators, as well as loading docks, supply and storage rooms and non-barrier laboratories. They may also include offices; conference, teaching and restrooms; and secure areas for central computer systems and file storage. There is limited opportunity to control potential animal contaminants in these areas and consequently they must be categorised as "dirty".

A loading dock area with dedicated bays for receiving animals and material from suppliers as well as removal of wastes is essential for the proper running of the facility. This area serves as a logistic hub and provides links to different parts of the facility including the various barrier elements. It should be planned and organised in a way that facilitates proper sanitation and pest control.

A dedicated route of transportation into the animal facility should be established and should include a reception area in which temperature and humidity are controlled in order to minimise climatic changes to the animals' environment as far as possible. Other features of this area could include a pass-through autoclave for materials to be sterilised on entry to the facility, and a UV sterilising port or a chamber for chemical surface disinfection of materials that cannot be autoclaved. This area may also be used for decontamination of the animal containers before entering the animal facility.

To prevent contamination of animal holding areas and other clean areas, entry and exit locks for personnel and/or material should be located as required.

2.3.4 ANIMAL HOLDING ROOMS

During the design of a laboratory animal facility, adequate space should be provided for animal holding rooms. These should be laid out in a flexible way to allow the possibility of holding a range of different animal species depending on current and perceived research needs. An efficient size of room for housing small laboratory animals such as mice, rats, guinea pigs or even rabbits could be 4×5 m approximately (Ruys 1991). The number of animals held in a room of this size depends on the caging system (conventional cages, filter top, individual ventilated cages or pens), and the size, the age and the sex of the animals as well as requirements of the project (Stakutis 2003). Whenever possible, all equipment likely to be required in the room should be defined at the beginning of the planning phase to allow optimisation of the room's dimensions. A detailed plan should be drawn up for all room types; size and composition of these, the number required and the work-pattern envisaged are major determinants of the eventual layout (CCAC 2003).

To maximise efficiency and flexibility, animal holding rooms should have the minimum fixed equipment—for example, a sink without a draining board. All racks, tables and cage change laminar flows should be on castors and able to pass through the doorway.

Floor materials should be seamless, turned up at the walls to form a cove base, non-slippery (especially when wet) and resistant to chemicals, particularly those used for regular sanitation. They should be capable of supporting racks, equipment and stored items without becoming gouged, cracked or pitted. Depending on the purpose to which the rooms will be assigned, floor drains might be necessary or undesirable. This is discussed further in Section 2.4.1.

Special consideration should be given to specifying the level of barrier/containment of the animal holding rooms.

For better control of the macroenvironment, windows should be avoided unless considered necessary for the housing of certain species (e.g., non-human primates).

The electrical supply system should be safe and provide appropriate lighting, a sufficient number of power outlets, and be of sufficient amperage for specialised equipment. An emergency power supply should be provided in case of power failure, and installed so as to maintain critical services such as heating, ventilation and air conditioning (HVAC), and isolators, ventilated racks, cabinets, and so on. Light fixtures, timers, switches and outlets should be sealed to prevent infestation by vermin and should be moisture resistant.

An adequate HVAC system should be capable of providing 15–20 air changes per hour to all of the animal rooms in use at the time. Positive or negative air pressure will be determined by the required barrier/containment level. A difference of 10–20 mm H_2O is usually sufficient to achieve an adequate air pressure differential. All technical data should be recorded daily to a data sheet and a regular maintenance programme should be in place (ILAR/NRC 1996; Wilkins and Waters 2004).

2.3.5 VENTILATION OF SECONDARY ENCLOSURES

The ventilation system plays an important role in the overall performance of the animal facility. Temperature, humidity, air changes and air velocity all contribute to this and should be designed and specified so as to take account of the level of barrier/containment, the type of animal rooms, and the type of caging system (Besch 1980; Wilkins and Waters 2004).

The ventilation system supplies the facility with fresh air and removes airborne contaminants such as ammonia and carbon dioxide and pheromones released by the animals; it compensates for varying thermal loads produced by animals' respiration, lights and equipment, regulates the humidity of room air and where appropriate, establishes static-pressure differentials between adjoining areas. Ventilation is also important to assure the health and safety of working personnel mainly by maintaining low levels of airborne allergens (ASHRAE 2001; Wilkins and Waters 2004).

The revised Appendix A of the European Convention ETS123 of the Council of Europe which was adopted by the European Commission (2007) requires that:

- Adequate ventilation should be provided in the holding room and the animal enclosures to satisfy the requirements of the animals housed. The purpose of the ventilation system is to provide sufficient fresh air of an appropriate quality and to keep down the levels and spread of odours, noxious gases, dust and infectious agents of any kind. It also provides for the removal of excess heat and humidity.
- The air in the room should be renewed at frequent intervals. A ventilation rate of 15–20 air changes per hour is normally adequate. However, in some circumstances, for example where stocking density is low, 8–10 air changes per hour may suffice. In some cases, natural ventilation may suffice and mechanical ventilation may not even be needed. Recirculation of untreated air should be avoided. However, it should be emphasised that even the most efficient system cannot compensate for poor cleaning routines.
- The ventilation system should be so designed as to avoid harmful draughts and noise disturbance.

The cost of investment in the ventilation system is considerable, so it is absolutely necessary that detailed technical specifications are available to inform the design process. It is important to know exactly how much space is to be ventilated, the structure of the facility, the number and kinds of animals that will be housed, the type of cages, the geographic location of the facility and the local climatic conditions.

A great deal of time and effort can be saved if computational fluid dynamics (CFD) software programmes are implemented during the design phase of a heating and ventilation air conditioning (HVAC) system (Morse et al. 1995; Hughes, Reynolds, and Rodriguez 1996; Hughes and Reynolds 1997; Memarzadeh et al. 2004).

2.3.6 Procedure Rooms

2.3.6.1 General Requirements for Furniture and Equipment

Furniture and equipment is likely to be subjected to heavy wear and tear, particularly in general service areas where it may be used by a number of different researchers. Equipment should be robust, accurate, reliable and regularly serviced and calibrated according to manufacturers recommendations. All persons using it must be acquainted with, and comply with the instructions. Furniture should be solidly constructed and well finished. Surfaces, especially worktops and cupboard/drawer faces, must be durable and able to be frequently and thoroughly cleaned and disinfected. High quality laminated finishes are commonly used. There is considerable advantage in using mobile furniture which can be removed from the room for disinfection, or when the role of the room changes.

2.3.6.2 Laboratories

A number of simple scientific procedures may be undertaken in the animal holding room, as long as the activity does not have any negative impact on other animals. More demanding procedures must be undertaken in rooms especially designed for the purpose and provided with an adequate supply of electricity and facilities for using anaesthetic gases, and so on. For most procedures, including animal handling, the use of a down-ventilated table minimises exposure to allergens and anaesthetic gases above the perforated table surface and affords a positive contribution to human health and safety.

In general, laboratories may be provided with daylight, but should have a facility for dimming the light or to exclude the influence of daylight. Temperature and humidity should be maintained within the same range as animal holding rooms to ensure that animals are not influenced by changes in the ambient environment during *in vivo* procedures.

2.3.6.3 Surgical Suites

Depending on the animal species, their number, and research needs, surgical facilities may vary from small and simple dedicated spaces in a laboratory, appropriately managed to minimise contamination during surgery from other activities in the room (e.g., by use of a laminar flow hood), to large and more complex facilities for large volume surgical procedures (Hessler 1991). The surgical facilities should be located away from high-traffic corridors and potential sources of contamination such as cage wash, necropsy and waste storage areas (Humphreys 1993). The relationship of surgical facilities to diagnostic laboratories, radiology facilities, animal housing, staff offices and so on should be considered in the overall context of the complexity of the surgical programme. Ideally, separate lockers, housekeeping and toilet facilities should be provided in close vicinity to the surgical suite.

Control of contamination and ease of cleaning should be key considerations in the design of a surgical facility. The interior surfaces should be constructed of materials that are seamless and impervious to moisture. Ventilation systems supplying HEPA filtered air at positive pressure can help to reduce the risk of postoperative infection (Bourdillon and Colebrook 1946; Schonholtz 1976; Ayscue 1986; Bartley 1993). Careful location of air supply and exhaust ducts and appropriate room-ventilation rates are also important to minimise airborne contamination (Ayliffe 1991; Bartley 1993; Holton and Ridgway 1993; Humphreys 1993). To facilitate cleaning, operating rooms should incorporate as little fixed equipment as possible (Schonholtz 1976). The surgical theatre should also be equipped with surgical lights to provide adequate illumination (Ayscue 1986), sufficient and suitable electrical power outlets for support equipment, and gas-scavenging capability (Poole 1999).

Surgical facilities should be back-up equipped with standby power, such as an electric generator with automatic switchover in case of power failure and ideally an uninterruptible power supply (UPS).

Functional areas for surgery should include a surgical support area including facilities for instrument preparation, a changing area for personnel, animal preparation and recovery rooms, surgeon preparation area (scrub room), operating theatres and intensive care and treatment rooms.

The areas that support those functions should be located and designed so as to minimise traffic flow and to separate unrelated, non-surgical activities from the surgical procedures in the operating theatre. The separation is best achieved by physical barriers (AORN 1982), but might also be realised by locating at some distance from other areas. In some cases, the necessary separation of non-surgical and surgical activities in the same area might be achieved by strict timetabling and thorough cleaning and disinfection between different activities. The number of personnel and their level of activity have been shown to be directly related to the level of bacterial contamination and the incidence of postoperative wound infections (Fitzgerald 1979). Traffic in the operating room itself can be reduced by installing an observation window and a communication system (such as an intercom system) in conjunction with judicious location of doors.

2.3.6.4 Necropsy

The necropsy room should be located remotely from the animal holding area and away from the general traffic flow of personnel working in the facility. Dangerous pathogens and hazardous chemicals may be present in the necropsy room, depending on the nature of investigations carried out in the facility. Because this area may be one of the most hazardous in the whole facility, it is recommended that the same standards are adopted for planning this area as for a human autopsy suite.

2.3.7 TRANSGENIC CORE FACILITIES

At the present time, most genetically modified mammals are mice, and this section will consider requirements for this species, although the general principles have wider applicability; in broad terms these are no different from any other kind of small rodent breeding and holding programme (see Chapter 9). Any breeding programme requires heightened health and barrier standards. The presence of newborn mice introduces a risk of infection, occurrence of which can be difficult to manage. Newborn mice are particularly at risk during the time of immune transfer; that is, when the passive immunity conferred by the mother decreases and the young animal's own immune system takes over. This makes it necessary for the breeding facility to be isolated and barrier-protected to ensure that unwanted infection is not introduced at this time (see Figure 2.2). The breeding of transgenic animals is also regulated in many countries and requires some specialised resources not normally associated with non-transgenic breeding.

When effective barrier conditions are in place, a transgenic breeding facility has the same requirements as a conventional one, but with two important differences. There are legal requirements to be met that vary depending on which country the facility is located. Most European and North American laws require that transgenic mice be considered as *genetically modified organisms* and as such, their inadvertent release into the environment must be prevented. Usually a designated person is appointed and assumes responsibility for ensuring that inadvertent release does not occur. The barrier area and any laboratories or other facilities into which genetically modified animals may be moved are usually considered to be areas dedicated to transgenic breeding and procedures; appropriate signage is posted to ensure that all persons entering are identified and informed of the hazard. The barrier also needs to include extra protection, such as rodent barriers (traps across doors), drain traps, and so on to prevent the risk of escapes. An additional restriction may be necessary if the transgenic animals are being created by use of a viral vector as the insertion mechanism of the DNA construct. This imposes a potential biosafety risk for humans working with the mice. In this case the area is restricted also in accordance with biosafety risk assessment criteria (Directive 2000/54/EC of the European Parliament and of the Council of 18 September 2000 on the protection

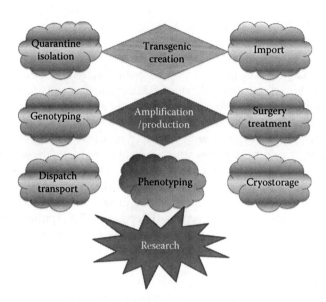

FIGURE 2.2 Components to be considered when planning a transgenic breeding facility.

of workers from risks related to exposure to biological agents at work (seventh individual directive within the meaning of Article 16(1) of Directive 89/391/EEC).

The genetic constitution of parent mice lines and of their offspring must be tracked and recorded (see Chapter 9). Consequently, there is a need to be able to identify each animal that is selected as a candidate for the transgenic characteristic. This places demands on available space as well as requiring access to a genotyping laboratory (in-house or external).

As work proceeds, the number of animals associated with each line increases. Space is at a premium particularly in a creation facility. This can be offset by providing cryogenic storage resources for lines that are not needed in the short-term, but may be needed at a later date (see Chapter 9). Cryogenic storage is also a way of providing backup in the event of a mishap with the breeding programme, to facilitate sharing of genetic resources, or simply as an archive. Creation of new transgenic lines represents a significant investment in time and resources (see Chapter 8) and discarding them should be discouraged. However, the establishment and maintenance of a cryogenic storage facility is demanding and requires highly trained and experienced staff as well as a reliable supply of liquid nitrogen if low temperature electric freezer cabinets are not available.

The process of line creation, stabilisation, and amplification is highly resource intensive. If it is not carefully controlled, numbers of mice can increase exponentially and buildings can rapidly become overstocked. Individual animals can be tracked by using simple databases or spreadsheets. There are also specialised computer applications that fill this role and facilitate the monitoring of programmes and tracing of individual, and use of these should be considered when setting up a breeding programme.

A plethora of additional resources may be needed if full or even partial phenotyping is to be carried out. These may include access to blood chemistry, histology/histopathology, and behavioural test apparatus to mention a few. The facility can arrange for testing to be done internally or external resources should be sought. There are many commercial phenotyping specialist companies and this approach often proves to be more efficient than trying to establish an in-house resource.

2.3.8 TEACHING AND TRAINING ROOMS

Training is a prerequisite for personnel working with laboratory animals. Central animal facilities should provide classrooms and laboratories for providing training for category A, B and C personnel in accordance with the FELASA recommendations (see Chapter 16). These laboratories should be spacious enough to accommodate students and supervisors and all necessary equipment, especially for practical training. Animals for these training courses should be housed in a separate room preferably in ventilated cabinets or IVCs. Training rooms should have sufficient natural and/or artificial light and should be air conditioned to provide a comfortable environment for participants.

2.3.9 Architectural Finishes and Materials

All finishes and materials must be free of cracks, impervious to moisture and have smooth surfaces that are easy to clean. There should be no places in which detergent liquids (cleaning water) could collect and no sharp angles, surface roughness, sharp corners or edges that could make cleaning difficult. Spaces that cannot be accessed for cleaning, disinfection and sterilisation must be completely and permanently sealed. All materials and finishes must be resistant to commonly used detergents and disinfectants, including formaldehyde, ammonia and hydrogen peroxide if it is possible that these will be used in the rooms. All equipment should be either easily accessible for personnel or removable for cleaning purposes.

2.4 DISTRIBUTION SYSTEMS

2.4.1 Drains

Depending on the purpose to which the room might be put, floor drains might or might not be necessary or desirable. However, for maximum possible flexibility it is recommended that drains of at least 10 cm of diameter be provided and where used, the floors should be sloped towards them and drain traps kept filled with liquid. For times when they are not required, they should be equipped with lockable stainless steel covers (rodent traps) that are removed only during room cleaning. In this case, frequent regular maintenance of the floor drain system is recommended (Ruys 1991). To minimise fluctuations in room humidity when wet-cleaning, drainage should encourage rapid removal of water and drying of surfaces (Gorton 1974). Drainage pipes should be at least 10 cm in diameter. In some areas, such as pen-housed rabbits, dog runs, and farm animal facilities, larger diameters are recommended. A rim-flush drain or heavy-duty disposal unit set in the floor may be useful for the disposal of solid waste. When drains are not in use for long periods, they should be capped and sealed to prevent backflow of sewer gases and vermin.

2.4.2 Animal Watering Systems: Bottle or Line Delivery

Animals must normally have access to fresh water at all times. This can be achieved by providing each cage with one or more water bottles, or disposable water bags, or equipping the facility with an automated water line delivery system with a nipple in each cage.

Handling water bottles is one of the most labour-intensive activities in a laboratory animal facility. There are multiple repetitive steps associated with capping, washing, filling, recapping and so on depending on species and stocking levels, these must be repeated several times each week. With the exception of delivery and connection to the cage, much of this process can be robotised. The main advantage of supplying water in bottles is that it allows monitoring of water consumption at every cage. It is also possible to add medication or test compounds to individual bottles. The disadvantage is the labour-intensive nature of the operation and the need to store considerable numbers of spare bottles, sipper tubes and so on.

Automatic watering systems are a labour saving way of distributing water to animals in the facility. If fitted correctly and maintained properly, they are reported to be relative problem free. However there is always a risk that the drinking nipples will malfunction and the animals will either not have access to water or the nipple may leak and flood the cage. The nipples are often not easily accessible and a considerable period of time may elapse before it is realised that there is a problem in an individual cage. The water lines require a considerable amount of maintenance to prevent the growth of pathogens, such as *Pseudomonas spp*. Ease of use must be weighed against the risk of blockage and needs for system maintenance.

2.5 BUILDING MANAGEMENT SYSTEMS (ENVIRONMENTAL CONTROL) AND MONITORING

Modern laboratory animal science dictates that animals used for scientific purposes shall be kept in an environment that is as uniform as possible. This includes maintaining temperature and relative humidity (RH) at predefined values, which depend on the species housed. As a rule of thumb, set points have been chosen that coincide approximately with the thermoneutral zone of the species concerned (but see Chapter 3). When selected, the room environment is then maintained within a predetermined range of this point by means of the building management system (BMS).

Temperature and RH are normally controlled via a BMS that uses sensors located in the air extract ducts or in the room itself. It is therefore essential that the sensors are correctly calibrated and maintained. Temperature should be regulated separately in each room, even if the same species of animals are kept in different rooms, to allow for differences in stocking density, age, and so on. In addition this arrangement makes the system more flexible; even species that require similar temperature and RH values but have different metabolic heat emissions can be switched easily from room to room. If multiple rooms will be always used to house the same species and at a constant stocking density, a single sensor is sometimes used to regulate several rooms or a selection of rooms can be monitored and a mean figure used to regulate the zone as whole.

Animals are kept in rooms with a preset lighting control that regulates duration and sometimes intensity. By convention, lighting is preset to 300 lux measured at 1 metre above the floor and in most cases the light cycle set to 12 hours dark and 12 hours light each 24 hour period, or 10 hours dark and 14 hours light.

Best practice in laboratory animal care requires that the room itself be monitored independently of the BMS system. This reduces the possibility of a sensor failure going unnoticed that might result in the BMS sensor deriving a value that is too high or too low and thus incorrectly regulating the ventilation system. In this case, because the same sensor is used to regulate and trigger an alarm, if control limits were to be exceeded, no alarm would be generated because it is the input of the sensor that is incorrect. The same applies to room lighting. The BMS turns the room lights on or off, but a faulty light bulb or fluorescent tube will not be registered. In this case the room may be set to have lights on, but the actual situation is poor illumination or darkness because of fixture failure.

The installation of independent room sensors increases the likelihood that an alarm would be generated in these cases because it acts independently of the BMS system. Individual room sensors are also valuable for monitoring rooms where zone control systems are used to regulate multiple rooms in conjunction with a single BMS sensor. Separate room sensors can also capture local phenomena such as water leakage (increased humidity), variations in room animal load (high or low stocking density), and defects in lighting. Closed circuit camera systems are sometimes used to monitor the animals but they can never replace direct inspection of animals by skilled persons, which should be performed on a daily basis including weekends and holidays.

2.5.1 MAINTENANCE PROGRAMMES OF THE FACILITY AND EQUIPMENT

To reduce the welfare and research consequences of malfunction or breakdown of essential equipment such as HVAC, lighting, and water supply, it is essential that maintenance programmes and emergency procedures are established and monitored.

2.6 OTHER FUNCTIONAL AREAS

2.6.1 WASHING AREA

The washing area (WA) is probably the most important and critical part of the animal facility. The WA is divided into "clean" and "dirty" parts, with a barrier separating them to avoid any cross contamination. Depending on the caging capacity of the animal facility, these areas may be physically separated by a double door washer and a double door bottle washer, if available. The location of the WA should be decided by considering the disposition of the animal rooms, waste disposal and the storage areas. Traffic flow should be arranged to guarantee separation between dirty and clean materials and personnel.

There must be sufficient space to facilitate movement of equipment and personnel and clear areas should be maintained around equipment that needs to be maintained and serviced. Door openings must be wide enough to allow equipment to enter the area. Utilities such as hot and cold water, steam, adequate floor drainage, and waterproof electric power outlets must be available. Walls and ceilings should be hard wearing, resistant and sealed. In addition, the ventilation should be sufficient to remove the heat-load created in this area. Prominently displayed and clearly legible signs should be in place, emphasising the occupational health precautions necessary for personnel working in that area.

2.6.1.1 Robot Systems

The use of robotic cage wash equipment is being considered increasingly by animal facilities of sufficient size to justify the initial outlay, as a means of reducing hazards to staff and decreasing the impact of staff shortages. Part or all of the process of emptying cages and bottles, washing and refilling them may be automated. Although normally the expense of installing and maintaining such equipment can only be justified if a sufficient number of units are handled each day, there are other, less quantifiable benefits such as reducing exposure of staff to allergens, ergonomic improvement, creation of a more interesting workplace and elimination of repetitive work routines. In practical terms automation is advantageous if most cages and accessories are of the same type, or at least dimensions. Also, it is usually necessary to have a skilled engineer available to maintain the equipment. Consideration of all the above factors should be the basis for a decision about whether or not to robotise.

2.6.2 STORAGE (CLEAN AND DIRTY AREA)

A common characteristic of all animal facilities is lack of adequate storage. Storage needs must be determined in relation to species, activity, and need for buffer space in the cage-clean area. Dedicated separate storage must be planned to accommodate feed, bedding and hazardous chemicals such as disinfectants. The size of the storage area also depends on the logistics of delivery, distance to suppliers and nature and shelf life of the materials.

All animal facilities will sooner or later have to store different types of equipment (cages, racks, etc.) that are not in use. Much of this requires significant volume. A rough rule of thumb is that provision should be made for approximately 20% of the floor area to be used for storage, depending on the type of facility. The consequence of not incorporating this into planning is that it may become necessary to hire external storage space in the future. Material brought in from external storage must usually be fully decontaminated before it can be used.

Adequate storage must also be provided on the clean side of the barrier for materials used in daily husbandry. Feed and bedding stocks should be maintained for at least one week of activity. Bedding and food should be stored separately from materials that pose a risk of contamination from toxic or hazardous substances. Corridors are not appropriate storage areas.

2.6.3 COLD ROOM

Depending on the size of the facility and level and type of activity, provision must be made for short-term storage for the loading dock; in addition necropsy rooms require facilities for storing carcasses/tissues for examination and prior to disposal. For smaller units, a refrigerator (typically set to 4°C) or freezer will be sufficient, but for larger facilities with a high level of activity or in situations where large animal models are used, a walk-in cold storage room (down to –10°C) is the solution of choice.

2.6.4 WASTE AREA: WASTE DISPOSAL

A waste disposal programme based on national legal requirements must be implemented in the facility. Dirty bedding could be transferred to a dedicated area or to a part of a WA on the dirty side. Special precautions should be taken to protect personnel during the disposal procedure, especially from allergens but also from dirty bedding infected with infectious agents or polluted with chemical hazards.

2.6.5 SECURITY: ALARM SYSTEMS—CONTINGENCY PLANS

For security reasons, access into the animal facility should be permitted only to authorised personnel. Access control may take many forms such as locks, card keys, and guard stations. A monitored alarm system should also be installed to guarantee early warning in the event of physical threats or risks.

In order to avoid panic and to ensure effective crisis management, a contingency plan addressing different kinds of threats (fire, earthquake, flood, etc.) should be drawn up. It is recommended that personnel are given frequent training in accordance with this for emergencies. A step-by-step procedure for dealing with emergencies should be clearly displayed on posters at key places within the animal facility.

2.7 COMMUNICATIONS

Communication systems are used in laboratory animal facilities to monitor important information such as environmental conditions necessary for the maintenance of a constant environment for the animals and the efficient running of the facility as well as fire sensors, alarm points, and sounders that are necessary for personnel safety. Environmental monitoring systems, including fire alarm systems, may be connected to automatic telephone notification equipment in order to notify strategic personnel of an emergency.

Fire alarm systems should include visual as well as audible signals and be heard/seen at all locations in the facility independent of activity. Sound may induce seizures and/or stress in susceptible animals, so most audible alarms for laboratory animal facilities operate between 430 and 470 Hz, which is clearly audible to humans but apparently has no adverse effect on the animals.

Equipment, such as laminar air flow cabinets, ultra freezers and water supply systems may be equipped with alarms that ensure notification when they do not function properly. A wireless telephone system combined with a room to office intercom system gives optimal flexibility to contact individuals as well as being an important tool in emergency situations. The intercom system may also be used to provide background music throughout the facility. It is increasingly common to install wireless computing networks within laboratory animal facilities and these can be of great help to facility managers in terms of housekeeping as well as providing flexibility for users of the facility.

A new development is the use of software that makes it possible to track animals and activities within the animal facility. This uses real-time registration of data at the level of the cage and enables communication with working stations outside the facility and between barriers by means of wireless local area networks. Staff in barrier areas may use small handheld wireless terminals (PDAs) or similar to register events and activities and initiate tasks that are notified to people outside the barrier. This is rapidly becoming a tool that ensures information is immediately accessible by all persons associated with an individual animal. For example, a technician may register that an animal is ill; this is transmitted to a clinical veterinarian who uses the system to create a treatment plan. At the same time the responsible researcher is notified and is able to evaluate and approve or suggest modification of the treatment. The treatment plan is transmitted to the technician or a veterinarian who is working in the barrier and is able to identify the room, the animal, and the cage. The treatment is carried out and logged. These systems greatly facilitate rapid responses to situations within the animal facility as well as tracking who is responsible for any given procedure at any given time of the day.

2.8 BARRIERS: LEVELS

2.8.1 Bioexclusion (Microbiological Standardisation)

Microorganisms constitute an important part of the human and animal environment. Although pathogens have the capacity to cause diseases in animals, other microbes that have not been shown to cause any clinical signs of disease can alter the animal in subtle ways, for example, by stimulating or down-regulating the immune system. Animals thus affected react in a different way to the same stimulus during an experiment (Morris 1999). Moreover, if immunocompromised animals are housed in the same barrier, they could be adversely affected by microorganisms that are not normally pathogenic. Because of this, it is important to standardise the microbiological environment in which the animals are maintained. Different barrier regimes are adopted depending on the type of experiments conducted and the characteristics of the animals used (e.g., immunocompetent, immunocompromised, experimentally infected, etc.).

2.8.2 Health Monitoring

The purpose of health monitoring is to determine and standardise the microbiological quality of animals; it is an important prerequisite for reproducible animal experiments. FELASA (Nicklas et al. 2002) has published recommendations for standardising and reporting health monitoring programmes. Although it is not a requirement of these recommendations that animals tested should be free from all of the microorganisms listed, it should be recognised that latent infections can have a considerable impact upon the outcome of animal experiments (behaviour, growth rate, relative organ weight, immune response; Nicklas et al. 1999). Depending on the area of research, it might be necessary for further agents to be excluded from the barrier. If a pathogen from the FELASA-list is identified in the colony, then careful consideration has to be given to the most appropriate course of action. The decision should be made on a case-by-case basis and recorded alongside all arguments including the risk to other studies in the same facility. Depending on the nature of the facility, the general rule is to protect the most sensitive studies.

In the FELASA recommendations the term *unit* describes a self-contained microbiological entity; this might be: (a) the total facility, animal rooms within different buildings that are attended by the same group of people (without special precautionary measures); (b) a classical barrier facility with one or more rooms (irrespective of how many species or strains are maintained within it); (c) an animal room that is protected by preventive measures, such as changing clothing; (d) an isolator or isolators between which animals are freely transferred with no special preventive measures, using procedures that are appropriate to the use of isolators; and (e) a filter top cage or an individually ventilated cage (IVC), which is opened only within a laminar flow cabinet using appropriate biosecurity procedures.

Taken together, not only does the facility structure have an impact on health monitoring and status, but also the cage and rack systems (see Chapter 3). The smaller the microbiological entity (hygiene unit) the greater is the effort needed for health monitoring and the more difficult it is to detect potential pathogens, especially on a single cage basis.

Colonies should be monitored at least quarterly, or more frequently if it is required to identify the presence of some widespread organisms that could seriously interfere with the research. A sample size of at least 10 animals per microbiological (breeding and experimental) unit is recommended (Rehbinder et al. 1996). In microbiological units consisting of two or more rooms or subunits, the sample should comprise animals from as many rooms or subunits as possible.

2.8.3 Immunocompetent Animals

Several methods and barrier systems are used to standardise the microbiological environment (see Section 2.3.1). All equipment and materials used for the animals must be hygienically controlled either by choosing items known already to be pathogen-free or by sterilising or disinfecting them. Personnel entering the barrier unit use an entry lock system, are decontaminated (i.e., water shower, air shower, and/or removal of street clothing), and dress in protective clothing.

2.8.4 Immunocompromised Animals

There are different reasons why an animal may be immunocompromised:

- Genetic models (e.g., nude, SCID, RAG mice, RNU rats)
- Induced (e.g., by irradiation, chemicals, etc.)
- Arising from experimental stress (e.g., surgery, stress models)

Depending on the degree of immune incompetence, special provisions are necessary such as sterilisation of all equipment and materials, special housing systems (i.e., IVC-systems, isolators, etc.).

2.8.5 Experimentally Infected Animals

Microbiological standardisation is at least as important in the case of infection studies as for other experiments. The difference is that the microorganism of interest is added to a preexisting microbiological environment. Special attention has to be given to avoid cross contamination between animals in different experimental groups (e.g., control groups, animals from other research projects in the same room or unit). To avoid cross contamination, special housing systems must be used and, depending on the experimental organism concerned, additional safety precautions must be considered; special containment systems might be necessary if there might otherwise be a risk for humans and/or the environment.

2.9 BIOCONTAINMENT

Research activities might involve work with biological agents in the animal facility. Where it is known that these agents have the potential to be hazardous, dedicated biocontainment systems must be used. The type of containment and the necessary safety measures depend on the risk classification of the agents.

2.9.1 Health and Safety S 0–4

Biological agents are classified into four hazard groups according to their pathogenicity.

Group 1
Biological agents that are very unlikely to cause human disease.

Group 2

European definition: Biological agents that may cause human disease and that might be a hazard to workers but are unlikely to spread to the community. Effective prophylaxis or effective treatment is usually available.

United States definition (NIH guidelines): Risk Group 2 (RG2) agents are associated with human disease, which is rarely serious and for which preventive or therapeutic interventions are often available.

Group 3

European definition: Biological agents that may cause severe human disease and present a serious hazard to workers. They may present a risk of spread to the community but there is usually effective prophylaxis or treatment available.

United States definition: Biological agents that may cause severe human disease and present a serious hazard to workers. Preventive measures/treatment may be available (April 2008 S509: Tribio Classification Standard 5/23).

Group 4

Biological agents that cause severe human disease and are a serious hazard to workers. They may present a high risk of spread to the community and there is usually no effective prophylaxis or treatment.

2.10 QUARANTINE

The objectives of quarantine are to protect other animals in the facility from the introduction of new pathogens and to protect man against zoonotic infection. To achieve this, incoming or reintroduced animals must be isolated from other animals in the facility using special containment systems, dedicated animal rooms or separate buildings. In addition to this level of isolation, material flow and access of people has to be controlled also to avoid cross contamination via material and man. Quarantine rules must be adapted to each situation depending on the risk posed by the introduction and the consequences of a possible outbreak of infection. These can range from isolating the newly arrived animals and closely observing them for clinical signs of disease, using serology, microbiology, and other health monitoring techniques or using sentinels. Depending on the species and microorganisms concerned, it is important to recognise that very often no clinical signs are apparent for uncomplicated infections, especially in rodents. An alternative strategy is to routinely introduce new strains or lines of animals using embryo transfer or by caesarean rederivation, so-called wet hysterectomy. This may be particularly relevant when introducing transgenic animals from outside sources, because often their microbiological status cannot be guaranteed (see Chapter 9).

For rodents and rabbits, the quarantine period can serve also for acclimatisation. This allows the animals to recover from the stress of transport and to become accustomed to the new environment, husbandry, and care practices. The time required for acclimatisation depends on the species, the age of the animals, the duration of transport, disturbance of the diurnal rhythm, the extent of change in environment, and the experiments the animals are intended for. It can range from one week up to a month or more.

2.11 CONCLUSIONS

- Planning and building a laboratory animal facility requires the input of a broadly based design team including representatives of the client and the professional team as described in Section 2.1.1.1 right from the beginning.
- If cost cutting is necessary, it must be in line with the facility concept.

- Specific knowledge and experience are crucial for a successful building programme and advice should be sought from persons with previous experience in planning and building animal facilities.
- Great benefit may be obtained by visiting other facilities and participating in special courses in laboratory animal facility planning.
- For proper planning, all different types of research and the specific requirements for each have to be identified and incorporated. Flexibility has to be built-in to cover possible changes in demand in the future.
- Standardisation of the microbiological environment is essential to secure animal welfare, to facilitate high-quality research and to minimise between-animal variability, which increases the numbers of animals required for an experiment or invalidates the results. The environmental monitoring required depends on the research topic and the animal model.
- Whenever possible a test room should be set up in the initial phase of construction to validate different engineering solutions and the compatibility of the various components before full-scale building commences.
- Before using the new facility, enough time must be allowed for a test run (proving period), which is crucial to ensure that all functions work correctly. This time can be used for testing and validation and to obtain all necessary licences from accreditation bodies and authorities.
- Training and recruitment strategies should be implemented to ensure that there are sufficient experienced and well-educated staff as described in Section 2.2 to run the facility.
- Special attention should be given to overall management of the facility. The head of a laboratory animal facility must possess the specialised competencies needed to support scientists by providing the highest quality of laboratory animals with respect to welfare (assured by high standards of husbandry within a well-defined environment), genetic integrity, and microbiological status.
- Building and running a laboratory animal facility is a total concept that includes the quality of the building and the competence and continuing education of the whole staff.
- Animals should always be kept and used under the best possible conditions in order to assure a maximal output of valid and precise data.

2.12 QUESTIONS UNRESOLVED

- Little is known about the impact of different facility concepts and housing systems on the running costs (i.e., IVC versus open cage system for mice).
- Not all elements, structures or equipment deployed have been tested and certified for use in laboratory animal facilities.

REFERENCES

AORN. 1982. Recommended practices for traffic patterns in the surgical suite. From the AORN recommended practice subcommittee. *AORN Journal* 35:750–58.
ASHRAE. 2001. *Fundamentals handbook.* Atlanta: American Society of Heating, Refrigeration, and Air-Conditioning Engineers.
Ayliffe, G. A. 1991. Role of the environment of the operating suite in surgical wound infection. *Reviews of Infectious Diseases* 13 Suppl. 10: S800–S804.
Ayscue, D. 1986. Operating room design. Accommodating lasers. *AORN Journal* 43:1278–87.
Bartley, J. M. 1993. Environmental control: Operating room air quality. *Today's OR Nurse* 15:11–18.
Besch, E. L. 1980. Environmental quality within animal facilities. *Laboratory Animal Science* 30:385–406.

Bourdillon, R. B., and L. Colebrook. 1946. Air hygiene in dressing-rooms for burns or major wounds. *The Lancet* 247:601–5.

CCAC. 2003. CCAC (Canadian Council on Animal Care) guidelines on laboratory animal facilities, characteristics, design and development. [Cited December 31 2009]. Available from http://www.ccac.ca/en/CCAC_Main.htm

European Commission. 2007. *Commission recommendation of 18 June 2007 on guidelines for the accommodation and care of animals used for experimental and other scientific purposes* (2007/526/EC).

Fitzgerald, R. H., Jr. 1979. Microbiologic environment of the conventional operating room. *Archives of Surgery* (Chicago, IL: 1960) 114:772–75.

Gorton, R. L. and E. L. Besch. 1974. Air temperature and humidity response to cleaning water loads in laboratory animal storage facilities. *ASHRAE Trans.* 80:37–52.

Hessler, J. R. 1991. Facilities to support research. In *Laboratory animal facilities*, ed. T. Ruys, Vol. 2, 34–55. New York: Van Norstrand.

Holton, J. and G. L. Ridgway. 1993. Commissioning operating theatres. *The Journal of Hospital Infection* 23:153–60.

Hughes, H. S. Reynolds, and M. Rodriguez. 1996. Designing animal rooms to optimize airflow using computational fluid dynamics. *Pharmaceutical Engineering* 16:44–65.

Hughes, H. C., and S. Reynolds. 1997. The influence of position and orientation of racks on airflow dynamics in a small animal room. *Contemporary Topics in Laboratory Animal Science* 36:62–67.

Humphreys, H. 1993. Infection control and the design of a new operating theatre suite. *The Journal of Hospital Infection* 23:61–70.

ILAR/NRC. 1996. *Guide for the care and use of laboratory animals*. Washington, DC: National Academy Press.

Memarzadeh, F., P. C. Harrison, G. L. Riskowski, and T. Henze. 2004. Comparison of environment and mice in static and mechanically ventilated isolator cages with different air velocities and ventilation designs. *Contemporary Topics in Laboratory Animal Science* 43:14–20.

Morris, T. 1999. Laboratory animal health monitoring. *Laboratory Animals* 33:1.

Morse, B. C., S. D. Reynolds, D. G. Martin, A. J. Salvado, and J. A. Davis. 1995. Use of computational fluid dynamics to assess air distribution patterns in animal rooms. *Contemporary Topics in Laboratory Animal Science* 34:65–69.

Nevalainen, T., E. Berge, P. Gallix, B. Jilge, E. Melloni, P. Thomann, B. Waynforth, and L. F. van Zutphen. 1999. FELASA guidelines for education of specialists in laboratory animal science (category D). Report of the Federation of Laboratory Animal Science Associations Working Group on Education of Specialists (Category D) accepted by the FELASA board of management. *Laboratory Animals* 33:1–15.

Nicklas, W., P. Baneux, R. Boot, T. Decelle, A. A. Deeny, M. Fumanelli, B. Illgen-Wilcke, and FELASA (Federation of European Laboratory Animal Science Associations Working Group on Health Monitoring of Rodent and Rabbit Colonies). 2002. Recommendations for the health monitoring of rodent and rabbit colonies in breeding and experimental units. *Laboratory Animals* 36:20–42.

Nicklas, W., F. R. Homberger, B. Illgen-Wilcke, K. Jacobi, V. Kraft, I. Kunstyr, M. Mahler, H. Meyer, G. Pohlmeyer-Esch, and GVSOLAS Working Grp Hyg. 1999. Implications of infectious agents on results of animal experiments—Report of the working group on hygiene of the Gesellschaft fur Versuchstierkunde—Society for Laboratory Animal Science (GV-SOLAS). *Laboratory Animals* 33:39–87.

Poole, T. B., ed. 1999. *The UFAW handbook on the care and management of laboratory animals*. Vol. 1. Oxford: Blackwell Scientific.

Rehbinder, C., P. Baneux, D. Forbes, H. van Herck, W. Nicklas, Z. Rugaya, and G. Winkler. 1996. FELASA recommendations for the health monitoring of mouse, rat, hamster, gerbil, guinea pig and rabbit experimental units. Report of the Federation of European Laboratory Animal Science Associations (FELASA) Working Group on Animal Health accepted by the FELASA board of management, November 1995. *Laboratory Animals* 30:193–208.

Ruys, T. 1991. *Handbook of facilities planning. Laboratory animal facilities*. Vol. 2. New York: Van Nostrand Reinhold.

Schonholtz, G. J. 1976. Maintenance of aseptic barriers in the conventional operating room: General principles. *The Journal of Bone and Joint Surgery. American Volume* 58:439–45.

Stakutis, R. E. 2003. Cage RACK ventilation options for laboratory animal facilities. *Lab Animal* 32:47–52.

Wilkins, C. K. and B. A. Waters. 2004. HVAC design in animal facilities. *ASHRAE Journal* 46:35–40.

3 Housing and Care of Laboratory Animals

Hanna-Marja Voipio, Finland; Ping-Ping Tsai, Germany; Heinz Brandstetter, Germany; Marcel Gyger, Switzerland; Hansjoachim Hackbarth, Germany; Axel Kornerup Hansen, Denmark; and Thomas Krohn, Denmark

CONTENTS

Objectives 30
Key Factors 30
3.1 Routine Care of Laboratory Animals 31
3.2 Primary Enclosure: Cages and Housing 31
    3.2.1 Open Cages 32
    3.2.2 Filter-Top Cages 33
    3.2.3 Individually Ventilated Cages (IVCs) 34
    3.2.4 Metabolic Cages 35
    3.2.5 Disposable Cages 35
    3.2.6 Complex Systems 36
    3.2.7 Pens 36
3.3 Secondary Enclosure and Rack System 36
    3.3.1 Open Rack 36
    3.3.2 Ventilated Cabinet 38
    3.3.3 IVC Rack System 39
    3.3.4 Isolator 40
    3.3.5 Special Housing 41
3.4 The Macro- and Micro-Environment 41
    3.4.1 Carbon Dioxide (CO_2), Ammonia and Relative Humidity 41
    3.4.2 Temperature in the Secondary Enclosure 42
    3.4.3 Odour and Housing 44
    3.4.4 Acoustic Environment 45
        3.4.4.1 Auditory Perception 45
        3.4.4.2 Vocalisation 45
        3.4.4.3 Effects of Sound on Physiology and Behaviour 46
        3.4.4.4 Acoustic Environment in Animal Rooms 46
    3.4.5 Illumination 48
        3.4.5.1 Visual Perception 48
        3.4.5.2 Vision and Housing 49
    3.4.6 Food 51
        3.4.6.1 Supply and Treatments of Diets 51
        3.4.6.2 The Effect of Diet on Experimental Results 52

    3.4.7 Water Supply...52
        3.4.7.1 Water Dispensing...52
        3.4.7.2 Water Treatment..53
        3.4.7.3 The Effect of Water Treatment on Experimental Results...............53
    3.4.8 Bedding Material..54
        3.4.8.1 The Effect of Bedding on Experimental Results............................55
        3.4.8.2 The Effect of Bedding on Animals...55
        3.4.8.3 The Effect of Bedding on Humans...55
3.5 Transport of Animals..56
3.6 Health Care Programme..57
    3.6.1 Preventive Veterinary Medicine for Recently Arrived Animals..................57
    3.6.2 Veterinary Surveillance..57
3.7 Engineering Control and Maintenance of Biosecurity...57
3.8 Conclusions..58
3.9 Questions Unresolved...59
References...60

OBJECTIVES

Recent decades have witnessed substantial developments in laboratory animal science in the fields of hygiene control, genetics and environment. This has led to major improvements in the way animals are housed and cared for in laboratory animal facilities.

Laboratory animals are dependent entirely on the care given to them, including the choice or design of primary and secondary enclosures (housing), routine/daily care, transportation, suitable macro- and micro-environmental conditions, veterinary care and biosecurity management.

A variety of primary (cage or pen) and secondary enclosures (rack systems) is available for rodents and rabbits, suitable for a range of different purposes; many of them are commercially available. In order to meet species-specific environmental needs, laboratory animal scientists have paid particular attention to the macro- and micro-environment including temperature, CO_2 and ammonia concentration, relative humidity, light intensity and cycle, bedding, nutrition, water, ventilation and sounds. Initiatives in these areas aim to enhance animals' well-being by improving their physiological and mental health. Other studies have focused on the impact that macro- and micro-environment exerts on the quality of experimental results.

This chapter reviews the literature underpinning our current knowledge and summarises areas of uncertainty relating to the housing and care of laboratory rodents (mice and rats) and rabbits.

KEY FACTORS

1. The characteristics of primary and secondary enclosures for mice, rats and rabbits should be determined by the purpose of housing and condition of facility.
2. Routine/daily and veterinary care procedures must be carefully planned, implemented and verified with a view to maintaining the health and well-being of animals and the validity of scientific findings.
3. Arrangements for transportation must take account of the type of transport and duration of the journey.
4. Macro- and micro-environmental factors, such as: CO_2 and ammonia concentrations, humidity, temperature, sound, light, food, water and bedding can markedly affect the welfare of animals and science.
5. The macro- and micro-environment can influence experimental results and the extent of interference depends on the nature of the environment and the parameter of interest.

3.1 ROUTINE CARE OF LABORATORY ANIMALS

Laboratory animals rely totally on the care they receive from staff. In addition to providing basic needs such as feed and water, laboratory animals must be observed daily, 7 days a week (Council of Europe 2006); the practise of providing sufficient feed and water to last for a weekend without subsequently checking the animals is not acceptable. Caging should be designed and racks loaded so as to permit animals to be observed without unduly disturbing them. Water bottles should be transparent or translucent so that staff can easily observe the amount of water in the bottle and automatic watering systems must be tested regularly. Staff should be trained to recognise when animals are deprived of water, for example, because water nipples or valves have become blocked. Cages should always be checked for flooding if bottles are found unexpectedly empty.

Animals may be disturbed by inexperienced handling, or even simply by a change of handler. It is good management practise to ensure that all animal care staff remain confident and competent at handling the range of animals with which they work routinely, and also those they may be required to care for periodically, for example, at weekends or to provide cover for vacations or sickness.

When animals are under study and subject to procedures that could compromise their normal behaviour or ability to eat and drink, the frequency and closeness of observation should be increased. Staff responsible for these duties must be trained to observe deviations from normal behaviour and to report concerns promptly to the appropriate person (see Chapter 14). It is also essential that staff have access to information about the procedure that has been carried out to help them correctly interpret the animal's condition. An animal that is recovering from anaesthesia or has been exposed to a microbiological challenge may react in a manner that causes concern if the observer was not adequately informed in advance, or does not have access to instructions about the correct action to take. Wherever possible, procedures should be planned so as to avoid recovery after hours or during the weekend. Should this not be possible, arrangements must be made to ensure adequate care and observation during the recovery period. There is an obligation to keep staff, including contract and holiday temporary staff, fully informed about ongoing procedures. Provision should be made to enable the responsible veterinarian and principle investigator (or his/her representative) to be contacted should this be necessary; telephone numbers should be accessible at all times, preferably in the facility or sometimes via the institution's security personnel. Crisis plans must be developed and tested to ensure adequate care of the animals in the event of power or plant failure, industrial action, fire, flooding and so on.

3.2 PRIMARY ENCLOSURE: CAGES AND HOUSING

Primary and secondary enclosures determine the immediate environment of laboratory animals, and should be arranged to support their species-specific requirements. The term "animal enclosure" such as cage, pen, run and stall has been defined in the revised Appendix A of the European Convention for the protection of vertebrate animals used for experimental and other scientific purposes (ETS 123; Council of Europe 2006) as the primary accommodation in which the animals are confined. For rodents this usually means cages, although for rabbits it can also be pens. The minimum cage dimensions and floor areas for mice, rats and rabbits are given in Tables 3.1 through 3.3 adapted from Appendix A.

A cage is a permanently fixed or movable container that is enclosed by solid walls and, on at least one side or the top, by bars or meshed wire or sometimes nets, and in which one or more animals are kept or transported. Depending on the stocking density and the size of the container, the freedom of animals to move is relatively restricted (Council of Europe 2006).

TABLE 3.1
Minimum Enclosure Dimensions and Space Allowances for Mice

	Body Weight (g)	Minimum Floor Area (cm²)	Floor Area Per Animal (cm²)	Minimum Cage Height (cm)
In stock and during procedure	≤20	330	60	12
	>20–25	330	70	12
	>25–30	330	80	12
	>30	330	100	12
Breeding		330		12
		For a monogamous pair (outbred/inbred) or trio (inbred) For each additional female plus litter 180 cm² should be added		

TABLE 3.2
Minimum Enclosure Dimensions and Space Allowances for Rats

	Body Weight (g)	Minimum Floor Area (cm²)	Minimum Cage Height (cm)	Floor Area Per Animal (cm²)
In stock and during procedure	≤200	800	18	200
	>200–300	800	18	250
	>300–400	800	18	350
	>400–600	800	18	450
	>600	1500	18	600
Breeding		800	18	
		Mother and litter; for each additional adult animal permanently added to the enclosure add 400 cm²		

3.2.1 OPEN CAGES

In the early 1960s a systematic approach was developed to the size and construction of cages for rodents (cage types I–IV; Spiegel and Gunnert 1961). The cage is covered with a metal grid, in which a food hopper and water bottle are placed. For laboratory rodents, cages are commonly fabricated of stainless steel and polycarbonate (PC). However, polysulfone (PSU) and polyetherimide (PEI) are also used and are said to be more resistant to autoclaving and most alkaline or acid detergents. Compared to PC, cages fabricated from PSU and PEI are claimed to reduce light transmission by 65–80% without affecting the ability to see into the cage and to have a longer life span because of their higher heat resistance. At the present time the cost of PSU and PEI cages are still higher than PC. There should be a programme in place to replace cages that are cracked or crazed. Some authors (Howdeshell et al. 2003; Koehler et al. 2003) have reported that such cages release bisphenol-A, a compound used in the production of plastic cages, which has oestrogen-like characteristics and that may impair reproduction and interfere with experimental results. In addition, it is difficult to observe animals in cracked cages. In most European countries four different standard cage sizes (Type I, II, III and IV) were commonly provided for rodents (Figure 3.1). Because the minimum floor area for mice has been set at 330 cm² by the Council of Europe, Type I cages (floor area 200 cm²)

TABLE 3.3
Minimum Enclosure Dimensions and Space Allowances (Cages and Pen) for Rabbits

Age	Body Weight (kg)	Minimum Enclosure Size (cm²)	Minimum Floor Area Per Animal (cm²)	Minimum Height (cm)	Raised Area[a] (Size/High, cm)
7 weeks (weaning)		4000	800	40	25 × 25/25
8–10 weeks		4000	1200	40	
> 10 weeks	< 3	3500		45	55 × 25/25
	3–5	4200		45	55 × 30/25
	> 5	5400		60	60 × 35/30
		Minimum floor area for one or two socially harmonious animals			
	Doe weight		Addition for nestbox (cm²)		
Doe plus litter	< 3	3500	1000	45	
	3–5	4200	1200	45	
	> 5	5400	1400	60	

[a] If there are scientific or veterinary justifications for not providing a raised area then the floor area should be 33% larger for a single rabbit and 60% larger for two rabbits.

are no longer being sold. Currently additional cage types, such as Type IL, IIL or IVS are available. The floor area of these cage types varies from one company to another; for example, the floor area of Type IL cages can vary between 330 cm² and 426 cm².

Many laboratory animal facilities use both wire-grid and solid floor cages. Solid floor cages are provided with bedding material that absorbs urine and desiccates faecal pellets and can be utilised by animals to manipulate their local micro-environment. In contrast, wire-grid floors can be mildly stressful and may lead to foot lesions during prolonged (>1 year) experiments (Heidbreder et al. 2000; Peace et al. 2001; Krohn, Hansen, and Dragsted 2003b; Eriksson et al. 2004; Sauer et al. 2006).

Rabbit housing should incorporate a raised area on which the animal can sit and beneath which it can lie, sit and move (see Chapter 4); this should cover no more than 40% of the floor area (Council of Europe 2006).

Commercial rabbit cages contain a solid or perforated floor; the latter is usually provided with a waste pan or tray suspended under the cage (Figure 3.2). All cages are designed to permit the rabbit to engage in caecotrophy. Cages with floor areas ranging from 3600 cm², 4200 cm² and 5400 cm² are available commercially.

Besides the traditional stainless steel, other cage materials are available that can resist temperatures up to 90°C or 121°C and that are resistant to saline solutions, lye and non-oxidising mineral acids.

3.2.2 Filter-Top Cages

The static filter-top cage was developed originally to prevent the spread of viruses from mice to the surroundings (Kraft 1958) and was subsequently modified to utilise a standard cage. The filter was placed on top of the cage with an overhang along the cage edge (Sedlacek et al. 1980). Different styles of filter tops are marketed commercially (examples are shown in Figure 3.3).

Measurements inside filter-top cages reveal that filter tops increase temperature and humidity as well as CO_2 and NH_3 concentrations in the cage (Simmons et al. 1968; Murakami 1971; Serrano 1971;

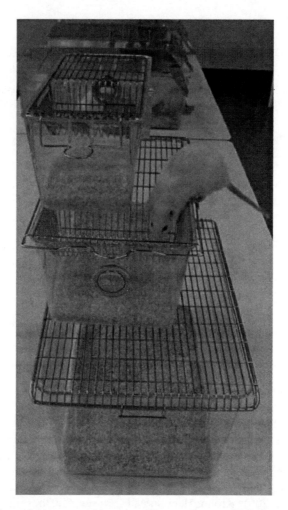

FIGURE 3.1 Standard open laboratory rodent cages. (Photo courtesy of T. P. Rooymans.)

Keller et al. 1989; Corning and Lipman 1991; Memarzadeh et al. 2004). This can be overcome by using individually ventilated cages (Lipman, Corning, and Coiro 1992; Memarzadeh et al. 2004). However, filter-top cages, also named microisolators, are not as effective as flexible film isolators because normally there is no seal between the filter top and cage and moreover the filter is coarser and does not exclude particles as fine as those trapped by HEPA filters that protect film isolators. Husbandry and handling of animals requires lids to be removed from cages so breaking the hygiene barrier; this is overcome by staff being appropriately trained and carrying out such tasks inside a sterile hood, which must be properly maintained (efficiency monitored, regular servicing and filter changing and so on). Currently, filter-top cages are often used for in-house transport of animals.

3.2.3 INDIVIDUALLY VENTILATED CAGES (IVCs)

As described above, IVCs were developed to overcome the limitations of the filter-top cage. A variety of systems are available commercially. The reduced ventilation within filter-top cages is overcome by providing a controlled supply of filtered air to each cage, the arrangements for which are incorporated into the IVC rack. When access to animals, cleaning and so on is carried out only beneath laminar flow cabinets, the level of containment is higher than that afforded by filter-top cages but

FIGURE 3.2 Rabbit cage with a shelter and perforated floor. (Photo courtesy of T. P. Rooymans.)

FIGURE 3.3 Two different types of filter-top cages. (Photos courtesy of T. P. Rooymans and J. Kirchner.)

less than that of isolator systems (for more information about IVC rack systems see Section 3.3.3, "IVC Rack System").

3.2.4 METABOLIC CAGES

A metabolic cage for rodents is one in which faeces and urine excreted by an individual animal can be collected separately without significant loss. It is also possible to monitor food and water intake (Scheline 1965; Weigelt 1970; Engellenner et al. 1982). The standard metabolic cage for rodents is circular and the animal stands on a grid floor above a separation system for faeces and urine (Figure 3.4). Typical sizes are small with a floor area of 320 cm^2 and a height of 14 cm or a large with a floor area of 450 cm^2 and a height of 18 cm. Because of their size and construction, containment in metabolic cages can be stressful (Eriksson et al. 2004) and they should be used only when necessary for sampling faeces and urine during experimental procedures (see Chapter 4).

3.2.5 DISPOSABLE CAGES

Disposable conventional cages as well as IVCs for rodents are now available commercially. Cages that incorporate pre-filled water bottles and bedding may simplify operation of a facility and save

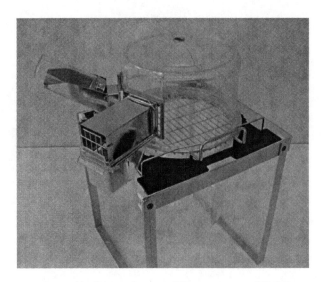

FIGURE 3.4 An example of a metabolic cage for rats. (Photo courtesy of T. P. Rooymans.)

labour; possibly even building costs; purchase, operation, and maintenance of washing room or autoclave and so on. Because these are relatively recent introductions it will take some time before the pros and cons of such cages compared with reusable, conventional caging can be evaluated in terms of workload, cost, environmental impact (can the material be recycled?) and animal welfare.

3.2.6 COMPLEX SYSTEMS

Highly complex enriched housing systems are sometimes used to house experimental animals for neuroscientific experiments. More structured and complex housing systems have also been developed and made available commercially for rodents and rabbits in order to improve animal welfare (see Chapter 4).

3.2.7 PENS

As an alternative to open cages, pens provide a satisfactory way of group-housing rabbits. "Pen" is defined (Council of Europe 2006) as an area enclosed, for example, by walls, bars or meshed wire in which one or more animals are kept. Depending on the size of the pen and the stocking density; the freedom of movement of animals is usually less restricted than in a cage and appropriate environmental enrichment items can be provided (see Chapter 4).

3.3 SECONDARY ENCLOSURE AND RACK SYSTEM

Secondary enclosures can be holding rooms, open racks or containment systems, such as isolators, ventilated cabinets or IVC systems, within which primary enclosures may be placed. The choice of rack system depends on the barrier level (see Chapter 2 for more general description of barrier levels), the condition of the animal room and the intended function or purpose. An overview of different rack systems is given in Table 3.4.

3.3.1 OPEN RACK

The open rack is a popular, well-established system used in different levels of barriers; it provides multiple shelves or guiderails that can support open or filter-top cages. Open racks may be fitted either with swivel casters and brakes or fixed on the wall; the former is easier to move in the animal

TABLE 3.4
An Overview of Different Rack Systems

System	Description	Advantage	Disadvantage	Hygiene Unit
Open rack	A rack contains multiple shelves, in which open or filter cages can be arranged and stored.	Easy to access Easy to observe animal Easy to perform health monitoring	No protection against infection Micro-environment communicates directly with macro-environment	Animal room
Ventilated cabinet	A closed cabinet offers a constant air exchange rate with HEPA filtered air Open or filter cages can be arranged and stored in cabinet	Relatively easy to access Easy to observe animals Easy to perform health monitoring Protects against transmission of infection Reduces dust in animal rooms Reduces exposure to allergens	No protection, when cabinet is opened (unless filter-top cages are used or working within a laminar airflow cabinet) Microclimate depends on the macro-environment Cage changing is more labour intensive Requires more space More expensive	Cabinet If filter top cages are used, each single cage is a hygiene unit
IVC (individual ventilated cage system)	A rack system providing assisted ventilation, in which special filter-top cages are used Exhaust air may go through a HEPA filter back to animal room or directly out of the room Cages can be maintained at positive or negative pressure to surroundings	Good hygienic protection (no cross-contamination) if a laminar airflow cabinet is used Needs less space Reduces dust in animal rooms Reduces the risk to staff of infection or allergy May enable the air exchange rate of the animal room to be reduced	Difficult to observe animal High workload, when working with laminar airflow bench The performance of health monitoring is not totally solved Very high cost	Each single cage
Isolator	Air-tight transparent container, precludes direct contact with the outside environment (everything is handled using built-in sleeves with attached gloves) Both air supply and exhaust can pass through a HEPA filter Incorporate a double lock system for passing animals or materials in and out Possibility of operation at positive or negative pressure to the room	Highest degree of hygienic protection Gnotoniotic and germ-free animal can be maintained Can be also used for biocontainment (i.e., infection studies) purposes Easy to perform health monitoring Reduces dust in animal rooms Reduces the risk to staff of infection or allergy	Very labour intensive Difficult to manipulate animals with attached gloves Highest cost	Isolator

room and more flexible for the animal care staff. Open racks for rabbits contain doors in which food hoppers and water bottles can be fixed. Open racks are relatively inexpensive and are easy to maintain and use on a routine basis.

3.3.2 VENTILATED CABINET

The ventilated cabinet could be considered as a secondary enclosure for rodents, which are placed in their cages into the cabinet (Figure 3.5). The cabinet offers a constant flow of HEPA filtered air and makes it possible to reduce the allergen load of air in the room if the air exhaust is coupled to the room ventilation extract or filtered. Both open and filter-top cages can be used with this system. Airborne cross-contamination is still possible and if animals are housed in open cages they will be exposed to room air when the cabinet is opened. This can be avoided by connecting the whole cabinet directly to a laminar airflow cabin during cage changing or experimental procedures. A combination of filter-top cages and ventilated cabinets together with cage changing and conduct of procedures on animals under sterile conditions—for example, in a laminar flow cabinet-can constitute a hygiene barrier. The effectiveness of the hygiene barrier depends mainly on the individual persons using the equipment and its proper maintenance (control, change of filters, etc.). Most ventilated cabinets do not offer heating and ventilation air conditioning (HVAC) facilities. The ventilation system for the cabinets must be backed up by emergency power and if a centralised ventilation system is used, some redundancy must be included to accommodate mechanical failure or power failures, and their consequences. A comparison of the characteristics of different rack systems is provided in Table 3.4.

FIGURE 3.5 Ventilated cabinet.

Housing and Care of Laboratory Animals

3.3.3 IVC Rack System

The microclimate within cages may be improved by actively ventilating them, so-called IVC systems. However, if direct air inlets are provided for each cage, localised areas of high air velocity could occur and this could compromise the well-being of animals (see also Chapter 4). For this reason air velocity in space occupied by animals should be validated, as described in "Performance Evaluation of IVC Systems" (Brandstetter et al. 2005). The IVC systems have been commercially available since the early 1980s and are widely used currently to protect animals and/or staff (Figure 3.6a and b). To protect the animals from infectious agents in the in-going air,

FIGURE 3.6 Examples of different IVC systems. (Photos courtesy of T. P. Rooymans.)

the latter must be ventilated through a HEPA-filter with an efficiency of up to 99.97% (Mrozek et al. 1994). Several experiments have shown that a cage operating at positive pressure is able to protect animals against infections (McGarrity and Coriell 1973; Lipman, Corning, and Saifuddin 1993; Mrozek et al. 1994; Clough et al. 1995). When the pressure in the cage is set to be negative to that in the room and exhaust air is filtered through a HEPA filter or ducted directly out of the room, rodent-derived allergens can be prevented from contaminating the workplace (Sakaguchi et al. 1990; Renström, Bjoring, and Höglund 2001); this also prevents spread of microorganisms from infected animals. Like filter-top cages, IVCs are sometimes called "microisolators" but this term can be misleading because they are still not as efficient as flexible film isolators. The effectiveness as a hygiene barrier depends mainly on cages being opened and the animal handled only within sterile handling facilities such as laminar flow cabinets that have been properly maintained (monitored, checked periodically, filters changed, etc.). There are many different systems on the market and it is important to properly evaluate the IVC system with respect to its intended application (i.e., bioexclusion, biocontainment, protection against allergens; Brandstetter et al. 2005). Most IVC systems do not have facilities to control temperature and humidity of the air supplied to cages, although a few do. As with ventilated cabinets, back-up power must be available in case of emergency.

3.3.4 Isolator

Isolators provide the highest degree of microbiological exclusion or containment. They provide an absolute barrier between animals kept in open cages within the isolator and the room in which they are located (Figure 3.7). An isolator can be constructed of either flexible or rigid material (often plastic). Incoming air is passed through a HEPA filter. Staff have no direct contact with the inside environment. Animals, cages and so on are handled using built-in sleeves to which impervious gloves are attached. All equipment and material is sterilised; for example, by autoclaving in special

FIGURE 3.7 Flexible film isolator for rodents, an example. (Photo courtesy of T. P. Rooymans.)

microbiologically secure containers that are subsequently connected to the isolator by means of a special sealed port (or "lock"). The interior of the port is chemically sterilised (e.g., with peracetic acid or other oxidising agents) before the contents of the container can be transferred. Alternatively, the surfaces of materials can be sterilised directly with suitable vapours or mists in the port itself, although if there are viable organisms underneath the surface they may not be killed by this method. Waste materials are removed also through the port. The frequency with which materials are passed into or out of isolators should be minimised because each penetration of the barrier constitutes a contamination risk. Consequently, supplies including equipment and consumable materials are stored inside isolators for longer periods, depending on their capacity. Isolators can be used for bioexclusion (e.g., to house gnotobiotic animals) or biocontainment (e.g., studies with infective micro-organisms, quarantine of newly arrived animals, etc.). For bioexclusion the isolator is operated under positive air pressure and incoming air is HEPA filtered. For biocontainment, the isolator is operated at negative pressure and out-going air (sometimes also in-going) is HEPA filtered. As with IVCs and ventilated cabinets, the ventilation system must be provided with emergency power. The integrity of the isolator should be regularly assessed, both visually and by leak detection methods. When isolators are emptied and rebuilt for use they must be thoroughly cleaned and re-sterilised.

3.3.5 SPECIAL HOUSING

Special types of housing are sometimes needed in order to compensate for biological deficiencies in animals—for example, for certain phenotypes of mutant and genetically modified mice or particular or toxicological models. Special housing for these purposes is designed to meet the needs of the animals and on the uses to be made of them, so they should be chosen on an individual basis.

3.4 THE MACRO- AND MICRO-ENVIRONMENT

A major factor in decreasing variation between animals and consequently the number of animals used in experimental research ("Reduction") has been achieved by standardising their macro- and micro-environment, and improving their health status ("Refinement"). In view of this it is very important for animal facilities to closely monitor and control the macro- and micro-environment, to keep appropriate records and to periodically review these as a management tool.

3.4.1 CARBON DIOXIDE (CO_2), AMMONIA AND RELATIVE HUMIDITY

Relative humidity (RH) and the concentrations of gases such as carbon dioxide and ammonia, are important factors in an animal's physical environment. The characteristics of the micro-environment differ from those of the macro-environment, principally because of thermal exchange and uptake and release of gases by animals in the cage. For example RH and CO_2/ammonia concentration tends to be higher within cages than in macro-environment, especially when filter tops are being used or the air exchange rate is low (Weihe 1965; Simmons et al. 1968; Murakami 1971; Serrano 1971). Increasing the rate of ventilation decreases humidity and CO_2/ammonia concentration both in animal rooms and inside cages (Fujita et al. 1981; Hasenau, Baggs, and Kraus 1993; Choi et al. 1994; Huerkamp and Lehner 1994; Reeb et al. 1997; Memarzadeh et al. 2004; Ooms et al. 2008; more detailed information about the importance of ventilation is provided in Chapter 2).

Relative humidity plays a major role in animals' thermoregulation. An ambient RH of between 50 and 60% is often recommended for breeding and maintenance of rats, mice, and rabbits. Low humidity decreases the viability of airborne micro-organisms and their potential for transmission, and reduces the release of ammonia by microbial degradation of urine; it has also been shown to induce earlier puberty (Clough 1984; Drickamer 1990). However, if RH falls below 30%, and especially at temperatures above 22°C, the incidence of ringtail disease increases in some strain of rats (Njaa, Utne, and Braekkan 1957; Flynn 1967) or pouched mice (Ellison and Westlin-van

Aarde 1990). Although in mice, RH of 60–70% has been shown to result in improved breeding performance (survival to weaning) and body weight gain or in reduction of airborne allergen levels (Anderson, Werboff, and Les 1968; Ellendorff et al. 1970; Donnelly and Saibaba 1989; Drickamer 1990; Jones et al. 1995), higher humidity, combined with high temperature, may result in decreased whole body and sexual organ weight (Ellendorff et al. 1970) as well as increased ammonia production in the cage (Gamble and Clough 1976). Elevated temperatures associated with high RH may lead to mortality (Batchelor 1999).

Intra-cage ammonia and carbon dioxide concentration is influenced by numerous factors: strain, gender and age of animals, diet, bedding and cage material, the presence and type of filter top, internal air movements, ambient temperature and RH, animal density within the cage, and frequency and time after bedding changes (Murakami 1971; Serrano 1971; Gamble and Clough 1976; Stolpe and Sedlag 1976; Raynor, Steinhagen, and Hamm 1983; White and Mans 1984; Hirsjärvi and Väliaho 1987; Corning and Lipman 1991; Eveleigh 1991; Eveleigh 1993; Hasenau, Baggs, and Kraus 1993; Choi et al. 1994; Potgieter and Wilke 1996; Lipman 1999; Smith et al. 2004a; Smith et al. 2004b; Smith et al. 2005; Burn and Mason 2005; Burn et al. 2006b; Ooms et al. 2008; Silverman et al. 2008). Schaerdel and others (1983) exposed rats to air containing different concentrations of ammonia and showed that up to 100 ppm the only measurable effect was a very small change in blood ammonia concentration. Ammonia release from faeces and urine inside soiled cages or pens plays an important role in the pathogenesis of diseases (i.e., murine respiratory mycoplasmosis) and may alter metabolic and physiological processes (Broderson, Lindsey, and Crawford 1976; Vesell et al. 1976; Gamble 1982; Schaefer 1982; Schoeb, Davidson, and Lindsey 1982). There are few reported studies of the impact of intra-cage CO_2 concentration and most were performed before 1970; they failed to show substantial toxicological effects on the animals (Krohn and Hansen 2000). Lipman, Corning, and Coiro (1992) reported that CO_2 concentration inside open cages was 334 ± 40 ppm (281 ± 20 ppm in holding room), comparable to the concentration in atmospheric air. It has been reported that ventilation failure can result in a significant increase in intra-cage CO_2 concentration (5000 up to 20,000–80,000 ppm) within 2 hours, depending on the type of cage (Krohn and Hansen 2002). Concentrations above 30,000 ppm may affect the welfare of the animals (Krohn, Hansen, and Dragsted 2003a) as well as experimental results (Schaefer 1982; Krohn and Hansen 2000). There is still a need to collect more information about acceptable levels of intra-cage CO_2 and ammonia concentrations.

3.4.2 TEMPERATURE IN THE SECONDARY ENCLOSURE

The recommended temperatures for housing laboratory rodents are between 20 and 24°C; it is in the range 16–21°C for rabbits.

In general, the ambient temperature should be maintained within the animal's thermoneutral zone (TNZ). The TNZ for an animal can be defined as the ambient temperature at which thermal stress is minimal (for review see Romanovsky, Ivanov, and Shimansky 2002). Gordon (1985) identified it as the temperature at which the animal spends most of its time when given the choice along a thermogradient. This approach has been applied to mice, rats (Gordon 1987), hamsters and guinea pigs (Gordon 1986). Szymusiak and Satinoff (1981), using maximal REM sleep time, showed that REM sleep time in rat increases regularly between 23 and 29°C followed by a rapid fall between 31 and 33°C. They concluded the TNZ in Long Evans rats is set around 29°C.

Romanovsky, Ivanov, and Shimansky (2002) emphasised that TNZ is influenced by several factors including age, health status, air humidity and velocity, barometric pressure, body contact with surfaces of the primary enclosure and the effective radiant field. Many other studies (Szymusiak and Satinoff 1981) have shown that the TNZ lies well above the temperature range of 20–24°C recommended for housing several laboratory rodent species. In the rabbit, Gekhman (1974) has determined that the TNZ ranges between 22 and 28°C and the recommended temperature range for housing is between 16 and 20°C.

Gordon, Becker, and Ali (1998) have examined the behavioural thermoregulatory responses of individually caged mice versus mice of different ages housed in groups; singly housed aged mice showed a significant preference for warmer places (around +1°C) compared to grouped mice, although this difference was not apparent in 2-month-old mice.

Gordon, Becker, and Ali (1998) estimated the heat loss of a "phantom" mouse in cages with different inclusions; the greatest contribution to thermal protection was bedding comprising of wood shavings in which the animal can burrow: the gain was around 6°C above the ambient housing temperature. In the case of bedding that was sparely distributed or consisted of wood chips, when burying was more difficult, the gain of temperature was minimal (Gordon 2004). The presence of a top filter and group living improved protection by 2°C and 1°C, respectively; under such conditions, mice, during daytime, would reach their specific TNZ at an ambient temperature between 20 and 24°C. During the night when they are active, mice are much less dependent on ambient temperature for their thermal comfort (Table 3.5; Gordon 2004). Moreover, Gaskill and others (2009) used preference testing for C57BL mice giving the animals a choice between cages maintained at different temperatures; they found that preference depends on the sex of the animal, its body condition, circadian rhythm and behaviour. The authors concluded that it is impossible to recommend a single optimal room temperature valid under all conditions for housing mice.

The inside temperature of a cage on a rack depends on its location and there are differences between cages and also between cages and room (Clough 1984).

In summary, although the temperature range currently recommended for housing laboratory rodents is significantly lower than the TNZ, it is satisfactory from a welfare viewpoint, conversely higher temperatures would adversely impact on personnel working with the animals. On the other hand, animals should be provided with nesting material and/or shelters for thermal insulation to help them avoid thermal stress.

The preceding discussion highlights the possibility that ambient temperature may influence experimental results. Laboratory rodents housed in metabolic cages with grid floor and without bedding or nesting material will be subject to mild thermal stress if maintained in rooms between 20 and 24°C, although the impact on metabolism and scientific results remains to be established. Swoap and others have demonstrated that small changes in ambient temperature have a significant impact on cardiovascular variables in rats and mice (Swoap, Overton, and Garber 2004; Swoap et al. 2008). Changes in ambient temperature also have marked effects on immune responses and can interfere with research into both infective conditions (Rudaya et al. 2005) and toxicology (Gordon

TABLE 3.5
Neutral Thermal Zones of Laboratory Rodents

Species/Strain	NTZ	References
Wistar rat	29.0–30.5°C	Romanovsky, Ivanov, and Shimansky 2002
BDIX rat	29.5–31.0°C	Romanovsky, Ivanov, and Shimansky 2002
Long Evans rat	28.0–30.5°C	Romanovsky, Ivanov, and Shimansky 2002
Long Evans and F344 rat	around 28°C (day) 22–24°C (night)	Gordon (1993)
Zucker lean rat	28.0–31.0°C	Romanovsky, Ivanov, and Shimansky 2002
Zucker fatty rat	28.0–29.0°C	Romanovsky, Ivanov, and Shimansky 2002
Sprague Dawley rat	24.9 ± 0.4°C	Gordon (1987)
Mouse	29.5–31.9°C	Gordon (1985)
Hamster	32–33°C (day) 26–28°C (night)	Gordon (1993)
Guinea pig	29–32°C	Gordon (1986)

2005; Fawcett 2008; Gordon et al. 2008). In conclusion, the thermal environment of animals undergoing experimental procedures should be carefully regulated and monitored and should be consistent between subjects and studies.

3.4.3 ODOUR AND HOUSING

Olfaction is an important means of social communication for mice and rats (see Chapter 4) and for this reason cage cleaning can be problematic (Hurst 2005). There are two conflicting pressures: the need to clean cages to maintain hygiene and health and to avoid too frequent disturbance of scent-marking patterns. Cage cleaning, in which scent marks are removed from only parts of the cage and the substrate, can promote aggression in male mice. In view of this, it is recommended that mice be transferred into completely clean cages with fresh bedding (Gray and Hurst 1995; Jennings et al. 1998). If it is anticipated that aggression might be a problem, transfer of animals together with their own nesting material into a clean cage has been shown to reduce aggression of male BALB/c mice, while transfer of sawdust containing urine and faeces seem to intensify aggression (Van Loo et al. 2000).

It has also been suggested that unfamiliar odours—for example, those associated with humans such as perfumes and deodorants—can induce stress responses in laboratory mice (Dhanjal 1991). This should be taken into account when cleaning cages and handling the animals; it is an important precaution to change clothing and wash hands after handling predator species, such as rats and cats, or their bedding to avoid causing fear reactions in mice.

Mice show fear responses when they encounter anaesthetised rats (Blanchard et al. 1998), possibly because rats are natural predators; if both species are housed in the same room, mice might become aware of rats by olfaction, even if they cannot see them. It is recommended therefore that mice and rats should always be kept in separate rooms (Jennings et al. 1998). However, a current study (Meijer, van Loo, and Baumans 2009) has demonstrated that heart rate (HR), body temperature (BT), activity (AC) and urinary corticosterone were not elevated in female C57BL/6 mice during and after Wistar rats were introduced into their room. Housing rats and mice in the same room appeared to be no more distressing than cage cleaning, which caused a temporary increase of HR, BT and AC.

Because olfaction plays a critical role in the social behaviour of mice, concern has been expressed that inbreeding of laboratory mice might lead to male mice becoming unable to discriminate between their own scent marks and those of other males (Nevison et al. 2000). This could have obvious consequences for behaviours such as agonistic and aggressive encounters and could easily confound responses in experiments. In addition, the lack of ability to discriminate between olfactory signals at an early age might influence performance in behavioural studies in which this sensory modality is involved (Forestell et al. 2001).

In European rabbits (*Oryctolagus cuniculus*) olfactory stimuli play an important role in regulating many aspects of their life. In addition to pheromone deposition in urine and faeces, chin marking is one of the most conspicuous forms of olfactory communication. Secretion of the chin gland is used in maintaining social status (Hayes et al. 2002) and the frequency of chinning correlates with other behavioural indicators of dominance (Black-Cleworth and Verberne 1975; Albonetti, Dessí-Fulgheri, and Farabollini 2006; Shimozuru et al. 2006; Arteaga et al. 2008). Chinning activity is also positively correlated with oestrus in females, as a form in sexual advertisement (Gonzalez-Mariscal et al. 1990; Hudson and Vodermayer 1992; Gonzalez-Mariscal et al. 1997). Marking behaviour was observed more frequently in group-penned rabbits than those individually caged (Podberscek, Blackshaw, and Beattie 1991). Group-housed rabbits appeared to mark their pens by rubbing their chins on the edges of various structures (Love 1991) and would often perform marking in the course of agonistic activities as a sequel to approach and olfactory investigation, particularly rabbits that were unfamiliar with both environment and group-mates (Albonetti, Dessí-Fulgheri, and Farabollini 2006).

3.4.4 Acoustic Environment

3.4.4.1 Auditory Perception

Auditory sensitivity varies between species, strains or stocks, and even ages. Unlike human beings, rodents can hear ultrasounds (above 20 kHz). In mice, the hearing range is from 2 kHz to about 80 kHz; the most sensitive area is between 8 and 25 kHz (Heffner and Masterton 1980; Zheng, Johnson, and Erway 1999). In mice especially, strain differences in hearing sensitivity have been reported and many strains have genetically impaired hearing ability (Willott et al. 1995; Zheng, Johnson, and Erway 1999). Furthermore, hearing loss has been shown in aged mice, with clear gender differences in the age of onset (Henry 2004).

The hearing range of albino rats is between 250 Hz and 80 kHz at 70 dB with a hearing threshold for most frequencies of 10–15 dB; sensitivity is highest between 8 and 38 kHz (Gourevitch and Hack 1966; Kelly and Masterton 1977; Borg 1982). Although albinism has been said to predispose to auditory defects, no differences have been observed between the audiograms of albino and pigmented Norway rats (Heffner et al. 1994).

Auditory evoked nerve-brainstem responses can first be detected in rat pups 7–8 days old (Geal-Dor et al. 1993). The auditory threshold decreases at the age of 12–13 days when the meatus opens and by day 20 it resembles that of adult rats (Crowley and Hepp-Reymond 1966). At the same time, the animals develop behavioural responses, such as orientation to sound (Kelly, Judge, and Fraser 1987). Hearing loss occurs in rats from the age of 20–25 months and is most severe between 24 and 40 kHz (Backoff and Caspary 1994; Keithley, Lo, and Ryan 1994).

Rabbits hear sounds from about 500 Hz up to about 30 kHz, and are most sensitive to frequencies between 2 and 16 kHz; the hearing threshold is between 10 and 20 dB (Heffner and Masterton 1980; Borg and Engström 1983).

3.4.4.2 Vocalisation

Mice and rats emit sounds within both audible and ultrasonic ranges. In mice, ultrasonic vocalisations are less common than in rats (Portfors 2007; Scattoni, Crawley, and Ricceri 2009) and occur during female–male (Maggio, Maggio, and Whitney 1983), female–female (Maggio and Whitney 1985) and mother–pup interactions (Noirot 1972). There have been only a few descriptions of audible vocalisations in rats, for instance when they are in pain (Levine et al. 1984; Voipio 1997) and during mother–pup contact, especially when the mother manipulates the pups (Ihnat, White, and Barfield 1995). There is a positive correlation between audible squeaks and non-aggressive skirmishing in conjunction with cage changing (Burn, Peters, and Mason 2006a).

In contrast, there are many publications about ultrasonic vocalisation in rats. Most ultrasounds are emitted at frequencies between 21 and 32 kHz (Adler and Anisko 1979), although sometimes higher frequencies are produced. During fighting, male rats emit short high-pitched calls at 40–70 kHz, or long calls between 23 and 30 kHz (Sales 1972). Females attacked by lactating females emit both high (33–60 kHz) and low (20–32 kHz) frequency calls. Compared to males, female ultrasounds at low frequencies are shorter in duration and higher in frequency (Haney and Miczek 1993). During oestrus, the frequency of 50 kHz vocalisation increases (Matochik, White, and Barfield 1992). Several calls have been recorded, during mating, including 22 kHz post-ejaculatory songs (Barfield and Geyer 1972) and short pre-ejaculatory calls (Barfield et al. 1979).

Isolated infant rats emit calls between 40 and 50 kHz before they reach 3 weeks of age (Noirot 1968) and also in response to cold (Okon 1971). Vocalisation studies suggest that social contact and olfactory cues are more important than warmth to the welfare of rat pups (Hofer and Shair 1978; Carden and Hofer 1992). Unusual tactile stimulation, such as being picked up or rolled, induces newborn pups to emit audible sounds, later audible and ultrasounds and after 10 days of age only ultrasounds (Ihnat, White, and Barfield 1995).

Adult rats may emit ultrasounds (20–35 kHz) when they are in pain (Kaltwasser 1991), or in response to a startle reaction induced by acoustic stimuli (Kaltwasser 1990). They also vocalise during ordinary handling (Kock 1986; Brudzynski and Ociepa 1992), isolation (Francis 1977) and in stressful situations (Blanchard et al. 1991); there are differences between females and males (Blanchard et al. 1992). In fact, vocalisation responses to stress have been used widely as an experimental tool in pharmacological investigations. The characteristics and functions of ultrasonic vocalisation in rats and mice have been reviewed by Portfors (2007).

In their commentary on communication, Arch and Narins (2008) conclude that even if ultrasonic vocalisation by rat pups is only the by-product of laryngeal braking as proposed by Blumberg and Alberts (1990), it still has a signalling role and leads to more or less predictable behaviours, thus justifying use of the term communication.

3.4.4.3 Effects of Sound on Physiology and Behaviour

Sound has been shown to have several physiological effects in laboratory animals. In rats, development of the auditory cortex is delayed if the animal is exposed to continuous moderate intensity noise (Chang & Merzenich 2003). One of the earliest behavioural findings was that exposure to sound increases washing and grooming activity (Anthony, Ackerman, and Lloyd 1959). It may affect also growth rate, learning ability, blood parameters, locomotor activity, the circulatory system and reproduction, and can evoke audiogenic seizures (Zondek and Tamari 1964; Pfaff 1974; Algers, Ekesbo, and Strömberg 1978; Peterson 1980; Gamble 1982; Sales 1991; Soldani et al. 1999; Prior 2006). Sounds can induce specific reactions, such as acoustic startle (Hoffman and Fleshler 1963; Fleshler 1965; Ison and Russo 1990), freezing (Anthony, Ackerman, and Lloyd 1959) and orientation (Kelly, Judge, and Fraser 1987; Brudzynski and Chiu 1995). These reactions have been combined into a classification system by Voipio (1997), who showed that white-noise sounds are capable of evoking strong behavioural and fear responses even at low sound pressure levels (SPL), whereas wave type sounds such as a whistle are effective only at a high SPL although animals can adapt to particular sounds. Turner, Bauer, and Rybak (2007) have reviewed the effects of sounds on laboratory animals.

3.4.4.4 Acoustic Environment in Animal Rooms

The acoustic environment in laboratory animal facilities has been studied and characterised by several researchers. Sounds are produced, especially during working hours, by animal care activities (Fletcher 1976; Pfaff and Stecker 1976; Sales, Milligan, and Khirnykh 1989; Milligan, Sales, and Khirnykh 1993; Sales, Milligan, and Khirnykh 1993). Sound intensity varies between the day and night time, and between weekdays and weekends; for example background noise in animal rooms may exceed 100 dB(A) during working hours, while at night it can be below 50 dB(A) (Peterson 1980; Sales, Milligan, and Khirnykh 1993). Maintenance tasks and routine care procedures generate environmental sounds of varying frequency and intensity. Noise is generated by (1) environmental control systems including ventilation, lighting and humidity control; (2) maintenance and husbandry procedures such as feeding, watering and transport of animals; (3) operation of cleaning equipment, for example, cage and bottle washers, vacuum and floor cleaning; and (4) other equipment such as tattoo guns, centrifuges and computer equipment. The intensity and frequency of sounds, including ultrasound, above 20 kHz present in animal facilities have been measured by Sales and co-workers. One concern frequently expressed is that environmental ultrasound, may disrupt communication between animals and it is argued that special care should be taken to detect and minimise such sounds in the animal's environment. Routine care activities including cleaning, changing cage grids, addition of food pellets, running tap water and animal transport caused SPLs between 60 and 90 dB. In most of these procedures both low (10 Hz–12.5 kHz) and high (12.5–70 kHz and up to 100 kHz) frequencies were measured. The ventilation system contributed only a small part of the background noise but the SPL varied between 35 and 50 dB, being lower at night-time than during working hours; only low levels of high frequency sounds were detected (Sales et al. 1988; Sales, Milligan, and Khirnykh 1989; Milligan, Sales, and Khirnykh 1993; Sales, Milligan, and Khirnykh 1999).

The amount of noise produced during cage change is influenced by the material from which cages are made and working practises; metal cages cause about 10–20 dB(R) higher sound intensity compared to plastic cages. When procedures are performed in a hurried way, they generate significantly higher sound levels than during calm working. Even the sequence in which animals and feed are removed and placed into a new cage is important. When changing rats into clean cages, addition of diet pellets to the clean feed hopper first reduced sound intensity by about 15 dB(R)* compared to pouring in pellets after the rats have been transferred. Plastic animal cage material causes less noise than steel, which should be considered when advantages and disadvantages of cage materials are assessed (Voipio et al. 2006). Furthermore, regular care maintenance of the equipment (e.g., oiling the wheels of cage racks) is an easy way to reduce adventitious environmental sound intensity. Using suitable insulation materials on the animal room surfaces helps to moderate noise level.

Animals also generate noise themselves and there is considerable variation between species. In rodent rooms, overall sound levels are higher during working hours than the rest of the day, and it is animal care activities that cause most of the noise during daytime. Mice and rats appear to be quiet because they generate sound levels at 40–50 dB, whereas rabbits are noisier. These measurements are affected by material the cages are constructed from: rabbits maintained in metal cages generate sound levels of 80 dB whilst those in plastic cages produce only 60 dB (Sales, Milligan, and Khirnykh 1989). In animal rooms, especially during working hours, often sudden loud sound peaks are produced and the combined noise caused by animal care tasks and animals themselves can exceed 90–100 dB(A) briefly (Peterson 1980; Sales, Milligan, and Khirnykh 1989; Sales, Milligan, and Khirnykh 1999); these may cause fear responses (Voipio 1997).

The acoustic environment differs in each animal facility, depending on the species and activities undertaken. It is not possible to give clear recommendations regarding sound levels for laboratory animals, in part because it can be difficult to avoid temporary loud noises while working, although care should be taken to minimise them. It is important to be aware that sounds influence animal physiology, behaviour and welfare, and consequently experimental results, so it is important to monitor sound intensity, including high frequency sounds audible to laboratory rodents and rabbits, in facilities and individual animal rooms. Such measurements are also necessary to protect the health of staff; in washing areas especially, sound intensity may exceed limits set down in regulations and in these cases, ear plugs or other ear protection should be worn. In contrast with species such as dogs and pigs, noise in rodent and rabbit rooms is seldom sufficiently loud to require the use of hearing protectors. There are many practical and often easy means to minimise the intensity of sound generated by equipment or husbandry practises. Some equipment, for example computers, may generate ultrasound and, if it is necessary to bring such devices into animal rooms, it is advisable to enclose them in a cupboard behind closed doors and some distance from animals because ultrasound attenuates within a short distance and does not penetrate thick materials.

The responses and reactions of animals depend on the characteristics of the sound. Rats show fear reactions when they hear white-noise sounds even at low intensity. Some sources of these sounds, such as tearing paper products close to the animals, can be avoided (Voipio 1997). In the same study, it was observed that rats showed fear reactions when exposed without warning to short bursts of high intensity whistling sounds, but not if the intensity was low. To some extent animals can adapt to sounds—the response decreases if the sound is played repeatedly—however after a rest period, the response returns to that of naïve animals. It is common experience that rodents eat their pups after exposure to sudden or continuous loud noises, for example during construction work in

---

* dB = Decibel: A ratio scale for the measurement of sound intensity. dB(lin): unweighted linear sound pressure level (SPL); when linear scale is used, the SPL is calculated to include the sound intensity over the whole spectrum. dB(A): The A frequency weighting is based on a person's hearing and it compensates for the error between unweighted SPL and the sound level experienced in the human ear. dB(R): The R frequency weighting is correspondingly based on rat hearing.

a facility, although this has not been verified experimentally. It is a wise precaution to avoid sudden noises in animal rooms.

In some facilities attempts are made to mask or offset the effect of exposure to sudden or loud sounds by playing radio music or white noise in the background; this has stimulated considerable debate and a few publications (Pfaff 1974; Pfaff and Stecker 1976; BVAAWF/FRAME/RSPCA/UFAW 1993; Sales, Milligan, and Khirnykh 1999). Sales, Milligan, and Khirnykh (1999) observed that the effect of sound from a radio depended on the volume set by the listener, being around 70 dB. She concluded that because the loudspeakers of most radios limit the frequency response to below 16 kHz, they fail to cover the full hearing spectrum of most laboratory animals, so any masking benefit is unclear and any direct effect on animals seems unlikely. The effectiveness of playing music or generating white-noise type sounds as a way of mitigating the effect of background noise in animal rooms is unproven, and more studies are needed to identify the advantages and disadvantages to animals and research.

When assessing acoustic environments or individual sounds and their effects on animals it is important to bear in mind the difference in auditory sensitivity between different species. For example, in many routine cage changing procedures, the apparent intensity of sound perceived by humans was 10–20 dB louder than would have been perceived by the rat, when corrected for its auditory sensitivity (Voipio et al. 2006). The importance of such differences in hearing range has been highlighted by Sales, Milligan, and Khirnykh (1999) and by Turner, Bauer, and Rybak (2007). Despite these hearing differences, results of most sound studies are expressed as dB(lin), a measure of the total loudness of sounds, or adjusted by using an A-weighting filter based on human hearing. Neither expression correctly represents the perception of sound by an animal, which makes interpretation of results difficult. This shortcoming can be avoided by using species-specific weighting filters for sound, for example the R-weighting that takes account of the rats' auditory spectrum (Björk et al. 2000; Voipio et al. 2006).

3.4.5 ILLUMINATION

3.4.5.1 Visual Perception

Many rodents are crepuscular or nocturnal and are active under low light conditions. Their retina is afoveate and there is a high predominance of rods making them sensitive to low light intensities as well as to slight contrast and slow movements. Rodents have laterally placed eyes and a large visual field (Burn 2008). Visual acuity in laboratory rodents is much poorer than in diurnal species such as humans and therefore laboratory rodents have an indistinct visual image of the world (Prusky, West, and Douglas 2000).

Although mouse and rats have been assumed to lack colour vision, their retina bears dispersed cones (1% versus 99% rod photoreceptors) and of these many are sensitive to red–green (93% of cones 510–530 nm) and the remainder to ultraviolet (UV; 7% of cones, 360 nm) wavelengths (Burn 2008). The significance of UV photoreceptors in rodents is not known, although mice do not seem to use UV cues for foraging or in environmental interactions (Sherwin 2007; Honkavaara, Aberg, and Viitala 2008). Other hypotheses need to be explored—for example, it has been suggested that UV perception might assist in spatial orientation.

There is continuing debate about whether the laboratory rabbit (*Oryctolagus cuniculus*) should be classified as a nocturnal, crepuscular, or diurnal species (Jilge 1991). The retina is characterised by horizontal strips enriched with cones (Juliusson et al. 1994) and in contrast to the mouse, which has a ventral streak of cones (Szél et al. 1996), the rabbit bears two retinal streaks, a medial horizontal one rich in green cones and a ventral one rich in blue cones. It is believed that the first of these serves to scan the horizon for predators and the ground with its rich green colours, whereas the second is more devoted to detection of aerial predators as in the mouse (Famiglietti 2005). The green visual pigment of rabbit has peak absorbance at 509 nm (Radlwimmer and

Yokoyama 1998) and the blue one around 430 nm (Nuboer, Vannuys, and Wortel 1983). Rabbits are prey animals and rely on a large visual field and sensitivity to movement in order to rapidly identify potential danger. Together, both eyes cover the whole animal surrounding except a small blind zone behind the head. The binocular visual field is approximately 24° and may serve visual tasks such as food recognition and prehension (Collewijn, Martins, and Steinman 1981). The provision of windows at the rear of cages and a platform on which the rabbit can rest may provide opportunities for the rabbit to scan and interact with its surroundings (BVAAWF/FRAME/RSPCA/UFAW 1993).

3.4.5.2 Vision and Housing

Many guidelines recommend that secondary enclosures for rodents and lagomorphs should be illuminated by light with a spectrum that mimics daylight as closely as possible (CCAC 2003). However, this is difficult because the spectrum of daylight varies during the day and cage furniture and construction materials also interfere with provision of the full light spectrum. For example, PC and PSU, the two materials most commonly used for fabricating cages, cut off short wavelengths below 400 nm. It is particularly important to recognise this in relation to housing rodents (which are able to detect light in the range of 360 nm) in completely contained primary enclosures such as filter-top cages and IVC systems. Currently, it is not known whether a distorted light spectrum affects the health, behaviour, or welfare of laboratory rodents. The literature carries conflicting reports of consequences for dental health (Sharon, Feller, and Burney 1971; Feller et al. 1974), growth and development (Wurtmann and Weisel 1969; Ozaki and Wurtman 1979; Saltarelli and Coppola 1979).

The most widely used light sources in animal holding rooms are discharge lamps (fluorescent tubes). A maintenance programme should be established to ensure that fluorescent tubes are regularly changed before they fail completely, because their spectral quality and light intensity degrade over time. If the flicker frequency of a discharge lamp is below the critical fusion frequency of the animal species housed, there could theoretically be a welfare issue, although Sherwin (2007) has shown that the flicker of conventional fluorescent tubes is not aversive. Despite this, perceptible low frequency flickering of tubes in animal housing rooms should be avoided and lamps replaced immediately. Lalitha, Suthanthirarajan, and Namasivayam (1988) have documented that a low frequency flickering light source of 80 lux for 30 minutes can cause stress to rats.

The best established effect of high light intensity on the health of laboratory animals is retinal damage, principally reported in albino rats (O'Steen, Shear, and Anderson 1972; Weihe 1978; Clough 1982; Semple-Rowland and Dawson 1987a, 1987b; Blom 1993). Stotzer and others (1970) and Williams, Howard, and Williams (1985) showed that light intensities as low as 60 lux over a period of 13 weeks or of 133 lux for 3 days were sufficient to induce retinal damages in albino rats. Assuming the most important parameter is retinal irradiance (intensity × time), then hyperthermia, intermittent light exposure and pre-exposure environment may accelerate retinal damages (Organisciak and Winkler 1994). However, retinal degeneration is a consequence not only of environmental conditions, but is strongly influenced by genetic factors, and in particular the Pde6brd1 mutation (Anonymous 2002). Recently, albino outbred mice have also been found with this mutation (Serfilippi et al. 2004; Chia et al. 2005; Clapcote et al. 2005). Clearly, environmental conditions are not the only factor predisposing to blindness. However, low intensity lighting should be provided for laboratory rodents, especially albinos.

A distinction should be made between recommendations about the level of light inside cages and in the room, which is shared by both animals and animal caretakers. Schlingmann, Pereboom, and Remie (1993b) have demonstrated that albino rats avoid areas in the cage brighter than 25 lux and melanic strains above 60 lux. Juvenile rats play significantly less in an environment with a light intensity around 570 lux compared to a darker environment (Vandershuren et al. 1995). Rat sleep patterns are also affected by light intensity. Van Betteray, Vossen, and Coenen (1991) found that rats, housed at a light intensity over 500 lux, sleep in a curled position in contrast to the extended body posture adopted at light intensities of less than 10 lux.

The exposure of animals to light depends not only on the illumination in the room but also the position of the rack and location of the cage on this (Kupp et al. 1989). A screen should be installed on top of racks in order to protect animals in the uppermost row of cages from high light intensity.

A conflict of interest exists between the dim light intensity required for housing laboratory rodents, especially albino strains, and the higher intensity required to provide animal house staff with a visually acceptable and safe environment. Schlingmann and others (1993a) proposed that the minimal intensity of room lighting should be 210 lux at working height, whereas CCAC (2003) recommends provision of two levels of room illumination, one for when people are working and a lower one when animals are not attended.

Light affects not only vision but also impacts endocrine functions and particularly affects reproduction; it is recommended that light intensity inside cages for rat breeding should be not be lower than 20–25 lux (GV-SOLAS 2004).

A regular circadian light cycle is of utmost importance for maintaining the temporal organisation of biological systems, from the molecular to entire organism levels. Constant illumination has important consequences for physiological processes (see, for a review, Cos et al. 2006). Disturbance of the light–dark cycle for 1 week markedly increases stress and aggressive behaviours in mice (Van der Meer, Van Loo, and Baumans 2004). Even a short period of illumination during the dark phase of the circadian cycle impacts drastically on pineal melatonin metabolism of laboratory rodents (Illnerova et al. 1979; Lerchl 1995). Melatonin has multiple actions on physiology and metabolism. It is not uncommon for people working in animal houses after office hours to use flash lights, switch on room lights or work under illuminated laminar flow for short periods of time. It is important that the consequences of short bouts of illumination on animal welfare and scientific data are considered not only in relation to the investigator's own animals but also others housed with them.

Usually, animals are maintained under a 12:12 or 14:10 light/dark cycle, year around (Jennings et al. 1998). A regular circadian light cycle such as this provides consistency throughout the year for experimental subjects. However, even in a controlled and consistently illuminated laboratory animal facility, hormonal profiles of male laboratory rat show circa annual and semi-annual fluctuations (Wong et al. 1983); usually, no adjustment is made for winter time/summer time.

Sunset and sunrise duration may be less important for ground dwelling species such as mice and rats than for arboreal species engaged in motor activity at the time the light switches off. However, the biological impact of sunrise and sunset on the temporal organisation of behaviour and physiology has not been clearly established, and the matter is not usually addressed in studies of circadian rhythms (Jud et al. 2005).

The consequences of prolonging dusk and dawn in rooms housing birds have been investigated and shown to improve their welfare. On the other hand, dim illumination as low as 0.25 lux during the night phase enhances tumour growth in rats (Dauchy et al. 1999; Cos et al. 2006).

In order to perform functional tests on nocturnal rodents during their active phase, it is usual to reverse day–night cycles. Caution has to be taken to avoid light pollution during the dark phase and to ensure that there is sufficient time during the light phase for staff to provide adequate care for the animals. Traditionally, a red light of the type used in photographic darkrooms is used, but this can be uncomfortable to work in and the spectral range of some cheaper types of lamps can be quite broad and provide visual cues to the animals. Sodium lamps are preferable because they emit over a very narrow spectrum, which is outside the photopigment sensitivity of laboratory rodents; the illumination is more comfortable for researchers and allows animal care staff to perform straightforward housing duties (McLennan and Taylor 2004). As an alternative, low-level illumination has been used during the reversed nocturnal active phase to allow observation of animals.

Beeler et al. (2006) tested this concept with mice on a reversed 15 lux, 12 hours/100 lux, 12 hours cycle and, although the animals' circadian rhythms were maintained, some aspects of metabolism were disturbed (Dauchy et al. 1999; Beeler, Prendergast, and Zhuang 2006).

When practising surgery and using high intensity or cold lights (6000–20,000 lux) it is of utmost importance to avoid directing illumination into the eyes of subjects, particularly of albino animals (GV-SOLAS 2004).

3.4.6 Food

A nutritionally balanced diet is important to maintain the health of laboratory animals and to ensure that experimental results are not biased by unintended nutritional factors. Consequently, laboratory diets should be formulated according to a limited selection of raw materials and manufactured in accordance with a standardised process; raw materials should be sourced with a view to assuring quality and avoiding potential contaminants and there should be a comprehensive monitoring and quality control programme. For these reasons, food should be sourced only from reputable suppliers who are subject to periodic audits. This is discussed further in Chapter 7.

Most commercially available laboratory animal diets are based on a fixed formula, designed to meet the requirements of the respective species (Clarke et al. 1977). The basic components of valid commercial diets are: carbohydrate, protein, fat, fibre, vitamins, and minerals. Although even purified ingredients may contain small quantities of various micronutrients, resourceful manipulation of the ingredients enable special diets to be formulated that adequately supply all known essential nutrients except one of interest.

Various formulations are available for different laboratory animal species at different stages of their lives (e.g., breeding, growing, or maintenance). The adequacy of a diet is determined by the animal's physiological status including its age, genetic constitution (Fenton and Cowgill 1947; Fenton et al. 1950; Lee, King, and Visscher 1953; Hoag and Dickie 1965; Knapka, Smith, and Judge 1977; Mizushima et al. 1984; Nishikawa et al. 2007), physical status (e.g., room temperature or humidity; Treichler and Mitchell 1941; Hegsted and McPhee 1950; Collins, Schreiber, and Elvehjem 1953; Hartsook and Nee 1976), and the nature of the experiment. Growth and reproductive performance are key indicators of dietary adequacy. Apparent strain/species differences in nutritional requirements have been observed and with the development of increasing numbers of transgenic strains, it may become necessary to perform more research on the nutrient requirements of each individual.

3.4.6.1 Supply and Treatments of Diets

Food for laboratory animals can be provided in different forms, the commonest of which for laboratory rodents and rabbits is the pelleted diet. Suppliers are responsible for monitoring all incoming ingredients, the manufacturing process, and all finished diets. Formulation usually makes allowance for possible nutrient losses during manufacturing and storage.

Within an animal facility, food should be stored under clean conditions, at a low temperature, low humidity, and in an enclosed area to minimise the nutrition degradation and contamination and to avoid contamination by disease agents. In general, diets based on natural ingredients and stored in an air-conditioned environment should be used within 180 days of manufacture, and diets containing vitamin C within 90 days of manufacture.

To ensure the safety of food, it is often sterilised, for example by irradiation or heat treatment (frequently within animal facilities). The consequences of sterilisation on nutritional quality depend on the nature of the diet (Porter and Festing 1970). Autoclaving may increase the hardness of pelleted diet and the efficiency of food utilisation, but can cause destruction of many vitamins and denature proteins (Zimmerman and Wostmann 1963; Porter and Lane-Petter 1965; Ford 1976, 1977). To counteract such effects, commercial autoclavable diets contain extra nutrients to compensate for degradation during steam sterilisation. Diets packed under vacuum or nitrogen can be sterilised by irradiation with less damage to nutrients than by heat sterilisation (Coates et al. 1969; Ley et al. 1969; Adamiker 1976), although there is some loss of vitamin activity and, in same instances, this can lead to vitamin deficiency (Diehl 1991; Hirayma et al. 2007).

Ad libitum feeding is normal practise for rodents, but not always for rabbits. Feeding concentrate diets *ad libitum* to laboratory rodents (especially rats) and rabbits is a common cause of obesity.

3.4.6.2 The Effect of Diet on Experimental Results

Food restriction is beneficial in terms of life span, the incidence and severity of degenerative diseases such as nephropathy and cardiomyopathy, and so on as considered at length in Chapter 10. If restricted feeding schedules are introduced, they should take account of species-specific needs, for example with respect to the timing and duration of any deprivation. The biological consequences also need to be recognised; for example, chronic food restriction in rats can lead to elevated blood corticosterone concentration and modified open field activity (Heiderstadt, Mclaughlin, and Wright 2000) and has been shown to differentially alter secretory patterns of prolactin, GH, and ACTH in group-housed Lewis and Wistar rats (Lopez-Varela et al. 2004). Unlike food restriction, fasting is not used during routine care (see Chapter 4). Other factors, including the presence of dietary fats, appear to produce subtle changes in the physiology and behaviour of mice and may account for some differences in the outcomes of investigations (Ritskes-Hoitinga et al. 1992, 1998; Brain, Maimanee, and Andrade 2000). It is good practise to request batch-analysis reports on a regular basis.

Currently, a few researchers (Marsman et al. 1991; Ritskes-Hoitinga et al. 1991; Jensen and Ritskes-Hoitinga 2007) have examined variation in the composition of commercial diets to identify possible influences on experimental results.

3.4.7 Water Supply

Laboratory animals require access to potable and uncontaminated water 24 hours a day. Water quality can vary from domestic drinking water, demineralised/partial desalinated water, and sterilised water. Depending on the animal species and management needs, it is appropriate to set different qualities of water for different purposes and to establish the way in which it will be dispensed.

The definition of domestic potable water may vary with locality. The person responsible for management of the animal facility should monitor local water quality as documented by the local water supply companies (waterworks) or arrange for in-house or outside testing of water leaving the taps. Moreover, the quality and material of fabrication of pipes and the layout in a building must be taken into consideration to avoid contamination by leaching, the inadvertent establishment of lengths of pipework through which water rarely flows (with subsequent local microbiological colonisation) susceptibility to blockages and leakages and so on.

3.4.7.1 Water Dispensing

Individual drinking bottles made from PC, PSU, or PEI are the most popular way of providing water and are suitable for most laboratory animals (up to 3 kg body weight). Automatic drinking systems based on bowls or nipple drinkers are also popular for all laboratory animal species (see Chapter 2). It is important that drinkers and drinking bottles are cleaned or replaced regularly to avoid possible contamination from the oral flora of animals.

Recently, the "single use water bag" comprising a disposable pouch and valve has been introduced as an alternative way of supplying drinking water to caged mice, rats, guinea pigs and hamsters during transportation. Alternatively, a moist diet such as soaked food pellets, or gel preparations can be provided to animals that have been recently weaned or undergone surgery, some genetically modified strains and sometimes during transport.

3.4.7.2 Water Treatment

Laboratory animals maintained under conventional conditions can be supplied with drinking water from approved domestic water supply sources although, because drinking water is frequently a source of microbial contamination, it is necessary to monitor it regularly (see Chapter 2) or, depending on the study protocol, to perform further sanitisation. If it is necessary to purify the water several methods are available, including reverse osmosis, ion exchange and distillation that produce either partially desalinated or demineralised water. Such purified water is not necessarily free of microbial contamination, depending on the method used, the cleanliness of the treatment system and the conditions and duration of storage before it is used. In this case also, routine microbiological monitoring is advisable. More rigorous treatment such as acidification, chlorination, UV disinfection, ozone treatment or sterilisation (heat or sterile filtration) can be used to minimise or eliminate all microbiological contamination, depending on the application.

In animal facilities acidification (pH 2.5–3.0) is commonly used to disinfect drinking water and to control most bacterial contaminants (Tober-Meyer and Bieniek 1981). Several acids can be added to water (Juhr, Klomburg, and Haas 1978; Hall, White, and Lang 1980), but currently the commonest are citric or hydrochloric acids; it is important to ensure that drinking nipples and cage fittings, that water may contact, are made of corrosion-resistant materials. Chlorination (10–13 ppm) is another simple, inexpensive, and proven way of reducing the microbiological content of drinking water, but it can lead to offensive odours and, if the chlorine concentration is too high, it may irritate the mucous membranes of both animals and humans (Thunert and Heine 1975). Both strategies greatly reduce or prevent growth and colonisation of potentially harmful bacteria capable of colonising nasopharynx and intestines of mice, but will not eliminate an established infection (NRC 1991; NIH 1994). Chlorination is sometimes performed in conjunction with other methods such as acidification, although this is not recommended because a pH below 4 facilitates release of chlorine from the water (Orcutt 1980).

Ultraviolet disinfection is advisable as an environmentally friendly way of disinfecting drinking water without changing its chemical properties. The sterility achieved depends on the wavelength and power output of the emitters and the characteristics of the installation and its operation (e.g., pipework, flow rate, water quality, etc.). It remains to be established whether UV radiation is sufficiently reliable to serve as a sole method of treating drinking water for specified pathogen-free (SPF) animals.

The treatment of water with ozone requires special equipment that can be expensive. It is recommended that the method is used in combination with further treatment such as acidification (Bienieck and Remmers 1981). Residual ozone has to be eliminated from the water, either using residual ozone converters (activated carbon or catalysts) or by UV irradiation (Thunert and Heine 1975; Bienieck and Remmers 1981).

Heat sterilisation is the method of choice for supplying germ-free water to gnotobiotic-, germ-free-, and SPF animals. Water can be sterilised in the drinking bottles or in larger containers at high temperatures and an extended processing time (e.g., 121°C for 15–30 minutes). Sterile filtration is only practical and effective when the filters are located at the end of a run of pipework. Such systems require a great deal of maintenance, but are reliable and economical if regularly serviced.

Demineralisation and desalination of water can be important pre-treatments for subsequent water processing, especially in the case of calcium and magnesium rich (so-called hard) water. In such cases, pre-treatment avoids mineral precipitation during subsequent autoclaving, enable other disinfection procedures to work more effectively and to prevent drinking nipples from becoming blocked.

3.4.7.3 The Effect of Water Treatment on Experimental Results

It has been reported that acidification (pH 2.3–2.5) and chlorination of water improved reproductive performance of two strains of mice (Les 1968). Although disinfection of the water does not alter growth curves or clinical chemistry parameters of rats and rabbits (Tober-Meyer and Bieniek 1981), it can influence physiological characteristics, modify microflora populations or interfere with experimental results (Fidler 1977; Hall, White, and Lang 1980; Hermann, White, and Lang 1982; Homberger, Pataki, and Thomann 1993). In both normal and irradiated mice, 6 weeks of treatment with acidified water led to reduced body weight and the mean number of bacterial species isolated from the terminal ileum of acid-treated mice was less than that found in control animals (Hall, White, and Lang 1980). Moreover, when water is acidified (pH 2.0–3.0) the tooth enamel of rodents may be attacked and they are predisposed slightly to proteinuria and reduced urine output (Karle, Gehring, and Deerberg 1980; Clausing and Gottschalk 1989).

3.4.8 Bedding Material

Solid floored cages supplied with bedding are recommended for housing laboratory rodents. Many different materials used as bedding have been subjected to evaluation including cedar, pine, aspen, corncob, cotton-based cellulose fibres and papers. Bedding material is considered to provide animals with a comfortable substrate that allows them to nest, dig and rest comfortably.

Some authors have reported that animals prefer solid floors covered with bedding to wire floors (Arnold and Estep 1994; Blom et al. 1996; Manser et al. 1995). The presence of bedding and nesting material enables rats to fulfil their needs to hide and manipulate (see Chapter 4). Although a deep layer of wood shavings may enable animals to reduce their heat loss (Gordon 2004), the most appropriate depth has not been established.

Bedding used with laboratory animals may contain biologically active compounds that may be released into the air, and the dustiness of different products has a large influence on the amount and effect of exposure of care staff and others to airborne contaminants (Port and Kaltenbach 1969; Kaliste et al. 2004). Bedding can also affect reproductive performance of mice (Iturrian and Fink 1968; Odynets, Simonova, and Kozhuhov 1991; Potgieter and Wilke 1997). Most bedding materials absorb urine, and by controlling its availability to urease-producing micro-organisms alter the rate of ammonia generation (Raynor, Steinhagen, and Hamm 1983); this can have a significant impact on the concentration of ammonia in cages (Raynor, Steinhagen, and Hamm 1983; Lipman 1999; Smith et al. 2004b).

For these and other reasons, the characteristics of the bedding material can have a major influence on the animal's well-being and validity as an experimental model. Factors to consider include: absorbency (urine/faeces), ammonia/hydrogen sulphide binding capacity, odour, microbiological contamination, dustiness, chemical composition and potential for toxicity (Schoental 1973; Kraft 1980; Raynor, Steinhagen, and Hamm 1983; Wirth 1983; Weichbrod et al. 1986; Hämäläinen and Tirkkonen 1991; Potgieter and Wilke 1992; Smith et al. 2004b; Burn and Mason 2005).

Rats and mice are reported to prefer bedding based on aspen shavings to other tested materials (Mulder 1975; Ras et al. 2002; Krohn and Hansen 2008), and Burn and others (2006b) have shown that this material is relatively inert compared with other wood products, but nevertheless is more biologically active than paper.

There have been few comparisons of preferences for different bedding materials. Blom and others (1996) showed that rats prefer bedding consisting of large particles; Ago and others (2002) concluded that of four different paper-based bedding materials, ICR male mice preferred one that was soft and allowed them to hide and build nests. There are fewer recommended bedding materials for rabbits than for rats and they include straw, shredded paper and non-toxic wood.

The micro-environment within primary enclosures can vary from cage to cage with respect to temperature, RH, and olfactory clues; it is important that care routines preserve this microenvironment whilst ensuring that extreme divergences between cage climates do not arise. Except for animals housed in ventilated cage systems, it is usual to change bedding once each week for mice and twice for rats and rabbits. The frequency of clean-out depends on the species, the age of the animal, the number of animals and cage size, the type of bedding and the cage/rack system.

In any case, bedding must be replaced before there is a noticeable accumulation of ammonia. Höglund and Renström (2001) reported that the ammonia concentration in IVC cages was generally below 10 ppm after 2 weeks and the frequency of cage changing can be reduced to every 2 weeks for mice kept in IVC rack under higher ventilation rates (at least 60 air changes per hour, ACH; Höglund and Renström 2001; Reeb-Whitaker et al. 2001). However caution is needed in adopting a rule-of-the-thumb approach because the optimal ACH can vary for different situations. The frequency of bedding change should be checked for each system before it is adopted as a routine.

3.4.8.1 The Effect of Bedding on Experimental Results

There has been considerable interest in the influence of different bedding materials on experimental results. Materials based on pine products have been reported to influence hepatic drug metabolism, aspects of endocytosis, mucosal immune responses in the intestines, tissue antioxidant concentration or cytotoxic effects (Pick and Little 1965; Vesell 1967; Sabine 1975; Cunliffe-Beamer, Freeman, and Myers 1981; Nielsen, Andersen, and Svendsen 1986; Weichbrod et al. 1988; Törrönen, Pelkonen, and Kärenlampi 1989; Potgieter and Wilke 1992; Potgieter, Torronen, and Wilke 1995; Potgieter et al. 1996; Pelkonen and Hänninen 1997; Sanford et al. 2002; Buddaraju and Van Dyke 2003; Davey et al. 2003). Other studies failed to demonstrate differences in rabbit hepatic microsomal enzyme activity or cytokine/chemokine expression in mammary gland tumours (Heston 1975; Vlahakis 1977; Määttä et al. 2006; Ruben et al. 2007). Many bedding materials contain variable concentrations of volatile organic compounds, some of which can interfere with experimental results, as well as animals' well-being. The effect of these compounds is significantly reduced if the bedding is autoclaved (Nevalainen and Vartiainen 1996). Other chemically mediated effects include an influence on drug disposition, the nature of which is strain dependent (Vesell et al. 1976). Finally, the texture of bedding can affect the development of specific stimulus modalities of neuropathic pain behaviour following peripheral nerve injury (Robinson, Dowdall, and Meert 2004).

Frequent changing of cage bedding and/or nesting material disrupts the olfactory environment and may alter the age of puberty in mice (Drickamer 1990) and the ability to metabolise drugs in rats (Vesell et al. 1973).

3.4.8.2 The Effect of Bedding on Animals

Bedding and nesting materials may contain contaminants that have the potential to exert harmful effects on animal health; these include bacteria, mycobacteria, fungi and endotoxins (Gale and Smith 1981; Odynets, Simonova, and Kozhuhov 1991; Mayeux et al. 1995; Royals, Getzy, and VandeWoude 1999; Ewaldsson et al. 2002). Lawton, Taylor, and Perks (2006) further reported that bedding type can influence aggression of male nude mice.

An unexplained incident reported in a population of nude rats was a decrease in the incidence of blepharitis when they were transferred to a paper-based bedding from one based on hardwood (Zahorsky-Reeves et al. 2005). On another occasion, commercial cotton nest material was reported as a predisposing factor in the development of conjunctivitis in nude mice (Bazille et al. 2001).

3.4.8.3 The Effect of Bedding on Humans

Animal care staff are at high risk of exposure to potent allergens and subsequently developing occupational allergy (Edwards, Beeson, and Dewdney 1983; Hunskaar and Fosse 1990; Gordon and Preece 2003; Portengen et al. 2003; Elliott et al. 2005; Krakowiak et al. 2007; Krop et al. 2007); particular hazards can occur when cleaning dirty cages, pouring clean bedding into the cages, cleaning floors or handling animals during experiments (Yamauchi et al. 1989; Ohman et al. 1994; Lieutier-Colas et al. 2002; Thulin et al. 2002; Kaliste et al. 2004; Pacheco et al. 2006; Ooms et al. 2008).

The use of a centralised vacuum system for cleaning, and of a ventilated bench when changing cages or handling animals substantially reduces exposure to airborne particles including allergens

(Kaliste et al. 2002; Thulin et al. 2002). Hair caps may be worn to further prevent the spread of occupational allergens (Krop et al. 2007).

The risk of developing laboratory animal allergy is known to increase with the duration of exposure to animals and work in animal related tasks (Elliott et al. 2005), but little is known about the precise relationship between levels of exposure and the risk of developing allergy (Hollander et al. 1997). Atopy and smoking predispose to sensitisation to laboratory animals and the development of bronchial asthma and allergic rhinitis (Krakowiak, Szulc, and Gorski 1997). Exposure-response relationships in humans are very dependent on genetic factors, and interaction between genetically determined traits, occupational allergens and other cofactors in the environment, such as endotoxin, are all important risk factors in sensitisation and the development of asthma (Jones 2008).

3.5 TRANSPORT OF ANIMALS

Transport is any movement of animals to another animal room within the same facility, to a procedure room, to another animal facility on the campus and, of course, conveyance from one institution or establishment to another. Transportation is always a stressful experience for the animals (see Chapter 2). There are two reasons for that: the transport itself, including conditions during the transport and the change of environment when they reach the new area. The following principles apply to all type of transports, but to different degrees (Council of Europe 2006).

- All animals must be fit to undertake the journey proposed.
- Animals that are sick or injured should not be considered fit for transport, except for slightly injured or sick animals whose transport would not cause additional suffering.
- Sick or injured animals may also be transported for experimental or other scientific purposes, if the illness or injury is part of the research programme and when it is approved by the person/committee responsible for animal welfare issues.
- Transport boxes must be suitable for the species, the age and number of animals; should maintain the microclimate; and should avoid cross-contamination by micro-organisms. They must prevent escape of animals.
- For animals that may present a hazard to human health, special measures have to be taken on a case-by-case basis to avoid any risk for humans and the environment. For external transport the containers must comply with relevant national and international legislation.
- Planning transport of genetically modified animals must also take account of regulations and restrictions concerning the transport of genetically modified organisms and, if created with a viral vector, restrictions regarding the transport of potentially biohazardous material. Staff should be trained appropriately and dedicated space allocated to the activities needed to carry out all steps in compliance with applicable laws and regulations.
- For both internal and external transport, the sender and recipient should agree with the conditions of transport, departure and arrival times to ensure that full preparation can be made for the animals' arrival. For in-house transports these points should be well-defined.
- There must be a transport plan for the whole journey. The route should be so planned as to ensure the best environmental conditions, short transportation time and to minimise the risk of any disturbances and possible delays.
- The area or animal room to which the animals will go to must be determined in advance. All animals arriving from outside should be quarantined immediately after arrival.
- Transport must be performed only by experienced and competent persons, who are responsible for the welfare of the animals during the transport itself.
- On arrival, the animals should be removed from their transport boxes as soon as possible, examined by a competent person, and if there are no abnormalities the animals should be

transferred to the area/animal room that was defined before. In case of in-house transportation to a procedure room, the animals should be used as soon as possible; however, in some cases, the stress associated with transport may interfere with procedures and a period of acclimatisation and recovery should be considered.
- All incoming animals should be kept under close observation for a suitable period after arrival.

During transportation, rodents and rabbits should be provided with food and water in a form that minimises the likelihood of it becoming contaminated. Transport boxes should be leak-proof and secure but should also have provision to allow inspection of animals without unduly disturbing them or endangering their health status (LASA 2005). Responsibility for veterinary care lies with the originating establishment.

3.6 HEALTH CARE PROGRAMME

A health care programme, including a comprehensive and relevant health monitoring programme, should be implemented by the veterinarian responsible for the facility. Veterinary care for laboratory animals should start at the time of their arrival from the host facility at the new facility; it comprises the care of animal during experiments including before and after surgical interventions, establishing and implementing humane end points and policy for dealing with expected/unexpected situations. These issues are discussed in greater detail in Chapters 7 and 14. The health monitoring programme is reviewed next.

3.6.1 Preventive Veterinary Medicine for Recently Arrived Animals

The entry of the animals into the facility should be performed in accordance with specific and well-defined procedures that include inspection of all health certificates and positive identification of the animals based on their certificates and records, their marking and tags. Isolation (quarantine) of the animals is recommended for a time corresponding with current legislation or, in other cases, with the institution's own policies. During the quarantine period animals should be acclimatised to conditions at the receiving establishment and clinically monitored; where appropriate laboratory tests may be carried out to confirm their health status.

3.6.2 Veterinary Surveillance

The occurrence of disease in laboratory animals requires immediate identification, diagnosis, isolation (where appropriate) and treatment, if possible. A regular programme of veterinary surveillance (health monitoring programme) consisting of microbiological, serological and parasitological testing of the animals makes possible early recognition of health problems, which could compromise the animals' well-being, interfere with scientific findings, and may have health and safety implications for employees (Nicklas et al. 2002).

3.7 ENGINEERING CONTROL AND MAINTENANCE OF BIOSECURITY

Equipment used to prevent animals from becoming infected, such as that for steam and irradiation sterilisation for food or water, and cage and water bottle washers, should be checked regularly and rigorously. A planned maintenance programme should be set up to ensure their correct operation. Other equipment such as ventilated cabinets, IVCs and isolators should also be checked regularly to ensure they continue to operate to their specified performance standards.

The objectives of biocontainment are to protect men, animals and the outside environment from potential biological hazards (see Chapter 2). These hazards can be infective or other agents brought in with imported animals and agents to which the animals are exposed during the course of research.

To avoid the spread of microbiological agents associated with experiments or illness of animals or staff (zoonotic infections), all imported animals should be immediately isolated for quarantine reasons (see Chapter 2). For occupational health and safety reasons biological agents used for research purposes have been classified into four hazard groups, depending on the risk involved and possible consequences. Animals are maintained in different degrees of barrier in accordance with the relevant regulations (BMBL 2007).

Biocontainment is the term that includes all safety practises and techniques applied to reduce or eliminate exposure of personnel and the outside environment to potentially hazardous biological agents (see Chapter 2 for details). These practises include:

Personal hygiene:	Hand washing, skin disinfection, changing clothing
Protective clothing:	Gowns, shoe covers or boots, gloves, bonnets, goggles, face masks or respirators
Barrier facilities:	Clear separation between clean and potentially contaminated areas to avoid entry or escape of biological material, cross-contamination and inadvertent contact exposure through contamination of the work environment; barrier elements can include entrance locks, airlocks, disinfection and sterilisation locks or ports, specialised ventilation and air treatment systems and so on
Sanitation of cages, pens and so on:	Special washing area and procedures
Disposal of wastes:	Waste disposal programme based on risk assessment and legal requirements
Safety equipment:	Biological safety cabinets (class I, II or III), enclosed containers and so on
Training of personnel:	Practises and techniques required for safe handling of dangerous material
Standard operation procedures (SOPs):	Describing all methods and safety equipment used
Monitoring of effectiveness:	Applied at all stages and to all measures including routine health screening of animals

Maintenance of biosecurity requires the exclusion of all microbiological contaminants that could be introduced each time the barrier is breached, either intentionally or accidentally. There is no such thing as an absolute and perfect barrier, but standard operation procedures are a crucial element of an effective containment programme, because they take into consideration that people may make mistakes and that failure of technical equipment will happen. The selection and combination of equipment and methods to provide assured containment should be based on a risk assessment for all agents concerned.

3.8 CONCLUSIONS

The type of cage and rack or other enclosure chosen for laboratory mice, rats and rabbits may differ depending on the species and characteristics of the animals, the purpose of housing, and the role of the facility. However pens, cages and racks should be designed and stocked so as to meet the animals' natural behavioural needs and to allow them to be observed with minimal disturbance. Moreover, carefully planned routine, daily care and veterinary procedures, as well as carefully planned transportation assure the health status and reduce the stress of laboratory animals.

Other important macro- and micro-environmental factors, such as carbon dioxide/ammonia concentration, humidity, temperature, olfaction, sound, light, food, water and bedding can affect the animals and lead to welfare problems in a variety of ways, for example, retinal damage resulting from high light intensity or inappropriate behavioural responses and fear reactions caused by white noise. Not only are there adverse welfare consequences from such disturbances, but relatively subtle changes in the macro- and micro-environment can confound some experimental results, and their elimination should be part of the experimental design process.

Although books and literature offer many recommendations for providing suitable macro- and micro-environments for different laboratory species, there remain many unanswered questions. Considerably more information is required before we can be certain that we are providing species-specific environments suitable for the laboratory animals housed and that do not interfere with experimental findings.

3.9 QUESTIONS UNRESOLVED

Care
- Handling: It is not entirely clear what characteristics of a handler are of greatest importance to an animal in relation to the stress it experiences.

Enclosures
- There is very limited information about the effects of different cage materials or rack systems on animals or experimental results.
- A better understanding is needed of the influences of the location of the cage within a rack and its relation to air inlets, exhausts and the doorway.

Relative Humidity and CO_2
- More knowledge is required about the effect of RH on animal health and welfare and the relationship the animal perceives between RH and ambient temperature.
- There are very few reports in the literature about the consequences of exposing healthy laboratory animals to different, naturally developed ammonia and/or CO_2 concentrations.
- There is still a need to collect more information about the variability of CO_2 and ammonia concentrations between cages.

Temperature
- The relationship between IVC air circulation systems and the TNZ of laboratory rodents should be assessed with a view to revising the temperature recommended in the secondary enclosure.

Audition and vocalisation
- More information is required about the hearing range and development of hearing in laboratory rabbits.
- Vocalisation by laboratory rabbits and its role in communication is unclear.
- The role of vocalisation in rodents remains unclear and there have been few studies of audible vocalisation.

Sounds
- What level of sound intensity and frequency is critical for well-being of the different species?
- What are the advantages and disadvantages of attempting to mask environmental sounds in animal rooms and what, if any, measures are appropriate and when?
- Little is known about the noise levels caused by individual care procedures. This information would be valuable in trying to improve the animal's environment and working routines.

- What sound levels and frequencies are present in isolators and IVCs, resulting from ventilation and what are their effects on the animals?
- What is the impact on animals of the differing sound levels in animal rooms between noisier weekdays and quiet weekends and does this need to be considered in the study protocol?

Vision
- Little is known about the significance of ultraviolet light for the biology of laboratory rodents; the literature appears to give conflicting views, varying from no significance (Honkavaara, Aberg, and Viitala 2008) to potentially enriching cues (Olsson et al. 2003) and deleterious effects of short wavelengths on animal physiology (Tong and Goh 2000, cited in Burn 2008).

Food
- Suitable and reliable techniques are needed for feeding group-housed rats and mice on a restricted diet, whilst avoiding individual variations resulting from hierarchy within the group.
- There is still very limited information about implementing restricted feeding schedules in accordance with species-specific needs.

Bedding materials
- Although there is a wide range of commercial bedding materials available, little is known about the animals' preferences for these or their impact on their physiology.
- There is a lack of information about the most appropriate amount of bedding and nesting material to provide and whether the latter can substitute in part for the former.
- The factors to consider in establishing a suitable cage change interval need further evaluation.

REFERENCES

Adamiker, D. 1976. Irradiation of laboratory animal diets. A review. *Zeitschrift Für Versuchstierkunde* 18:191–201.
Adler, N., and J. Anisko. 1979. Behavior of communicating—Analysis of the 22-khz call of rats (*Rattus norvegicus*). *American Zoologist* 19:493–508.
Ago, A., T. Gonda, M. Takechi, T. Takeuchi, and K. Kawakami. 2002. Preferences for paper bedding material of the laboratory mice. *Experimental Animals* 51:157–61.
Albonetti, M. E., F. Dessí-Fulgheri, and F. Farabollini. 2006. Organization of behavior in unfamiliar female rabbits. *Aggresive Behavior* 17:171–78.
Algers, B., I. Ekesbo, and S. Strömberg. 1978. The impact of continuous noise on animal health. *Acta Veterinaria Scandinavica* 68:1–26.
Anderson, A., J. Werboff, and E. P. Les. 1968. Effects of environmental temperature-humidity and cage density on body weight and behavior in mice. *Experientia* 24:1022–23.
Anonymous. 2002. Genetic background: can your mice see? Jax notes N° 485:2. Available from http://jaxmice.jax.org/jaxnotes/archive/485.pdf, Accessed May 18, 2010.
Anthony, A., E. Ackerman, and J. A. Lloyd. 1959. Noise stress in laboratory rodents. 1. Behavioral and endocrine response of mice, rats, and guinea pigs. *The Journal of the Acoustical Society of America* 31:1430–37.
Arch, V. S., and P. M. Narins. 2008. 'Silent' signals: Selective forces acting on ultrasonic communication systems in terrestrial vertebrates. *Animal Behaviour* 76:1423–28.
Arnold, C. E., and D. Q. Estep. 1994. Laboratory caging preferences in golden hamsters (*Mesocricetus auratus*). *Laboratory Animals* 28:232–38.
Arteaga, L., A. Bautista, M. Martinez-Gomez, L. Nicolas, and R. Hudson. 2008. Scent marking, dominance and serum testosterone levels in male domestic rabbits. *Physiology & Behavior* 94: 510–15.

Backoff, P. M., and D. M. Caspary. 1994. Age-related changes in auditory brainstem responses in Fischer 344 rats: Effects of rate and intensity. *Hearing Research* 73:163–72.

Barfield, R. J., and L. A. Geyer. 1972. Sexual behavior: Ultrasonic postejaculatory song of the male rat. *Science* 176:1349–50.

Barfield, R. J., P. Auerbach, L. A. Geyer, and T. K. Mcintosh. 1979. Ultrasonic vocalizations in rat sexual-behavior. *American Zoologist* 19:469–80.

Batchelor, G. R. 1999. The laboratory rabbit. In *UFAW handbook on the care and management of laboratory animals*, ed. T. B. Poole, Vol. 7, 395–408. Oxford: Blackwell Science.

Bazille, P. G., S. D. Walden, B. L. Koniar, and R. Gunther. 2001. Commercial cotton nesting material as a predisposing factor for conjunctivitis in athymic nude mice. *Laboratory Animals* 30:40–42.

Beeler, J. A., B. Prendergast, and X. Zhuang. 2006. Low amplitude entrainment of mice and the impact of circadian phase on behavior tests. *Physiology & Behavior* 87:870–80.

Bienieck, H. J., and C. H. Remmers. 1981. Ozonierung des tränkewassers fur das versuchstier. *GIT: Labor-Fachzeitschrift* 25:383–86.

Björk, E., T. Nevalainen, M. Hakumäki, and H.-M. Voipio. 2000. R-weighting provides better estimation for rat hearing sensitivity. *Laboratory Animals* 34:136–44.

Black-Cleworth, P., and G. Verberne. 1975. Scent-marking, dominance and flehmen behavior in domestic rabbits in an artificial laboratory territory. *Chemical Senses & Flavor* 1:465–94.

Blanchard, R. J., D. C. Blanchard, R. Agullana, and S. M. Weiss. 1991. Twenty-two kHz alarm cries to presentation of a predator, by laboratory rats living in visible burrow systems. *Physiology & Behavior* 50:967–72.

Blanchard, R. J., M. A. Hebert, P. F. Ferrari, P. Palanza, R. Figueira, D. C. Blanchard, and S. Parmigiani. 1998. Defensive behaviors in wild and laboratory (Swiss) mice: The mouse defense test battery. *Physiology & Behavior* 65:201–9.

Blanchard, R. J., R. Agullana, L. McGee, S. Weiss, and D. C. Blanchard. 1992. Sex differences in the incidence and sonographic characteristics of antipredator ultrasonic cries in the laboratory rat (*Rattus norvegicus*). *Journal of Comparative Psychology* 106:270–77.

Blom, H. 1993. *Evaluation of housing conditions for laboratory mice and rats*. Thesis. Utrecht, Netherlands: Rijksuniversiteit.

Blom, H. J., G. Van Tintelen, C. J. Van Vorstenbosch, V. Baumans, and A. C. Beynen. 1996. Preferences of mice and rats for types of bedding material. *Laboratory Animals* 30:234–44.

Blumberg, M. S., and J. R. Alberts. 1990. Ultrasonic vocalizations by rat pups in the cold: An acoustic by-product of laryngeal braking? *Behavioral Neuroscience* 104:808–17.

BMBL. 2007. Biosafety in microbiological and biomedical laboratories. Available from http://www.cdc.gov/biosafety/publications/BMBL_5th_Edition.pdf, Accessed 26 June 2010.

Borg, E. 1982. Auditory thresholds in rats of different age and strain. A behavioral and electrophysiological study. *Hearing Research* 8:101–15.

Borg, E., and B. Engström. 1983. Hearing thresholds in the rabbit. A behavioral and electrophysiological study. *Acta Otolaryngologica* 95:19–26.

Brain, P. F., T. A. Maimanee, and M. Andrade. 2000. Dietary fats influence consumption and metabolic measures in male and female laboratory mice. *Laboratory Animals* 34:155–61.

Brandstetter, H., M. Scheer, C. Heinekamp, C. Gippner-Steppert, O. Loge, L. Ruprecht, B. Thull, R. Wagner, P. Wilhelm, and H. P. Scheuber. 2005. Performance evaluation of IVC systems. *Laboratory Animals* 39:40–44.

Broderson, J. R., J. R. Lindsey, and J. E. Crawford. 1976. The role of environmental ammonia in respiratory Mycoplasmosis of rats. *The American Journal of Pathology* 85:115–30.

Brudzynski, S. M., and D. Ociepa. 1992. Ultrasonic vocalization of laboratory rats in response to handling and touch. *Physiology & Behavior* 52:655–60.

Brudzynski, S. M., and E. M. Chiu. 1995. Behavioural responses of laboratory rats to playback of 22 kHz ultrasonic calls. *Physiology & Behavior* 57:1039–44.

Buddaraju, A. K., and R. W. Van Dyke. 2003. Effect of animal bedding on rat liver endosome acidification. *Comparative Medicine* 53:616–21.

Burn, C. C. 2008. What is it like to be a rat? Rat sensory perception and its implications for experimental design and rat welfare. *Applied Animal Behavior Science* 112:1–32.

Burn, C. C., A. Peters, and G. J. Mason. 2006a. Acute effects of cage cleaning at different frequencies on laboratory rat behaviour and welfare. *Animal Welfare* 15:161–71.

Burn, C. C., A. Peters, M. J. Day, and G. J. Mason. 2006b. Long-term effects of cage-cleaning frequency and bedding type on laboratory rat health, welfare, and handleability: A cross-laboratory study. *Laboratory Animals* 40:353–70.

Burn, C. C., and G. J. Mason. 2005. Absorbencies of six different rodent beddings: Commercially advertised absorbencies are potentially misleading. *Laboratory Animals* 39:68–74.

BVAAWF/FRAME/RSPCA/UFAW. 1993. Refinements in rabbit husbandry. Second report of the BVAAWF/FRAME/RSPCA/UFAW joint working group on refinement. British Veterinary Association Animal Welfare Foundation. Fund for the Replacement of Animals in Medical Experiments. Royal Society for the Prevention of Cruelty to Animals. Universities Federation for Animal Welfare. *Laboratory Animals* 27:301–29.

Carden, S. E., and M. A. Hofer. 1992. Effect of a social companion on the ultrasonic vocalizations and contact responses of 3-day-old rat pups. *Behavioral Neuroscience* 106:421–26.

CCAC. 2003. CCAC (Canadian Council on Animal Care) guidelines on laboratory animal facilities, characteristics, design and development. [Cited December 31, 2009]. Available from http://www.ccac.ca/en/CCAC_Main.htm

Chang E. F., and M. M. Merzenich. 2003. Environmental noise retards auditory cortical development. *Science* 300: 498–502.

Chia, R., F. Achilli, M. F. W. Festing, and E. M. C. Fisher. 2005. The origins and uses of mouse outbred stocks. *Nature Genetics* 37:1181–86.

Choi, G. C., J. S. McQuinn, B. L. Jennings, D. J. Hassett, and S. E. Michaels. 1994. Effect of population size on humidity and ammonia levels in individually ventilated microisolation rodent caging. *Contemporary Topics in Laboratory Animal Science* 33:77–81.

Clapcote, S. J., N. L. Lazar, A. R. Bechard, G. A. Wood, and J. C. Roder. 2005. NIH Swiss and black Swiss mice have retinal degeneration and performance deficits in cognitive tests. *Comparative Medicine* 55:310–16.

Clarke, H. E., M. E. Coates, J. K. Eva, D. J. Ford, C. K. Milner, P. N. O'Donoghue, P. P. Scott, and R. J. Ward. 1977. Dietary standards for laboratory animals: Report of the laboratory animals centre diets advisory committee. *Laboratory Animals* 11:1–28.

Clausing, P., and M. Gottschalk. 1989. Effects of drinking water acidification, restriction of water supply and individual caging on parameters of toxicological studies in rats. *Zeitschrift Für Versuchstierkunde* 32:129–34.

Clough, G. 1982. Environmental effects on animals used in biomedical research. *Biological Reviews of the Cambridge Philosophical Society* 57:487–523.

Clough, G. 1984. *Environmental factors in relation to the comfort and well-being of laboratory rats and mice.* Paper presented at Standards in Laboratory Animal Management. 7–24. UFAW: London.

Clough, G., J. Wallace, M. R. Gamble, E. R. Merryweather, and E. Bailey. 1995. A positive, individually ventilated caging system: A local barrier system to protect both animals and personnel. *Laboratory Animals* 29:139–51.

Coates, M. E., J. E. Ford, M. E. Gregory, and S. Y. Thompson. 1969. Effects of gamma-irradiation on the vitamin content of diets for laboratory animals. *Laboratory Animals* 3:39–49.

Collewijn, H., A. J. Martins, and R. M. Steinman. 1981. Natural retinal image motion: Origin and change. *Annals of the New York Academy of Sciences* 374:312–29.

Collins, R. A., M. Schreiber, and C. A. Elvehjem. 1953. The influence of relative humidity upon vitamin deficiencies in rats. *The Journal of Nutrition* 49:589–97.

Corning, B. F., and N. S. Lipman. 1991. A comparison of rodent caging systems based on microenvironmental parameters. *Laboratory Animal Science* 41:498–503.

Cos, S., D. Mediavilla, C. Martinez-Campa, A. Gonzalez, C. Alonso-Gonzalez, and E. J. Sanchez-Barcelo. 2006. Exposure to light-at-night increases the growth of DMBA-induced mammary adenocarcinomas in rats. *Cancer Letters* 235:266–71.

Council of Europe. 2006. *European convention 123, appendix A revised 2006, European convention for the protection of vertebrate animals used for experimental and other scientific purposes, guidelines for accommodation and care of animals.* Available from http://www.coe.int/t/e/legal_affairs/legal_cooperation/biological_safety,_use_of_animals/laboratory_animals/2006/Cons123%282006%293AppendixA_en.pdf, Accessed 26 June 2010.

Crowley, D., and M.-C. Hepp-Reymond. 1966. Development of cochlear function in the ear of the infant rat. *Journal of Comparative and Physiological Psychology* 62:427–32.

Cunliffe-Beamer, T. L., L. C. Freeman, and D. D. Myers. 1981. Barbiturate sleeptime in mice exposed to autoclaved or unautoclaved wood beddings. *Laboratory Animal Science* 31:672–75.

Dauchy, R. T., D. E. Blaska, L. A. Sauer, G. C. Brainard, and J. A. Krause. 1999. Dim light during darkness stimulates tumor progression by enhancing tumor fatty acid uptake and metabolism. *Cancer Letters* 144:131–36.

Davey, A. K., J. P. Fawcett, S. E. Lee, K. K. Chan, and J. C. Schofield. 2003. Decrease in hepatic drug-metabolizing enzyme activities after removal of rats from pine bedding. *Comparative Medicine* 53:299–302.

Dhanjal, P. 1991. *The assessment of stress in laboratory mice due to olfactory stimulation with fragranced odours.* Birmington, UK: University of Birmingham.

Diehl, J. F. 1991. Nutritional effects of combining irradiation with other treatments. *Food Control* 1:20–24.

Donnelly, H., and P. Saibaba. 1989. *Effects of humidity on breeding success in laboratory mice.* Paper presented at Laboratory animal welfare research—Rodents. 17–24, London: UFAW.

Drickamer, L. C. 1990. Environmental factors and age of puberty in female house mice. *Developmental Psychobiology* 23:63–73.

Edwards, R. G., M. F. Beeson, and J. M. Dewdney. 1983. Laboratory animal allergy: The measurement of airborne urinary allergens and the effects of different environmental conditions. *Laboratory Animals* 17:235–39.

Ellendorff, F., D. Smidt, M. Monzavifar, and P. de Boer. 1970. Temperature, humidity and reproduction in mice. II. Influence of 35 degrees C, 45 and 91 per cent relative humidity on body weight and genital tract. *International Journal of Biometeorology* 14:85–93.

Elliott, L., D. Heederik, S. Marshall, D. Peden, and D. Loomis. 2005. Incidence of allergy and allergy symptoms among workers exposed to laboratory animals. *Occupational and Environmental Medicine* 62:766–71.

Ellison, G. T. H., and L. M. Westlin-van Aarde. 1990. Ringtail in the pouched mouse (*Saccostomus campestris*). *Laboratory Animals* 24:205–6.

Engellenner, W. J., L. Rozboril, V. P. Perdue, R. G. Burright, and P. J. Donovick. 1982. A simple and inexpensive metabolic cage for mice. *Physiology & Behavior* 28:177–79.

Eriksson, E., F. Royo, K. Lyberg, H. E. Carlsson, and J. Hau. 2004. Effect of metabolic cage housing on immunoglobulin A and corticosterone excretion in faeces and urine of young male rats. *Experimental Physiology* 89:427–33.

Eveleigh, J. R. 1991. Cage ammonia levels during murine reproduction. *Journal of Experimental Animal Science* 34:236–38.

Eveleigh, J. R. 1993. Murine cage density: Cage ammonia levels during the reproductive performance of an inbred strain and two outbred stocks of monogamous breeding pairs of mice. *Laboratory Animals* 27:156–60.

Ewaldsson, B., B. Fogelmark, R. Feinstein, L. Ewaldsson, and R. Rylander. 2002. Microbial cell wall product contamination of bedding may induce pulmonary inflammation in rats. *Laboratory Animals* 36:282–90.

Famiglietti, E. V. 2005. "Small-tufted" ganglion cells and two visual systems for the detection of object motion in rabbit retina. *Visual Neuroscience* 22:509–34.

Fawcett, A. 2008. ARRP guideline 22: Guidelines for the housing of mice in scientific institutions. In *Animal Research Review Panel.* Available from http://www.animalethics.org.au/__data/assets/pdf_file/0004/249898/draft-guidelines-housing-mice.pdf, Accessed May 23, 2010.

Feller, R. P., E. J. Edmonds, I. L. Shannon, and K. O. Madsen. 1974. Significant effect of environmental lighting on caries incidence in the cotton rat. *Proceedings of the Society for Experimental Biology and Medicine* 145:1065–68.

Fenton, P. F., and G. R. Cowgill. 1947. Reproduction and lactation in highly inbred strains of mice on synthetic diets. *Journal of Nutrition* 33:703–12.

Fenton, P. F., G. R. Cowgill, M. A. Stone, and D. H. Justice. 1950. The nutrition of the mouse. VIII. Studies on pantothenic acid, biotin, inositol and paminobenzoic acid. *The Journal of Nutrition* 42:257–69.

Fidler, I. J. 1977. Depression of macrophages in mice drinking hyperchlorinated water. *Nature* 270:735–36.

Fleshler, M. 1965. Adequate acoustic stimulus for startle reaction in the rat. *Journal of Comparative and Physiological Psychology* 60:200–207.

Fletcher, J. L. 1976. Influence of noise on animals. In *Control of the animal house environment,* ed. T. McSheehy, 51–62. London: Laboratory Animals Ltd.

Flynn, R. J. 1967. *Notes on ringtail in rats.* Paper presented at Husbandry of laboratory animals: International Symposium of the International Committee on Laboratory Animals. 285–88, London: Academic Press.

Ford, D. J. 1976. The effect of methods of sterilization on the nutritive value of protein in a commercial rat diet. *The British Journal of Nutrition* 35:267–76.

Ford, D. J. 1977. Effect of autoclaving and physical structure of diets on their utilization by mice. *Laboratory Animals* 11:235–39.

Forestell, C. A., H. M. Schellinck, S. E. Boudreau, and V. M. LoLordo. 2001. Effect of food restriction on acquisition and expression of a conditioned odor discrimination in mice. *Physiology & Behavior* 72:559–66.

Francis, R. L. 1977. 22-kHz calls by isolated rats. *Nature* 265:236–38.

Fujita, S., T. Obara, I. Tanaka, and C. Yamauchi. 1981. Distribution of environmental temperature and relative humidity according to the number of conditioned air changes in laboratory animals rooms. *Experimental Animals* 30:21–29.

Gale, G. R. and A. B. Smith. 1981. Ureolytic and urease-activating properties of commercial laboratory animal bedding. *Laboratory Animal Science* 31:56–59.
Gamble, M. R. 1982. Sound and its significance for laboratory animals. *Biological Reviews of the Cambridge Philosophical Society* 57:395–421.
Gamble, M. R., and G. Clough. 1976. Ammonia build-up in animal boxes and its effect on rat tracheal epithelium. *Laboratory Animals* 10:93–104.
Gaskill, B. N., S. A. Rohr, E. A. Pajor, J. R. Lucas, and J. P. Garner. 2009. Some like it hot: Mouse temperature preferences in laboratory housing. *Applied Animal Behaviour Science* 116:279–85.
Geal-Dor, M., S. Freeman, G. Li, and H. Sohmer. 1993. Development of hearing in neonatal rats: Air and bone conducted ABR thresholds. *Hearing Research* 69:236–42.
Gekhman, B. I. 1974. Mechanisms of variations in heat emission in the thermoneutral zone in rabbits. *Bulletin of Experimental Biology and Medicine* 76:1393–95.
González-Mariscal, G., A. I. Melo, A. Zavala, and C. Beyer. 1990. Variations in chin-marking behavior of New Zealand female rabbits throughout the whole reproductive cycle. *Physiology & Behavior* 48:361–65.
González-Mariscal, G., M. E. Albonetti, E. Cuamatzi, and C. Beyer. 1997. Transitory inhibition of scent marking by copulation in male and female rabbits. *Animal Behaviour* 53:323–33.
Gordon, C. J. 1985. Relationship between autonomic and behavioral thermoregulation in the mouse. *Physiology & Behavior* 34:687–90.
Gordon, C. J. 1986. Relationship between behavioral and autonomic thermoregulation in the guinea pig. *Physiology & Behavior* 38:827–31.
Gordon, C. J. 1987. Relationship between preferred ambient temperature and autonomic thermoregulatory function in rat. *The American Journal of Physiology* 252: R1130–R1137.
Gordon, C. J. 1993. Twenty-four hour rhythms of selected ambient temperature in rat and hamster. *Physiology & Behavior* 53: 257–263.
Gordon, C. J. 2004. Effect of cage bedding on temperature regulation and metabolism of group-housed female mice. *Comparative Medicine* 54:63–68.
Gordon, C. J. 2005. *Temperature and toxicology: An integrative, comparative and environmental approach.* Boca Raton, FL: CRC Press.
Gordon, C. J., P. Becker, and J. S. Ali. 1998. Behavioral thermoregulatory responses of single-and group-housed mice. *Physiology & Behavior* 65:255–62.
Gordon, C. J., P. J. Spencer, J. Hotchkiss, D. B. Miller, P. M. Hinderliter, and J. Pauluhn. 2008. Thermoregulation and its influence on toxicity assessment. *Toxicology* 244:87–97.
Gordon, S., and R. Preece. 2003. Prevention of laboratory animal allergy. *Occupational Medicine* (Oxford, England) 53:371–77.
Gourevitch, G., and M. H. Hack. 1966. Audibility in the rat. *Journal of Comparative and Physiological Psychology* 62:289–91.
Gray, S., and J. L. Hurst. 1995. The effects of cage cleaning on aggression within groups of male laboratory mice. *Animal Behaviour* 49:821–26.
GV-SOLAS. 2004. *Tiergerechte haltung von laborratten.* Available from http://www.gv-solas.de/auss/hal/rattenhaltung.pdf, Accessed 26 June 2010.
Hall, J. E., W. J. White, and C. M. Lang. 1980. Acidification of drinking water: Its effects on selected biologic phenomena in male mice. *Laboratory Animal Science* 30:643–51.
Hämäläinen, A., and T. Tirkkonen. 1991. How to choose a good bedding material. *Laboratornye Zhyvotnye* 1:60–65.
Haney, M., and K. A. Miczek. 1993. Ultrasounds during agonistic interactions between female rats (*Rattus norvegicus*). *Journal of Comparative Psychology* 107:373–79.
Hartsook, E. W., and J. C. Nee. 1976. Effects in the rat of environmental temperature, diet dilution and treadmill running on voluntary food intake, body composition and endocrine organ mass: A multiple regression analysis. *The Journal of Nutrition* 106:1314–25.
Hasenau, J. J., R. B. Baggs, and A. L. Kraus. 1993. Microenvironments in microisolation cages using BALB/c and CD-1 mice. *Contemporary Topics in Laboratory Animal Science* 33:11–16.
Hayes, R. A., B. J. Richardson, S. C. Claus, and S. G. Wyllie. 2002. Semiochemicals and social signaling in the wild European rabbit in Australia: II. Variations in chemical composition of chin gland secretion across sampling sites. *Journal of Chemical Ecology* 28:2613–25.
Heffner, H., and B. Masterton. 1980. Hearing in glires—Domestic rabbit, cotton rat, feral house mouse, and kangaroo rat. *Journal of the Acoustical Society of America* 68:1584–99.
Heffner, H. E., R. S. Heffner, C. Contos, and T. Ott. 1994. Audiogram of the hooded Norway rat. *Hearing Research* 73:244–47.

Hegsted, D. M., and G. S. McPhee. 1950. The thiamine requirement of the adult rat and the influence on it of a low environmental temperature. *Journal of Nutrition* 41:127–36.

Heidbreder, C. A., I. C. Weiss, A. M. Domeney, C. Pryce, J. Homberg, G. Hedou, J. Feldon, M. C. Moran, and P. Nelson. 2000. Behavioral, neurochemical and endocrinological characterization of the early social isolation syndrome. *Neuroscience* 100:749–68.

Heiderstadt, K. M., R. M. Mclaughlin, and D. C. Wright. 2000. The effect of chronic food and water restriction on open-field behavior and serum corticosterone levels in rats. *Laboratory Animals* 34:20–28.

Henry, K. R. 2004. Males lose hearing earlier in mouse models of late-onset age-related hearing loss; females lose hearing earlier in mouse models of early-onset hearing loss. *Hearing Research* 190:141–48.

Hermann, L. M., W. J. White, and C. M. Lang. 1982. Prolonged exposure to acid, chlorine, or tetracycline in the drinking water: Effects on delayed-type hypersensitivity, hemagglutination titers, and reticuloendothelial clearance rates in mice. *Laboratory Animal Science* 32:603–8.

Heston, W. E. 1975. Testing for possible effects of cedar wood shavings and diet on occurrence of mammary gland tumors and hepatomas in C3H-A-vy and C3H-avy-fB mice. *Journal of the National Cancer Institute* 54:1011–14.

Hirayma, K., K. Uetsuka, Y. Kuwabara, M. Tamura, and K. Itoh. 2007. Vitamin K deficiency of germfree mice caused by feeding standard purified diet sterilized by gamma-irradiation. *Experimental Animals* 56:273–78.

Hirsjärvi, P. A., and T. U. Väliaho. 1987. Microclimate in two types of rat cages. *Laboratory Animals* 21:95–98.

Hoag, W. G., and M. M. Dickie. 1965. Nutrition. In *Biology of the laboratory mouse*, ed. E. L. Green. Bar Harbor, ME: The Jackson Laboratory.

Hofer, M. A., and H. Shair. 1978. Ultrasonic vocalization during social interaction and isolation in 2-weeek-old rats. *Developmental Psychobiology* 11:495–504.

Hoffman, H. S., and M. Fleshler. 1963. Startle reaction: Modification by background acoustic stimulation. *Science* (New York, NY) 141:928–30.

Höglund, A. U., and A. Renström. 2001. Evaluation of individually ventilated cage systems for laboratory rodents: Cage environment and animal health aspects. *Laboratory Animals* 35:51–57.

Hollander, A., P. Van Run, J. Spithoven, D. Heederik, and G. Doekes. 1997. Exposure of laboratory animal workers to airborne rat and mouse urinary allergens. *Clinical and Experimental Allergy* 27:617–26.

Homberger, F. R., Z. Pataki, and P. E. Thomann. 1993. Control of *Pseudomonas aeruginosa* infection in mice by chlorine treatment of drinking water. *Laboratory Animal Science* 43:635–37.

Honkavaara, J., H. Aberg, and J. Viitala. 2008. Do house mice use UV cues when foraging? *Journal of Ethology* 26:339–45.

Howdeshell, K. L., P. H. Peterman, B. M. Judy, J. A. Taylor, C. E. Orazio, R. L. Ruhlen, F. S. Vom Saal, and W. V. Welshons. 2003. Bisphenol A is released from used polycarbonate animal cages into water at room temperature. *Environmental Health Perspectives* 111:1180–87.

Hudson, R., and T. Vodermayer. 1992. Spontaneous and odor-induced chin marking in domestic female rabbits. *Animal Behaviour* 43:329–36.

Huerkamp, M. J., and N. D. Lehner. 1994. Comparative effects of forced-air, individual cage ventilation or an absorbent bedding additive on mouse isolator cage microenvironment. *Contemporary Topics in Laboratory Animal Science* 33:58–61.

Hunskaar, S., and R. T. Fosse. 1990. Allergy to laboratory mice and rats: A review of the pathophysiology, epidemiology and clinical aspects. *Laboratory Animals* 24:358–74.

Hurst, J. L. 2005. *Making sense of scents: Reducing aggression and uncontrolled variation in laboratory mice.* Available from http://www.nc3rs.org.uk/downloaddoc.asp?id=230&page=164&skin=0

Ihnat, R., N. R. White, and R. J. Barfield. 1995. Pup's broadband vocalizations and maternal behavior in the rat. *Behaviour Processes* 33:257–452.

Illnerova, H., J. Vanecek, J. Krecek, L. Wetterberg, and J. Saaf. 1979. Effect of one minute exposure to light at night on rat pineal serotonin N-acetyltransferase and melatonin. *Journal of Neurochemistry* 32:673–75.

Ison, J. R., and J. M. Russo. 1990. Enhancement and depression of tactile and acoustic startle reflexes with variation in background-noise level. *Psychobiology* 18:96–100.

Iturrian, W. B., and G. B. Fink. 1968. Comparison of bedding material: Habitat preference of pregnant mice and reproductive performance. *Laboratory Animal Care* 18:160–64.

Jennings, M., G. R. Batchelor, P. F. Brain, A. Dick, H. Elliott, R. J. Francis, R. C. Hubrecht, et al. 1998. Refining rodent husbandry: The mouse—Report of the rodent refinement working party. *Laboratory Animals* 32:233–59.

Jensen, M. N., and M. Ritskes-Hoitinga. 2007. How isoflavone levels in common rodent diets can interfere with the value of animal models and with experimental results. *Laboratory Animals* 41:1–18.

Jilge, B. 1991. The rabbit: A diurnal or a nocturnal animal? *Journal of Experimental Animal Science* 34:170–83.

Jones, M. G. 2008. Exposure-response in occupational allergy. *Current Opinion in Allergy and Clinical Immunology* 8:110–14.

Jones, R. B., J. B. Kacergis, M. R. MacDonald, F. T. McKnight, W. A. Turner, J. L. Ohman, and B. Paigen. 1995. The effect of relative humidity on mouse allergen levels in an environmentally controlled mouse room. *American Industrial Hygiene Association Journal* 56:398–401.

Jud, C., I. Schmutz, G. Hampp, H. Oster, and U. Albrecht. 2005. A guideline for analyzing circadian wheel-running behavior in rodents under different lighting conditions. *Biological Procedures Online* 7: 101–16.

Juhr, N. C., S. Klomburg, and A. Haas. 1978. Sterilization of drinking water using peracetic acid. *Zeitschrift Für Versuchstierkunde* 20:65–72.

Juliusson, B., A. Bergstrom, P. Rohlich, B. Ehinger, T. van Veen, and A. Szel. 1994. Complementary cone fields of the rabbit retina. *Investigative Ophthalmology & Visual Science* 35:811–18.

Kaliste, E., M. Linnainmaa, T. Meklin, and A. Nevalainen. 2002. Airborne contaminants in conventional laboratory rabbit rooms. *Laboratory Animals* 36:43–50.

Kaliste, E., M. Linnainmaa, T. Meklin, E. Torvinen, and A. Nevalainen. 2004. The bedding of laboratory animals as a source of airborne contaminants. *Laboratory Animals* 38:25–37.

Kaltwasser, M. T. 1990. Startle-inducing acoustic stimuli evoke ultrasonic vocalization in the rat. *Physiology & Behavior* 48:13–17.

Kaltwasser, M. T. 1991. Acoustic startle induced ultrasonic vocalization in the rat: A novel animal model of anxiety? *Behavioural Brain Research* 43:133–37.

Karle, E. J., F. Gehring, and F. Deerberg. 1980. Acidifying of drinking water and its effect on enamel lesions of rat teeth. *Zeitschrift Für Versuchstierkunde* 22:80–88.

Keithley, E. M., J. Lo, and A. F. Ryan. 1994. 2-deoxyglucose uptake patterns in response to pure tone stimuli in the aged rat inferior colliculus. *Hearing Research* 80:79–85.

Keller, L. S., W. J. White, M. T. Snider, and C. M. Lang. 1989. An evaluation of intra-cage ventilation in three animal caging systems. *Laboratory Animal Science* 39:237–42.

Kelly, J. B., and B. Masterton. 1977. Auditory sensitivity of the albino rat. *Journal of Comparative and Physiological Psychology* 91:930–36.

Kelly, J. B., P. W. Judge, and I. H. Fraser. 1987. Development of the auditory orientation response in the albino rat (*Rattus norvegicus*). *Journal of Comparative Psychology* (Washington, DC: 1983) 101:60–66.

Knapka, J. J., K. P. Smith, and F. J. Judge. 1977. Effect of crude fat and crude protein on reproduction and weaning growth in four strains of inbred mice. *Journal of Nutrition* 107:67–71.

Kock, N. 1986. Ultrasound production and stress in rodents. *The Veterinary Record* 118:588.

Koehler, K. E., R. C. Voight, S. Thomas, B. Lamb, T. Hassold, and P. A. Hunt. 2003. When disasters strikes—Rethinking caging materials. *Laboratory Animals* 32:24–27.

Kraft, L. M. 1958. Observations on the control and natural history of epidemic diarrhea of infant mice (EDIM). *The Yale Journal of Biology and Medicine* 31:121–37.

Kraft, L. M. 1980. The manufacture, shipping and receiving and quality control of rodent bedding materials. *Laboratory Animal Science* 30:366–76.

Krakowiak, A., B. Szulc, and P. Gorski. 1997. Occupational respiratory diseases in laboratory animal workers: Initial results. *International Journal of Occupational Medicine and Environmental Health* 10:31–36.

Krakowiak, A., M. Wiszniewska, P. Krawczyk, B. Szulc, T. Wittczak, J. Walusiak, and C. Palczynski. 2007. Risk factors associated with airway allergic diseases from exposure to laboratory animal allergens among veterinarians. *International Archives of Occupational and Environmental Health* 80:465–75.

Krohn, T. C., and A. K. Hansen. 2000. The effects of and tolerances for carbon dioxide in relation to recent developments in laboratory animal housing. *Scandinavian Journal of Laboratory Animal Science* 27:173–81.

Krohn, T. C., and A. K. Hansen. 2002. Carbon dioxide concentrations in unventilated IVC cages. *Laboratory Animals* 36: 209–12.

Krohn, T. C., and A. K. Hansen. 2008. Evaluation of corncob as bedding for rodents. *Scandinavian Journal of Laboratory Animal Science* 35:231–36.

Krohn, T. C., A. K. Hansen, and N. Dragsted. 2003a. The impact of low levels of carbon dioxide on rats. *Laboratory Animals* 37:94–99.

Krohn, T. C., A. K. Hansen, and N. Dragsted. 2003b. Telemetry as a method for measuring the impact of housing conditions on rats' welfare. *Animal Welfare* 12:53–62.

Krop, E. J., G. Doekes, M. J. Stone, R. C. Aalberse, and J. S. van der Zee. 2007. Spreading of occupational allergens: Laboratory animal allergens on hair-covering caps and in mattress dust of laboratory animal workers. *Occupational and Environmental Medicine* 64:267–72.

Kupp, R., C. A. Pinto, L. F. Rubin, and H. E. Griffin. 1989. Effects of ambient lighting on the eyes of rats. *Laboratory Animals* 18:32–37.

Lalitha, R., N. Suthanthirarajan, and A. Namasivayam. 1988. Effect of flickering light stress on certain biochemical parameters in rats. *Indian Journal of Physiology and Pharmacology* 32:182–86.

LASA. 2005. Guidance on the transport of laboratory animals. *Laboratory Animals* 39:1–39.

Lawton, S., V. Taylor, and V. Perks. 2006. Evaluation of five bedding types on male nude mouse health and aggression. *Animal Technology and Welfare* 5:163–64.

Lee, Y. C., J. T. King, and M. B. Visscher. 1953. Strain difference in vitamin E and B12 and certain mineral trace-element requirements for reproduction in A and Z mice. *The American Journal of Physiology* 173:456–58.

Lerchl, A. 1995. Sustained response of pineal melatonin synthesis to a single one-minute light-pulse during night in Djungarian hamsters (*Phodopus sungorus*). *Neuroscience Letters* 198:65–67.

Les, E. P. 1968. Effect of acidified-chlorinated water on reproduction in C3H-HeV and C57BL-5J mice. *Laboratory Animal Care* 18:210–13.

Levine, J. D., M. Feldmesser, L. Tecott, N. C. Gordon, and K. Izdebski. 1984. Pain-induced vocalization in the rat and its modification by pharmacological agents. *Brain Research* 296:121–27.

Ley, F. J., J. Bleby, M. E. Coates, and J. S. Paterson. 1969. Sterilization of laboratory animal diets using gamma radiation. *Laboratory Animal Care* 3:221–54.

Lieutier-Colas, F., P. Meyer, F. Pons, G. Hedelin, P. Larsson, P. Malmberg, G. Pauli, and F. De Blay. 2002. Prevalence of symptoms, sensitization to rats, and airborne exposure to major rat allergen (rat n 1) and to endotoxin in rat-exposed workers: A cross-sectional study. *Clinical and Experimental Allergy* 32:1424–29.

Lipman, N. S. 1999. Isolator rodent caging systems (state of the art): A critical view. *Contemporary Topics in Laboratory Animal Science* 38:9–17.

Lipman, N. S., B. F. Corning, and M. A. Coiro. 1992. The effects of intracage ventilation on microenvironmental conditions in filter-top cages. *Laboratory Animals* 26:206–10.

Lipman, N. S., B. F. Corning, and M. Saifuddin. 1993. Evaluation of isolator caging systems for protection of mice against challenge with mouse hepatitis virus. *Laboratory Animals* 27:134–40.

Lopez-Varela, S., F. Chacon, P. Cano, A. Arce, and A. I. Esquifino. 2004. Differential responses of circulating prolactin, GH, and ACTH levels and distribution and activity of submaxillary lymph node lymphocytes to calorie restriction in male Lewis and Wistar rats. *Neuroimmunomodulation* 11:247–54.

Love, J. A. 1991. Group housing rabbits. *Laboratory Animals* 20:37–43.

Määttä, J., R. Luukkonen, K. Husgafvel-Pursiainen, H. Alenius, and K. Savolainen. 2006. Comparison of hardwood and softwood dust-induced expression of cytokines and chemokines in mouse macrophage RAW 264.7 cells. *Toxicology* 218:13–21.

Maggio, J. C., and G. Whitney. 1985. Ultrasonic vocalizing by adult female mice (*Mus musculus*). *Journal of Comparative Psychology* 99:420–36.

Maggio, J. C., J. H. Maggio, and G. Whitney. 1983. Experience-based vocalization of male mice to female chemosignals. *Physiology & Behavior* 31:269–72.

Manser, C.E., T.H. Morris, and D.M. Broom. 1995. An investigation into the effects of solid or grid cage flooring on the welfare of laboratory rats. *Laboratory Animals* 29: 353–63.

Marsman, G., F. J. Pastoor, J. N. Mathot, H. M. Theuns, and A. C. Beynen. 1991. Vitamin D, within its range of fluctuation in commercial rat diets, does not influence nephrocalcinogenesis in female rats. *Laboratory Animals* 25:330–36.

Matochik, J. A., N. R. White, and R. J. Barfield. 1992. Variations in scent marking and ultrasonic vocalizations by Long-Evans rats across the estrous cycle. *Physiology & Behavior* 51:783–86.

Mayeux, P., L. Dupepe, K. Dunn, J. Balsamo, and J. Domer. 1995. Massive fungal contamination in animal care facilities traced to bedding supply. *Applied and Environmental Microbiology* 61:2297–2301.

McGarrity, G. J., and L. L. Coriell. 1973. Mass airflow cabinet for control of airborne infection of laboratory rodents. *Applied Microbiology* 26:167–72.

McLennan, I. S., and J. J. Taylor . 2004. The use of sodium lamps to brightly illuminate mouse houses during their dark phases. *Laboratory Animals* 38:384–92.

Meijer, M. K., P. L. P. van Loo, and V. Baumans. 2009. There's a rat in my room! Now what? Mice show no chronic physiological response to the presence of rats. *Journal of Applied Animal Welfare Science* 12:293–305.

Memarzadeh, F., P. C. Harrison, G. L. Riskowski, and T. Henze. 2004. Comparison of environment and mice in static and mechanically ventilated isolator cages with different air velocities and ventilation designs. *Contemporary Topics in Laboratory Animal Science* 43:14–20.

Milligan, S. R., G. D. Sales, and K. Khirnykh. 1993. Sound levels in rooms housing laboratory animals: An uncontrolled daily variable. *Physiology & Behavior* 53:1067–76.

Mizushima, Y., T. Harauchi, T. Yoshizaki, and S. Makino. 1984. A rat mutant unable to synthesize vitamin C. *Experientia* 40:359–61.

Mrozek, M., U. Zillmann, W. Nicklas, V. Kraft, B. Meyer, E. Sickel, B. Lehr, and A. Wetzel. 1994. Efficiency of air filter sets for the prevention of airborne infections in laboratory animal houses. *Laboratory Animals* 28:347–54.

Mulder, J. B. 1975. Bedding preferences of pregnant laboratory-reared mice. *Behavior Research Methods & Instrumentation* 7:21–22.

Murakami, H. 1971. Differences between internal and external environments of the mouse cage. *Laboratory Animal Science* 21:680–84.

Nevalainen, T., and T. Vartiainen. 1996. Volatile organic compounds in commonly used beddings before and after autoclaving. *Scandinavian Journal of Laboratory Animal Science* 23:101–4.

Nevison, C. M., C. J. Barnard, R. J. Beynon, and J. L. Hurst. 2000. The consequences of inbreeding for recognizing competitors. *Proceedings of the Royal Society of London Series B-Biological Sciences* 267:687–94.

Nicklas, W., P. Baneux, R. Boot, T. Decelle, A. A. Deeny, M. Fumanelli, and B. Illgen-Wilcke. 2002. Recommendations for the health monitoring of rodent and rabbit colonies in breeding and experimental units. *Laboratory Animals* 36:20–42.

Nielsen, J. B., O. Andersen, and P. Svendsen. 1986. Hepatic O-deethylase activity in mice on different types of bedding. *Zeitschrift Für Versuchstierkunde* 28:69–75.

NIH. 1994. *Manual of microbiologic monitoring of laboratory animals*. National Institutes of Health. 151–54, Washington, DC: NIH Publication.

Nishikawa, S., A. Yasoskima, K. Doi, H. Nakayama, and K. Uetsuka. 2007. Involvement of sex, strain and age factors in high fat diet-induced obesity in C57BL/6J and BALB/cA mice. *Experimental Animals* 56:263–72.

Njaa, L. R., F. Utne, and O. R. Braekkan. 1957. Effect of relative humidity on rat breeding and ringtail. *Nature* 180:290–91.

Noirot, E. 1968. Ultrasounds in young rodents. 2. Changes with age in albino rats. *Animal Behaviour* 16:129–34.

Noirot, E. 1972. Ultrasounds and maternal behavior in small rodents. *Developmental Psychobiology* 5:371–87.

NRC. 1991. *Infectious diseases of mice and rats*. 141–45. National Research Council. Washington DC. USA: National Academy Press.

Nuboer, J. F. W., W. M. Vannuys, and J. F. Wortel. 1983. Cone systems in the rabbit retina revealed by ERG-null-detection. *Journal of Comparative Physiology* 151:347–52.

Odynets, A., O. Simonova, and A. Kozhuhov. 1991. Beddings for laboratory animals: Criteria of biological evaluation. *Laboratornye Zhyvotnye* 1:70–76.

Ohman, J. L., Jr., K. Hagberg, M. R. MacDonald, R. R. Jones, Jr., B. J. Paigen, and J. B. Kacergis. 1994. Distribution of airborne mouse allergen in a major mouse breeding facility. *The Journal of Allergy and Clinical Immunology* 94:810–17.

Okon, E. E. 1971. Temperature relations of vocalization in infant golden hamsters and Wistar rats. *Journal of Zoology* 164:227–37.

Olsson, I. A. S., C. M. Nevison, E. G. Patterson-Kane, C. M. Sherwin, H. A. Van de Weerd, and H. Wurbel. 2003. Understanding behaviour: The relevance of ethological approaches in laboratory animal science. *Applied Animal Behaviour Science* 81: 245–64.

Ooms, T. G., J. E. Artwohl, L. M. Conroy, T. M. Schoonover, and J. D. Fortman. 2008. Concentration and emission of airborne contaminants in a laboratory animal facility housing rabbits. *Journal of the American Association for Laboratory Animal Science* 47:39–48.

Orcutt, R. P. 1980. Bacterial diseases: Agents, pathology, diagnosis, and effects on research. *Laboratory Animals* 9:21–43.

Organisciak, D. T., and B. S. Winkler. 1994. Retinal light damage: Practical and theoretical considerations. *Progress in Retinal and Eye Research* 13:1–29.

O'Steen, W. K., C. R. Shear, and K. V. Anderson. 1972. Retinal damage after prolonged exposure to visible light. *The American Journal of Anatomy* 134:5–22.

Ozaki, Y., and R. J. Wurtman. 1979. Spectral power distribution of light sources affects growth and development of rats. *Photochemistry and Photobiology* 29:339–41.

Pacheco, K. A., C. McCammon, P. S. Thorne, M. E. O'Neill, A. H. Liu, J. W. Martyny, M. Vandyke, L. S. Newman, and C. S. Rose. 2006. Characterization of endotoxin and mouse allergen exposures in mouse facilities and research laboratories. *The Annals of Occupational Hygiene* 50:563–72.

Peace, T. A., A. W. Singer, N. A. Niemuth, and M. E. Shaw. 2001. Effects of caging type and animal source on the development of foot lesions in Sprague Dawley rats (*Rattus norvegicus*). *Contemporary Topics in Laboratory Animal Science* 40:17–21.

Pelkonen, K. H., and O. O. Hänninen. 1997. Cytotoxicity and biotransformation inducing activity of rodent beddings: A global survey using the hepa-1 assay. *Toxicology* 122:73–80.

Peterson, E. A. 1980. Noise and laboratory animals. *Laboratory Animal Science* 30:422–39.

Pfaff, J. 1974. Noise as an environmental problem in the animal house. *Laboratory Animals* 8:347–354.

Pfaff, J., and M. Stecker. 1976. Loudness level and frequency content of noise in the animal house. *Laboratory Animals* 10:111–17.

Pick, J. R., and J. M. Little. 1965. Effect of type of bedding material on thresholds of pentylenetetrazol convulsions in mice. *Laboratory Animal Care* 15:29–33.

Podberscek, A. L., J. K. Blackshaw, and A. W. Beattie. 1991. The behaviour of group penned and individually caged laboratory rabbits. *Applied Animal Behavior Science* 28:353–63.

Port, C. D., and J. P. Kaltenbach. 1969. The effect of corncob bedding on reproductivity and leucine incorporation in mice. *Laboratory Animal Care* 19:46–49.

Portengen, L., A. Hollander, G. Doekes, G. de Meer, and D. Heederik. 2003. Lung function decline in laboratory animal workers: The role of sensitisation and exposure. *Occupational and Environmental Medicine* 60:870–75.

Porter, G., and M. Festing. 1970. A comparison between irradiated and autoclaved diets for breeding mice, with observations on palatability. *Laboratory Animals* 4:203–13.

Porter, G., and W. Lane-Petter. 1965. Observations on autoclaved, fumigated and irradiated diets for breeding mice. *The British Journal of Nutrition* 19:295–305.

Portfors, C. V. 2007. Types and functions of ultrasonic vocalizations in laboratory rats and mice. *Journal of American Association for Laboratory Animal Science* 46:28–34.

Potgieter, F. J., and P. I. Wilke. 1992. Laboratory animal bedding: A review of wood and wood constituents as a possible source of external variables that could influence experimental results. *Animal Technology* 43:65–88.

Potgieter, F. J., and P. I. Wilke. 1996. The dust content, dust generation, ammonia production, and absorption properties of three different rodent bedding types. *Laboratory Animals* 30:79–87.

Potgieter, F. J., and P. I. Wilke. 1997. Effect of different bedding materials on the reproductive performance of mice. *Journal of the South African Veterinary Association* 68:8–15.

Potgieter, F. J., P. I. Wilke, H. van Jaarsveld, and D. W. Alberts. 1996. The *in vivo* effect of different bedding materials on the antioxidant levels of rat heart, lung and liver tissue. *Journal of the South African Veterinary Association* 67:27–30.

Potgieter, F. J., R. Torronen, and P. I. Wilke. 1995. The *in vitro* enzyme-inducing and cytotoxic properties of South African laboratory animal contact bedding and nesting materials. *Laboratory Animals* 29:163–71.

Prior, H. 2006. Effects of the acoustic environment on learning in rats. *Physiology & Behavior* 87:162–65.

Prusky, G. T., P. W. West, and R. M. Douglas. 2000. Behavioral assessment of visual acuity in mice and rats. *Vision Research* 40:2201–9.

Radlwimmer, F. B., and S. Yokoyama. 1998. Genetic analyses of the green visual pigments of rabbit (*Oryctolagus cuniculus*) and rat (*Rattus norvegicus*). *Gene* 218:103–9.

Ras, T., M. van de Ven, E. G. Patterson-Kane, and K. Nelson. 2002. Rats' preferences for corn versus wood-based bedding and nesting materials. *Laboratory Animals* 36:420–25.

Raynor, T. H., W. H. Steinhagen, and T. E. Hamm, Jr. 1983. Differences in the microenvironment of a polycarbonate caging system: Bedding *vs* raised wire floors. *Laboratory Animals* 17:85–89.

Reeb, C. K., R. B. Jones, D. W. Bearg, H. Bedigian, and B. Paigen. 1997. Impact of room ventilation rates on mouse cage ventilation and microenvironment. *Contemporary Topics in Laboratory Animal Science* 36:74–79.

Reeb-Whitaker, C. K., B. Paigen, W. G. Beamer, R. T. Bronson, G. A. Churchill, I. B. Schweitzer, and D. D. Myers. 2001. The impact of reduced frequency of cage changes on the health of mice housed in ventilated cages. *Laboratory Animals* 35:58–73.

Renström, A., G. Bjoring, and A. U. Höglund. 2001. Evaluation of individually ventilated cage systems for laboratory rodents: Occupational health aspects. *Laboratory Animals* 35:42–50.

Ritskes-Hoitinga, J., J. N. Mathot, L. F. Van Zutphen, and A. C. Beynen. 1992. Inbred strains of rats have differential sensitivity to dietary phosphorus-induced nephrocalcinosis. *The Journal of Nutrition* 122:1682–92.

Ritskes-Hoitinga, J., J. N. Mathot, L. H. Danse, and A. C. Beynen. 1991. Commercial rodent diets and nephrocalcinosis in weanling female rats. *Laboratory Animals* 25:126–32.

Ritskes-Hoitinga, J., P. M. Verschuren, G. W. Meijer, A. Wiersma, A. J. van de Kooij, W. G. Timmer, C. G. Blonk, and J. A. Weststrate. 1998. The association of increasing dietary concentrations of fish oil with hepatotoxic effects and a higher degree of aorta atherosclerosis in the ad lib.-fed rabbit. *Food and Chemical Toxicology* 36:663–72.

Robinson, I., T. Dowdall, and T. F. Meert. 2004. Development of neuropathic pain is affected by bedding texture in two models of peripheral nerve injury in rats. *Neuroscience Letters* 368:107–11.

Romanovsky, A. A., A. I. Ivanov, and Y. P. Shimansky. 2002. Selected contribution: Ambient temperature for experiments in rats: A new method for determining the zone of thermal neutrality. *Journal of Applied Physiology* 92:2667–79.

Royals, M. A., D. M. Getzy, and S. VandeWoude. 1999. High fungal spore load in corncob bedding associated with fungal-induced rhinitis in two rats. *Contemporary Topics in Laboratory Animal Science* 38:64–66.

Ruben, D., N. Muratore, S. Pin, and K. Gabrielson. 2007. Effects of bedding substrates on microsomal enzymes in rabbit liver. *Journal of the American Association for Laboratory Animal Science* 46:8–12.

Rudaya, A. Y., A. A. Steiner, J. R. Robbins, A. S. Dragic, and A. A. Romanovsky. 2005. Thermoregulatory responses to lipopolysaccharide in the mouse: Dependence on the dose and ambient temperature. *American Journal of Physiology. Regulatory, Integrative and Comparative Physiology* 289:R1244–R1252.

Sabine, J. R. 1975. Exposure to an environment containing the aromatic red cedar, *Juniperus virginiana*: Procarcinogenic, enzyme-inducing and insecticidal effects. *Toxicology* 5:221–35.

Sakaguchi, M., S. Inouye, H. Miyazawa, H. Kamimura, M. Kimura, and S. Yamazaki. 1990. Evaluation of countermeasures for reduction of mouse airborne allergens. *Laboratory Animal Science* 40:613–15.

Sales, G. D. 1972. Ultrasound and aggressive behaviour in rats and other small mammals. *Animal Behaviour* 20:88–100.

Sales, G. D. 1991. The effect of 22 kHz calls and artificial 38 kHz signals on activity in rats. *Behavioural Processes* 24:83–93.

Sales, G. D., K. J. Wilson, K. E. Spencer, and S. R. Milligan. 1988. Environmental ultrasound in laboratories and animal houses: A possible cause for concern in the welfare and use of laboratory animals. *Laboratory Animals* 22:369–75.

Sales, G. D., S. R. Milligan, and K. Khirnykh. 1989. *The acoustic environment of laboratory animals*. Potters Bar: Universities Federation for Animal Welfare South Mimms.

Sales, G. D., S. R. Milligan, and K. Khirnykh. 1999. Sources of sound in the laboratory animal environment: A survey of the sounds produced by procedures and equipment. *Animal Welfare* 8:97–115.

Saltarelli, C. G., and C. P. Coppola. 1979. Influence of visible light on organ weights of mice. *Laboratory Animal Science* 29:319–22.

Sanford, A. N., S. E. Clark, G. Talham, M. G. Sidelsky, and S. E. Coffin. 2002. Influence of bedding type on mucosal immune responses. *Comparative Medicine* 52:429–32.

Sauer, M. B., H. Dulac, S. Clark, K. M. Moffitt, J. Price, D. Dambach, H. Mosher, D. Bounous, and L. Keller. 2006. Clinical pathology laboratory values of rats housed in wire-bottom cages compared with those of rats housed in solid-bottom cages. *Journal of American Association for Laboratory Animal Science* 45:30–35.

Scattoni, M. L., J. Crawley, and L. Ricceri. 2009. Ultrasonic vocalizations: A tool for behavioural phenotyping of mouse models of neurodevelopmental disorders. *Neuroscience and Biobehavioral Reviews* 33:508–15.

Schaefer, K. E. 1982. Effects of increased ambient CO_2 levels on human and animal health. *Experientia* 38:1163–68.

Schaerdel, A. D., W. J. White, C. M. Lang, B. H. Dvorchik, and K. Bohner. 1983. Localized and systemic effects of environmental ammonia in rats. *Laboratory Animal Science* 33:40–45.

Scheline, R. R. 1965. A restraining cage for metabolic studies in rats. *The Journal of Pharmacy and Pharmacology* 17:52–53.

Schlingmann, F. S., H. L. M. de Rijk, W. J. Pereboom, and R. Remie. 1993a. Avoidance as a behavioural parameter in the determination of distress amongst albino and pigmented rats at various light intensities. *Animal Technology* 44:87–96.

Schlingmann, F. S., H. L. M. de Rijk, W. J. Pereboom, and R. Remie. (1993b) Light intensity in animal rooms and cages in relation to the care and management of albino rats. *Animal Technology* 44: 97–107.

Schoeb, T. R., M. K. Davidson, and J. R. Lindsey. 1982. Intracage ammonia promotes growth of *Mycoplasma pulmonis* in the respiratory tract of rats. *Infection and Immunity* 38:212–17.

Schoental, R. 1973. Carcinogenicity of wood shavings. *Laboratory Animals* 7:47–49.

Sedlacek, R. S., H. D. Suit, K. A. Mason, and E. R. Rose. 1980. Development and operation of a stable limited defined flora mouse colony. Paper presented at Animal Quality and Models in Biomedical Research. Stuttgart, New York 197–201.

Semple-Rowland, S. L., and W. W. Dawson. 1987a. Cyclic light intensity threshold for retinal damage on albino rats raised under 6 lux. *Experimental Eye Research* 44:643–61.

Semple-Rowland, S. L., and W. W. Dawson. 1987b. Retinal cyclic light damage threshold for albino rats. *Laboratory Animal Science* 37:289–98.

Serfilippi, L. M., D. R. S. Pallman, M. M. Gruebbel, T. J. Kern, and C. B. Spainhour. 2004. Assessment of retinal degeneration in outbred albino mice. *Comparative Medicine* 54:69–76.

Serrano, L. J. 1971. Carbon dioxide and ammonia in mouse cages: Effect of cage covers, population, and activity. *Laboratory Animal Science* 21:75–85.

Sharon, I. M., R. P. Feller, and S. W. Burney. 1971. The effects of lights of different spectra on caries incidence in the golden hamster. *Archives of Oral Biology* 16:1427–32.

Sherwin, C. M. 2007. Validating refinements to laboratory housing: Asking the animals. In NCR3RS (database online). Available from http://www.nc3rs.org.uk/downloaddoc.asp?id=603&page=759&skin=0, Accessed July 5, 2010.

Shimozuru, M., T. Kikusui, Y. Takeuchi, and Y. Mori. 2006. Scent-marking and sexual activity may reflect social hierarchy among group-living male Mongolian gerbils (*Meriones unguiculatus*). *Physiology & Behavior* 89:644–49.

Silverman, J., D. W. Bays, S. F. Cooper, and S. P. Baker. 2008. Ammonia and carbon dioxide concentrations in disposable and reusable ventilated mouse cages. *Journal of the American Association for Laboratory Animal Science* 47:57–62.

Simmons, M. I., D. M. Robie, J. B. Jones, and L. J. Serrano. 1968. Effect of filter cover on temperature and humidity in a mouse cage. *Laboratory Animals* 2:113–20.

Smith, A. L., S. L. Mabus, J. D. Stockwell, and C. Muir. 2004a. Effects of housing density and cage floor space on C57BL/6J mice. *Comparative Medicine* 54:656–63.

Smith, E., J. D. Stockwell, I. Schweitzer, S. H. Langley, and A. L. Smith. 2004b. Evaluation of cage microenvironment of mice housed on various types of bedding materials. *Contemporary Topics in Laboratory Animal Science* 43:12–17.

Smith, A. L., S. L. Mabus, C. Muir, and Y. Woo. 2005. Effects of housing density and cage floor space on three strains of young adult inbred mice. *Comparative Medicine* 55: 368–76.

Soldani, P., M. Gesi, P. Lenzi, G. Natale, F. Fornai, A. Pellegrini, M. P. Ricciardi, and A. Paparelli. 1999. Long-term exposure to noise modifies rat adrenal cortex ultrastructure and corticosterone plasma levels. *Journal of Submicroscopic Cytology and Pathology* 31:441–48.

Spiegel, A., and R. Gunnert. 1961. Neue käfige fur mäuse und ratten. *Zeitschrift Für Versuchstierkunde* 1:38–46.

Stolpe, J., and R. Sedlag. 1976. Single and complex effect of ammonia and hydrogen sulfide in the air on small laboratory animals (rats) under various environmental conditions. 1. Effect of ammonia. *Archiv Für Experimentelle Veterinärmedizin* 30:533–39.

Stotzer, H., I. Weisse, F. Knappen, and R. Seitz. 1970. Die retina-degeneration der ratte. *Arzneimittel-Forschung* 20: 811–17.

Swoap, S. J., C. Li, J. Wess, A. D. Parsons, T. D. Williams, and J. M. Overton. 2008. Vagal tone dominates autonomic control of mouse heart rate at thermoneutrality. *American Journal of Physiology. Heart and Circulatory Physiology* 294:H1581–H1588.

Swoap, S. J., J. M. Overton, and G. Garber. 2004. Effect of ambient temperature on cardiovascular parameters in rats and mice: A comparative approach. *American Journal of Physiology. Regulatory, Integrative and Comparative Physiology* 287:R391–R396.

Szél, A., P. Rohlich, A. R. Caffeé, and T. vanVeen. 1996. Distribution of cone photoreceptors in the mammalian retina. *Microscopy Research and Technique* 35:445–62.

Szymusiak, R., and E. Satinoff. 1981. Maximal REM sleep time defines a narrower thermoneutral zone than does minimal metabolic rate. *Physiology & Behavior* 26:687–90.

Thulin, H., M. Björkdahl, A. S. Karlsson, and A. Renström. 2002. Reduction of exposure to laboratory animal allergens in a research laboratory. *The Annals of Occupational Hygiene* 46:61–68.

Thunert, A., and W. Heine. 1975. The water supply of SPF animal houses. III. Heating and acidification of drinking water. *Zeitschrift Für Versuchstierkunde* 17:50–52.

Tober-Meyer, B. K., and H. J. Bieniek. 1981. Studies on the hygiene of drinking water for laboratory animals. 1. The effect of various treatments on bacterial contamination. *Laboratory Animals* 15:107–10.

Törrönen, R., K. Pelkonen, and S. Kärenlampi. 1989. Enzyme-inducing and cytotoxic effects of wood-based materials used as bedding for laboratory animals. Comparison by a cell culture study. *Life Sciences* 45:559–65.

Treichler, R., and H. H. Mitchell. 1941. The influence of plane of nutrition and of environmental temperature on the relationship between basal metabolism and endogenous nitrogen metabolism subsequently determined. *Journal of Nutrition* 22:333–43.

Turner, J. G., C. A. Bauer, and L. P. Rybak. 2007. Noise in animal facilities: Why it matters. *Journal of American Association for Laboratory Animal Science* 46:10–13.

van Betteray, J. N. F., J. M. H. Vossen, and A. M. L. Coenen. 1991. Behavioural characteristics of sleep in rats under different light/dark conditions. *Physiology & Behavior* 50:79–82.

Van der Meer, E., P. L. P. Van Loo, and V. Baumans. 2004. Short-term effects of a disturbed light-dark cycle and environmental enrichment on aggression and stress-related parameters in male mice. *Laboratory Animals* 38:376–83.

Vandershuren, L. J. M. J., R. J. M. Niesink, B. M. Spruijt, and J. M. V. R. Ree. 1995. Influence of environmental factors on social play behavior of juvenile rats. *Physiology & Behavior* 58:119–23.

Van Loo, P. L., C. L. J. Kruitwagen, L. F. M. van Zutphen, J. M. Koolhaas, and V. Baumans. 2000. Modulation of aggression in male mice: Influence of cage cleaning regime and scent marks. *Animal Welfare* 8:281–95.

Vesell, E. S. 1967. Induction of drug-metabolizing enzymes in liver microsomes of mice and rats by softwood bedding. *Science (New York, N.Y.)* 157:1057–58.

Vesell, E. S., C. M. Lang, W. J. White, G. T. Passananti, R. N. Hill, T. L. Clemens, D. K. Liu, and W. D. Johnson. 1976. Environmental and genetic factors affecting the response of laboratory animals to drugs. *Federation Proceedings* 35:1125–32.

Vesell, E. S., C. M. Lang, W. J. White, G. T. Passananti, and S. L. Tripp. 1973. Hepatic drug metabolism in rats: Impairment in a dirty environment. *Science (New York, NY)* 179:896–97.

Vlahakis, G. 1977. Possible carcinogenic effects of cedar shavings in bedding of C3h-avy fb mice. *Journal of the National Cancer Institute* 58:149–50.

Voipio, H.-M. 1997. How do rats react to sound? *Scandinavian Journal of Laboratory Animal Science* 24:1–80.

Voipio, H.-M., T. Nevalainen, P. Halonen, M. Hakumaki, and E. Björk. 2006. Role of cage material, working style and hearing sensitivity in perception of animal care noise. *Laboratory Animals* 40:400–409.

Weichbrod, R. H., C. F. Cisar, J. G. Miller, R. C. Simmonds, A. P. Alvares, and T. H. Ueng. 1988. Effects of cage beddings on microsomal oxidative enzymes in rat liver. *Laboratory Animal Science* 38:296–98.

Weichbrod, R. H., J. E. Hall, R. C. Simmonds, and C. F. Cisar. 1986. Selecting bedding material. *Laboratory Animals* 15:25–29.

Weigelt, O. 1970. Metabolism cage for rats on the building block principle. *Zeitschrift Für Versuchstierkunde* 12:68–76.

Weihe, W. H. 1965. Temperature and humidity climatograms for rats and mice. *Laboratory Animal Care* 15:18–28.

Weihe, W. H. 1978. Richtwerte der belegungsdichte von standardkäfigen. *Zeitschrift Für Versuchstierkunde* 20:305–9.

White, W. J., and A. M. Mans. 1984. Effect of bedding changes and room ventilation rates on blood and brain ammonia levels in normal rats and rats with portacaval shunts. *Laboratory Animal Science* 34:49–52.

Williams, R. A., A. G. Howard, and T. P. Williams. 1985. Retinal damage in pigmented and albino rats exposed to low levels of cyclic light following a single mydriatic treatment. *Current Eye Research* 4:97–102.

Willott, J. F., L. C. Erway, J. R. Archer, and D. E. Harrison. 1995. Genetics of age-related hearing loss in mice. II. Strain differences and effects of caloric restriction on cochlear pathology and evoked response thresholds. *Hearing Research* 88:143–55.

Wirth, H. 1983. Criteria for the evaluation of laboratory animal bedding. *Laboratory Animals* 17:81–84.

Wong, C. C., K. D. Dohler, M. J. Atkinson, H. Geerlings, R. D. Hesch, and A. von zur Muhlen. 1983. Circannual variations in serum concentrations of pituitary, thyroid, parathyroid, gonadal and adrenal hormones in male laboratory rats. *Journal of Endocrinology* 97:179–85.

Wurtmann, R. J., and J. Weisel. 1969. Environmental lighting and neuroendocrine function: Relationship between spectrum of light source and gonadal growth. *Endocrinology* 85:1218–21.

Yamauchi, C., T. Obara, N. Fukuyama, and T. Ueda. 1989. Evaluation of a one-way airflow system in an animal room based on counts of airborne dust particles and bacteria and measurements of ammonia levels. *Laboratory Animals* 23:7–15.

Zahorsky-Reeves, J. L., A. A. Schimmel, C. E. Suckow, M. A. Couto, and P. T. Lawson. 2005. Paper-based bedding dramatically decreased the incidence of blepharitis in a nude rat population. *Contemporary Topics in Laboratory Animal Science* 44:62.

Zheng, Q. Y., K. R. Johnson, and L. C. Erway. 1999. Assessment of hearing in 80 inbred strains of mice by ABR threshold analyses. *Hearing Research* 130:94–107.

Zimmerman, D. R., and B. S. Wostmann. 1963. Vitamin stability in diets sterilized for germfree animals. *The Journal of Nutrition* 79:318–22.

Zondek, B., and I. Tamari. 1964. Effect of audiogenic stimulation on genital function and reproduction. iii. Infertility induced by auditory stimuli prior to mating. *Acta Endocrinologica* 45:227–33.

4 Animal Needs and Environmental Refinement

Vera Baumans, The Netherlands; Hanna Augustsson, Sweden; and Gemma Perretta, Italy

CONTENTS

Objectives 76
Key Factors 76
4.1 Animal Needs 76
    4.1.1 Introduction 76
    4.1.2 The Five Freedoms 77
        4.1.2.1 Hunger and Thirst 77
        4.1.2.2 Discomfort 78
        4.1.2.3 Pain, Injury, Disease 79
        4.1.2.4 Expressing Normal Behaviour 81
        4.1.2.5 Fear and Distress 81
4.2 Motivation and Needs 82
    4.2.1 Nesting and Hiding 83
    4.2.2 Social Behaviour 83
    4.2.3 Exploration 85
    4.2.4 Foraging 85
    4.2.5 Locomotory Activity and Movement 86
4.3 Environmental Refinement 86
    4.3.1 Definition and Principles of Environmental Refinement for
           the Welfare of the Animals 86
    4.3.2 Refinement of Physical Environment 87
        4.3.2.1 Cage Structure and Furnishings 88
        4.3.2.2 Feeding 89
        4.3.2.3 Opportunities for Physical Exercise and Exploration 89
    4.3.3 Refinement of Social Environment 90
    4.3.4 Validation of Environmental Refinement 91
    4.3.5 The Impact on Scientific Results 92
    4.3.6 Animal Welfare Assessment and the Future 92
4.4 Conclusions 92
4.5 Questions Unresolved 93
References 93

OBJECTIVES

Laboratory animals are maintained in standardised laboratory environments that are usually impoverished and monotonous. Although the basic physiological needs of the animals (i.e., food, water, protection from climatic extremes, injury and disease etc.) are fulfilled, their is often compromised because of the low level of stimulation and the lack of resources for performance of certain species-typical behaviours. Such poor conditions may result in a range of physiological or psychological changes that add unintentional variables to the experimental design and so might affect the reliability of scientific results.

This chapter will discuss the applicability of the concept of the "Five Freedoms" to the field of laboratory animal science, in relation to behavioural needs and potential environmental refinement. It is essential to have knowledge of the species-typical behaviour in order to assess whether housing conditions meet the needs of an individual animal in accordance with its species, and when appropriate, also its age, sex and previous experiences.

Appropriate programs for environmental refinement should be developed to promote the psychological well-being of laboratory animals. These programs should include and document criteria for evaluating the effectiveness of the program; they should also be validated and regularly assessed for their continuing positive effect on animal welfare and impact on scientific results.

KEY FACTORS

The concepts of the Five Freedoms should be taken into account in laboratory animal care and use.

1. The ability of an animal to express its species-specific behaviour pattern is essential to ensuring its well-being.
2. Exploration, foraging, nesting and digging, as well as social interaction, are essential needs in laboratory rodents and rabbits.
3. Environmental provisions should aim to fulfil the animal's specific needs and be validated to ensure animal welfare and scientific integrity.
4. Structural complexity of the home environment increases the amount of usable space and promotes behavioural diversity and physical activity.
5. Social company is a continuous source of stimulation and variation and may help to reduce stress and enhance recovery.
6. Nesting material and nest boxes enable the animal to exert some degree of environmental control and gives opportunities for shelter and hiding without compromising experimental results.
7. Subtle differences in environmental provisions may have a marked impact on the value of refinements and success with their implementation.

4.1 ANIMAL NEEDS

4.1.1 INTRODUCTION

Standardisation of laboratory animal housing has the objective of reducing individual variation between animals and thereby the number of animals needed for each experiment. For laboratory animal housing, standardisation commonly means control and regulation within narrow limits of the temperature, ventilation, light schedule and air humidity. Although current standards for the laboratory environment already ensure provision of resources necessary for physical survival, issues related to animal behaviour and well-being are receiving increasing attention.

The traditional approach to animal welfare is based on assuring longevity, growth and reproductive performance and absence of stereotypies or other abnormal behaviours as well as physiological stress markers such as glucocorticoids. Although it is relatively simple to identify a potential

Animal Needs and Environmental Refinement

stressor and to measure the animal's response to it using appropriate methodology, interpretation of the results might be difficult. Functional parameters frequently indicate the absence rather than the presence of welfare. In addition, they are sometimes considered crude on the grounds that they may not be manifested until the welfare of the animal has been severely compromised.

One of the first and most widely accepted definitions of welfare is the concept of the Five Freedoms (Brambell 1965), published as minimal standards for farm animals and later reformulated by the UK Farm Animal Welfare Council (FAWC 1993).

4.1.2 THE FIVE FREEDOMS

- *Freedom from Hunger and Thirst:* By ready access to fresh water and a diet to maintain full health and vigour
- *Freedom from Discomfort:* By providing an appropriate environment including shelter and a comfortable resting area
- *Freedom from Pain, Injury, or Disease:* By prevention or rapid diagnosis and treatment
- *Freedom to Express Normal Behaviour:* By providing sufficient space, proper facilities, and company of the animal's own kind
- *Freedom from Fear and Distress:* By ensuring conditions and treatment that avoid mental suffering

Although the Five Freedoms are as relevant to laboratory animals as they are to farm animals in relation to general husbandry and to refining procedures, the context in which laboratory animals are used places certain constraints and introduces some special considerations that need to be taken into account in terms of animal welfare. Experimental procedures sometimes necessitate withholding food and water from animals or inflicting pain, disease and/or stress as part of the experimental protocol. When serious conflicts arise between the Five Freedoms and the requirements of animal experiments, justification of those experiments might be questioned. Good quality animal research should be restricted to normal, healthy animals unless the illness is itself the subject of investigation (Poole 1997). Unavoidable pain and distress that is a consequence of the research should be minimised by applying refined experimental procedures and husbandry techniques. Animals are adapted by evolution to natural variations in their environment and may have difficulty acclimatising to the uniformity of the artificial environment of a laboratory animal facility. Total freedom from environmental challenges may therefore not be favourable for animal welfare (Korte, Olivier, and Koolhaas 2007). Nonetheless, the general concepts promoted in the Five Freedoms are still valid and serve well as a basis for our discussion.

4.1.2.1 Hunger and Thirst

The first freedom, freedom from hunger and thirst, is not a problem in the general husbandry of most laboratory species but rather the opposite. Laboratory rodents have *ad libitum* access to nutritionally well-defined diets and water. This, in combination with limited space and opportunities for exercise, often leads to animals, especially laboratory rats and rabbits, becoming overweight and sedentary. Moreover, single housing may also cause rats to overeat, possibly as a response to boredom and frustration (Fiala, Snow, and Greenough 1977). Maintaining full health and vigour might therefore be difficult from this perspective. Even low levels of exercise are effective in counteracting the gain of body mass in rats (Suzuki and Machida 1995). Increased activity levels may also enhance physical fitness and thereby improve health and potentially also, well-being (Spangenberg et al. 2005). Restricted feeding, as discussed in Chapter 10, has been shown to benefit the animal's health and to prolong its life span.

Deprivation of food is a common part of some experimental procedures. Sometimes, possibly because of lack of knowledge, fasting is continued longer than necessary, posing an important animal welfare issue. For example, when a procedure requires gastric emptying in a rat, it is not necessary to fast the animal longer than 6 hours because its stomach will be empty after that time. When fasted overnight (generally more than 12 hours) rats will be overactive during the dark period searching for food and will be exhausted in the morning. One strategy is to estimate the food intake between 5 p.m. and midnight and give the rats that amount of food that will have been consumed by midnight so that their stomachs will be empty in the morning. Fasting for a longer period leads to increased locomotory (Vermeulen et al. 1997) and grooming behaviour and 18 hours fasting causes a body weight loss of at least 10%. If it necessary to completely empty the intestinal tract, fasting up to 22 hours is required in the rat (Ritskes-Hoitinga and Chwailibog 2003). In that case rats and mice should be given sucrose cubes or a 10% solution of sucrose or maltose for the time required for the gastrointestinal tract to become empty, provided that this does not interfere with the experiment. In general, mice and rats should never be fasted longer than 12 and 24 hours, respectively. See Chapter 10 for more information on nutrition and feeding.

4.1.2.2 Discomfort

The second freedom, freedom from discomfort, referred originally to relieving physical discomfort during resting and shelter from bad weather or other environmental disturbances. In laboratory animal facilities discomfort from adverse climatic conditions is generally not encountered because environmental parameters are maintained within relatively narrow optimal ranges. On the other hand small enclosures or cage sizes impose restrictions on movement and locomotion and may produce a state of discomfort in animals. Paragraph 4.5.1 of the revised Appendix A of the CoE Convention (ETS123, Council of Europe 2006) states that:

> All animals should be allowed adequate space to express a wide behavioural repertoire. Animals should be socially housed wherever possible and provided with an adequately complex environment within the animal enclosure to enable them to carry out a range of normal behaviours. (p. 14)

Housing rodents and rabbits in small cages that restricts their ability to run, climb and jump in a way that they have evolved to do, can cause both frustration and the appearance of stereotypies as well as excessive weight gain and apathy. This in turn may lead to other health problems that not only impair welfare, but also interfere with experimental results. Increasing the complexity of the cage is generally more important than enlarging the floor area, because inclusion of structures provides more opportunities for activity and increases the usable space. Although laboratory animals are commonly housed in cages or pens that are only a tiny proportion of the areas natural to their habitat, cages should at least allow animals to stretch fully in all directions when resting and also allow for a range of locomotory activities. It is difficult to objectively specify adequate cage sizes for maintaining laboratory animals because much depends on the strain, group size and age of the animals; their familiarity with each other; and their reproductive condition. In consequence, minimum requirements, including recommendations for enriched environments and social housing, are based on both experimental results and good/best practical experience. A great deal of background information has also been made available in what is generally referred to as "Appendix A, Part B" by working groups of experts in the Working Party for the preparation of the fourth Multilateral Consultation of Parties to the European Convention for the Protection of Vertebrate Animals used for Experimental and other Scientific Purposes (ETS 123; Stauffacher and Peters 2002).

Both rats and mice rear or stand erect to sniff and explore their surroundings. In rats, upright positions are also used in play behaviour and in the agonistic boxing position. Even though it is possible for mice to stand fully erect, in the standard cages used today this is not the case for rats (Brain et al. 1993). Adult rats need a height of at least 30 cm to rear fully (Lawlor 2002), although, the minimum cage height for rats in Appendix A has been raised only to 18 cm. However, another preference study found that rats have a preference for cages of 18–20 cm height, and it has been suggested that a cage with a low roof may be perceived as a safer shelter than one with a high roof, when other shelter is lacking (Blom et al. 1995; Buttner 1993). In a trade-off, shelter may be more

FIGURE 4.1 Housing system for rats. (Photo courtesy of Novo Nordisk.)

important than the ability to rear but, as a minimum, the cage should be high enough to give rats the opportunity to groom fully and to play and interact with cage mates. Nowadays, alternative housing systems for rats (e.g., elevated cage tops and two-storey cages that allow climbing, rearing, and shelter) are available commercially (Figure 4.1).

Traditionally, it has been common practise to keep rabbits in cages, although fortunately many facilities are replacing this approach by pen housing. Housing rabbits in small cages limits their freedom of movement and prevents normal movements such as hopping and rearing; one consequence of this is the development of pathological changes such as osteoporosis (Morton 1993). One common solution to this problem is to connect two cages together so that rabbits can be pair-housed, or at least can jump between cages. Pen housing might be an even better practise and is widely used for female rabbits and has been used also to house castrated males.

Sometimes it is necessary to reduce the floor area for experimental reasons, for example when measuring the animals' metabolism. Metabolic cages are generally smaller, have grid floors that allow faeces and urine to be collected and require animals to be housed singly. All these features are known to reduce welfare in laboratory rodents and therefore the time animals spend in metabolic cages should be kept as short as possible; when an experiment lasts for a few days or even weeks, larger cages should be used. Depending on the experimental protocol, a period of acclimatisation may be needed; it has been shown that the urinary concentration of metabolites of stress hormones and other physiological parameters might require about 4–5 days to return to base levels, depending on the parameter measured (Gomez-Sanches and Gomez-Sanches 1991).

A recently introduced source of potential discomfort may be individually ventilated cage systems (IVCs). Humans consider air speed greater than 0.2 m/s to be draughts, and this is generally agreed to be an upper limit also for rodents (Lipman 1999). High intra-cage ventilation rates could be perceived as draughts and cause chronic stress and heat loss. Preference tests have shown that rats choose cages with fewer than 80 air changes per hour (Krohn, Hansen, and Dragsted 2003). The physiology and behaviour of mice were not adversely affected by fewer than 80 air changes per hour when the air inlet was located at the top of the cage and nesting material was provided (Baumans et al. 2002). In summary, the location of the air supply to the cage (from the wall or from the top), the ventilation rate and the presence of nesting material are important considerations when assessing the impact of IVC housing on the well-being of mice. Evidence that animals are reacting to draught could be a change of location of the nest and the building barriers of bedding.

4.1.2.3 Pain, Injury, Disease

The construction of standard animal cages largely precludes them as a source of physical injury. Instead, aggression in connection with dominance and territorial defence is the most frequent cause of injury. The problem is greatest with male mice and rabbits, especially when they are grouped after sexual maturation. In mice, bite wounds are usually located around the back and tail area and, in rabbits, around the neck and genitalia. Although social housing is preferable to isolation, aggression may be a potential source of pain and distress if managed poorly. It is essential to maintain group stability to allow a stable dominance structure to become established so that groups are socially harmonious. An extensive study of the effect of cage stocking density in rats (Hurst et al. 1999) found that group composition had a greater impact on animal welfare than group size.

In nature, animals are able to retreat from the territory of a resident animal, but in captivity the only options are aggressive conflict or subordination. This situation may constitute a socially stressful environment (Blanchard, McKittrick, and Blanchard 2001) and lead to physical injury and reduced welfare (Van Loo 2001). In rats, repeated defeat is known to induce behavioural syndromes indicative of depression and helplessness (Willner 1991). Frustration and frequent escape attempts are also common in rats housed in socially stressful environments (Hurst and Barnard 1997) and physiological and immunological changes can be demonstrated if the situation is prolonged. To minimise the occurrence of fighting in male mice, the number of individuals per cage should be no more than three animals, nesting material (not soiled bedding) should be transferred to the new cage at clean-out, and animals should be disturbed as little as possible (Van Loo, Van Zutphen, and Baumans 2003). In general, litter mates fight less than unrelated mice (Olsson and Westerlund 2007).

Barbering or whisker trimming is an example of an abnormal behaviour commonly seen in laboratory mice occupying inadequate living conditions. The individual responsible for barbering plucks the fur of its cage mates and removes the vibrissae, which are a very important source of sensory input in rodents and play a key part in establishing their behavioural repertoire. In fact, rodents use whiskers to identify and distinguish objects, to control balance and for orienting and exploration. Whisker trimming may result in sensory deprivation in the affected animal and disrupt its behaviour (Sarna, Dyck, and Whishaw 2000; see Figure 4.2). The problem is not fully understood as yet but recent studies indicate that it is not a dominance behaviour, as previously thought (Long 1972), but is more likely to be a form of abnormal repetitive behaviour related to obsessive compulsive disorders. There are reports that barbering may be associated with endogenous biological components (Garner et al. 2004b) and several husbandry and social factors (Garner et al. 2004a) such as diet, weaning age and enrichment, suggesting that it may represent a stress-evoked behavioural response, for example coping with inappropriate housing (van den Broek and Omtzigt 1993).

There is ample evidence that sub-clinical infections have a deleterious effect on experimental outcomes, so it is important to maintain laboratory animals free from disease; therefore,

FIGURE 4.2 Barbering in mice. (Photo courtesy of T. P. Rooymans.)

comprehensive health monitoring procedures should be established (Nicklas et al. 2002; see Chapter 3). Furthermore, humane endpoints for euthanasia should be defined (see Chapter 14).

4.1.2.4 Expressing Normal Behaviour

An animal needs adequate space, proper facilities, and the company of conspecifics; that is, the possibility for social interaction if it is to express its normal behavioural repertoire. However, the identification of a catalogue of all behaviours that a certain species "needs" to perform and a further list of those that they don't need to is virtually impossible and not necessarily helpful in this context (Jensen and Toates 1993). The needs of a laboratory rodent or rabbit are not different from their wild counterparts; however, not all naturally occurring behaviours are beneficial for animal health and well-being. Some behaviours, such as infanticide or intense physical fighting and so on, occur as natural responses to a challenging environment and may occur also in captivity, but should not be considered normal. On the other hand social behaviours, foraging activities and behaviours relating to maintaining body condition, are usually considered to be needs in most animal species. The Rodent Refinement Working Party (Jennings et al. 1998) attempted to define the behavioural needs of mice and suggested that ideally, cages should allow for "resting, grooming, exploring, hiding, searching for food, gnawing, social interaction, nesting, digging and going into retreats".

Domestication can be defined as an evolutionary process by which the original natural and sexual selection pressures, acting upon the species, are modified to favour adaptation to a captive environment and cohabitation with a human population (Price 1999). In general, changes in behaviour brought about by domestication are quantitative rather than qualitative. The behavioural repertoire seen in wild animals has not been lost in domesticated animals. Instead, it has been argued that the most important effects of domestication are the reduced responsiveness and sensitivity to environmental change (Price 1999; Berdoy 2002). Domestication involves genetic changes and increases taming through reduction of flight responses and decreased aggressiveness towards humans.

An informed scientific approach is crucial to avoid misguided attempts to improve the husbandry and care of captive animals. Only by taking into account knowledge of the natural habitat and behaviour of the species concerned may informed improvements be made to the captive environment (Brain 1992; Newberry 1995). For comprehensive reviews of the natural behaviour of laboratory rodents and rabbits and for valuable references to further reading, the following articles and books are recommended: for mice (Crowcroft 1966; Mackintosh 1981; Latham and Mason 2004), for rats (Barnett 1963; Lore and Flannelly 1977; Berdoy 2002; Burn 2008), and for rabbits (Stauffacher 1996; RSPCA/UFAW 2008).

4.1.2.5 Fear and Distress

Emotions such as fear have evolved as adaptations to alert the animal to threats or challenges in its environment, prompting it to modify its behaviour, thereby preventing it from being injured in a potentially dangerous situation (Boissy 1995). The existence of such innate modifiers of behaviour also increases the likelihood of survival, because they result in a more flexible and suitable behavioural response to environmental stimuli. It has been argued that suffering occurs 'when unpleasant subjective feelings are acute or continue for a long time because the animal is unable to carry out the actions that would normally reduce risks to life and reproduction in those circumstances' (Dawkins 1990, p. 2). Hence it is important, to enable natural defence reaction patterns towards many stressors to facilitate successful coping strategies.

Defensive strategies include escape or avoidance, aggressive defence, freezing or immobility and submission if the threat comes from a conspecific (Marks and Nesse 1994). The most successful strategy depends on situational factors, such as the proximity of the threat (defensive distance), environmental constraints and individual factors (Blanchard, Griebel, and Blanchard 2001). In mice, reactions to clearly defined threat stimuli include flight if an escape route is available, hiding if a shelter can be found, defensive burying if substrate is available, and freezing if it perceives it is still undetected or if neither escape nor concealment is possible (Rodgers 1997). Defensive burying is a behaviour consisting of digging and pushing substrate so as to cover an aversive stimulus, which thereby could be avoided (De Boer and Koolhaas 2003). In a natural environment, defensive burying might lead to the closing of a tunnel entrance or building a wall to protect against predators or intruders. In the laboratory, defensive burying is shown by both rats and mice in response to novel objects, draught and for concealment. Being able to hide in nesting material or a nest-box, burrowing in bedding or fleeing behind a shelter resembles what a rat or mouse would do in the wild if threatened. When possible, risk assessment, freezing, and defensive burying appear to occur from a place of concealment because they enable the animal to investigate the threat while being minimally visible to it (Rodgers and Dalvi 1997; Yang et al. 2004). Ideally, the animal should feel secure in a complex, challenging environment that it can control. A sense of security can be achieved by providing materials suitable for constructing a nest, hiding places and compatible cage mates.

The human–animal interaction is an important factor in animal experimentation both for the welfare of the animal and the outcome of the experiment (Davies and Balfour 1992). Human interactions that are perceived by the animal as positive may facilitate the development of a cooperative relationship based on mutual confidence, whereas absence of human interaction, negative or inconsistent experiences of human contact may lead the animal to develop a fearful relationship (Hemsworth, Barnett, and Hansen 1987). Many laboratory species such as rodents and rabbits are highly susceptible to predators, and are likely to show strong fear responses in unfamiliar situations if they cannot find shelter. This is shown by attempts to flee, biting when handled or sudden immobility to avoid being detected. The literature contains many studies of human–animal interactions in laboratory rats (McCall and Lester 1969; Gartner et al. 1980; van Bergeijk and van Herck 1990; Augustsson et al. 2002; Maurer et al. 2008), and rabbits (Podberscek, Blackshaw, and Beattie 1991). It is clear from these studies and from practical experience that it is of the highest importance that researchers are competent and take time to handle the animals before the start of the experiment so as to reduce fearful responses (Morton 1993; Jennings et al. 1998; RSPCA/UFAW 2008). Careful handling, together with conditioning to experimental and husbandry procedures, reduce stress responses and improve the reliability of scientific results.

4.2 MOTIVATION AND NEEDS

Another way of looking at animal welfare is to consider what motivates an animal and its needs instead of what should be avoided (Dawkins 1990). A behavioural need may vary with situational motivation, which in turn depends on age, sex, environment and so on. Another important influence is whether performance of the behaviour is rewarding in itself or is it the goal of the behaviour that is important. An example of this might be animals preferring access to nesting material to perform nesting behaviour rather than to a prefabricated nest box. In the case of hiding the display of behaviour in itself is not rewarding but the possibility of concealment is (van der Harst, Baars, and Spruijt 2003). Techniques have been developed for measuring choice and motivation such as preference tests (Baumans, Stafleu, and Bouw 1987; Sherwin 1996; Van de Weerd et al. 1998) and operant conditioning (Manser et al. 1998; Mason, McFarland, and Garner 1998; Sherwin 2004) and these have been used frequently during recent decades to identify preferences and assess their strength for different resources.

At least five important underlying motivations can be identified within the normal behavioural repertoire of mice, rats and rabbits, both in the wild and under laboratory conditions:

- Nesting and hiding
- Social behaviour
- Exploration

- Foraging
- Locomotory activity and movement

4.2.1 Nesting and Hiding

Burrows and nests are used by mice, rats and rabbits as shelters for hiding, resting and breeding. These animals are all prey species and flee when they feel threatened and, if possible, huddle with other conspecifics in a safe place. An avoidance reaction that occurs in both rodents and rabbits is the taking of cover under structures or in burrows.

Rats are proficient diggers and can build complex burrows (Boice 1977; Lore and Shultz 1988) in which they usually sleep during the daytime, either in a heap or separately depending on the ambient temperature. Mature animals settle down to sleep by tucking their heads between the forepaws while in the quadrupedal position. As the level of sleep deepens they suddenly keel over onto one side, typically at full length with tail extended (Lawlor 2002). In captivity, about 70% of the rat's time budget is reported to be spent in shelters (Eskola et al. 1999; Sörensen and Ottesen 2004). Burrows of house mice may vary between 10 cm and over 8 metre in length and include several entrance holes, tunnel segments and larger cavities functioning as nesting burrows and food caches (Adams and Boice 1981; Schmid-Holmes et al. 2001). Some wild mice nest in preexisting cavities such as stone crevices or walls while others live in fields or other open areas where they may dig burrows and tunnel systems (Latham and Mason 2004). Rabbits exhibit burrow digging for shelter and nesting and use raised areas as look outs to observe their environment (Stauffacher 2000; Baumans 2005).

Both male and female wild house mice exhibit elaborate nest building behaviour not only as part of their parental behaviour but also to create a suitable resting space and to facilitate thermoregulation by better controlling the microclimate. Nests may vary from bowl shaped to spherical, depending on the genetic background of the mice and the habitat in which they reside (Brown 1953; Van Oortmerssen 1971; Sluyter and Oortmerssen 2000) and is strongly influenced by the material available (Sluiter et al. 1985). The commonest nesting materials are paper, rags, grass, hair and feathers (Berry 1970). Laboratory mice show a strong preference for paper material to construct nests (Van de Weerd et al. 1998; Figure 4.3).

4.2.2 Social Behaviour

Laboratory rodents and rabbits are social species. They communicate using behavioural displays, vocalisations, scent marks and other olfactory cues in many situations relating to mating,

FIGURE 4.3 Mice social housing and nesting material. (Photo courtesy of T. P. Rooymans.)

mother–young interactions, aggressive interactions and so on. Dominance relationships are established within stable groups of animals that interact with each other on a frequent basis. The structure of the dominance hierarchy may differ, depending on species and population density but its common function is to reduce aggressive interactions arising from competition for limited resources. In a stable harmonious group, aggression is uncommon. Dominance displays preside over social interactions and subdominant animals seldom display aggression towards each other. In the wild, changes in group structure arising from dispersal, death or other reasons generally result in increased aggression within the group because of disruption of the established social hierarchy. Social isolation is a very unnatural situation for these species.

Wild rats generally live in large colonies of both males and females. At low densities one male may defend a small group of females but at high densities the dominant male tolerates subordinate males within the colony and will also allow them to mate with females in the group. A common type of social structure is a breeding deme comprising a group of females and one or several males inhabiting the same burrow system (McClintock 1987). The websites http://www.ratbehavior.org/rats.html (Hanson 2008) and http://www.ratlife.org (Berdoy 2002) are valuable illustrative resources about many aspects of rat behaviour.

Social encounters between rats commonly involve nose-to-nose contact and sniffing (of the body and the anogenital region). If one animal attacks the other (offensive aggression), it often involves a characteristic sequence of behaviours involving lateral attack, chasing, standing on top and back attack. In defence, the attacked rat responds by flight, freezing or defensive aggression. Defensive behaviours involve assumption of an upright boxing posture, lying-on-back or defensive bites aimed at the face of the attacker (Blanchard et al. 1977; Blanchard and Flannelly 1986). In mice, similar behaviours such as: approach, chase, lateral attack and bite chase have been identified as offensive behaviours whereas flight, freezing, boxing and defensive attack (towards the face) have been identified as defensive behaviours (Blanchard and O'Donnell 1979; Blanchard, Griebel, and Blanchard 2001). Play fighting serves a social function and is commonly seen in juvenile rats and also, to a lesser degree, in adult animals. It can be distinguished from aggressive fighting by its target: play fighting is not directed to the back but, instead, involves less aggressive attack of the back of the neck, nape, which is not bitten but rubbed gently with the snout (Pellis et al. 1987). Other socially cohesive behaviours in rats and mice include mutual grooming and huddling to sleep close together.

As in rats, the house mouse establish social structures that are influenced by factors such as environmental constraints (Mackintosh 1981), colony size (Poole and Morgan 1973) and the degree of male and female aggressiveness (Brain and Parmigiani 1990; Parmigiani, Ferrari, and Palanza 1998). Social organisation in mice can also be either commensal or feral. Territories vary in size from the area adjacent to the nesting area to home ranges extending 25–30 metre or even more (Brown 1953). Male house mice leave their natal territory and establish their own territory nearby. The resident territory holder defends the home range from conspecific intruders but may allow subordinate males to reside within the territory when unoccupied space is scarce. Female mice often stay within their natal territory, and mate mainly with the dominant male. Females may engage in territorial defence especially when pregnant (Mackintosh 1981). After parturition they protect the nest and pups from infanticidal attacks (Parmigiani, Ferrari, and Palanza 1998). There have been recent reviews both of maternal (Weber and Olsson 2008) and social organisational behaviour (Latham and Mason 2004) in wild mice and their implications for husbandry of laboratory mice.

Wild rabbits live in large colonies subdivided into smaller breeding groups consisting of 1–4 males and up to nine females (Meredith 2000). Territorial size varies with food resources and the number of individuals in the group but may be up to 500 metres in diameter (Donnelly 1997). Rabbits usually establish a strict linear dominance hierarchy that is maintained by scent marking (glandular secretions and urine) and a range of behavioural signals (Lehmann 1991). Males are very aggressive towards male intruders but are tolerant towards subordinate males and females.

Females fight mainly over nest sites but may also attack offspring of females within the same breeding group.

4.2.3 EXPLORATION

Attention to novel stimuli is essential for adaptation to a changing environment. Exploratory behaviour can be defined as the search for increased information about the environment with or without any evident need or incentive (Barnett and Cowan 1976). Exploratory motivation might be provoked by novelty or environmental change and results in the acquisition of information and learning about spatial and topographical aspects of the environment or of resources within it (Renner 1988; Pierre et al. 2001). There are also evolutionary benefits of exploration such as discovery of novel territories, food and partners. Because exploratory behaviour is such an important aspect of survival in the wild, it is highly conserved and all mammals are highly motivated to carry it out. Lack of exploratory opportunities may result in frustration and boredom, manifested in behavioural abnormalities such as redirected behaviours (aggression, self-injury, etc.), stereotypies and inactivity/apathy (Wemelsfelder 1990).

Exploration of the environment involves a trade-off between the chance of finding and utilising the resources necessary for survival and reproduction and the risk of being killed by a predator or an aggressive conspecific. The trade-off between risk taking and potential gain may vary in different situations and is influenced by motivational or emotional state. When undertaking exploration, rodents make short excursions from a safe area, avoid open areas, stay close to vertical structures and prefer sites with overhead cover and complex ground level structure (Plesner Jensen and Gray 2003). The actual behaviours involved in exploration include risk assessment (stretched attend), approach, rearing, sensory investigation (sniffing, chewing and biting) and scent marking. It may also involve defensive actions such as burying or withdrawal. Novel objects commonly elicit scent-marking behaviour (both gland rubbing and urine), which may serve as territorial marking or as identification of investigated objects. The extent of neophobic object avoidance depends on characteristics of the object, previous experience of similar objects and of the environment in which it is located.

Several laboratory studies have shown that exploratory behaviour and emotional reactivity vary between strains in both rats and mice (Trullas and Skolnick 1993; Ramos and Mormede 1998; Kopp, Vogel, and Misslin 1999; Griebel et al. 2000; Crusio 2001). Some studies also indicate that males and females may differ in their perception of novel and social events and use different approaches when exposed to the same task depending on context and reward (Bimonte et al. 2000; Palanza, Gioiosa, and Parmigiani 2001; Augustsson and Meyerson 2004; Augustsson, Dahlborn, and Meyerson 2005). This may be reflected in their requirements in a captive environment. Individually housed male mice tend to show a higher propensity for exploration and a lower level of anxiety compared with group-housed males. Individually housed female mice have higher risk assessment and less risk taking than group housed mice (Palanza, Gioiosa, and Parmigiani 2001).

4.2.4 FORAGING

Rodents spend a great part of their time exploring for and processing food in their natural environments. Rats and mice are nocturnal feeders and they carry a piece of food by their teeth to a suitable spot where they adopt a squatting posture and hold the food in their forepaws to nibble at it. Rabbits also consume most of their food during the dark period. Animals preferentially search for food even when it is readily available, and it has been proposed that this behaviour might have an information-gathering function in that it provides them with information about the location and quality of potential foraging sites (Mench 1998). For example, rats readily work (press a lever) for food even if identical food is freely available to them (Neuringer 1969). This behavioural response is called "contrafreeloading" (Osborne 1977; Inglis, Forkman, and Lazarus 1997). The tendency for animals to work for food in the presence of easily accessible food has been interpreted as evidence of a strong requirement for confined animals to engage in additional activity (Chamove 1989) or in foraging activities (Dawkins 1990).

4.2.5 Locomotory Activity and Movement

Motor activity involves an integrated sequence of events that are coordinated and regulated by the nervous system and take place in response to perceived changes in the relationship between the animal and its physical and social environment. The initiation and adaptation of motor activity takes place in response to summed sensory inputs from the environment, whether it involves changes to posture, equilibrium or locomotion (running, climbing or walking activity), or specific actions such as ocular, head or limb movements. Depending on the need or the aim, motor activity can be functionally described as orientation, exploration, approach, pursuit, manipulation, consumption, hiding, fighting and so on.

In the natural habitat, locomotory activity is generally goal-directed and, apart from juvenile animals that show frequent play behaviour, adult animals are active physically mainly to find food, patrol a territory or escape from predators. Depending on their home range, rats may travel up to several kilometres per day (Barnett 1963) patrolling the region and its surroundings. Both mice and rats generally move rapidly between sites following the course of regular runways (Boice 1977; Berdoy 2002; Latham and Mason 2004). They are excellent climbers and may run along wires and ropes as well as swim and jump to reach a position, although rats are more likely to swim than mice. In the laboratory, voluntary activity in rodents has been studied mainly using running wheels; there is great individual variation in running distance, which appears to be linked partly to hormonal status and dominance. However, the same great variety of movement seen in wild rodents and rabbits such as running, jumping, climbing and hopping can take place if enough space and a suitable environmental structure is provided (Brain 1992; Morton 1993; Berdoy 2002; Spangenberg et al. 2005).

4.3 ENVIRONMENTAL REFINEMENT

4.3.1 Definition and Principles of Environmental Refinement for the Welfare of the Animals

Baumans has defined environmental enrichment as "any modification in the environment of captive animals that seeks to enhance their physical and psychological well-being by providing stimuli meeting the animals' species-specific needs" (Baumans 2000, p. 1251). Neuroscience also utilises the term "environmental enrichment" although research protocols are based mainly on novelty-induced stimulation and regularly changing items, primarily to measure the effects on brain neuronal plasticity (for a review see: Adams, Korytko, and Blank 2001) in contrast to the enhancement of welfare by means of appropriate enrichment focused on animal needs (Baumans 2005).

We here propose changing the term "environmental enrichment", when applied to laboratory animals, to "environmental refinement". Nevertheless, in order to avoid confusion we still use "enrichment" when quoting results from "enrichment" studies.

In laboratory animal facilities, rodents and rabbits are commonly housed in standard-sized cages that provide barely any structuring and generally include items that satisfy only minimum requirements such as food, water and substrate; nesting material and refuges are often now provided but have only recently been endorsed by the revised Appendix A of the European Convention (Council of Europe 2006). It can be argued that what is generally termed "standardised" should be considered "impoverished" in that the behaviour of the animals is severely restricted. Chronic thwarting of motivated behaviours has been shown to produce functional modifications to the nervous system, which raises questions about the appropriateness of such animals as models for research on the normal functioning of the nervous system (Wurbel 2001).

As a matter of fact, neuroscience studies of enrichment often use the term "impoverished" to signify conditions of individually housed subjects in conventional cages (Mohammed et al. 2002) whereas "standard conditions" refers to socially housed subjects in conventional cages. "Enriched conditions" describes larger cages with many different kinds of enrichment objects and a larger number of animals.

The term "environmental refinement" is preferred here to the more widely and commonly used "environmental enrichment" for several reasons. Firstly, environmental refinement implies an improvement in quality of the life of animals, while the term and the concept of environmental enrichment are often misused and are applied whenever changes are made to the environment, or when the environment is made more complex, without any demonstrable benefit to the animals (Nevison, Hurst, and Barnard 1999). Secondly, it is important to distinguish between the different approaches and aims of environmental enrichment in neuroscience research compared with improvements of animal welfare. Thirdly, within the Three Rs principle, "refinement" is the generally accepted term for measures aiming to improve animal welfare.

Environmental refinement is an ongoing process and we should aim to provide stimuli beyond the satisfaction of basic needs, normally catered for in standard housing conditions.

The impact of changes in the environment may differ depending on the animal's strain, sex, weight, age, group composition, cage dimensions, reproductive state, and so on. Moreover, in the animal welfare-oriented approach, the types of environmental improvement may be quite dissimilar, raising the critical issue of what should be considered the basic cage environment.

Components for environmental refinement should be selected on the basis that they meet the animal's needs (see Section 4.1), which means that a thorough knowledge of the species-typical behaviour is essential. Any manipulation to improve the environment of laboratory animals should be evaluated in terms of its impact on their welfare; the assessments should be based on multiple indicators including: behaviours in response to, or associated with introduced changes, physiological parameters, physical health and so on. In addition, consideration should be given to the practicality of routine husbandry practises and the effects on scientific results, such as biologically significant increase in variation in experimental data or alterations of baseline values.

It is a moral and, in many countries, a legal obligation for those who maintain animals for experimental purposes to provide the best possible environment; improvement must result in enhanced animal welfare without compromising the scientific purpose for which the animals are being used.

Refinements of the animal's environment focus on its physical environment and associated sensory stimuli (auditory, visual, olfactory and tactile) and nutritional aspects (supply and type of food), and its social environment.

4.3.2 Refinement of Physical Environment

The basic constituents of the physical environment are the cage structure (i.e., size, shape and design), and the components within it (i.e., flooring, bedding and furnishing). Strategies to improve the physical environment of laboratory animals should include provision of stimuli (objects/devices) that are biologically meaningful to them, with which they can choose to interact or not (depending on individual differences) and that are not harmful to them. Stimuli that are too artificial or too unfamiliar will probably be of very little benefit to the animal. On the other hand, providing objects that simulate aspects of the animal's natural habitat often give rise to lasting and diverse species-specific behaviours. This does not mean that it is necessary to simulate "natural conditions" in animals' environments because these may not necessarily be the optimum. Not only does the immediate housing environment affect laboratory animals lives, but also the perception of this afforded by their senses, especially in rodents that have very sensitive auditory pathways (see Chapter 3). In order to provide auditory stimuli, some laboratories play radio music in animal holding rooms.

Very few studies of the effects of music on these species are available and the physiological and behavioural consequences remain unclear. Further research is needed to provide more scientifically based information relating to such changes in the auditory environment (Patterson-Kane and Farnworth 2006).

4.3.2.1 Cage Structure and Furnishings

Refinement of the physical environment can be achieved by ensuring the dimensions and structuring of the primary enclosure are adequate and by providing sufficient furnishing to satisfy physiological and ethological needs (i.e., nesting, grooming, exploring, hiding, searching for food, gnawing and social contacts; see Section 4.1). Cage equipment allows the animal to interact with, to have control over and, in some cases, to manipulate its environment, thereby prompting the animal to psychologically appraise its environment and thereby improve its well-being (Baumans 2005).

The revised Appendix A of the European Convention ETS 123 (Council of Europe 2006) requires that bedding and nesting material and refuges should be provided for rodents in breeding, stock or under procedure unless there is a justification on veterinary or welfare grounds against doing so. For rabbits, a raised area should be provided on which the animal can lie and sit and easily move underneath.

Bedding provides a comfortable surface and enables behaviours such as digging, chewing and nesting (Boyd 1988). Commonly used bedding materials include: wood chips, cellulose-based chips and shredded filter paper. Preference tests have shown that when mice are given a choice between combinations of a substrate of wood chips, shredded filter paper or sawdust, they show a preference for paper (Sherwin and Nicol 1997).

Nesting material for rodents may consist of paper tissue, hay or wood-wool and appear to be successful in that they enable the animals to build a nest to rest, get a shelter and hide and retreat from their cage mates or excessive light (Van de Weerd and Baumans 1995).

Rats prefer paper strips that can be used to form more cohesive nests and nest boxes for hiding and are willing to work for it (Ras et al. 2002). However, in contrast to mice, rats don't readily use nesting material if they have not had experience of it from birth onwards (Van Loo and Baumans 2004). Nest-boxes of opaque or semi-opaque materials are particularly suitable shelters (Manser et al. 1998). The use of shredded paper, which incidentally allows construction of a nest resembling that of the wild rat, allows the female with her young to burrow and insulate themselves from disturbing environmental influences, thus enhancing the mother's feeling of security. In addition, access to shredded paper has been shown to greatly reduce infant mortality in rodents (Nolen and Alexander 1966). However, some types of paper increase pup mortality, probably due to hygroscopic properties of the material leading to dehydration of the pups (Norris and Adams 1976).

Suitable structuring of cages for rodents and rabbits allows them to identify separate areas for feeding, resting, defecation and urination, thereby facilitating their natural tendency to use different areas for different behaviours.

This can be achieved by providing surfaces that enable the animals to better utilise the space and control their environment. Such divisions may be facilitated by the inclusion of shelves, shelters, nest boxes, nesting material, tubes and platforms (Baumans 2005). Rats with access to an appropriate shelter are more explorative and less timid than those in barren cages (Townsend 1997).

Rabbits benefit from housing in pens rather than cages because they provide more space and height to perform natural behaviours such as sitting in an upright posture with their ears erect. If floor pens are not a feasible option, an alternative strategy is to provide extra space by joining cages together in a modular system. Suitable environmental refinement objects for rabbits include roughage, hay blocks or chew sticks. Access to a raised area, such as a platform, enables rabbits to jump on and off, providing valuable exercise to strengthen bones and making it possible for them to withdraw but to look and observe the surroundings and probably reducing anxiety (RSPCA/

FIGURE 4.4 Rabbit pen. (Photo courtesy of Novo Nordisk.)

UFAW 2008; Figure 4.4). When floor pens are used to group house rabbits, visual barriers should be provided (Baumans et al. 2006). For breeding does, nesting material and a nest box or another refuge should be provided.

4.3.2.2 Feeding

Under laboratory conditions food is most commonly provided *ad libitum* in fixed dispensers and is easily obtainable by the animals. A varied diet and scattering food (e.g., sunflower seeds) in the substrate encourage activity and natural behaviours such as foraging and food storing. Mice, rats and hamsters quickly learn to search and apparently welcome this variation in their feed (Scharmann 1991). Leach and Ambrose (1999) have designed and evaluated floor feeding where, in addition to the standard diet present in the food hopper, other food was incorporated into the cage substrate. Floor feeding appeared to promote natural feeding, foraging and exploratory behaviours without affecting growth rate.

Providing items of food within the cage can encourage foraging behaviours and allow rats to adopt normal postures for eating. Foraging behaviours and processing of food commonly occur also under laboratory conditions, thus suggesting that the animals may benefit from opportunities to engage in physical activity and manipulation of objects (Hutchinson, Avery, and Vandewoude 2005).

Hay, straw or grass cubes can satisfy the need of rabbits for roughage and for chewing (Baumans 1997); these materials can readily be presented in such a way that the animals have to work to retrieve them and so to engage in foraging activities (Prowse 2002; Banjanin and Barley 2004). A significantly reduced frequency of stereotypies was reported when a restricted amount of food was fed just before the dark period instead of providing *ad libitum* or restricted diet in the morning (Krohn and Ritskes-Hoitinga 1999).

4.3.2.3 Opportunities for Physical Exercise and Exploration

Mice and rats are usually kept in cages that do not allow them to express the amount of physical activity that occurs in wild rodents. Lack of locomotor and exploratory opportunities may result in frustration and boredom, manifested in behavioural abnormalities such as redirected behaviours (aggression, self-injury, etc.), stereotypies and inactivity/apathy (Wemelsfelder 1990).

Provision of an adequately large area and the inclusion of tubes, shelters and platforms encourage locomotor activity and provide the animal with opportunities for exercise. Numerous studies have demonstrated that exercise and/or behavioural enrichment can increase neuronal survival and resistance following brain insult (Stummer et al. 1994; Carro et al. 2001), promote brain vascularisation

(Black et al. 1990; Isaacs et al. 1992), stimulate neurogenesis, enhance learning (van Praag et al. 1999; Young et al. 1999) and contribute to the maintenance of cognitive function during aging (Escorihuela, Tobena, and Fernandezteruel 1995). On the other hand it has been argued that the restriction of physical activity imposed on conventionally caged laboratory animals affects their CNS gene expression levels, the interactions among gene products and, consequently diminishes the suitability of these animal models for the development of new CNS medicines (Gurwitz 2001).

Access to properly sized and designed running wheels promotes locomotor activity in rodents (Sherwin 1998), decreases the amount of body fat, lowers blood pressure, improves resistance to oxygen toxicity, increases immunocompetence (Coleman and Rager 1993; Suzuki and Machida 1995) and increases life expectancy (Holloszy 1992). Single-caged hamsters display less stereotypical bar-mouthing when running wheels are provided (Baumans and Coke 2007). However, under some circumstances wheel-running is apparently a stereotypical and maladaptive behaviour arising as an artefact of experimental/impoverished conditions (Barnett and Cowan 1976; Hogan and Roper 1978; Mason 1991). Howerton and Garner (2008) have recently shown that competition among group-housed male CD-1 mice for access to running wheel-igloo enrichment increased the amount of potentially injurious aggression and disturbed hierarchy linearity, suggesting a disruption in social structure. It remains unclear whether the decrease in stereotypies, observed in cages equipped with wheels results from a redirection of activity towards stereotypic wheel-running, or increased opportunities for exploratory locomotion or is caused by a shift in the behavioural time-budget.

Several studies have investigated the effects of enriched housing on exploratory behaviour in novel environments. Some found that environmental enrichment may have a positive impact on diversity (Widman and Rosellini 1990) and efficiency of exploration (Zimmermann et al. 2001) as well as on emotional reactivity (Prior and Sachser 1995; Chapillon et al. 1999). Operant conditioning techniques have also shown that mice will work for access to greater space/novel areas both when singly housed (Sherwin 1996; Sherwin and Nicol 1997) and housed in groups (Sherwin 2004).

4.3.3 REFINEMENT OF SOCIAL ENVIRONMENT

Laboratory rodents and rabbits are social, gregarious species for which social contact is a stimulating part of their environments. Social interactions are important contributors to animal welfare provided that the group composition is appropriate for the species' natural demography. Group housed animals are able to engage in social exploration; the behavioural activities of one animal, such as scent marking or digging, may also be a valuable source of novelty that elicits exploration by the other individuals (Olsson and Westerlund 2007).

Social housing is beneficial only if the pairs or groups are harmonious and stable. The successful establishment of groups requires the grouping of individuals that are compatible, a characteristic that is strongly influenced by age, sex and hierarchical rank; some of these can be managed by good husbandry practises. The establishment of groups must be handled with care, particularly in the case of male mice and rabbits, adult hamsters or gerbils, which can exhibit severe conspecific aggression. The provision of visual barriers and refuges that allow animals to withdraw out of sight when a threat occurs can minimise aggression. Disruption of established stable and harmonious groups can be very stressful and should be minimised (Van Loo et al. 2003). One option for preventing aggression in male rabbits is castration; an aggressive male mouse could be housed together with an ovariectomised female as companion (Wersinger and Martin 2009) when the animals have to be kept for long periods, although consideration needs to be given to whether the negative impact of the surgical procedure is outweighed by the benefit of social housing.

Although social contacts are an important component of animals' life, it may be necessary sometimes to house animals individually, for example, if adverse effects or injury are likely to occur, or for scientific reasons (e.g., metabolic studies and infectious disease research). In these cases, animals should have visual, olfactory or auditory contact with conspecifics. Laboratory rabbits benefit from minimal tactile contacts, which can be facilitated by providing perforated mesh panels between pens or cages (RSPCA/UFAW 2008).

In a recent study (Van Loo et al. 2007), several physiological and behavioural parameters were assessed after surgery for implantation of a telemetry transmitter; postoperative mice benefited most from being socially housed with respect to postoperative recovery and distress; however, when social housing is not possible, individual caging appears to be a better option than separating mice by a grid partition, possibly because they are frustrated by not being able to huddle together.

4.3.4 Validation of Environmental Refinement

Improving of the welfare of laboratory animal housing depends on meeting the physiological and behavioural needs of the occupants. This can be achieved through environmental refinement programs that entail modification of the animals' environment. The consequences of such changes need to be carefully evaluated in order to establish whether the animal's welfare really has been improved and to determine the impact on experimental results.

The assessment of the impact of environmental changes on animal well-being is based on a complex of behavioural and physiological parameters comprising:

- Animals' ability to cope with the changes in their social and physical environment
- Increase of their behavioural repertoire
- Balanced display of socio-positive and negative behaviours
- Reduction or absence of maladaptive or abnormal behaviours
- Absence of signs of chronic fear and distress
- Activity of the hypothalamo-pituitary-adrenal axis and the sympatho-adrenal system
- Body temperature
- Heart rate
- Immunocompetence
- Body weight/food intake

Other hormonal and physiological measures are included also (Manser 1992; Broom and Johnson 1993; Clark, Rager, and Calpin 1997; Mench 1997; Terlouw and Schouten 1997; ILAR 1998; Olsson and Dahlborn 2002). The measures selected to evaluate the effects of environmental alteration on animal welfare depend on the objective of the environmental refinement program and should also take into account the specific type of research for the animals required (Olsson and Dahlborn 2002).

Some environmental changes may impact negatively on animal welfare. For example some types of cage supplementation may induce aggression in male mice as a consequence of object monopolisation (Marashi et al. 2003). Other authors have reported that providing cages of male mice with a shelter increased the level of aggression and was accompanied by physiological indicators of stress (Van Loo et al. 2002). In contrast, the provision of nesting material reduced aggressive behaviour as well as stress hormones and adrenal activity (Van Loo et al. 2002, 2003, 2004).

In an extensive review of the effects of environmental enrichment in mice, Olsson and Dahlborn (2002) pointed out that for some species not all types of enrichment clearly benefit an animal's well-being and that more systematic research is needed to define those resources that are important to mice. Any housing supplementation or change of the animals' physical or social environment should be tested, validated and documented. Although many enrichment items are available commercially, only a few have been scientifically evaluated and validated (Van Loo et al. 2005; Wurbel and Garner 2007).

Critical to developing environmental refinement programs and assessing their value is the participation of all personnel involved in the care and use of the animals, who must be able to recognise the normal behaviours of the species concerned and to evaluate the outcomes of a change in housing conditions.

4.3.5 THE IMPACT ON SCIENTIFIC RESULTS

Animals that are well adapted to their physical environment are better animal models than those that lack environmental stimulation and may show signs of chronic stress, maladaptive behaviours and stereotypies. Garner and Mason described cage stereotypies as repetitive sequences of motor behaviour that are topographically and morphologically invariant, often rhythmical and apparently purposeless. They also observed that such "cage stereotypies" are typically elicited by the frustration of specific motivated behaviours, in combination with stress and reduced behavioural competition caused by low environmental complexity (2002). As a consequence, captive environments may affect the validity and replicability of experiments by altering animals' behavioural and physiological parameters; increased inter-individual variation might also reduce the reliability of results (Garner 2005). Mason and Clubb (2007) argued that, rather than genetic selection, pharmacological treatment and so on, environmental enrichment is the preferred approach to tackle the problems underlying stereotypic behaviours and to improve both welfare and behaviour.

In making any modification to the complexity of the environment, the potential for this to introduce variability into a study and so confound the experimental results should be considered. Furthermore, the provision of enrichment should be evaluated in the context of the health of the animal and research goals on a case-by-case basis (Bayne 2005). Experimental variability has been reported to be unchanged, increased or decreased, depending on the type of enrichment, strain of animal, parameter measured and statistical method used and so on (Baumans 1997; Stauffacher 1997; Tsai et al. 2002; Van de Weerd et al. 2002; Augustsson et al. 2003; Wolfer et al. 2004).

It is reasonable to assume that relative refinements that focus on specific needs of the animals, will introduce much less variability than complex cage enrichments such as those used to induce changes in brain structure and learning and memory abilities. More detailed information about the type of enrichment used should be included in the materials and methods section of a scientific paper in order to render experimental results more reproducible (Baumans 2005).

4.3.6 ANIMAL WELFARE ASSESSMENT AND THE FUTURE

Many physiological stress markers, preference tests and operant conditioning techniques have been used to study the effects on animal welfare of environmental conditions. More recently, methods have been developed to measure the anticipation of reward (Spruijt, van den Bos, and Pijlman 2001; van der Harst, Baars, and Spruijt 2003) and for assessing subjective emotional states (Desire, Boissy, and Veissier 2002). Moreover, there is evidence that ambiguous stimuli may be subjected to differential cognitive evaluation based on previous experience, leading to reduced expectation of positive events in rats, which are housed in stressful, unpredictable conditions (Harding, Paul, and Mendl 2004). The concept of affective consciousness (Panksepp 2005) and the development of tools for assessing positive emotions (Boissy et al. 2007) are also receiving increasing attention and have the potential to increase our understanding of animal welfare beyond just the absence of negative experiences.

4.4 CONCLUSIONS

- The concept of the Five Freedoms, developed originally for farm animals, is applicable also to laboratory animals and can be used to assess the welfare of these animals.
- Animals should be housed in an environment, which meets their needs, where they feel secure and where they can express species-typical behaviour.
- It is essential to have knowledge of the species-typical behaviour in order to assess whether housing conditions meet the needs of an individual animal based on its species, and when appropriate, also on its age, gender and previous experiences.

- Knowledge about the motivation and needs of animals is required to enable assessment of animal welfare.
- The term "environmental refinement" implies providing stimuli beyond the basic needs, which are normally catered for in animals under standard housing conditions.
- Refinement programs should include and document the criteria used to evaluate the effectiveness of the program; they should also be validated and regularly assessed for their lasting improvement of animal welfare and lack of impact on scientific results.

4.5 QUESTIONS UNRESOLVED

- How long does it take for the stomach and intestinal tract, respectively, to become empty following food deprivation? How long a period of food deprivation is acceptable for rabbits?
- Is the provision of running wheels an environmental refinement? Does their use indicate a stereotypy or a benefit?
- What is the best way to manage group-housed male mice and rabbits so as to minimise aggression? Under what conditions is castration justified to allow group housing? Are there other strategies for environmental refinement or husbandry that help minimise aggression?
- In IVC housing, how does air speed and noise affect the animals? How can opportunities for observation of animals be maximised? How does the reduced frequency of cleaning affect the animals (e.g., habituation to being handled vs. not disturbing scent marks)?
- How important is cage height to rats in permitting them to display normal behaviours such as rearing and climbing?
- What basic cage structure or furniture should be recommended for housing rodents and rabbits?
- How can we use positive welfare assessment (e.g., anticipation) on a practical day-to-day basis?

REFERENCES

Adams, C. S., A. I. Korytko, and J. L. Blank. 2001. A novel mechanism of body mass regulation. *The Journal of Experimental Biology* 204:1729–34.

Adams, N., and R. Boice. 1981. Mouse (mus) burrows: Effects of age, strain, and domestication. *Animal Learning & Behavior* 9:140–44.

Augustsson, H. and B. J. Meyerson. 2004. Exploration and risk assessment: A comparative study of male house mice (*Mus musculus musculus*) and two laboratory strains. *Physiology & Behavior* 81:685–98.

Augustsson, H., H. A. van de Weerd, C. L. Kruitwagen, and V. Baumans. 2003. Effect of enrichment on variation and results in the light/dark test. *Laboratory Animals* 37:328–40.

Augustsson, H., K. Dahlborn, and B. J. Meyerson. 2005. Exploration and risk assessment in female wild house mice (*Mus musculus musculus*) and two laboratory strains. *Physiology & Behavior* 84:265–77.

Augustsson, H., L. Lindberg, A. U. Hoglund, and K. Dahlborn. 2002. Human-animal interactions and animal welfare in conventionally and pen-housed rats. *Laboratory Animals* 36:271–81.

Banjanin, S., and J. Barley. 2004. Environmental enrichment for guinea pigs: A discussion by the laboratory animal refinement & enrichment forum. *Animal Technology and Welfare* 3:161–63.

Barnett, S. A. 1963. *The rat—A study in behaviour.* London: Methuen Co.

Barnett, S. A. and P. E. Cowan. 1976. Activity, exploration, curiosity and fear—Ethological study. *Interdisciplinary Science Reviews* 1:43–62.

Baumans, V. 1997. Environmental enrichment: Practical applications. In *Animal alternatives, welfare and ethics*, eds. L. F. M. van Zutphen and M. Balls, 187–97. Amsterdam, New York: Elsevier.

Baumans, V. 2000. *Environmental enrichment: A right of rodents!* Paper presented at Progress in the Reduction, Refinement and Replacement of animal experimentation, 3rd world congress on Alternatives and animal use in the life sciences.

Baumans, V. 2005. Environmental enrichment for laboratory rodents and rabbits: Requirements of rodents, rabbits, and research. *ILAR Journal/National Research Council, Institute of Laboratory Animal Resources* 46:162–70.

Baumans, V., F. R. Stafleu, and J. Bouw. 1987. Testing housing system for mice—The value of a preference test. *Zeitschrift Fur Versuchstierkunde* 29:9–14.

Baumans, V., F. Schlingmann, M. Vonck, and H. A. Van Lith. 2002. Individually ventilated cages: Beneficial for mice and men? *Contemporary Topics in Laboratory Animal Science* 41:13–19.

Baumans, V., P. Clausing, R. Hubrecht et al. 2006. FELASA working group on standardization of enrichment. Available from http://www.lal.org.uk/index.php?option=com_content&view=article&id=56&Itemid=60

Baumans, V., C. Coke, J. Green et al. 2007. Feeding enrichment making lives easier for animals in research labs. Washington, DC: Animal Welfare Institute.

Bayne, K. 2005. Potential for unintended consequences of environmental enrichment for laboratory animals and research results. *ILAR Journal/National Research Council, Institute of Laboratory Animal Resources* 46:129–39.

Berdoy, M. 2002. *The laboratory rat: A natural history.* Available from http://www.ratlife.org

Berry, R. J. 1970. The natural history of the house mouse. *Field Studies* 3:219–62.

Bimonte, H. A., L. A. Hyde, B. J. Hoplight, and V. H. Denenberg. 2000. In two species, females exhibit superior working memory and inferior reference memory on the water radial-arm maze. *Physiology & Behavior* 70:311–17.

Black, J. E., K. R. Isaacs, B. J. Anderson, A. A. Alcantara, and W. T. Greenough. 1990. Learning causes synaptogenesis, whereas motor-activity causes angiogenesis, in cerebellar cortex of adult-rats. *Proceedings of the National Academy of Sciences of the United States of America* 87:5568–72.

Blanchard, D. C., G. Griebel, and R. J. Blanchard. 2001. Mouse defensive behaviors: Pharmacological and behavioral assays for anxiety and panic. *Neuroscience and Biobehavioral Reviews* 25:205–18.

Blanchard, R. J., C. R. McKittrick, and D. C. Blanchard. 2001. Animal models of social stress: Effects on behavior and brain neurochemical systems. *Physiology & Behavior* 73:261–71.

Blanchard, R. J., D. C. Blanchard, T. Takahashi, and M. J. Kelley. 1977. Attack and defensive behavior in albino-rat. *Animal Behaviour* 25:622–34.

Blanchard, R. J., and K. J. Flannelly. 1986. Defensive behavior of laboratory and wild *Rattus norvegicus*. *Journal of Comparative and Physiological Psychology* 100:101–7.

Blanchard, R. J., and V. O'Donnell. 1979. Attack and defensive behaviors in the albino mouse. *Aggressive Behaviour* 5:341–52.

Blom, H. J. M., G. Vantintelen, V. Baumans, J. Vandenbroek, and A. C. Beynen. 1995. Development and application of a preference test system to evaluate housing conditions for laboratory rats. *Applied Animal Behaviour Science* 43:279–90.

Boice, R. 1977. Burrows of wild and albino-rats—Effects of domestication, outdoor raising, age, experience, and maternal state. *Journal of Comparative and Physiological Psychology* 91:649–61.

Boissy, A. 1995. Fear and fearfulness in animals. *Quarterly Review of Biology* 70:165–91.

Boissy, A., G. Manteuffel, M. B. Jensen et al. 2007. Assessment of positive emotions in animals to improve their welfare. *Physiology & Behavior* 92:375–97.

Boyd, J. 1988. Mice. *Humane Innovations and Alternatives in Animal Experimentation* 3:98–99.

Brain, P. F. 1992. Understanding the behaviours of feral species may facilitate design of optimal living conditions for common laboratory rodents. *Animal Technology* 43:99–105.

Brain, P. F., D. Buttner, P. Costa et al. 1993. *Rodents*. Paper presented at The Accommodation of Laboratory Animals in Accordance with Animal Welfare Requirements.

Brain, P. F. and S. Parmigiani. 1990. Variation in aggressiveness in house mouse-populations. *Biological Journal of the Linnean Society* 41:257–69.

Brambell, F. W. 1965. *Report of the technical committee to enquire into the welfare of animal kept under intensive livestock husbandry system.* London: HMSO.

Broom, D. M. and K. G. Johnson. 1993. *Stress and animal welfare.* London: Chapman & Hall.

Brown, R. Z. 1953. Social behavior, reproduction, and population changes in the house mouse (*Mus musculus* L.). *Ecological Monographs* 23:217–40.

Burn, C. C. 2008. What is it like to be a rat? Rat sensory perception and its implications for experimental design and rat welfare. *Applied Animal Behaviour Science* 112:1–32.

Buttner, D. 1993. Upright standing in the laboratory rat—Time expenditure and its relation to locomotor activity. *Journal of Experimental Animal Science* 36:19–26.

Carro, E., J. L. Trejo, S. Busiguina, and I. Torres-Aleman. 2001. Circulating insulin-like growth factor I mediates the protective effects of physical exercise against brain insults of different etiology and anatomy. *The Journal of Neuroscience: The Official Journal of the Society for Neuroscience* 21:5678–84.
Chamove, A. S. 1989. Environmental enrichment: A review. *Animal Technology* 40:155–78.
Chapillon, P., C. Manneche, C. Belzung, and J. Caston. 1999. Rearing environmental enrichment in two inbred strains of mice: 1. Effects on emotional reactivity. *Behavior Genetics* 29:41–46.
Clark, J. D., D. R. Rager, and J. P. Calpin. 1997. Animal well-being. II. Stress and distress. *Laboratory Animal Science* 47:571–79.
Coleman, K. J. and D. R. Rager. 1993. Effects of voluntary exercise on immune function in rats. *Physiology & Behavior* 54:771–74.
Council of Europe. 2006. *Appendix of the European convention for the protection of vertebrate animals used for experimental and other scientific purposes (ETS123). Guidelines for accommodation and care of animals (Article 5 of the convention) approved by the multilateral consultation.* Trans. Council of Europe. Vol. Cons 123 (2006) 3. Strasbourg.
Crowcroft, P. 1966. *Mice alla over.* London: Foulis & Co Ltd.
Crusio, W. E. 2001. Genetic dissertation of mouse exploratory behaviour. *Behavioral and Brain Research* 125:127–32.
Davies, H. and D. Balfour, eds. 1992. *The inevitable bond: Examining scientist-animal welfare.* Cambridge: Cambridge University Press.
Dawkins, M. S. 1990. From an animal's point of view—Motivation, fitness, and animal-welfare. *Behavioral and Brain Sciences* 13:1–61.
De Boer, S. F. and J. M. Koolhaas. 2003. Defensive burying in rodents: Ethology, neurobiology and psychopharmacology. *European Journal of Pharmacology* 463:145–61.
Desire, L., A. Boissy, and I. Veissier. 2002. Emotions in farm animals: A new approach to animal welfare in applied ethology. *Behavioural Processes* 60:165–80.
Donnelly, T. M. 1997. Ferrets, rabbits and rodents—Clinical medicine and surgery. In *Basic anatomy, physiology and husbandry*, eds. E. W. Hillyer and K. E. Quesenberry, 147–59. London: WB Saunders.
Escorihuela, R. M., A. Tobena, and A. Fernandezteruel. 1995. Environmental enrichment and postnatal handling prevent spatial-learning deficits in aged hypoemotional (Roman high-avoidance) and hyperemotional (Roman low-avoidance) rats. *Learning & Memory* 2:40–48.
Eskola, S., M. Lauhikari, H. M. Voipio, and T. Nevalainen. 1999. The use of aspen blocks and tubes to enrich the cage environment of laboratory rats. *Scandinavian Journal of Laboratory Animal Science* 26:1–10.
FAWC. 1993. *Second report on priorities for research and development in farm animal welfare.* Tolworth, UK: Farm Animal Welfare Council.
Fiala, B., F. M. Snow, and W. T. Greenough. 1977. Impoverished rats weigh more than enriched rats because they eat more. *Developmental Psychobiology* 10:537–41.
Garner, J. P. 2005. Stereotypies and other abnormal repetitive behaviors: Potential impact on validity, reliability, and replicability of scientific outcomes. *ILAR Journal/National Research Council, Institute of Laboratory Animal Resources* 46:106–17.
Garner, J. P., B. Dufour, L. E. Gregg, S. M. Weisker, and J. A. Mench. 2004a. Social and husbandry factors affecting the prevalence and severity of barbering ('whisker trimming') by laboratory mice. *Applied Animal Behaviour Science* 89:263–82.
Garner, J. P., and G. J. Mason. 2002. Evidence for a relationship between cage stereotypies and behavioural disinhibition in laboratory rodents. *Behavioural Brain Research* 136:83–92.
Garner, J. P., S. M. Weisker, B. Dufour, and J. A. Mench. 2004b. Barbering (fur and whisker trimming) by laboratory mice as a model of human trichotillomania and obsessive-compulsive spectrum disorders. *Comparative Medicine* 54:216–24.
Gartner, K., D. Buttner, K. Dohler, R. Friedel, J. Lindena, and I. Trautschold. 1980. Stress response of rats to handling and experimental procedures. *Laboratory Animals* 14:267–74.
Gomez-Sanches, E. P., and Gomez-Sanches, C. E. 1991. 19-nordeoxycorticosterone aldosterone and corticosterone excretion in sequential urine samples from male and female rats. *Steroids* 56:451–54.
Griebel, G., C. Belzung, G. Perrault, and D. J. Sanger. 2000. Differences in anxiety-related behaviours and in sensitivity to diazepam in inbred and outbred strains of mice. *Psychopharmacology* 148:164–70.
Gurwitz, D. 2001. Are drug targets missed owing to lack of physical activity? *Drug Discovery Today* 6:342–43.
Hanson, A. 2008. *Rat behaviour and biology* [Cited January 22, 2009]. Available from http://www.ratbehavior.org/rats.html

Harding, E. J., E. S. Paul, and M. Mendl. 2004. Animal behaviour: Cognitive bias and affective state. *Nature* 427:312.

Hemsworth, P. H., J. L. Barnett, and C. Hansen. 1987. The influence of inconsistent handling by humans on the behavior, growth and corticosteroids of young-pigs. *Applied Animal Behaviour Science* 17:245–52.

Hogan, J. A. and T. J. Roper. 1978. A comparison of the properties of different reinforcers. *Advances in the Study of Behavior* 8:155, 156–255.

Holloszy, J. O. 1992. Exercise and food restriction in rats. *Journal of Nutrition* 122:774–77.

Howerton, C. and J. P. Garner. 2008. Effects of a running wheel-igloo enrichment on aggression, hierarchy linearity, and stereotypy in group-housed male CD-1 (ICR) mice. *Applied Animal Behaviour Science* 115:90–103.

Hurst, J. L. and C. J. Barnard. 1997. Well by design: The natural selection of welfare criteria in laboratory rats. In *Animal alternatives, welfare and ethics*. Utrecht: Elsevier.

Hurst, J. L., C. J. Barnard, U. Tolladay, C. M. Nevision, and C. D. West. 1999. Housing and welfare in laboratory rats: Effects of cage stocking density and behavioural predictors of welfare. *Animal Behaviour* 58:563–86.

Hutchinson, E., A. Avery, and S. Vandewoude. 2005. Environmental enrichment for laboratory rodents. *ILAR Journal/National Research Council, Institute of Laboratory Animal Resources* 46:148–61.

ILAR, ed. 1998. *The psychological well-being of nonhuman primates*. Washington, DC: National Academy Press.

Inglis, I. R., B. Forkman, and J. Lazarus. 1997. Free food or earned food? A review and fuzzy model of contrafreeloading. *Animal Behaviour* 53:1171–91.

Isaacs, K. R., B. J. Anderson, A. A. Alcantara, J. E. Black, and W. T. Greenough. 1992. Exercise and the brain—Angiogenesis in the adult-rat cerebellum after vigorous physical-activity and motor skill learning. *Journal of Cerebral Blood Flow and Metabolism* 12:110–19.

Jennings, M., G. R. Batchelor, P. F. Brain et al. 1998. Refining rodent husbandry: The mouse. Report of the rodent refinement working party. *Laboratory Animals* 32:233–59.

Jensen, P. and F. M. Toates. 1993. Who needs behavioral needs—Motivational aspects of the needs of animals. *Applied Animal Behaviour Science* 37:161–81.

Kopp, C., E. Vogel, and R. Misslin. 1999. Comparative study of emotional behaviour in three inbred strains of mice. *Behavioural Processes* 47:161–74.

Korte, S. M., B. Olivier, and J. M. Koolhaas. 2007. A new animal welfare concept based on allostasis. *Physiology & Behavior* 92:422–28.

Krohn, T. C., A. K. Hansen, and N. Dragsted. 2003. The impact of cage ventilation on rats housed in IVC systems. *Laboratory Animals* 37:85–93.

Krohn, T. C., and J. Ritskes-Hoitinga. 1999. The effects of feeding and housing on the behaviour of the laboratory rabbit. *Laboratory Animals* 33:101–7.

Latham, N., and G. Mason. 2004. From house mouse to mouse house: The behavioural biology of free-living *Mus musculus* and its implications in the laboratory. *Applied Animal Behaviour Science* 86:251–89.

Lawlor, M. M. 2002. Comfortable quarters for rats in research institution. In *Comfortable quarters for laboratory animals*, eds. V. Reinhardt and A. Reinhardt, 27–33. Washington, DC: Animal Welfare Institute.

Leach, M. C. and N. Ambrose. 1999. Practical rodent enrichment. *Animal Technology* 50:177–79.

Lehmann, M. 1991. Social-behavior in young domestic rabbits under seminatural conditions. *Applied Animal Behaviour Science* 32:269–92.

Lipman, N. S. 1999. Isolator rodent caging systems (state of the art): A critical view. *Contemporary Topics in Laboratory Animal Science* 38:9–17.

Long, S. Y. 1972. Hair-nibbling and whisker-trimming as indicators of social hierarchy in mice. *Animal Behaviour* 20:10–12.

Lore, R. and K. Flannelly. 1977. Rat societies. *Scientific American* 236:106.

Lore, R. K. and L. A. Shultz. 1988. The ecology of wild rats: Application in the laboratory. In *Ethoexperimental approaches to the study of behavior*, eds. R. J. Blanchard, P. F. Brain, D. C. Planchard, and S. Parmigiani. Dortrecht: Kluver Academic Publishers.

Mackintosh, J. H. 1981. Behaviour of the house mouse. In Biology of the house mouse. *Symposium of the Zoological Society London* 47: 337–65. London: Academic Press.

Manser, C. E. 1992. *The assessment of stress in laboratory animals*. RSPCA. Horsham, West Sussex, UK.

Manser, C. E., D. M. Broom, P. Overend, and T. H. Morris. 1998. Operant studies to determine the strength of preference in laboratory rats for nest-boxes and nesting materials. *Laboratory Animals* 32:36–41.

Marashi, V., A. Barnekow, E. Ossendorf, and N. Sachser. 2003. Effects of different forms of environmental enrichment on behavioral, endocrinological, and immunological parameters in male mice. *Hormones and Behavior* 43:281–92.

Marks, I. M. and R. M. Nesse. 1994. Fear and fitness—An evolutionary analysis of anxiety disorders. *Ethology and Sociobiology* 15:247–61.
Mason, G. J. 1991. Stereotypies—A critical-review. *Animal Behaviour* 41:1015–37.
Mason, G. J., D. McFarland, and J. Garner. 1998. A demanding task: Using economic techniques to assess animal priorities. *Animal Behaviour* 55:1071–75.
Mason, G. J. and J. N. Clubb. 2007. How and why should we use enrichments to tackle stereotypic behaviour? *Applied Animal Behaviour Science* 102:163–88.
Maurer, B. M., D. Doering, F. Scheipl, H. Kuechenhoff, and M. H. Erhard. 2008. Effects of a gentling programme on the behaviour of laboratory rats towards humans. *Applied Animal Behaviour Science* 114:554–71.
McCall, R. B., and M. L. Lester. 1969. Caretaker effect in rats. *Developmental Psychology* 1:771.
McClintock, M. K. 1987. Psychobiology of reproductive behavior: An evolutionary perspective. In *A functional approach to the behavioral endocrinology of rodents*, ed. D. Crews, 176–203. Englewood Cliffs, NJ: Prentice Hall.
Mench, J. A. 1997. Behaviour. In *Animal welfare*, eds. N. C. Appleby and B. O. Hughes, 127–42. Oxon: CAB International.
Mench, J. A. 1998. Environmental enrichment and the importance of exploratory behaviour. In *Second nature: Environmental enrichment for captive animals*, eds. D. J. Shepherdson, J. D. Mellen, and M. Hutchins, 30–46. Washington, DC: Smithsonian Institution Press.
Meredith, A. 2000. General biology and husbandry. In *Manual of rabbit medicine and surgery*, ed. P. Flecknell. London: British Small Animal Veterinary Association.
Mohammed, A. H., S. W. Zhu, S. Darmopil et al. 2002. Environmental enrichment and the brain. *Progress in Brain Research* 138:109–33.
Morton, D. B. 1993. Refinements in rabbit husbandry—Second report of the BVAAWF/FRAME/RSPCA/UFAW joint working group on refinement (JWGR). *Laboratory Animals* 27:301–29.
Neuringer, A. J. 1969. Animals respond for food in the presence of free food. *Science* (New York, NY) 166:399–401.
Nevison, C. M., J. L. Hurst, and C. J. Barnard. 1999. Strain-specific effects of cage enrichment in male laboratory mice (*Mus musculus*). *Animal Welfare* 8:361–79.
Newberry, R. C. 1995. *Environmental enrichment—Increasing the biological relevance of captive environments*. Paper presented at Applied Animal Behaviour Science.
Nicklas, W., P. Baneux, R. Boot, T. Decelle, A. A. Deeny, M. Fumanelli, B. Illgen-Wilcke 2002. Recommendations for the health monitoring of rodent and rabbit colonies in breeding and experimental units. *Laboratory Animals* 36: 20–42.
Nolen, G. A. and H. C. Alexander. 1966. Effects of diet and type of nesting material on the reproduction and lactation of the rat. *Laboratory Animal Care* 16:327–36.
Norris, M. L. and C. E. Adams. 1976. Incidence of pup mortality in the rat with particular reference to nesting material, maternal age and parity. *Laboratory Animals* 10:165–69.
Olsson, I. A. and K. Dahlborn. 2002. Improving housing conditions for laboratory mice: A review of "environmental enrichment". *Laboratory Animals* 36:243–70.
Olsson, I. A. S. and K. Westerlund. 2007. More than numbers matter: The effect of social factors on behaviour and welfare of laboratory rodents and non-human primates. *Applied Animal Behaviour Science* 103:229–54.
Osborne, S. R. 1977. Free food (contrafreeloading) phenomenon—Review and analysis. *Animal Learning & Behavior* 5:221–35.
Palanza, P., L. Gioiosa, and S. Parmigiani. 2001. *Social stress in mice: Gender differences and effects of estrous cycle and social dominance*. Paper presented at Physiology & Behavior.
Panksepp, J. 2005. Affective consciousness: Core emotional feelings in animals and humans. *Consciousness and Cognition* 14:30–80.
Parmigiani, S., P. F. Ferrari, and P. Palanza. 1998. An evolutionary approach to behavioral pharmacology: Using drugs to understand proximate and ultimate mechanisms of different forms of aggression in mice. *Neuroscience and Biobehavioral Reviews* 23:143–53.
Patterson-Kane, E. G. and M. J. Farnworth. 2006. Noise exposure, music, and animals in the laboratory: A commentary based on laboratory animal refinement and enrichment forum (LAREF) discussions. *Journal of Applied Animal Welfare Science* 9:327–32.
Pellis, S. M., V. C. Pellis, D. P. O'Brien, F. de la Cruz, and P. Teitelbaum. 1987. Pharmacological subtraction of the sensory controls over grasping in rats. *Physiology & Behavior* 39:127–33.
Pierre, P. J., A. Skjoldager, A. J. Bennett, and M. J. Renner. 2001. A behavioral characterization of the effects of food deprivation on food and nonfood object interaction: An investigation of the information-gathering functions of exploratory behavior. *Physiology & Behavior* 72:189–97.

Plesner Jensen, S. and J. Gray. 2003. How does habitat structure affect activity and use of space among house mice. *Animal Behaviour* 65:239–50.

Podberscek, A. L., J. K. Blackshaw, and A. W. Beattie. 1991. The effects of repeated handling by familiar and unfamiliar people on rabbits in individual cages and group pens. *Applied Animal Behaviour Science* 28:365–73.

Poole, T. 1997. Happy animals make good science. *Laboratory Animals* 31:116–24.

Poole, T. B., and H. D. Morgan. 1973. Differences in aggressive behaviour between male mice (*Mus musculus L.*) in colonies of different sizes. *Animal Behaviour* 21:788–95.

Price, E. O. 1999. Behavioral development in animals undergoing domestication. *Applied Animal Behaviour Science* 65:245–71.

Prior, H., and N. Sachser. 1995. Effects of enriched housing environment on the behaviour of young male and female mice in four exploratory tasks. *Journal of Experimental Animal Science* 37:57–68.

Prowse, L. 2002. Progression of environmental enrichment at Sequani Limited. *Animal Technology and Welfare* 1:119–21.

Ramos, A., and P. Mormede. 1998. Stress and emotionality: A multidimensional and genetic approach. *Neuroscience and Biobehavioral Reviews* 22:33–57.

Ras, T., M. van de Ven, E. G. Patterson-Kane, and K. Nelson. 2002. Rats' preferences for corn versus wood-based bedding and nesting materials. *Laboratory Animals* 36:420–25.

Renner, M. J. 1988. The role of behavioural topography during exploration in determining subsequent adaptive behaviour. *International Journal of Comparative Psychology* 2:43–56.

Ritskes-Hoitinga, M., and A. Chwailibog. 2003. Nutrient requirements, experimental design and feeding schedules in animal experimentation. In *Handbook of laboratory animal science*, eds. J. Hau and G. L. van Hoosier, Jr. Boca Raton, FL: CRC Press.

Rodgers, R. J. 1997. Animal models of 'anxiety': Where next? *Behavioural Pharmacology* 8:477, 496; discussion 497–504.

Rodgers, R. J., and A. Dalvi. 1997. Anxiety, defence and the elevated plus-maze. *Neuroscience and Biobehavioral Reviews* 21:801–10.

RSPCA/UFAW. 2008. *Refining rabbit care: A resource for those working with rabbits in research*, eds. P. Hawkins and R. C. Hubrecht.

Sarna, J. R., R. H. Dyck, and I. Q. Whishaw. 2000. The Dalila effect: C57BL6 mice barber whiskers by plucking. *Behavioural Brain Research* 108:39–45.

Scharmann, W. 1991. Improved housing of mice, rats and guinea-pigs—A contribution to the refinement of animal-experiments. *Alternatives to Laboratory Animals* 19:108–14.

Schmid-Holmes, S., L. C. Drickamer, A. S. Robinson, and L. L. Gillie. 2001. Burrows and burrow-cleaning behavior of house mice (*Mus musculus domesticus*). *American Midland Naturalist* 146:53–62.

Sherwin, C. M. 1996. Laboratory mice persist in gaining access to resources: A method of assessing the importance of environmental features. *Applied Animal Behaviour Science* 48:203–13.

Sherwin, C. M. 1998. The use and perceived importance of three resources which provide caged laboratory mice the opportunity for extended locomotion. *Applied Animal Behaviour Science* 55:353–67.

Sherwin, C. M. 2004. The motivation of group-housed laboratory mice, *Mus musculus*, for additional space. *Animal Behaviour* 67:711–17.

Sherwin, C. M. and C. J. Nicol. 1997. Behavioural demand functions of caged laboratory mice for additional space. *Animal Behaviour* 53:67–74.

Sluiter, W., E. Hulsing-Hesselink, I. Elzenga-Claasen, and R. Van Furth. 1985. Method to select mice in the steady state for biological studies. *Journal of Immunological Methods* 76:135–43.

Sluyter, F., and G. A. V. Oortmerssen. 2000. A mouse is not just a mouse. *Animal Welfare* 9:193–205.

Sörensen, D. B. and J. L. Ottesen. 2004. Consequences of enhancing environmental complexity for laboratory rodents—A review with emphasis on the rat. *Animal Welfare* 13:193–204.

Spangenberg, E. M., H. Augustsson, K. Dahlborn, B. Essen-Gustavsson, and K. Cvek. 2005. Housing-related activity in rats: Effects on body weight, urinary corticosterone levels, muscle properties and performance. *Laboratory Animals* 39:45–57.

Spruijt, B. M., R. van den Bos, and F. T. A. Pijlman. 2001. A concept of welfare based on reward evaluating mechanisms in the brain: Anticipatory behaviour as an indicator for the state of reward systems. *Applied Animal Behaviour Science* 72:145–71.

Stauffacher, M. 1996. Housing requirements: What ethology can tell us. In *Animal alternatives, welfare and ethics*. Utrecht: Elsevier Science B.V.

Stauffacher, M. 1997. Comparative studies in housing conditions. In *Harmonization of laboratory animal husbandry*, ed. P. N. O'Donoghue. London: Royal Soc. Med. Press.

Stauffacher, M. 2000. Refinement in rabbit housing and husbandry. In *Progress in reduction, refinement and replacement of animal experimentation, developments in animal and veterinary sciences*, eds. M. Balls, A.-M van Zeller, and M. E. Halder, 1269–77. Amsterdam: Elsevier Science.

Stauffacher, M., and A. G. Peters. 2002. *Background information for the proposals presented by the group of experts on rodent and rabbits, part B, working party for the preparation of the fourth multilateral consultation of parties to the European convention for the protection of vertebrate animals used for experimental and other scientific purposes (ETS 123)*. Strasbourg.

Stummer, W., K. Weber, B. Tranmer, A. Baethmann, and O. Kempski. 1994. Reduced mortality and brain damage after locomotor activity in gerbil forebrain ischemia. *Stroke; A Journal of Cerebral Circulation* 25:1862–69.

Suzuki, K. and K. Machida. 1995. Effectiveness of lower-level voluntary exercise in disease prevention of mature rats. 1. Cardiovascular risk factor modification. *European Journal of Applied Physiology and Occupational Physiology* 71:240–44.

Terlouw, E. M. C. and V. G. P. Schouten. 1997. Physiology. In *Animal welfare*, eds. M. C. Appelby and B. O. Hughes, 143–58. Oxon: CAB International.

Townsend, P. 1997. Use of in-cage shelters by laboratory rats. *Animal Welfare* 6:95–103.

Trullas, R. and P. Skolnick. 1993. Differences in fear motivated behaviors among inbred mouse strains. *Psychopharmacology* 111:323–31.

Tsai, P. P., U. Pachowsky, H. D. Stelzer, and H. Hackbarth. 2002. Impact of environmental enrichment in mice. 1: Effect of housing conditions on body weight, organ weights and haematology in different strains. *Laboratory Animals* 36:411–19.

Van Bergeijk, J. P., and H. van Herck. 1990. Effects of group size and gentling on behavior, selected organ masses and blood constituents in female rivm:TOX rats. *Zeitschrift Fur Versuchstierkunde* 33:85–90.

Van de Weerd, H. A., E. L. Aarsen, A. Mulder, C. L. Kruitwagen, C. F. Hendriksen, and V. Baumans. 2002. Effects of environmental enrichment for mice: Variation in experimental results. *Journal of Applied Animal Welfare Science* 5:87–109.

Van de Weerd, H. A., P. L. P. Van Loo, L. F. M. Van Zutphen, J. M. Koolhaas, and V. Baumans. 1998. Strength of preference for nesting material as environmental enrichment for laboratory mice. *Applied Animal Behaviour Science* 55:369–82.

Van de Weerd, H. A., and V. Baumans. 1995. Environmental enrichment in rodents. In *Environmental enrichment information resources for laboratory animals: 1965–1995: Birds, cats, dogs, farm animals, ferrets, rabbits, and rodents*, eds. C. P. Smith and V. Taylor. AWIC Resource Series No. 2, 145–212. Potters Bar, Hertfordshire, UK/Beltsville, MD: Universities Federation for Animal Welfare/United States Department of Agriculture.

Van den Broek, F. A. R., and C. M. Omtzigt. 1993. Whisker trimming behaviour in A2G mice is not prevented by offering means of withdrawal from it. *Laboratory Animals* 27:270–72.

Van der Harst, J. E., A. M. Baars, and B. M. Spruijt. 2003. Standard housed rats are more sensitive to rewards than enriched housed rats as reflected by their anticipatory behaviour. *Behavioural Brain Research* 142:151–56.

Van Loo, P. L. 2001. *Male management—Coping with aggression problems in male laboratory mice*. Utrecht, Utrecht University: Department of Laboratory Animal Science.

Van Loo, P. L., E. Van der Meer, C. L. Kruitwagen, J. M. Koolhaas, L. F. Van Zutphen, and V. Baumans. 2004. Long-term effects of husbandry procedures on stress-related parameters in male mice of two strains. *Laboratory Animals* 38:169–77.

Van Loo, P. L., H. J. Blom, M. K. Meijer, and V. Baumans. 2005. Assessment of the use of two commercially available environmental enrichments by laboratory mice by preference testing. *Laboratory Animals* 39:58–67.

Van Loo, P. L. P., C. L. J. J. Kruitwagen, J. M. Koolhaas, H. A. Van de Weerd, L. F. M. Van Zutphen, and V. Baumans. 2002. Influence of cage enrichment on aggressive behaviour and physiological parameters in male mice. *Applied Animal Behaviour Science* 76:65–81.

Van Loo, P. L. P., E. Van der Meer, C. L. J. J. Kruitwagen, J. M. Koolhaas, L. F. M. Van Zutphen, and V. Baumans. 2003. Strain-specific aggressive behavior of male mice submitted to different husbandry procedures. *Aggressive Behavior* 29:69–80.

Van Loo, P. L. P., L. F. M. Van Zutphen, and V. Baumans. 2003. Male management: Coping with aggression problems in male laboratory mice. *Laboratory Animals* 37:300–13.

Van Loo, P. L. P., N. Kuin, R. Sommer, H. Avsaroglu, T. Pham, and V. Baumans. 2007. Impact of 'living apart together' on postoperative recovery of mice compared with social and individual housing. *Laboratory Animals* 41:441–55.

Van Loo, P. L. and V. Baumans. 2004. The importance of learning young: The use of nesting material in laboratory rats. *Laboratory Animals* 38:17–24.

Van Oortmerssen, G. A. 1971. Biological significance, genetics and evolutionary origin of variability in behaviour within and between inbred strains of mice (*Mus musculus*). A behaviour genetic study. *Behaviour* 38:1–92.

Van Praag, H., B. R. Christie, T. J. Sejnowski, and F. H. Gage. 1999. Running enhances neurogenesis, learning, and long-term potentiation in mice. *Proceedings of the National Academy of Sciences of the United States of America* 96:13427–31.

Vermeulen, J. K., A. De Vries, F. Schlingmann, and R. Remie. 1997. Food deprivation: Common sense or nonsense? *Animal Technology* 48:45–54.

Weber, E. M. and I. A. S. Olsson. 2008. Maternal behaviour in *Mus musculus sp*: An ethological review. *Applied Animal Behaviour Science* 114:1–22.

Wemelsfelder, F. 1990. Boredom and laboratory animal welfare. In *The experimental animal in biomedical research*, ed. Rollin and Kesser, 243–72. Boca Raton, FL: CRC Press.

Wersinger, S. R. and L. B. Martin. 2009. Optimization of laboratory conditions for the study of social behavior. *ILAR Journal/National Research Council, Institute of Laboratory Animal Resources* 50:64–80.

Widman, D. R., and R. A. Rosellini. 1990. Restricted daily exposure to environmental enrichment increases the diversity of exploration. *Physiology & Behavior* 47:57–62.

Willner, P. 1991. Animal models of depression. In *Behavioral models in psychopharmacology: Theoretical, industrial and clinical perspectives*, ed. P. Willner, 521. Cambridge: Cambridge University Press.

Wolfer, D. P., O. Litvin, S. Morf, R. M. Nitsch, H. P. Lipp, and H. Wurbel. 2004. Laboratory animal welfare: Cage enrichment and mouse behaviour. *Nature* 432:821–22.

Wurbel, H. 2001. Ideal homes? Housing effects on rodent brain and behaviour. *Trends in Neurosciences* 24:207–11.

Wurbel, H. and J. P. Garner. 2007. Refinement of rodent research through environmental enrichment and systematic randomization. Available from http://www.nc3rs.org.uk, Accessed 26 June 2010.

Yang, M., H. Augustsson, C. M. Markham, D. T. Hubbard, D. Webster, P. M. Wall, R. J. Blanchard, and D. C. Blanchard. 2004. The rat exposure test: A model of mouse defensive behaviors. *Physiology & Behavior* 81:465–73.

Young, D., P. A. Lawlor, P. Leone, M. Dragunow, and M. J. During. 1999. Environmental enrichment inhibits spontaneous apoptosis, prevents seizures and is neuroprotective. *Nature Medicine* 5:448–53.

Zimmermann, A., M. Stauffacher, W. Langhans, and H. Wurbel. 2001. Enrichment-dependent differences in novelty exploration in rats can be explained by habituation. *Behavioural Brain Research* 121:11–20.

5 Ethical Evaluation of Scientific Procedures: Recommendations for Ethics Committees

Rony Kalman*, Israel; I. Anna S. Olsson*, Portugal; Claudio Bernardi, Italy; Frank van den Broek, The Netherlands; Aurora Brønstad, Norway; Istvan Gyertyan, Hungary; Aavo Lang, Estonia; Katerina Marinou, Greece; and Walter Zeller, Switzerland

CONTENTS

Objectives 102
Key Factors 103
5.1 Background 103
5.2 Checklist of Key Issues 104
    5.2.1 Description and Purpose of the Study 104
    5.2.2 Replacement 104
    5.2.3 Refinement 104
    5.2.4 Reduction 105
    5.2.5 Retrospective Information from Similar Studies 105
    5.2.6 Information about the Researcher and Institute 105
    5.2.7 Experts and Competent Persons 105
    5.2.8 General Aspects 106
    5.2.9 Composition and Dynamics of the Ethical Evaluation Body 106
5.3 Discussion Points 107
    5.3.1 Limited Benefit as a Reason for Denying a License 107
    5.3.2 Evaluation of Replacement Methods 107
    5.3.3 Selection of Species, Strain, Sex, Age and Considering Animals with Special Needs 108
        5.3.3.1 Additional Considerations 109
    5.3.4 Responsibility and Authority Including Humane Endpoints (HEP) 110
    5.3.5 Classification of the Severity of Procedures 111
    5.3.6 Efficient Study Design and the Need for Qualified Personnel 111
    5.3.7 Standardisation and Importance of Reduction 117
    5.3.8 Repeated Use of Animals 117
    5.3.9 Evaluation of Previous Studies 118
    5.3.10 Conflict of Interest 118
    5.3.11 Ensuring Scientific Validity as Part of the Ethics Evaluation 119
    5.3.12 Communication with the Public 120

---

* While all authors have contributed content, these authors have had the main responsibility for finalising the manuscript interms of content and style.

	5.3.13 Committee Composition	121
	5.3.14 The Social Responsibility of an Ethics Committee	123
	5.3.15 Project, Experiment and Protocol Evaluation	124
	5.3.16 General Discussion	125
5.4	Conclusions	126
5.5	Questions Unresolved	126
	Acknowledgements	127
	References	127

OBJECTIVES

Both the general public and the scientific community agree that the use of sentient animals for scientific research raises complex ethical issues. Ethical evaluation of animal experiments is an important way of assuring both these and other stakeholders that such experiments with animals are permitted only if they have been subjected to careful ethical scrutiny. Most countries in which biomedical research takes place have established a framework within which this ethical evaluation should be conducted, but perform this in different ways, guided by national and local regulatory requirements. The situation in Europe has been reviewed in detail by a FELASA working group (FELASA 2005; Smith et al. 2007) on which this chapter is partly based and which it complements. At the time of writing (November 2009), a revised European Directive is being prepared and is expected to require ethical evaluation of all animal investigations in Member States.* At present there are no detailed recommendations about how this review should be organised or conducted, although one objective of the Directive is to achieve a harmonised process throughout Europe.

It is possible to identify the major issues that should be addressed for effective ethical evaluation of scientific procedures involving animals. This chapter summarises these as a checklist of key issues to be considered during ethical evaluation. Key aspects of the checklist are addressed in discussion points that consider the extent to which they can be evaluated and how effectively they contribute to achieving responsible use of animals by respecting the principles of the Three Rs.

This document is based on existing literature and on the wide, international experience of the authors. Although the overall aim of this manual is to provide an evidence-based review of the various topics addressed, this has not proved possible in the present text because there has been almost no empirical research into the ethical evaluation process. Consequently, the present text relies to a great extent on the expertise and experience of members of a working group within the collaborative action COST B 24.

The issues considered include:

- Description and purpose of the study
- Replacement
- Refinement
- Reduction
- Retrospective information from similar studies
- Information about the researcher and institute
- Experts and competent persons
- General aspects

This checklist and accompanying text is presented as a working tool for ethics committees and individual committee members in carrying out an effective, efficient and robust evaluation process; in initial training of new members; and in continuing education programs for committee members, attending veterinarians, researchers, students and other concerned persons. It can also be used to

---

* http://www.europarl.europa.eu/sides/getDoc.do?pubRef=-//EP//TEXT+TA+P6-TA-2009-0343+0+DOC+XML+V0//EN&language=EN Accessed on 28 November 2009.

assess the quality of evaluation processes, offering guidance about reducing time in preparing and evaluating high-quality applications. We also address some general aspects of ethical evaluation of animal use, principally the social responsibility of the ethical review process towards the broad community in which it takes place, in promoting a culture of responsibility and care within the establishment, and the advantages and disadvantages of conducting evaluation at different levels.

We do not set out to provide guidance on *how* to conduct ethical evaluation or to provide or even suggest *right* answers to the questions raised. The process and dynamics of ethical evaluation should be continuously re-evaluated to take account of scientific developments and changing regulations.

KEY FACTORS

Experiments that may cause pain, suffering or distress to animals may provide important scientific information, but raise ethical questions. All such animal experiments should be evaluated from an ethical viewpoint using methods that are transparent, consistent and effective and that comply with national requirement and international standards. The process must take account both of scientific merits and the cost in terms of animal welfare.

The mechanism established for reviewing research programmes should have broad representation and be able to respond quickly and effectively to issues as they arise.

Proposals submitted for evaluation must explain the objective(s) clearly and concisely, identify key personnel who will be involved, and must state precisely what will happen to the animals.

The body responsible for evaluation must be satisfied that:

- The scientific objectives cannot be achieved without using animals.
- The model chosen is the most appropriate.
- The source of the animals, the quality of care provided, and the standards to which investigations will be conducted comply with the ethical values of the establishment.
- The fewest animals compatible with obtaining a valid scientific result are used.

Evaluation should include arrangements for limiting the severity of procedures, including the availability of suitable facilities and technical resources, analgesia and anaesthesia protocols, and humane endpoints.

Arrangements should be made to respond promptly to questions that may arise during a study and to gain experience from issues that may occur during or following it. This may include retrospective or on-going evaluation.

5.1 BACKGROUND

The ethical review process is a means of ensuring that all production and use of animals within laboratory animal facilities is carefully considered, adequately justified and carried out as humanely as possible, so that any adverse effects experienced by the animals are more than offset by the benefits that arise from the study. Wherever there are alternative ways of achieving the desired scientific outcomes without using sentient animals, these should be adopted; measures should be put in place to avoid unnecessary duplication of research/testing and fully implement the Three Rs from the moment it is recognised that an animal experiment will take place, through the period where animals are sourced and arrive at a facility, and up to the time they are either dead or have been re-homed. This includes optimising standards of animal husbandry and care and effective training, supervision and management of all personnel involved.

Ethical evaluation is the term used for the process of reviewing the justification for a particular scientific investigation, and may include local scrutiny of proposed investigations within a scientific establishment, application for approval of the study by the national or regional regulatory authority or grant awarding bodies, funding agencies or commercial sponsors, and so on or wider evaluation of projects submitted under supranational programmes such as the EU Framework Programme. The term includes assessment of progress of on-going investigations to establish whether the costs and benefits on which the original assessment had been made have been realised in practise and to recognise whether additional measures are appropriate.

At an institutional level, the ethical review process and/or ethical evaluation may be undertaken by an "Ethical Review Body" that may be an informal network of people or a more formal committee with specific terms of reference and a prescribed constitution, or by mechanisms designed for specific tasks. This chapter will address the process of evaluation and will provide guidance on the structure and activities appropriate to an ethical review body. The recommendations are summarised in a checklist of key issues that need to be considered; some of these issues require further clarification and this is given in the following discussion.

5.2 CHECKLIST OF KEY ISSUES

The following checklist introduces the key issues to consider during the ethical evaluation process, in the form of questions that should be asked to check whether sufficient information has been provided.

5.2.1 DESCRIPTION AND PURPOSE OF THE STUDY

1. Is there a description of the background and scientific reasoning of the general research and is the *hypothesis* to be tested specified (Section 5.3.1)?
2. Is it clear whether any of the procedures are required by regulatory, or public authorities?

5.2.2 REPLACEMENT

1. Is there a precise explanation why animal use cannot be replaced by non-animal approaches?
2. Is it stated that alternatives have been evaluated and which literature and databases have been consulted? (Section 5.3.2)

5.2.3 REFINEMENT

1. Is the origin of the animals to be used in the experiments clearly indicated? (Section 5.3.3)
   - In the case of wild-caught animals, are methods for capture, restraint, identity marking and transport described and is the acclimatisation period defined?
   - Is it specified whether the animals that will be used are naïve in regards to similar procedures or have been used previously?
   - Do the animals have special needs (specific congenital conditions, altered immunological status, etc.)?
2. Is it described how the animals will be prepared, handled and acclimatised prior to the experiment?
3. Are the time scale and types of procedures involved (acclimatisation phase, experimental phase, termination) clearly defined?
4. Are there clear protocols for anaesthesia and pain control? Are doses, volumes, administration routes and frequency, and treatment duration reasonable for the species and the actual procedure performed—and are they in accordance with relevant literature?
5. Have the humane endpoints (see Chapter 14) been considered and well defined and are the actions to be taken when reaching these clearly described? Has the person within the organisation, who will be ultimately responsible for making decisions, been identified (attending veterinarian, scientists, management, other)? (Section 5.3.4)
6. Are surgical procedures clearly described?

7. Have the experimental techniques to be performed been listed? What will be the impact of these techniques and the data collection on the animals?
8. Are follow-up inspections of the animals described, and do they identify potential complications in time for adequate actions?
9. Is the method and time of euthanasia clearly defined? If alternatives to euthanasia are available (i.e., sanctuary for primates, return to herd for farm animals), have they been indicated?
10. Have the housing, experimental conditions, and enrichment for the animals been described?
    - If special housing conditions (i.e., metabolic cages, single housing) are requested for the experiment, are they described?
    - If special husbandry conditions (i.e., moist diet, extra bedding) are required to safeguard animal welfare, are they described?
    - Is the social housing adequate to the animal species and the research design described?
    - Does the facility have all the requirements necessary to carry out the study (experimental areas, equipment for the experiment, rooms for post-operative care, special condition for microbiological containment, etc.)?
11. Has the experiment been evaluated and classified according to a severity classification system? (Section 5.3.5)
12. Has the review body considered whether it has the expertise and competency to evaluate technical and veterinary aspects of the application? If not, have experts in the specific area been consulted?

5.2.4 REDUCTION

1. Has the review body been assured that the optimal number of animals for the study has been requested? Was a proper analysis performed on these aspects by a qualified person? (Section 5.3.6; see also Chapter 6)
2. Has the potential conflict between minimising the number of animals and fairness to the individual animal been considered in the experiment? (Section 5.3.7)
3. Is the re-use of animals an option in the application? If yes, has the severity of the previous and of the proposed procedure been taken into consideration? (Section 5.3.8)
4. Has a pilot study been considered by the principal researcher, and if not, was this decision justified? (see Chapter 14)

5.2.5 RETROSPECTIVE INFORMATION FROM SIMILAR STUDIES

1. Has the researcher been asked to submit a retrospective review of the harm to the animals used in earlier studies and has the ethical review body evaluated such information? (Section 5.3.9)

5.2.6 INFORMATION ABOUT THE RESEARCHER AND INSTITUTE

1. Is the principal researcher in the project indicated?
2. Are his/her qualifications and competence clearly described and appropriate?
3. Is contact information reported?
4. Are other participants in the research project listed? Have education and training of personnel involved in the procedures been detailed?

5.2.7 EXPERTS AND COMPETENT PERSONS

1. Has any expert been consulted during design of the study and preparation of the application (statistician, veterinarian, animal welfare officer, other)?

5.2.8 GENERAL ASPECTS

1. Is there any potential for a conflict of interest affecting any person involved in the project or in associated administrative procedures? (Section 5.3.10)
2. Have scientific aspects of the research been subject to independent scientific appraisal? (Section 5.3.11)
3. Does the project description include a summary understandable by laypersons? (Section 5.3.12)
4. Is there any information to be concealed from the public? If yes, is this justified? (Section 5.3.12)
5. Has proper documentation of the entire ethical review process been kept for future evaluation? This includes:
   - the request for ethical evaluation submitted by the researcher and
   - the minutes of the committee meeting.

5.2.9 COMPOSITION AND DYNAMICS OF THE ETHICAL EVALUATION BODY

The competences necessary for a complete evaluation (Section 5.3.13) comprise:

- Scientists with knowledge and expertise in biological and biomedical research
- Veterinarian (preferably with qualifications in laboratory animal medicine)

Additional competencies relevant to the evaluation process are:

- Statistician
- A person without affiliation to the applicant organisation
- Lawyer
- Ethicist
- People with knowledge about animal welfare
- People with knowledge about alternatives to the use of animals
- Animal technician/caretaker
- Representative of animal protection organisations
- Representative of patient organisations

In the case of matters that need to be further addressed/clarified the following issues should be considered (Section 5.3.4):

- The review body should ensure that a mechanism for rapid communication and flexible interaction with the researcher is in place. A swift and non-bureaucratic dialogue is desirable between researcher and review body.
- Transcripts or records of discussions between committee members and the researcher
- Expert evaluations
- Decisions taken
- Summaries of audits performed during the course of the experiment
- Retrospective evaluations

In addition to the evaluation process itself, two additional key factors are identified and discussed further:

- The broad responsibility of the ethical review body towards society in promoting a culture of responsible use of animals (Section 5.3.14)
- An appropriate level of evaluation (Section 5.3.15)

# 5.3 DISCUSSION POINTS

The checklist highlights a number of issues, and while some are relatively self-evident, we expand on a few of them and discuss in more detail the potential and limitations for taking account of different perspectives. This discussion is based on the experience of members of and takes account of existing literature.

5.3.1 Limited Benefit as a Reason for Denying a License

Decision-making about the use of animals in experiments is often described as balancing animal harm against the expected benefits (e.g., APC 2003; Nuffield Council on Bioethics 2005). This presupposes that the potential benefit is relevant to the decision to be made; however, ethics committees generally focus on the harms. There may be several reasons for this: members of review bodies may perceive their mandate as being to safeguard the animals' interests only, or consider that they are not in a position to influence the benefit. There may also be an underlying assumption that if the investigation has received funding (whether from public sources or corporate funding for product development and testing), this is sufficient guarantee that it is reasonably likely to deliver a benefit. In addition, evaluating benefit generally involves more unknown factors and is therefore more difficult than evaluating animal harm; this is discussed further in the report of the UK Animal Procedures Committee (2003). Determining whether an applied study of a potential treatment is more likely to deliver benefit than a fundamental study of biological mechanisms, assuming both have been properly designed, is usually not perceived a useful exercise: progress in science and technology requires both fundamental and applied research. But it is possible to envisage situations when the purpose of the experiment would be highly unlikely to deliver more than a very limited benefit. Studies of public attitudes show that the purpose for which animals are used affects whether people find that use ethically acceptable (e.g., Aldhous, Coghlan, and Copley 1999; Lassen, Gjerris, and Sandoe 2006). For some purposes it is generally considered unacceptable to use animals, for example the European Union introduced a testing ban on finished cosmetic products on 11 September 2004 and on ingredients or combination of ingredients on 11 March 2009. Therefore, it is possible that an ethical review body may not approve an experiment on the grounds that its purpose is not sufficiently important. In fact, ethical review bodies rarely turn down applications: data from Sweden for the period 1989–2000 show that less than 3% of applications were turned down (Hagelin, Hau, and Carlsson 2003), and in The Netherlands fewer than 0.5% (ZoDoende 2007). For additional information, see Section 5.3.9.

5.3.2 Evaluation of Replacement Methods

It is important to promote the use of non-animal alternatives to meet public expectations, but ethical evaluation of individual projects or experiments may not be the most efficient mechanism for this. It may seem astonishing at first glance but in most cases when an animal model is proposed, there is no direct non-animal alternative. *In vitro* and *in vivo* systems have different characteristics and represent different levels of complexity (molecular, cellular, tissue level *vs.* whole-animal level) so they are not interchangeable. In basic research, *in vitro* studies are not alternatives to *in vivo* studies, rather the two complement each other. If the scientific question specifically targets the level of the whole living organism (e.g., regulation of blood pressure, cancer metastasis, learning and memory, or behaviour), then logically there are no *in vitro* alternatives (Balls 1998). In the course of drug discovery, *in vitro* studies usually precede *in vivo* experiments but the two approaches are complementary and the information provided by experiments on animals is considered critical to the drug development process.

Direct replacement methods exist in education (e.g., video materials, computer simulations), production of biological materials, for example, batch potency testing of tetanus and erysipelas vaccines has regulatory accepted alternatives and in a disappointingly limited number of drug safety tests (strictly speaking only four non-animal alternative safety testing strategies have received regulatory acceptance, all for dermatological applications: skin corrosivity, skin sensitivity, phototoxicity, and percutaneous absorption). Regulatory authorities are cautious about accepting non-animal alternatives to animal models in safety testing. The essential requirement is that the alternative method should have the same predictive power (in terms of sensitivity, specificity, accuracy, reproducibility) as the animal method. For an alternative method to be accepted by the regulatory authorities, a formal validation process should be completed successfully, involving international collaboration; this usually takes several years (Balls et al. 1995; Botham et al. 1998; Fentem et al. 1998; Hendriksen et al. 1998; Balls and Fentem 1999). On the other hand, it has also been argued that many currently accepted and traditionally used animal tests do not fulfil the requirements that are asked by candidate alternative methods (Balls and Combes 2005).

In general, research uses a combination of many different methods. Typically, the development of therapeutics involves three phases of research: non-animal models followed by animal models and then clinical trials. Only 30–40% of publications describe animal-based research and it is clear that non-animal methods play an important role in biological research. There may be scope for that proportion to increase, but that will depend on attitudes among researchers and the availability of funding rather than decisions of ethical review bodies.

5.3.3 Selection of Species, Strain, Sex, Age and Considering Animals with Special Needs

Animal experiments are usually conducted in order to understand a biological function, or to simulate human diseases or other pathological situations in order to predict their outcomes (Jokinen, Clarkson, and Prichard 1985; Conour, Murray, and Brown 2006). The selection of species, strain, sex, age and microbiological status is of crucial scientific importance. Although it is expected that scientists will make the appropriate choice for their research, in the interests of transparency and to allow thorough evaluation, they also need to provide the ethical evaluation body with a detailed explanation of their choice (for further discussion of model selection, see Chapter 7).

The selection of *species* involves a combination of scientific considerations and more practical issues, namely:

- The species should have the characteristics that best model the specific parameters under study (Xu 2004), including, when relevant, anatomical and physiological similarities (Svendsen and Gottrup 1998), which allow extrapolation of findings to the target species, typically humans (Jokinen, Clarkson, and Prichard 1985).
- The size and behaviour of the model should facilitate stress-free manipulation and experimental techniques.
- Where appropriate, it should be possible to obtain animals with well-defined genetic characteristics.
- There must be appropriate facilities and trained personnel for the housing and care of the model (Xu 2004).

When the species has been selected, it is important to choose and justify the most appropriate *strain*, stock or breed for the study. The main considerations for this selection should be its genetic characteristics (such as inbred or outbred), pathophysiological or behavioural characteristics relevant to the scientific question to be answered (Jokinen, Clarkson, and Prichard 1985).

The *sex* and *age* of the animal models will influence many physiological and behavioural parameters including growth and behaviour under different housing conditions (e.g., Laviola et al.; Tsai et al. 2003). In addition, age-related changes in morphology, physiology and metabolism may influence experimental results (Spagnoli et al. 1991; Yabuki et al. 2003). Restricting the sample in some way, for example using only one sex or a particular age group that makes extrapolation to the entire population difficult, must be carefully justified.

Microbiological status is important not only because there are welfare imperatives in minimising the incidence of disease but also to avoid the risk that subclinical infections affect research results (Morrell 1999). Housing and husbandry standards should be appropriate to the source of the animals (ILAR/NRC 1996).

5.3.3.1 Additional Considerations

5.3.3.1.1 Genetically Modified Animals

While many genetically modified animals appear normal, there is sometimes an impact on health status or welfare and they may be more sensitive to pain, distress or permanent harm, which should be identified immediately (Buehr et al. 2003; Wells et al. 2006; see Chapter 9).

Researchers should inform the ethical evaluation body of issues that may impact on the welfare of genetically modified animals. More specifically, they should provide reassurance that the research and animal care teams are able to immediately recognise animals experiencing discomfort and to apply well-defined procedures for each strain of genetically modified animal, including frequent observation at all stages of life and maintenance of relevant records within animal rooms and on a central database. Personnel at the establishment should be trained to identify animals in distress and act accordingly (Wells et al. 2006). If the genetic modification results in animals with distinctive needs, such as special diets (BVAAWF/FRAME/RSPCA/UFAW Joint Working Group on Refinement 2003), these must be taken into account.

5.3.3.1.2 Immunological Status of the Animals

The immunological status of laboratory animals directly affects their needs for housing and care. Severe combined immunodeficiency (SCID) and nude mice require more rigorous control of aspects such as environmental temperature and microbiological protection (Schuurman, Hougen, and van Loveren 1992; Hoang, Withers-Ward, and Camerini 2008). If special housing such as isolators or individually ventilated cages is required, these should provide the same welfare standards as conventional cages.

5.3.3.1.3 Animals of Non-Laboratory Species

The previous paragraphs apply principally to purpose-bred animals of species typically used in the laboratory. Within Europe, rodents are the most commonly used animals (up to 75%); fundamental science and "research and development of products and devices for human medicine and dentistry and for veterinary medicine" constitute about two-thirds of the reported scientific use of animals (European Commission 2007). Of the 12 million animals used in the EU, farm animals (ruminants and equidae) comprise around 1%; birds, including chickens, about 5% and fish nearly 15%. Farm animal research—in which animals commonly serve as a model for their own species—demands different competence and a different approach at several levels, and ethical evaluation procedures must take account of this. Furthermore, farm animals used in research are covered not only by legislation on the protection of laboratory animals, but also by regulations concerning transport, farms and animal trade and national veterinary health surveillance programs that are designed and overseen by the competent authorities (FASS 2010).

Wild-caught animals that are to be kept in the laboratory are a special cause of concern. The research establishment should be well acquainted with the behaviour of the species in their natural environment and competent at handling them (Schapiro and Everitt 2006). Special care must be provided whether an animal is particularly stressed by transportation, has needs that are difficult to satisfy in captivity, or is particularly sensitive to techniques that may be applied (Smith and Jennings 2003). Again, legislation other than research animal legislation may apply.

5.3.4 Responsibility and Authority Including Humane Endpoints (HEP)

Endpoints are predefined points beyond which an experiment is not allowed to proceed, and humane endpoints (HEPs) are determined with the objective of reducing animal suffering. They are discussed at greater length in Chapter 14. There are several definitions of humane endpoints (Bhasin et al. 1999; OECD 2000) all based on the idea that earlier, less severe clinical signs should trigger the end of an experiment, rather than awaiting death. The use of HEPs is an important refinement when planning experiments where animals are expected to develop progressively severe disease. Not only is this so for ethical reasons, but also for scientific ones because in their absence, animals may suffer and die as a result of secondary causes such as circulatory collapse, organ failure, dehydration or starvation, rather than direct consequences of the condition being investigated. Consequently, for all experiments where there is reason to expect that the welfare of (non-anesthetised) animals will become severely compromised, the ethical application should be accompanied by a detailed description of how HEPs will be applied. This applies to all progressive and potentially lethal conditions including monoclonal antibody production by ascites, cancer research, arthritis, sepsis, toxicity testing, aging studies, pain research, studies of infectious disease and vaccine trials. If a potential treatment is being studied, particular attention must be given to control animals which will develop the disease without any potentially preventive or therapeutic treatment.

Early recognition of humane endpoints and initiating appropriate action requires a clearly described routine for inspection and clinical evaluation of animals and clearly defined criteria of when to take action. Preliminary pilot studies provide guidance on how frequently animals should be observed and clearly defined responsibilities must be assigned to appropriately trained personnel. Actions may include stopping experiments or treating clinical symptoms, but in the case of rodents, euthanasia is the commonest action.

Score sheets are often used for decision-making and setting HEPs (see Chapter 14; ILAR 2000; Aldred, Cha, and Meckling-Gill 2002). Requirements for individual inspecting animals and using score sheets must be included when estimating human resource needs for the experiment. To avoid the risk that observers pay too much attention to completing the score sheet rather than assessing the overall clinical state of the animal, score sheets should include a space for "other remarks"; in addition, persons using them must understand disease mechanisms and be trained in clinical observation.

A conflict may arise between objectively defined endpoint criteria based on measurable parameters, such as body weight or blood characteristics and opinions based on clinical evaluation. This latter is based on experience and subjective judgements and is difficult to standardise, but bias can be reduced by designing experiments so that personnel working with the animals are blind to which treatment group the animals belong to. From the viewpoint of the animal welfare, the overall clinical picture is more relevant than a single parameter such as body weight, so when deciding whether or not to exclude an animal from the study, the most humane ethical approach is to take the overall clinical picture into consideration. When in doubt, a qualified veterinarian should be consulted and his/her advice carefully considered.

The time point and conditions under which animals are observed are relevant. Being able to observe nocturnal animals (such as rodents) in red light during their active period helps evaluation. New technology may make it easier to set earlier and accurate endpoints. For example, non-invasive imaging such as MRI and CT can be used to evaluate tumour sizes in inaccessible organs such as brain. Telemetry (Vlach, Boles, and Stiles 2000) and non-invasive disease markers (urine, faeces) are also useful in monitoring disease development in the animal.

The planning and application of HEPs is complex. There must be common understanding between all those involved in an experiment; there is potential for serious conflict between whether to apply a humane endpoint and possibly euthanise an animal or to maintain the animal in the experiment. The 18 responses to a request on the electronic list (Olsson, personal communication, 2008), all North American, affirmed that the veterinarian has ultimate authority to decide in this type of situation. This right is laid down in U.S. regulations or legislation and most respondents referred to the role of the animal ethics/animal care and use committee to adjudicate if disagreement remained and, most importantly, to establish clear protocols for humane endpoints so as to minimise the likelihood of practical conflicts arising. Protocols should also provide for the possibility that an animal becomes seriously ill in an experiment when this was not anticipated.

In Europe, ETS 123 prescribes that the decision to euthanise an animal should be taken by a competent person such as a veterinarian, or the person who is responsible for or has performed the procedure, whereas Directive 86/609 gives precedence to the veterinarian. Veterinary advice to euthanise an animal should prevail because failure to euthanise an animal experiencing severe and lasting suffering, means inflicting more harm than necessary and consequently may be a criminal offence. If this is due to lack of knowledge it may, depending on local legislation, be considered a minor offence. When euthanasia is declined against veterinary advice, this same act could be considered intentional and therefore culpable.

5.3.5 Classification of the Severity of Procedures

With increasing attention to harm-benefit evaluation, a growing number of countries have adopted schemes for severity classification. Five examples of such schemes are presented in Table 5.1. Classification is a valuable tool and if appropriately used can aid ethical evaluation, but there are also potential problems with use of a classification system in relation to its quality (reliability) and its purpose.

With regard to quality/reliability: A severity classification is an estimate of the harm inflicted on the animal. As with any estimate it should be as faithful, accurate and relevant as possible. However, this approach is currently confounded because there is poor equivalence of terms (e.g., moderate), based on differing and often quite obscure underlying 'algorithms', so that terms differ in their robustness and usefulness (e.g., an overall classification may be based solely on the most severe procedure of the experiment). Unfortunately, severity classification is a reductionist, simplifying approach that conceals many important details and aspects of an experiment. In addition, severity lists differ between countries and interpretation may vary between institutes. For example, a project involving animals bearing tumours is likely to be given a lower score in a cancer research institute than in an institute that does not routinely perform this type of study. Severity lists are particularly useful within a certain institute and especially for less experienced members of the ethical evaluation process.

Severity classification is mainly a practical tool: it speeds up the evaluation process and helps members of the ethical evaluation body in making their judgement. However, if its reliability is questionable this practical value may be lost. More importantly, there is the risk of bias when it is used to represent the harm in harm-benefit analysis. Severity classification has a simplifying and categorising nature that is at odds with a balanced, multifaceted, thorough, and more or less individually tailored approach to ethical evaluation. Severity classes constitute only one side of this balance (the harm) and if the corresponding benefits are not considered in parallel the review process can be misleading and counterproductive (see also APC 2003).

5.3.6 Efficient Study Design and the Need for Qualified Personnel

According to the Three Rs principle of Reduction, the minimum possible number of animals should be used to answer a scientific question. Using more animals than necessary is ethically unjustifiable, but using fewer animals even more so because an experiment yielding no information will have to be repeated and animals have suffered for no benefit at all. The minimal or rather optimal

TABLE 5.1
Five Examples of Severity Classifications

Israel	Switzerland	Canada	UK	Sweden
1. Collection of organs from animals that have not undergone any experimental process and were euthanised using a method acceptable for organ collection.	0 Interventions and manipulations in animals for experimental purposes as a result of which the animals experience no pain, suffering, injury, or extreme anxiety and no significant impairment of their general condition. Examples in veterinary practice include: withdrawal of blood samples for diagnostic purposes; subcutaneous injection of a drug.	A. Experiments on most invertebrates or on living isolated tissues or organs. Possible examples: the use of tissue culture and tissues obtained at necropsy or from the slaughterhouse; the use of eggs, protozoa, or other single-celled organisms; experiments involving containment, incision, or other invasive procedures on simple metazoan species.	Unclassified; performed entirely under general anaesthesia, from which the animal does not recover consciousness. This includes the preparation and use of decerebrated animals.	
2. Experiments that cause slight temporary discomfort or stress. Examples: IV IM IP SC injections, behavioural experiments that do not cause stress (but not including water maze or predator experiments), infliction of slight pain that the animal can avoid, withdrawing blood from peripheral vessels not requiring anaesthesia, feeding experiments that do not cause clinical manifestations, tail tip sampling.	1 Interventions and manipulations for experimental purposes that subject animals to a brief episode of mild stress (pain or injury). Examples in veterinary practice: injection of a drug requiring the use of restraint; castration of male animals under anaesthesia.	B. Experiments which cause little or no discomfort or stress Possible examples: maintenance of domestic flocks or herds under simulated or actual commercial production management systems; the short-term and skilful restraint of animals for purposes of observation or physical examination; blood sampling; injection of material in quantities that will not cause adverse reactions by the following routes: intravenous, subcutaneous, intramuscular, intraperitoneal, or oral, but not intrathoracic or intracardiac (Category C); acute non-survival studies in which the animals are completely anaesthetised and do not regain consciousness; approved methods of euthanasia in which there is rapid loss of consciousness, such as anaesthetic overdose, or decapitation preceded by sedation or light anaesthesia; short periods of food and/or water deprivation equivalent to periods that occur in nature.	Mild Procedures that, at worst, give rise to slight or transitory minor adverse effects. Examples include: withdrawal of small infrequent blood samples; skin irritation tests with substances expected to be non-irritant or only mildly irritant; minor surgical procedures such as small superficial tissue biopsies or cannulation of peripheral blood vessels where anaesthesia is necessary for restraint, not analgesia. However, if used in combination or repeated in the same animal, the cumulative severity may be increased beyond mild. Protocols may also be regarded as mild if they have the potential to cause greater suffering but contain effective safeguards to ensure effective symptomatic or specific treatment or the protocol is terminated before the animal experiences more than minor adverse effects.	Minor Experiments where animals do not risk being exposed to more than minor pain and/or other discomfort. Examples: • Restraint for physical examination • Gavage, or force-feeding • Skin tests with non-irritating substances • Injection of non-irritating substances • Blood sampling from peripheral blood vessels (artery and vein) • Sedation or anaesthesia to facilitate handling • Experiments under anaesthesia in animals that are euthanised without recovery • Anaesthesia for minor, superficial surgical interventions with recovery • Species-specific mild food or water deprivation • Euthanasia using accepted methods/techniques • Tail tissue sampling from rodents

3. Experiments that cause slight stress or short-term pain. Such experiments should not cause significant changes to the animal's appearance, to physiologic parameters such as heart rate or respiratory rate or to social behaviour. During and after such experiments animals should not exhibit signs of self-injury, anorexia, dehydration, anxiety, excessive recumbency, vocalisation, over aggressiveness or tendency for isolation. Examples: non-survival major surgery, cannulation, minor survival surgery, blood withdrawal under anaesthesia from the retro-orbital sinus or from the heart, restraint for short periods, water or food restriction for less than 12 hours a day.

C. Experiments that cause minor stress or pain of short duration

Possible examples: cannulation or catheterisation of blood vessels or body cavities under anaesthesia; minor surgical procedures under anaesthesia, such as biopsies, laparoscopy; short periods of restraint beyond that for simple observation or examination, but consistent with minimal distress; short periods of food and/or water deprivation, which exceed periods that occur in nature; behavioural experiments on conscious animals that involve short-term, stressful restraint; exposure to non-lethal levels of drugs or chemicals. Such procedures should not cause significant changes to the animal's appearance, to physiological parameters such as respiratory or cardiac rate, or faecal or urinary output, or to social responses.

Moderate

Experiments in which animals are not exposed to more than moderate pain and/or other discomfort that under normal circumstances is ameliorated by investigators with appropriate skills and knowledge. Examples include

- Permanent catheterisation of peripheral or central blood vessels (arteries and veins)
- Larger surgical interventions under anaesthesia involving the abdominal cavity, thorax, skeleton or central nervous system, with recovery and appropriate post-operative care and analgesia
- Blood sampling by retro-orbital puncture in small rodents
- Injection of irritating substances
- Housing in metabolic cages
- Disease models where animals are subject to pain or suffering that is minimised with appropriate care
- Immunisation using Freund's complete adjuvant
- Behavioural studies involving harmful stimuli but from which the animal can escape
- Toxicity tests without lethal endpoint
- Combined, repeated interventions or interventions of long duration, each of which are of minor severity

(*Continued*)

TABLE 5.1 (Continued)
Five Examples of Severity Classifications

Israel	Switzerland	Canada	UK	Sweden
4. Experiments that cause medium pain or distress that is alleviated by analgesics. Examples: major survival surgeries where animals receive analgesics, local non-metastatic tumours where animals receive analgesics, restraining animals for over 60 minutes, restriction of water or food for over 12 hours during the activity phase of the animal's day, significant change in environmental parameters (temperature, lighting), procedures that cause sensory or motor damage or severe and constant anatomical and/or physiological changes, the use of CFA-Complete Freund's Adjuvant.	2. Interventions and manipulations in animals for experimental purposes that subject the animals briefly to moderate stress, or for a moderately long to long-lasting episode of mild stress (pain, suffering, or injury, extreme anxiety, or significant impairment of general condition). Examples in veterinary practice: surgical treatment of a single leg-bone fracture; castration of female animals.	D. Experiments that cause moderate to severe distress or discomfort Possible examples: major surgical procedures conducted under general anaesthesia, with subsequent recovery; prolonged (several hours or more) periods of physical restraint; induction of behavioural stresses such as maternal deprivation, aggression, predator-prey interactions; procedures that cause severe, persistent or irreversible disruption of sensorimotor organisation; the use of Freund's Complete Adjuvant (FCA: see CCAC Guidelines on Acceptable Immunological Procedures). Other examples include induction of anatomical and physiological abnormalities that will result in pain or distress; the exposure of an animal to noxious stimuli from which escape is impossible: the production of radiation sickness; exposure to drugs or chemicals in quantities that impair physiological systems.	Moderate Experiments such as toxicity tests (which do not involve lethal endpoints) and many surgical procedures (provided that suffering is controlled and minimised by effective post-operative analgesia and care). Protocols that have the potential to cause greater suffering but include controls that minimise severity, or terminate the protocol before the animal shows more than moderate adverse effects, may also be classed within the moderate severity limit.	

		Substantial	Considerable
5. Experiments that cause severe and lasting pain or distress that are not alleviated by analgesics. Metastatic tumours or experiments in which the endpoint is death. In all such experiments the researcher is requested to justify why analgesics are not used.	E. Procedures that cause severe pain near, at, or above the pain tolerance threshold of unanaesthetised conscious animals This category of invasiveness is not necessarily confined to surgical procedures, but may include exposure to noxious stimuli or agents whose effects are unknown; exposure to drugs or chemicals in quantities that (may) markedly impair physiological systems and that cause death, severe pain or extreme distress; completely new biomedical experiments that have a high degree of invasiveness; behavioural studies in which the effects of the distress are not known; use of muscle relaxants or paralytic drugs without anaesthetics; infliction of burns or trauma on unanaesthetised animals; a method of euthanasia not approved by the CCAC; any procedures (e.g., the injection of noxious agents or the induction of severe stress or shock) that will result in pain, which approaches the pain tolerance threshold and cannot be relieved by analgesia (e.g., when toxicity testing and experimentally induced infectious disease studies have death as the endpoint).	that may result in a major departure from the animal's usual state of health or well-being. These include: acute toxicity procedures where significant morbidity or death is an endpoint, some efficacy tests of anti-microbial agents and vaccines, major surgery, and some models of disease, where animal welfare may be seriously compromised. If it is expected that even one animal would suffer substantial effects, the entire procedure would merit a 'substantial' severity limit.	Experiments where animals risk being subject to considerable pain and/or other discomfort, which cannot always be eliminated even with appropriate knowledge and techniques. Examples include • Major surgical interventions under anaesthesia but without adequate post-operative analgesia • Shock, burn, and radiation experiments where animals may be subject to considerable pain or suffering • Tumour biology experiments where the tumour growth must be allowed until advanced stages • Infectious biology experiments, including experiments for development, testing and control of vaccines, where animals can be expected to become seriously ill or with a lethal endpoint • Behavioural experiments involving harmful stimuli without the possibility of escape or using considerable restraint • Toxicity tests with a lethal endpoint • Antibody production using the ascites method • Induction of serious hypoxia to induce central nervous system injury • Induction of serious disease conditions without using alleviating treatment • Combined, repeated interventions or interventions of long duration, each of which is of moderate severity
3. Interventions and manipulations for experimental purposes, which cause the animals severe to very severe stress, or subject them to medium- to long-lasting moderate stress (severe pain, prolonged suffering or severe injury; extreme and persistent anxiety, or significant and persistent impairment of general condition). Examples in veterinary practice: induction of predictably lethal infectious and neoplastic disease without pre-emptive euthanasia.			

Source: The Israeli National Council for Animal Experimentation (adapted to English by Rony Kalman); the Swiss Federal Veterinary Office (http://www.tierversuch.ch/?show=AWLaw&nav_id=4104&lang=en); the Canadian Council for Animal Care (http://www.ccac.ca/en/CCAC_Programs/Guidelines_Policies/GUIDES/ENGLISH/V1_93/APPEN/APPXV.HTM#B); and the UK Home Office (Guidance on the Operation of the Animals (Scientific Procedures) Act 1986 http://www.archive.official-documents.co.uk/document/hoc/321/321-00.htm); and the Swedish Board of Agriculture (translated by Anna Olsson and Elisabeth Ormandy), respectively.

number of animals is usually based on a so-called power-analysis, and most review bodies require such an analysis in their application form (for more details see Chapter 6). A power-analysis is a reversed statistical analysis, where the desired outcome is entered in the equation along with foreseen biological variation, to calculate the number of animals required to obtain the desired results. To be able to utilise this equation, two parameters that may influence the outcome dramatically must be set. That is, the size of effect that is regarded as significant and the risk that the scientist is willing to take that the study will yield results that either falsely support the null hypothesis or falsely reject it.

Therefore four parameters (a–d) need to be set beforehand:

a. The wanted outcome (or Δ, the relevant size of the effect)
b. The estimation of biological variation (δ)
c. The acceptable risk that false-negative result is obtained (β)
d. The acceptable risk that false-positive result is obtained (α)

If the animal sample is too small, the intervention may indeed result in a biological effect though this is not detectable statistically. On the other hand, chance differences may occur between two experimental groups and investigators will want to minimise the risk that they wrongfully attribute such an effect to their intervention. There are no general rules for fixing the acceptability of false-positive or false-negative results, which greatly influences the number of animals needed. For some experiments 5 and 20% might be acceptable, respectively, whereas in others 1 and 10% are commonly accepted. Investigators are usually aware of acceptable ranges in their field. The relevant effect size must be estimated by the scientist and should be based on sound arguments. It is usually possible to predict the amount of biological variation from experience or literature sources. Standardisation of all amenable aspects of the experiment (including selection of animals, housing, care, and treatment) greatly reduces the number of animals needed. Furthermore, reliable power-analysis can be performed only when the scientist knows whether the parameter of interest will be normally distributed or not. Power-analysis regarding normal distribution is usually based on a reversed one- or two-sided Student's t-test, but when this is not the case, a completely different, non-parametric approach should be taken. Corrections are often necessary, for instance when more than one parameter or more than one intervention is tested simultaneously. Extra animals may have to be included to allow for withdrawal of animals from the experiment (for experimental or for humane reasons). For proper estimation of the optimal number of animals needed, all aforementioned numbers will have to be given and argued by the investigator.

The best way of optimising methodology is to educate investigators who design, perform and publish experiments; methodological knowledge is needed at each stage. Statistical analysis and methodology should form a major part of every scientific curriculum so the scientists are aware of the principles involved.

Despite this, it is generally accepted that efficient experimental design is very demanding, and most large institutions employ one or more specialists in this field to support their staff and in all institutions a case can be made for providing specialised advice in the appropriate scientific discipline. Then, as part of the ethical evaluation process, the scientist can be requested to consult such a specialist at an early stage so to ensure not only accurate prediction of the number of animals needed, but also a robust experimental design including, for instance, randomisation and proper control groups. Ethical review bodies can then be assured by the specialist's signature that the experimental design is efficient, so limiting the need for statistical expertise in the review body itself. Sometimes the composition of review bodies include a statistician; alternatively, protocols that appear weak may be submitted to a specialist and comments fed back to the investigator.

For a more extensive discussion about study design and optimisation of animal numbers see Chapter 6 as well as Festing and Altman (2002) and Festing (2006).

5.3.7 Standardisation and Importance of Reduction

There is considerable moral debate about whether it is imperative to minimise the number of animals in all circumstances or it is preferable to use more of them when this results in less suffering for individual animals. Public debates on the use of animals often emphasise reduction. However, often it is possible to plan an experiment so that harm to each animal is avoided or significantly reduced by doubling or tripling the number of animals. Drastic as this may seem, this might often be a more humane approach; to the individual animal, experiencing a high degree of discomfort, no benefit arises from a number of conspecifics being spared. There seems to be a morally relevant difference between a given amount of harm caused to one individual and the same amount of total harm distributed in smaller portions over a greater number of individuals. Quite apart from the possible increase of discomfort to individual animals, over-emphasis on reduction of animal numbers has yet another downside in that it may place the scientific rigour of the experiment at risk. In fact, an accumulating number of systematic reviews indicate that many animal studies use too few animals to provide reliable data (e.g., Sena et al. 2007).

As described elsewhere in this chapter, the number of animals used in an experiment can be greatly reduced if biological variation within or between animals is reduced. This can be achieved by a process called standardisation, which involves selecting experimental animals that are as similar as possible. Standardisation is often applied with respect to age, sex, body weight, genetic background and any other characteristics relevant to the study. The evaluation process must be sensitive to the need for standardisation, while also considering other aspects of the proposal. For example, efforts to standardise techniques over time may hinder innovation and neglect the implementation of refined experimental procedures. Another concern is that standardisation may reduce the possibility of extrapolation. For instance, if only male mice are used in a study, the number of animals required is reduced, but the outcome of the study may not be applicable to females. Whether or not this is a problem depends mainly on the parameters under study and the population to which extrapolation is proposed. The investigator should be able to justify the adoption or rejection of standardisation with respect to each aspect of the study. A different consequence of using only animals of the same sex is that breeders may cull high percentages of unwanted animals of the non-preferred sex. This is especially true for some inbred strains of mice, though culling as a result of under-demand for one sex has also been reported in other strains. Ethical review bodies should be aware of the possibility of this bias and question the choice of animals if this appears to limit the value of the experiment.

Imaging techniques are important tools for reduction and refinement because these methods are non-invasive and the same animals can be followed over a time period. However, since in most cases anaesthesia is required for image capture, the total burden on each animal, including the pathological condition, number of procedures performed under anaesthesia, and the intervals between them must be taken into consideration (see Chapter 12).

5.3.8 Repeated Use of Animals

The use of animals for repetitive studies or even for consecutive but completely different studies is a very delicate issue that deserves special consideration. It is in this area that the greatest conflicts between the two Rs of Reduction and Refinement arise (OLAW 2002).

Multiple surgical or other highly stressful repetitive procedures on a single animal are generally unacceptable and cannot be adequately justified by monetary cost savings (CCAC 1997). Nonetheless, they may be permissible on scientific ground and when adequately justified by the researcher; for example, it may be possible to justify multiple major survival surgical procedures if they are related components of a research project, if they will conserve scarce animal resources or if they are needed for clinical reasons (ILAR/NRC 1996).

When evaluating a request for repeated use of animals, the following additional issues should be further taken in consideration:

1. Whether using non-naive animals affects the scientific validity of the research
2. The benefit of Reduction compared with the impact on Refinement
3. The additional burden caused to the animal by the subsequent intervention(s) in combination with residual effects from the first one
4. The total length of the two experiments and the length of the interval between them

In circumstances where repeated use of animals is allowed by national law and is subject to local approval, one of the two following principles should be adopted (Israeli Ministry of Health 1994):

1. The study concerned can be approved if the first study was of minor severity (usually the first or second grade of the five-grade severity level scale shown in Table 5.1).
2. If the initial study was of more than minor severity, the repeated study can be approved, if it is terminal and animals are anesthetised throughout.

The ethical review body should record repeated use of animals separately from routine records for future evaluation and oversight. It is recommended that animals undergoing serial studies should receive additional veterinary attention and care. For additional information see Section 5.3.5.

5.3.9 Evaluation of Previous Studies

Information from similar previous projects is an important resource when evaluating the expected harm to animals in a given project. Although such information is rarely available in the scientific literature (Morton 1992), bodies that have established a retrospective review process (Jennings and Howard 2004) are well placed to take account of actual animal harm, so as to guide the evaluation of future projects. An increasing number of ethical review bodies are adopting retrospective review and a central database is urgently needed to integrate findings and to make such information available between ethical review bodies.

5.3.10 Conflict of Interest

A conflict of interest is a situation in which an employee, a professional person or a public official has a private or personal interest that appears to influence the objective exercise of his/her official duties.

In the case of ethical evaluation, conflicts of interest are most likely to arise in relation to the composition of the review body. Members, and in particular internal members, usually have multiple responsibilities within their institution, some of which have the potential to influence their decisions. For this reason, several organisations have established basic rules or internal policies to identify and, where possible, avoid conflicts of interest that may arise during the ethical evaluation process; examples are presented in the Australian *Code of Practice for the Care and Use of Animals for Scientific Purposes* (NHMRC 2004) or the UK *Code of Practice for Scientific Advisory Committees* (Government Office for Science 2007). Briefly, these guidelines underline the importance of the chairperson being independent of the research teams submitting protocols; he/she should advise any member of the review body involved directly with a research protocol to withdraw from the meeting while the protocol is being considered, so as to enable the committee to make (and be seen to make) an independent decision.

Similarly, a member in charge of managing the animal facility under evaluation by the review body should withdraw from the meeting while decisions are made that relate specifically to those facilities or their management.

It is recognised that these principles may be difficult to apply by ethical review bodies acting in smaller institutions where only one animal facility manager and a limited number of scientists using animals are present; in such cases consideration should be given to appointing a suitable person from outside the institution.

5.3.11 Ensuring Scientific Validity as Part of the Ethics Evaluation

Scientific validity is a prerequisite for any study to be ethically approvable. Although most—if not all—ethical review bodies on animal experimentation include scientists from a variety of fields, the body is not usually established with the intention of performing scientific scrutiny. Consequently, ethical evaluation is often carried out after scientific merit has been validated by other persons or groups. Typically, this takes place through the peer review process for funding scientific projects; in such cases, the ethical review body is usually presented only with the outcome of that application, and it would help the process if funding agencies made complete evaluation reports available. When this doesn't happen, the review body is left with the responsibility of estimating the value for science or society and there is a danger that it may attempt (partial) scientific evaluation itself (ZoDoende 2007). On the one hand, this is part of their remit: If it is judged that an animal experiment is invalid, it cannot be ruled to be ethically approvable. If it is perceived that the quality of the scientific process or animal welfare can be improved, the body is obliged to support this. Scientists, on the other hand, very often construe this as double scrutiny. An alternative risk is that the evaluation process omits the benefit altogether and assumes that a funded project has been proven to have sufficient scientific and societal merit to be justified. In this case, the process fails to evaluate harm and benefit together, something that funding evaluation is unlikely to do.

Experimental design is a critical component of the scientific validity. Although it is usually assumed that the scientific evaluation of funding applications ensures robust experimental approaches, funding applications are often written in more general terms and details of the experimental design are only finalised when the project has been funded. Consequently, a strong case can be made for the ethical review body to consider the experimental design, particularly when a proposed project or experiment has not been subject to previous scientific review. A basic assumption in the ethical evaluation process is that regulatory compliance and animal welfare considerations should be subject to the same rigorous scrutiny for both internally and externally funded (peer reviewed) projects. Peer review for such projects can be carried out via a mechanism established internally or involve independent consultants. Even if the peer review is less extensive and elaborate than that carried out during a competitive call, it must ensure that legal requirements relating to the expected benefit are met.

Whilst ethical review bodies often rely on the peer review process of funding agencies to determine the scientific value of proposed research, these funding committees, in turn, often rely on ethical review bodies to determine whether the pain and suffering induced by the experiment are acceptable. Moreover, because these review processes often occur concurrently, the only communication between the two committees may be a confirmation of approval status. There is a need to optimise this process, establishing where scientific and ethical evaluation should overlap and where they should be separated. For example, the process of ethical evaluation could be split into two levels: at a higher level—usually the funding agency—ethical and scientific evaluation could be carried out together. At the lower level—usually in the research institute—ethical evaluation could be restricted to those aspects that remain to be scrutinised, such as the Three Rs, especially Refinement and Reduction. In principle, this is the process adopted in European Union Framework Programmes, where ethical evaluation is part of the evaluation for funding in the case of projects involving non-human primates and proposals identified by the scientific review as being ethically problematic (Matthiessen, Lucaroni, and Sachez 2003).

5.3.12 Communication with the Public

Transparent administration is a fundamental principle in open modern democracy. In many countries, project licenses are required to be written in part or in full in a language intelligible to lay persons, in order to promote insight into and open debate about the use of animals in research (e.g., http://scienceandresearch.homeoffice.gov.uk/animal-research/publications-and-reference/001-abstracts/). This principle extends beyond the field of animal-based research into the issue of promoting public engagement with science in general. Decision-making about the acceptability of a research project may not depend on a lay summary being presented, although it can be argued that this may help non-scientist members of any evaluation committee. It is assumed that a policy of openness benefits acceptance of animal-based research by the public (e.g., Home Office 2005). The amount of information withheld from public should be minimised. This is discussed further in Chapter 17.

Within Europe, the extent to which information about animal experiments is made available to the public is influenced by national legislation, but two sorts of information are usually protected: personal data and intellectual property (IP) rights. The former means that personal data about those involved with the experiment should not be made public both as a matter of principle and for personal safety reasons (in which case the identification of the establishment where the experiments are done should be kept confidential also).

Intellectual property rights provide exclusivity for the owner to produce and sell newly invented medications or other inventions and this is the major driving force of the Western pharmaceutical industry. Industry has an interest in keeping results confidential until a patent has been obtained; this usually results in a delay of 5–6 years between the application to carry out the research and publication of the results. Interest in IP rights is not confined to industry; the academic sphere also wishes to protect innovation. An increasing number of academic institutions have established resource centres for innovation that assist research into patentable processes or devices with the objective of establishing commercial businesses. Consequently, an increased number of applications may be withheld from public release in order to protect IP. Any breach of confidentiality relating to IP rights may disrupt patentability, so very careful and prudent consideration is needed about what details of the experiment, if any, are to be made publicly available.

National security is another reason for making some information strictly confidential or even secret; national laws generally stipulate who has the right to declare information as secret for such reasons. Nevertheless, it is highly recommended that animal experiments, the results of which may influence national security, should be evaluated by an ethical review body, using the same criteria as for regular research. It may be necessary to restrict membership of ethical review bodies working in this area to persons operating under the appropriate security level, but they must be independent in their evaluation.

In recent years, the public has shown an increasing interest in obtaining information about animal experiments. In many countries, freedom of information (FOI) acts may be used to get access to information. Recently, FOI-rulings in The Netherlands required two universities to release ethical review committee decisions to the Political Party for Animals. Rather than awaiting use of FOI-acts to obtain information in a confrontational way, ethical review bodies (or institutions where animal experiments are performed) may pro-actively inform the public of their activities whilst taking care not to harm IP rights, national security or personal privacy. Several countries use the Internet to publish lay summaries (e.g., UK); this may play an important role in such communication, although it is not clear whether in their present format they give a complete and balanced picture (Phillips and Jennings 2008). In Denmark, the competent authority has opted to publish more extensive information about applications and approvals.

5.3.13 Committee Composition

- General aspects
  Although there is no clear reference to ethical evaluation in Directive 86/609/EEC, ethical evaluation committees and processes have been established in various countries of Europe and the Mediterranean region either on a national, legal, or voluntary basis. Guides describing best practice regarding their composition have already been published on the web and in peer-reviewed journals (e.g., Phillips and Jennings 2008).

  The revision of Directive 86/609/EEC is likely to introduce a requirement for ethical evaluation procedures; however, at the time of writing, there is no common European standard. To cover aspects of ethical evaluation outlined in this document, an ethics committee or ethical review process or body must include a number of competences. The minimal number of members required and their competencies differ between countries, as does the requirement for participation of non-institutional/public members. In our opinion, an ethical review committee should include at least the following competences:

- **Chairperson**
  The chair of the ethics committee must have independent status, authority and experience. An academic (preferably a full professor), a highly ranked governmental officer, or a judge (in some countries) are likely to possess sufficient experience in either laboratory animal science and/or ethical and legal aspects of animal experimentation for the role. The chairperson should have authority to implement decisions made by the ethical review committee and not be vulnerable to internal or external pressure. In the case of institutional or local committees it is debatable how formally independent the chairperson should be of the institution or company for which he/she operates. Although financial dependency may increase vulnerability, inside acquaintance with the institute's research activities and its particular problems and hints is likely to be an advantage.

- **Scientists with knowledge and expertise in biological and biomedical research**
  These might be scientists possessing specific knowledge about the field of the protocol being considered. Their role is to represent scientists as general stakeholders and also to serve as expert reviewers of the scientific content of the application(s) under review. Although it is not possible to involve experts in all fields of research likely to require evaluation, committees should aim to encompass a wide range of expertise.

- **Veterinarians**
  Veterinary participation in the ethical evaluation process is considered essential and in many European countries is mandated by law. Even when a veterinarian does not formally participate in the work of the ethical committees, he or she should always be consulted in advance regarding issues related to the Three Rs. The professional expertise of veterinarians gives them particular authority regarding animal health issues such as setting and interpreting humane endpoints, collection of body samples, animal health monitoring schemes and overview of animals' psychological condition (Joint Working Group on Veterinary Care et al. 2008). The veterinarian serving on a committee should preferably be a specialist in laboratory animal medicine (ECLAM diplomat or equivalent).

The inclusion of persons with experience in other relevant fields is encouraged. Although requirements for the composition of committees vary for each country or institution according to national legislation and the type of establishment, it is our opinion that persons with the following qualifications may enhance the evaluation process:

- **Statisticians**
  The participation of statisticians aims to ensure that the number of animals used is appropriate to achieving a valid scientific result.

- **Lay persons or persons without affiliation to the requesting organisation**
  While there is some controversy about involving lay members in the committee, this is already the practise in some countries. Arguments in favour are that they represent society and so can independently raise issues that might be overlooked by scientists taking a technical viewpoint. Their participation has been argued to be essential (House of Lords 2001–2002), especially because they may contribute to the transparency of the process and act in the animals' defence. If lay members are totally external to the institution they may be able to express their opinions more freely. They should be changed frequently and have a variety of backgrounds; their contribution may involve all proposed and on-going work, or they may be also asked to participate only in particularly sensitive projects. Opponents of involving lay members consider that some or even most of the times these are not able to fully understand the experimental process and may therefore not be able to make an informed, independent decision (Smith and Jennings 2003).

- **Lawyers**
  Persons with legal expertise are sometimes considered as "lay" persons as well. Their special contribution might be to ensure the integrity of the whole process (Smith et al. 2007), as well as providing expert advice on legal aspects of the evaluation and licensing process. One argument sometimes raised against including lawyers, especially if they have no previous experience or awareness of animal welfare legislation, is that they may contribute unnecessarily bureaucratic and "legislative" arguments to discussions.

- **Ethicists**
  Ethicists (persons with formal training in moral philosophy) bring in a wider ethical perspective regarding the use of animals (Smith et al. 2007). Because of their training in formal reasoning, they may also be invaluable in defining key issues and structuring debate.

- **Experts on animal welfare**
  Specialists in animal welfare, including animal welfare officers, contribute their expertise to concerns such as the level of pain or distress; they should always be available to advise researchers on issues regarding the welfare concept of the project.

- **Experts on replacements to the use of animals**
  Although there is a high priority to replace the use of live animals in research, scientists with sufficient specialist knowledge in particular research areas might not be always available, because the number of scientists primarily involved in developing replacement methods are limited.

- **Animal technicians**
  Their daily contact with the animals throughout their life within the establishment gives them unique insight into the consequences of different interventions on the animals. Moreover, they are entitled to participate in the ethical evaluation process as "defenders" of the animals (Smith et al. 2007).

- **Representatives of animal protection organisations**
  Representatives of animal welfare organisations may facilitate decisions of ethical review bodies because they acquire particular knowledge in fields of animal welfare (Smith and Jennings 2003). As with lay persons, their participation is an issue of debate though their presence is mandatory in some countries because they represent part of broader society.

- **Representatives of patient organisations**
  Patients have a particular interest in research about their specific illness. Like animal protection organisations, they represent a part of society with a particular interest in animal research where their voice can be helpful in evaluating experimental protocols because they are able to express opinions about aspects of therapy of their disease.

General aspects of committee dynamics and composition influence how it works. If the committee is composed primarily of scientists, there is a strong risk of bias towards defending the interests of research rather than those of animals. This is exacerbated if the climate causes non-scientist members to feel that they cannot participate fully (Schuppli and Fraser 2007).

All committee members, not just new members or lay persons, need (continuous) training to enable them to optimally fulfil their role in the ethical evaluation process. Topics might include regulations pertaining to the use of experimental animals or more specifically to the ethical evaluation of that use. It can be argued that committee members also need an appreciation of biomedical sciences and public attitudes towards them, in order to conduct a proper ethical evaluation. In some countries, these needs are met by conferences and workshops at which lay committee members (e.g., UK, organised by the RSPCA) or all committee members (The Netherlands, where the ethics committees have their own association) can meet and exchange experiences and views. Such meetings can also help harmonise procedures and decisions.

It is rare for all expert members, but particularly experts in replacement, to be knowledgeable about all aspects of topics considered, especially by a university-type committee. In addition, statisticians in particular may find that the amount of detail in a proposal is too limited for them to form a proper professional opinion. In conclusion, a committee cannot be expected to completely scrutinise every proposal, but rather it should aim to ascertain whether all possible means have been used to achieve the Three Rs principle. This might also involve more specialised people interacting directly with the researcher at the appropriate stage of the process, at which they can obtain all the necessary information.

5.3.14 The Social Responsibility of an Ethics Committee

Not only does ethical evaluation play an important role in reviewing projects but it is also one step in a chain of processes that influence the culture of the research community. Constructive dialogue with the researcher enables the ethical review body to assist in better planning, which is advantageous both for animals and research quality. However, after the project has been approved, ethical bodies have limited influence on how the experiments are performed; approval is not an adequate guarantee that animal experiments are performed in an ethically acceptable way, which depends largely on the qualifications and skills of the research group and compliance with the approved protocol. By promoting educational activities and debate, an ethics committee can contribute to development of a culture of responsible use of animals (sometimes referred to as a 'culture of care'; Smith et al. 2007).

Society expects that bodies such as ethical committees will take corporate social responsibility by acting as watchdogs for animal experiments and ensuring their decisions form a standard for acceptable or unacceptable behaviour. This goes beyond basic compliance with relevant laws and regulations that lay down a minimum level of conduct (Carroll 1998; Geva 2008). Moreover, laws may lag behind ethical thinking—as shown by the time taken to revise the already more than 20-year-old European Directive. Ethical committees can contribute actively to necessary changes in practice (Carroll 1998), which is likely to be what the public expects. An example of this was seen in Norway in 2007 when the National Animal Research Authority refused an application to use mice for testing the toxicity of batches of shellfish. Norwegian law clearly states that animals only should be used when there is no alternative, and the National Animal Research Authority argued that it was not within their mandate to approve an application where more humane alternatives exist, as in this case. The Norwegian Food Safety Authority overruled this decision because it was perceived to conflict with commercial interest with the shellfish industry in Norway. The National Animal Research Authority in Norway had set an example of what they regarded as unacceptable use of animals and wanted to put pressure on the authorities to change the regulations and speed the validation process of the alternative methods (Forsøksdyrutvalget 2007).

5.3.15 Project, Experiment and Protocol Evaluation

Ethical evaluation may be carried out at different stages or levels of projects involving animal use. Many countries use institutional, local or national ethical committees for evaluation of research proposals. Although the level of evaluation is usually determined by law and cannot be determined by individual committees, some general reflections are relevant.

Scientific experimentation has a hierarchical structure characterised by three levels, each of which has different functions:

1. Project level
2. Experiment/protocol level
3. Procedure level

It is crucial that the information and amount of details required to assess the levels defined above is clarified.

Project level: This is the level of research aimed at acquiring new knowledge or discovering new medicines or technologies in a specific field. It implies a broad approach and may or may not involve animal experiments, although usually there is a mixture of methods. Animal and non-animal experiments usually complement each other (e.g., research at a molecular as well as whole organism level of a disease) and there may be a hierarchical approach (for example in drug discovery, screening cascades where *in vitro* drug–receptor interaction assays precede *in vivo* efficacy and safety tests). It is at the project level that potential benefits of the research can be best estimated in the course of harm-benefit analysis. On the harm side, this level is especially suited for assessing *Replacement alternatives* and this is probably the most relevant aspect of ethical evaluation at this stage. The typical time-frame of a project spans from months to years. Administratively, ethical evaluation of projects may be carried out on national, regional or institutional levels.

Experiment level: An experiment is the basic methodological unit of research and involves highly regulated and elaborated methodologies, and requirement for reproducibility, controllability, reliability, validity and so on. An experiment seeks a concrete answer to a concrete question and, thus, represents a particular strategy within the general framework of its mother project. An experiment is also the smallest unit of scientific publication. One indispensable attribute of each experiment is the experimental protocol or design, which describes the conditions, circumstances and activities required, for example, experimental groups, timing of procedures, methods of data collection and evaluation and so on. A critical part of the protocol is the statistical design that largely determines the number of animals used in the experiment. Consequently, in the course of ethical evaluation, this is the most appropriate level for evaluating the realisation of *Reduction alternatives*. In comparison with the project level, balanced ethical judgement of an experiment requires more site-specific expertise/knowledge. In view of this, ethical evaluation of experiments may best be done at the institutional level, although it may happen on regional level, too. The typical time-frame of an experiment varies from days to months.

Procedure level: A procedure is a sequence of actions carried out on experimental animals with the aim of evoking a physiological or pathological state, or a physiological or behavioural response (often by creating changes in the environment), and including data collection—always in the context of an experiment. A procedure may involve invasive as well as non-invasive techniques. From an ethical point of view, a procedure is best understood if it is seen as the series of interventions involving a *single individual*. Replacement and reduction options are not relevant at the procedure level. Instead, ethical evaluation should focus on the implementation of *Refinement alternatives*. Evaluation of procedures requires a large amount of institution-specific, even laboratory-specific knowledge, so it is best carried out at the institutional level. Procedures typically last minutes to days.

Lifetime experience: Consideration needs to be given not only to suffering what animals might undergo during the course of the investigation, but to all negative experiences during their lives, including transport and possible shortcomings in care, husbandry and housing. Persons involved in animal use have a responsibility not only to minimise suffering, but also to take measures to ensure that animals benefit in a positive sense from their environment (see Chapter 4). Considerations about the animal's whole life experience extend to breeders and suppliers of laboratory animals who can influence the temperament and behaviour of animals by careful selection of breeding stock and adopting appropriate care routines in the initial stages of their life.

5.3.16 General Discussion

This document has been elaborated by a group of people with different backgrounds, drawn from different countries and cultures and working at different types of organisations. The ethical evaluation process was not seen in the same way by everybody. This led to extensive discussions about most points before an account was agreed that all members could accept. To name a few: the existence of actual, valid alternatives; the ideal composition of evaluation bodies or committees; the roles of the members and the detail with which applications should be scrutinised; and whether limited benefit is a reason to deny permission to proceed. Nevertheless, all members affirmed the importance and value of ethical evaluation and the need for a common understanding of the minimal requirements that should be included.

The resulting document is a synthesis of many issues that need to be evaluated by ethical evaluation process. It was not our aim to provide answers to complex ethical issues—this will be the task of the ethical review bodies. Rather, this document is intended to highlight questions and to establish reference points to help ensure that ethical evaluation is complete and best serves its purpose. Ethical evaluation can take place at various levels, each of which presents different aspects. Evaluation at the Project level is best placed to focus on the R of Replacement, at the Protocol/Experiment level it is particularly suited to consider the R of Reduction, and at the Procedure level to evaluate the R of Refinement. However, at each level, it is the duty of the ethical review body to ensure that essential aspects of the research are not overlooked and that both researchers and the public feel use of the animal was justified.

The composition of ethical review bodies or committees differs in various countries and they operate at different levels (national, regional, institutional). Some countries put more emphasis on technical or regulatory aspects while others concentrate on a more general harm-benefit evaluation; however, all act according to the same basic principles (Three Rs and harm-benefit balance). Even though scientists make up a large proportion of most committees, the committees reflect an aspect of public interest in the scientific work, and should look at research projects from a different perspective. One important criterion is that the evaluation process should be independent, strong, and unbiased. Ethical review committees are expected to act as watchdogs for animal experiments and their decisions should provide the basis of a standard of acceptable behaviour. In this context, we feel it is important to note that the evaluation process should never become formal. For example, electronic means of communication (e.g., e-mail) are convenient and may save time, but can never replace face-to-face discussion and dialogue.

Although regulation in all member countries is based on the principle of Russell and Burch's Three Rs, the working group wants to underline the importance of the fourth R—Responsibility. This stands for the personal responsibility of everyone involved in animal research to conduct research in the best possible way, in accordance with the Three Rs and aiming to maximise scientific benefit. The R of Responsibility places weight on the shoulders of the individual rather than the authorities and emphasises the importance of education and training, on-going explanation and openness rather than policing and punishing. It is the duty of the ethical committee, along with all other research, public and educational bodies to create a culture of care and responsibility among those who plan and perform animal experiments. This is particularly important because ethical review bodies and governmental authorities cannot control each step and aspect of the research process.

5.4 CONCLUSIONS

The use of animals in research is controversial, and the ethical evaluation of projects has an important function in ensuring that animals are used in the best possible way. This is of interest to both the general public and specialists, and most countries where biomedical research takes place have some system for ethical evaluation in place. A complete ethical evaluation has to take into account a number of issues, ranging from the scientific benefit of the research to technical matters. It is questionable whether the ethical evaluation process can actually address all these aspects without being over-bureaucratic for itself and the scientists. One other, equally important contribution that ethical evaluation can make is to the development of a culture of responsibility.

This chapter provides a working tool for ethical review bodies or committees and individuals involved in the process; it will be of value for training new members and in providing continuing education programs for committee members, researchers and students. Drawing from existing literature and the experience of working group members, several key considerations have been identified, including information about the purpose of the investigation, the research team and previous experience, and how the Three Rs are to be implemented.

Discussion points, corresponding to particular issues in the checklist described above, briefly discuss critical considerations and extent to which their evaluation is possible and assures responsible use of animals with respect to the Three Rs. Special attention has been given to the following topics: harm-benefit analysis, compliance with the Three Rs principle, composition and functioning of ethical committees, communication with society, tension between extensive control (resulting in a heavy administrative burden for scientists and committees), and trust in a culture of responsibility (with the risk of that responsibility not being taken seriously).

5.5 QUESTIONS UNRESOLVED

- Coordinating ethical and scientific review
  Currently, evaluations are carried out separately by ethical review bodies and grant awarding agencies. There is a need to identify areas where they overlap and to improving communication between the two bodies.
- Harmonisation
  Most European countries have introduced some form of ethical evaluation but there is very little consistency of process, constitution, or remit. Greater sharing of experiences would facilitate development of common standards for issues such as severity banding, re-homing policies, and consistent experimental design.
- Replacement techniques
  There may be scope for increasing the availability of replacement techniques but their uptake depends on attitudes among researchers and the availability of funding. A means is needed of making ethical review bodies more aware of opportunities for introducing these.
- Evaluating previous studies
  More and more ethical review bodies are undertaking retrospective review and there is an urgent need for a central database to help integrate and facilitate sharing of findings.

- Training for persons involved in ethical evaluation
  Although some countries already offer training opportunities, the wider availability of networking and training opportunities would facilitate harmonisation and improve the effectiveness of review in establishments.
- Balancing control and trust
  Use of a checklist that covers all topics is likely to be burdensome and impractical. Evaluation in more general terms places greater trust in the way scientists work but may weaken their involvement in maintaining the ethos of the establishment. There is a need to review the effectiveness of these different models.
- Social responsibility
  There is little evidence about whether ethical evaluation satisfies society's demand for more rigorous oversight of animal experimentation, and public debate about the role of ethics committees would be welcome.

ACKNOWLEDGEMENTS

Thanks to other members of the working group on ethics and harm-benefit evaluation of the COST B24, in particular Annie Reber and Jacques Serviere for useful input.

REFERENCES

Aldhous, P., A. Coghlan, and J. Copley. 1999. Animal experiments—Where do you draw the line? Let the people speak. *New Scientist (1971)* 162:26–31.

Aldred, A. J., M. C. Cha, and K. A. Meckling-Gill. 2002. Determination of a humane endpoint in the L1210 model of murine leukemia. *Contemporary Topics in Laboratory Animal Science* 41:24–27.

APC. 2003. *Review of cost-benefit assessment in the use of animals in research*. Available from http://www.apc.gov.uk/reference/reports.htm, Accessed 26 June 2010.

Balls, M. 1998. Why is it proving to be so difficult to replace animal tests? *Laboratory Animals* 27:44–47.

Balls, M., B. J. Blaauboer, J. H. Fentem, L. Bruner, R. D. Combes, B. Ekwall, R. J. Fielder, et al. 1995. Practical aspects of the validation of toxicity test procedures—the report and recommendations of ECVAM workshop-5. *Atla-Alternatives to Laboratory Animals* 23:129–47.

Balls, M., and J. H. Fentem. 1999. The validation and acceptance of alternatives to animal testing. *Toxicology in Vitro* 13:837–46.

Balls, M., and R. Combes. 2005. The need for a formal invalidation process for animal and non-animal tests. *Atla-Alternatives to Laboratory Animals* 33:299–308.

Bhasin, J., R. Latt, E. Macallum, K. McCutcheon, E. Olfert, D. Rainnie, and M. Schunk. 1999. *Canadian Council on Animal Care guidelines on choosing an appropriate endpoint in experiments using animals for research, teaching and testing*. Ottawa: CCAC.

Botham, P. A., L. K. Earl, J. H. Fentem, R. Roguet, and J. J. M. van de Sandt. 1998. Alternative methods for skin irritation testing: The current status—ECVAM skin irritation task force report 1. *Atla-Alternatives to Laboratory Animals* 26:195–211.

Buehr, M., J. P. Hjorth, A. K. Hansen, and P. Sandoe. 2003. Genetically modified laboratory animals—What welfare problems do they face? *Journal of Applied Animal Welfare Science* 6:319–38.

BVAAWF/FRAME/RSPCA/UFAW Joint Working Group on Refinement. 2003. Refinement and reduction in production of genetically modified mice. *Laboratory Animals* 37 Suppl 1: S1–S49.

Carroll, A. B. 1998. The four faces of corporate citizenship. *Business and Society Review* 100/101:1–7.

CCAC. 1997. *CCAC guidelines on: Animal use protocol review*. Ottawa: CCAC.

Conour, L. A., K. A. Murray, and M. J. Brown. 2006. Preparation of animals for research—issues to consider for rodents and rabbits. *ILAR Journal* 47:283–93.

European Commission. 2007. *Report from the commission to the council and the European Parliament; fifth report on the statistics on the number of animals used for experimental and other scientific purposes in the member states of the European Union*. Brussels: European Council, COM(2007)675 Final.

FASS. 2010. Guide for the care and use of agricultural animals in agricultural research and teaching. In *Federation of Animal Science Societies* [database online], 3rd ed. Champaign, IL, [Cited January 6, 2010]. Available from http://www.fass.org/page.asp?pageID = 357

FELASA. 2005. *Principles and practice in ethical review of animal experiments across Europe.* UK: Federation of European Laboratory Animal Science Associations.
Fentem, J. H., G. E. B. Archer, M. Balls, P. A. Botham, R. D. Curren, L. K. Earl, D. J. Esdaile, H. G. Holzhutter, and M. Liebsch. 1998. The ECVAM international validation study on in vitro tests for skin corrosivity. 2. Results and evaluation by the management team. *Toxicology in Vitro* 12:483–524.
Festing, M. F. 2006. Design and statistical methods in studies using animal models of development. *ILAR Journal* 47:5–14.
Festing, M. F., and D. G. Altman. 2002. Guidelines for the design and statistical analysis of experiments using laboratory animals. *ILAR Journal* 43:244–58.
Forsøksdyrutvalget. 2007. Case number S-2007/81689.
Geva, A. 2008. Three models of corporate social responsibility: Interrelationships between theory, research, and practice. *Business and Society Review* 113:1–41.
Government Office for Science. 2007. *Code of practice for scientific advisory committees.*
Hagelin, J., J. Hau, and H. E. Carlsson. 2003. The refining influence of ethics committees on animal experimentation in Sweden. *Laboratory Animals* 37:10–18.
Hendriksen, C., J. M. Spieser, A. Akkermans, M. Balls, L. Bruckner, K. Cussler, A. Daas, et al. 1998. Validation of alternative methods for the potency testing of vaccines—The report and recommendations of ECVAM workshop 31. *Atla-Alternatives to Laboratory Animals* 26:747–61.
Hoang, V., E. Withers-Ward, and D. Camerini. 2008. Nonprimate models of HIV-1 infection and pathogenesis. *Advances in Pharmacology* (San Diego, Calif.) 56:399–422.
Home Office. 2005. *Animals (scientific procedures) inspectorate report.* Available from http://tna.europarchive.org/20100413151426/http://scienceandresearch.homeoffice.gov.uk/animal-research/publications-and-reference/publications/reports-and-reviews/annual-report052835.pdf?view = Binary; Accessed on 26 June 2010.
House of Lords. 2001–2002. *Select committee on animals in scientific procedures.* House of Lords, Volume 1.
ILAR. 2000. Humane endpoints for animals used in biomedical research and testing. *ILAR Journal* 41:80–86.
ILAR/NRC. 1996. *Guide for the care and use of laboratory animals.* Washington, DC: National Academy Press, 1996.
Israeli Ministry of Health regulation. 2001. *Law on Animal Experimentation 1994.*
Jennings, M., and B. Howard. 2004. *Guidance notes on retrospective review. A discussion document prepared by the LASA ethics and training group.* Available from http://www.lasa.co.uk/Guidance%20notes%20RR%20%282004%29.pdf, Accessed 26 June 2010.
Joint Working Group on Veterinary Care, H. M. Voipio, P. Baneux, I. A. Gomez de Segura, J. Hau, and S. Wolfensohn. 2008. Guidelines for the veterinary care of laboratory animals: Report of the FELASA/ECLAM/ESLAV joint working group on veterinary care. *Laboratory Animals* 42:1–11.
Jokinen, M. P., T. B. Clarkson, and R. W. Prichard. 1985. Animal models in atherosclerosis research. *Experimental and Molecular Pathology* 42:1–28.
Lassen, J., M. Gjerris, and P. Sandoe. 2006. After Dolly—ethical limits to the use of biotechnology on farm animals. *Theriogenology* 65:992–1004.
Laviola, G., W. Adriani, S. Morley-Fletcher, and M. L. Terranova. 2002. Peculiar response of adolescent mice to acute and chronic stress and to amphetamine: Evidence of sex differences. *Behavioural Brain Research* 130:117–25.
Matthiessen, L., B. Lucaroni, and E. Sachez. 2003. Towards responsible animal research. Addressing the ethical dimension of animal experimentation and implementing the 'three R's' principle in biomedical research. *EMBO Reports* 4:104–7.
Morrell, J. M. 1999. Techniques of embryo transfer and facility decontamination used to improve the health and welfare of transgenic mice. *Laboratory Animals* 33:201–6.
Morton, D. 1992. A fair press for animals. *New Scientist* 134:28–30.
NHMRC. 2004. *Australian code of practice for the care and use of animals for scientific purposes.* 7th ed. Vol. EA16. Canberra: Australian Government Publishing Service.
Nuffield Council on Bioethics. 2005. *The ethics of research involving animals.*
OECD. 2000. *Guidance document of the recognition, assessment and use of clinical signs as humane endpoints for experimental animals used in safety evaluation.* Organization for Economic Cooperation and Development. OECD, Paris, France.
OLAW. 2002. *Institutional animal care and use committee guidebook.* Bethesda, MD: Office of Laboratory Animal Welfare.
Phillips, B., and M. Jennings. 2008. Home office licence abstracts—an assessment. *Atla-Alternatives to Laboratory Animals* 36:465–71.

Schapiro, S. J., and J. I. Everitt. 2006. Preparation of animals for use in the laboratory: Issues and challenges for the institutional animal care and use committee (IACUC). *ILAR Journal* 47:370–75.

Schuppli, C. A., and D. Fraser. 2007. Factors influencing the effectiveness of research ethics committees. *Journal of Medical Ethics* 33:294–301.

Schuurman, H. J., H. P. Hougen, and H. van Loveren. 1992. The rnu (rowett nude) and rnuN (nznu, New Zealand nude) rat: An update. *ILAR Journal* 34:1–14.

Sena, E., H. B. van der Worp, D. Howells, and M. Macleod. 2007. How can we improve the pre-clinical development of drugs for stroke? *Trends in Neurosciences* 30:433–39.

Smith, J. A., F. A. van den Broek, J. C. Martorell, H. Hackbarth, O. Ruksenas, W. Zeller, and FELASA Working Group on Ethical Evaluation of Animal Experiments. 2007. Principles and practice in ethical review of animal experiments across Europe: Summary of the report of a FELASA working group on ethical evaluation of animal experiments. *Laboratory Animals* 41:143–60.

Smith, J. A., and M. Jennings. 2003. *A resource book for lay members of local ethical review processes.* West Sussex: RSPCA.

Spagnoli, L. G., A. Orlandi, A. Mauriello, G. Santeusanio, C. de Angelis, R. Lucreziotti, and M. T. Ramacci. 1991. Aging and atherosclerosis in the rabbit. 1. Distribution, prevalence and morphology of atherosclerotic lesions. *Atherosclerosis* 89:11–24.

Svendsen, P., and F. Gottrup. 1998. Comparative biology of animals and man in surgical research. In *Animal modelling in surgical research*, ed. B. Jeppson, 1–15. Amsterdam: Harwood Academic Publishers.

Tsai, P. P., H. D. Stelzer, H. J. Hedrich, and H. Hackbarth. 2003. Are the effects of different enrichment designs on the physiology and behaviour of DBA/2 mice consistent? *Laboratory Animals* 37:314–27.

Vlach, K. D., J. W. Boles, and B. G. Stiles. 2000. Telemetric evaluation of body temperature and physical activity as predictors of mortality in a murine model of Staphylococcal enterotoxic shock. *Comparative Medicine* 50:160–66.

Wells, D. J., L. C. Playle, W. E. Enser, P. A. Flecknell, M. A. Gardiner, J. Holland, B. R. Howard, et al. 2006. Assessing the welfare of genetically altered mice: Laboratory environments and rodents' behavioural needs. *Laboratory Animals* 40:111–14.

Xu, Q. 2004. Mouse models of arteriosclerosis: From arterial injuries to vascular grafts. *The American Journal of Pathology* 165:1–10.

Yabuki, A., M. Matsumoto, H. Nishinakagawa, and S. Suzuki. 2003. Age-related morphological changes in kidneys of SPF C57BL/6Cr mice maintained under controlled conditions. *The Journal of Veterinary Medical Science* 65:845–51.

ZoDoende. 2007. *Statistical report on the use of animals for experimental and other scientific purposes in the Netherlands (in Dutch).* The Netherlands: Food and Consumer Product Safety Authority (VWA).

6 Reduction by Careful Design and Statistical Analysis

Michael Festing, United Kingdom

CONTENTS

Objectives 132
Key Factors 132
6.1 Introduction 132
    6.1.1 Improving the Design of Experiments 132
    6.1.2 The Three Rs 133
6.2 Main Principles of Design 133
    6.2.1 Purpose of the Experiment 133
    6.2.2 Types of Experiment 133
    6.2.3 Identifying the Experimental Unit 134
    6.2.4 Independent and Dependent Variables 134
        6.2.4.1 Independent Variables or Factors 134
        6.2.4.2 Dependent Variables 135
    6.2.5 The Five Requirements for a Good Experimental Design 135
        6.2.5.1 Absence of Bias: Randomisation and Blinding 135
        6.2.5.2 Power 136
        6.2.5.3 Range of Applicability: The Factorial Experiment 139
        6.2.5.4 Simplicity 139
        6.2.5.5 Amenable to Statistical Analysis 139
    6.2.6 Formal Designs 139
        6.2.6.1 The "Completely Randomised" or "Between Subject" Design 140
        6.2.6.2 The Randomised Block Design 140
        6.2.6.3 The Cross-Over, Repeated Measures or Within-Subject Design 141
        6.2.6.4 The Latin Square 142
        6.2.6.5 The Split-Plot Design 142
        6.2.6.6 Sequential Designs 143
        6.2.6.7 Factorial Designs 143
6.3 Statistical Analysis 145
    6.3.1 Examining the Data 145
    6.3.2 Parametric Statistical Analysis 145
    6.3.3 Non-Parametric Tests 146
    6.3.4 Discrete Data 146
    6.3.5 Interpreting and Reporting the Results 146
    6.3.6 A Note on Statistical Software 147
6.4 Conclusions 147
6.5 Questions Unresolved 147
References 148

OBJECTIVES

This chapter provides a broad overview of the design and statistical analysis of experiments. It does not instruct the investigator how to design and analyse an individual experiment, but aims more to give an understanding of the principles involved. Readers are urged to find a good statistical textbook that can be consulted for more details. A list of further reading is given at the end of the chapter.

KEY FACTORS

Before designing an experiment it is essential to establish a clear hypothesis and to select the most appropriate model; animals should be used only if there is no suitable alternative. Carefully establish the experimental unit and identify the dependant variables and independent variables that will be incorporated into the study design. This will usually suggest an appropriate experimental design. Key considerations in developing the design are:

- Ensure there is no bias by randomising the independent variables and/or blinding collection of experimental data.
- Select an appropriate power (the ability to detect a treatment effect).
- Ensure that findings will effectively test the hypothesis under consideration.
- Ensure the study is simple, easy to conduct and amenable to statistical analysis.

The methods of analysis to be applied should always be determined at the time the experiment is designed, to ensure appropriate group selection, data collection and group sizes. A poorly designed experiment may prove difficult or impossible to analyse in a meaningful way.

6.1 INTRODUCTION

6.1.1 Improving the Design of Experiments

Poorly designed and/or incorrectly analysed animal experiments can waste scientific resources and lead to the unethical use of animals (Altman 1982; Altman 1991; Festing et al. 2002). If experiments are unnecessarily large, then animals will be wasted. If too small, then important treatment effects may be missed and the wrong conclusion drawn and the animals will again be wasted. This may lead to consequential losses that could be of enormous significance. Failure to detect a subtle but important toxic effect of a chemical, for example, could lead to a medical or environmental disaster. Experiments that are biased because the treated and control groups are not treated the same way may also lead to the wrong conclusions. And, of course, a paper based on badly designed experiments may be rejected for publication.

Clearly it is in the interests of every investigator to ensure that their experiments are well designed. Yet there is good evidence that this is not always the case. Surveys of scientific papers using animals often show defects both in the design and in the resulting statistical analysis. In one study over 60% of 133 papers had obvious statistical errors. These were so serious in 5% of the papers that the conclusions were probably not supported by the data (McCance 1995). In another survey of 48 experiments published in two toxicology journals (Festing 1996) only 32% of the experiments were of an appropriate size, as estimated using the resource equation method of determining sample size (Mead 1988), with 27% of them using about twice as many animals as necessary. Many of the experiments were poorly designed. Only one used blocking to reduce variation, and this experiment was incorrectly analysed. Based on the criteria that use of t-test to analyse an experiment with more than two treatment groups, or using a one-way analysis of variance (ANOVA) to analyse a factorial design are incorrect, it was found that a third (16/48) of the experiments were incorrectly analysed.

More recently a meta-analysis of all animal papers that could be identified, which related to six interventions with a known outcome in humans (Perel et al. 2007) found that only in three cases did the animal experiments correctly predict the human outcome. Whether this was because the animal model was intrinsically inaccurate or because many of the papers were methodologically poor is not clear.

6.1.2 The Three Rs

Animals are capable of feeling pain and distress, and should not be subjected to experiments unless there is no alternative to their use, and the results are likely to improve the health and welfare of humans or other animals or improve understanding of biological mechanisms. Wherever possible, non-sentient Replacement (Russell and Burch 1959) alternatives such as tissue or cell culture should be used. If animals must be used, ones of the lowest possible sentience should be chosen such as insects rather than mice, or mice rather than non-human primates. Experiments should be Refined so that they cause the least possible pain and distress before, during, and after the experiment has been completed. Finally, the number of animals used should be Reduced to the smallest possible number consistent with achieving the scientific objectives of the experiment, remembering that the use of too few animals should be avoided as important treatment effects may be missed. If the number of animals can be reduced without reducing scientific output it will also save money, time, and scientific resources, so it is in the investigator's best interest to use well-designed experiments.

6.2 MAIN PRINCIPLES OF DESIGN

6.2.1 Purpose of the Experiment

The first step in designing any experiment is to have a clear idea of its purpose. This will determine the choice of experimental material, the dependent and independent variables (discussed below), and the method of statistical analysis of the results. For example, is the purpose of the experiment to determine the effect of some treatment on "rats" or "Sprague-Dawley" rats, or "female Sprague-Dawley" rats or humans, using rats as a model of humans? Are rats even a good model of humans in the specific case? Is the aim to compare the means of treatment groups, or to obtain a good estimate of the magnitude of a response, or to obtain a dose-response relationship? In every case the design of the experiment will depend on the question being asked. It is clearly ethically unacceptable to design an experiment that is incapable of answering the question of interest, as happens occasionally. A clear statement of the purpose of the experiment should be the first part of the experimental protocol.

6.2.2 Types of Experiment

The randomised controlled trial is the commonest type of experiment. Usually, the aim is to compare the means of two or more groups, one of which may receive no treatment, or a sham treatment (a placebo) and be designated the "control", and the others are "treated" groups that receive some intervention such as a drug. By comparing the group means it is possible to determine, subject to error caused largely by inter-individual variation, whether the treatment has any effect. An experiment to compare three newly formulated diets may just designate them A, B, and C, but it is still considered to be a controlled experiment. Occasionally positive controls are needed. If an experiment is expected to give largely negative results, such as when safety testing an intervention that is thought to be very safe, a positive control with a known response to one of the treatments may be needed to show that the experiment is capable of picking up a response when present.

There are three major subtypes of experiments:

- Pilot studies are small experiments used to explore the logistics of a proposed more substantial experiment, and to gather some preliminary data such as whether the treatments are excessively severe. The results are not usually published, unless they yield some interesting results. This, of course, introduces publication bias because the results are more likely to be published if they are statistically significant.
- Exploratory experiments are often used to produce data that can be used to generate ideas and formulate hypotheses. Often the aim is to measure many characters and to look at the pattern of response to the treatments. This may present problems in statistical analysis of the results as it may generate false positive results. If enough characters are studied some will differ "significantly" between the treatments just by chance. Where interesting results are obtained, a hypothesis should be formulated and explored using a confirmatory experiment.
- Confirmatory experiments are used to test some relatively simple formal hypothesis that must be specified before the experiment is started. Formulating the hypothesis after the experiment is completed and the results examined may lead to serious bias.

More rarely, an uncontrolled experiment is used in animal research. An acute toxicity study may involve administering a test substance to an animal to see whether it has any obvious adverse effect. In this case, controls can often be omitted because in a short period of time it can be safely assumed that an untreated animals will not show any adverse signs. However, this type of experiment is rare and is not discussed here.

6.2.3 Identifying the Experimental Unit

A critical step in designing any experiment is to identify correctly the subject of the experiment, known as the experimental unit (ExpU). The ExpU is defined as the smallest division of the experimental material such that any two ExpUs could receive different treatments (Cox 1958). It is the object that is randomised to the treatments. It is also the unit of the statistical analysis. An ExpU may be an individual animal if two animals can receive different treatments. However, it would be a cage of animals if the treatment is in the diet or water and several animals in the cage use the same diet hopper or water bottle. In this case, the statistical analysis is based on the mean of all animals in the cage (although if the data is collected separately on each animal it is then possible to find out whether there are important cage effects). In a teratology experiment the ExpU is the pregnant female that is assigned to a treatment, although it is the foetuses that are measured and assessed. So it is the mean characteristics of the pups, possibly weighted by litter size, which are subject to the statistical analysis. In a cross-over trial, treatments are given in a random sequential order to an individual or cage of individuals. In this case the ExpU is the animal or cage over a period of time. In a split-plot experimental design there are two different types of ExpUs, as discussed briefly below. *For convenience it will be assumed in the rest of this chapter the ExpU is a single animal.*

6.2.4 Independent and Dependent Variables

There are two types of variable in an experiment. The *independent variables* are the treatments or other conditions that are controlled by the researcher. The *dependent* variables are those which are measured or counted in response to the treatment.

6.2.4.1 Independent Variables or Factors

The investigator will usually wish to determine the effect of an independent variable that can be controlled, such as a drug, diet, or other variable on some property of the experimental subjects.

There will usually be two or more *levels* of the independent variable such as a control and one or more doses or types of some treatment. The aim may be just to compare means, or to estimate a dose/response relationship. In a factorial design two or more variables, or *factors,* such as drug treatment and sex, may be altered at the same time in a controlled manner.

Variables such as strain and sex are called classification variables because they cannot be assigned at random to the subjects in the same way as a drug or diet treatment can be. For these variables care needs to be taken to ensure that the things being compared really are comparable. For example, if the aim is to compare males and females they clearly should be the same age and from the same source. However, within the animal house the cages can be kept in random order to average out any within-room environment effect.

6.2.4.2 Dependent Variables

The response to the treatments must be measured in some way. Usually one or more *dependent variables* such as body weight, the concentration of some factor in a body fluid, the activity of an enzyme, an immune response, or the presence of a tumour are measured, counted, or scored (possibly subjectively) to obtain a numerical measure of the magnitude of the response. These dependent variables can also be called *traits* or *characters* or *outcomes*. In some experiments there will be a single outcome measurement, in others literally thousands of outcomes will be measured, as in a microarray experiment. The number of dependent variables have statistical implications if each of them is to be subjected to statistical analysis, particularly if they are correlated, because performing lots of statistical tests could result in many false positive results. In some cases this can be taken into account either by correcting for false positives or by doing a multivariate statistical analysis.

6.2.5 THE FIVE REQUIREMENTS FOR A GOOD EXPERIMENTAL DESIGN

There are five basic requirements for good experimental design (Cox 1958). These are: absence of bias, high power, wide range of applicability, simplicity, and being amenable to statistical analysis.

6.2.5.1 Absence of Bias: Randomisation and Blinding

When one or more of the treated groups has more favourable conditions than the other groups, differences between groups will be a consequence both of the effect of any treatment and the effect of the better environment. This is known as bias, and it can result in entirely misleading conclusions being drawn. Bias might occur, for example, if the animals of each treatment group are housed all together in one cage, and the animals in one of the cages are fighting so that their biological characteristics are altered. It could also occur when the animals are allocated to treatment groups if the groups differ in age, health, or weight. Bias can also occur if the investigator favours one group more than another. This is particularly likely where there is a subjective element to assessing the results. For example, when a pathologist assesses histological slides, or when an observer is assessing behaviour they may alter their assessment if they know which treatment group the subject belongs to.

Bias can be avoided by assigning the animals to the treatments using a formal method of randomisation and by blinding wherever this is possible. Note that randomisation should start with the assignment of the animals to treatment groups, but should continue throughout the time they are held in the animal house, the order in which they are treated, and the outcomes assessed. The exact method of randomisation will also depend on the experimental design. In a randomised block design, randomisation is done separately in each block (see Section 6.2.6.2).

Blinding is done by assigning to each animal a code number that does not indicate its treatment group. If treatment involves inclusion of a substance in the diet, then cage labels should be colour coded with the same colours as the diet containers. Persons feeding the animals should not know what colour is associated with which treatment. Outcomes should also be assessed using codes.

6.2.5.1.1 Methods of Randomisation

Many statistical textbooks contain tables of random numbers that can be used to assign animals to treatments, but often it is easier to use a physical method or to randomise using a computer program such as EXCEL. Suppose that, for example, the aim was to randomise 18 animals to three treatment groups. Six slips of paper numbered "1", six numbered "2" and six numbered "3" would be folded and placed in a receptacle, thoroughly shaken and then taken out one by one and the assignment noted on paper. This could then be taken to the animal house. Animals would then be assigned sequentially to groups in the order noted on the paper.

Spreadsheet software such as EXCEL usually includes a random number generator that produces a number between 0 and 1, to several decimal places. One column could be filled with six "1s", six "2s" and six "3s". Eighteen random numbers would then be placed in the next column. Both columns could then be sorted on the second column. This will put the first column in random order.

6.2.5.2 Power

The *power* of an experiment is its ability to detect the effect of a treatment, assuming it has an effect. It is usually expressed as a percentage. It is also sometimes called the *sensitivity* of the experiment. One minus the power is the probability of a false negative result, also known as a "type II error". Clearly, the aim should be to design a powerful/sensitive experiment. Methods of controlling power are discussed below.

6.2.5.2.1 Choice of Animals

The choice of species, sex, strain/genotype, age/weight are often determined by the nature of the experiment. It is important to use healthy, disease-free animals. Clinical or sub-clinical infection increases variability and may also alter the nature of a response to a treatment, giving misleading results. So-called Specific Pathogen Free or SPF animals should always be used if they are available. Sometimes it is not possible to obtain an adequate number of animals that are sufficiently uniform, so the experiment needs to be split. This should be done using a *randomised block* experimental design, discussed below.

Genetic variation in mice and rats can be controlled by using inbred strains. The case for using such strains rather than outbred stocks has been made repeatedly (Festing 1999), but its importance is still not recognised by many scientists. Toxicologists in particular have ignored the strong scientific case for using inbred strains in safety testing (Festing 1986, 1997; Chia et al. 2005).

6.2.5.2.2 Control of the Environment

Having chosen animals that are as uniform as possible, it is also important to ensure that the environment does not increase variability. A general rule should be that animals are given the opportunity to regulate their environment for themselves, rather than having this done for them in ways that may be inappropriate. The worst case would be to house animals individually in cages with a wire mesh floor without bedding. Such animals would have no opportunity to control their temperature, and would certainly be stressed. Groups of stressed animals are usually more variable, and may give different results from non-stressed animals. In contrast, mice in groups can live in a deep freeze if they are given sufficient quantities of nesting material. Where environmental enrichment is provided, it should be a true enrichment that decreases territorial aggression otherwise it will be counterproductive (Garner 2005).

The environment within the animal house can vary both spatially and temporally. It is well known that animals placed on the top shelf are subjected to higher light intensity, and possibly a higher ambient temperature than those on a lower shelf. It is sometimes worthwhile to take account of this using a randomised block design, discussed below. Similarly conditions may fluctuate over time due to uncontrollable factors such as variations in barometric pressure and seasonal trends. The randomised block design can also be used if the experiment needs to be split over time.

6.2.5.2.3 Sample Size

It is important to use an objective method to establish the required sample or group size. Relying on tradition, such as always using eight animals per group can lead to substantial waste of animals. In general, experiments with only a few treatment groups require larger sample sizes than those with many groups.

A *power analysis* is the preferred way of determining sample size, because it takes into account all of the important variables. It is almost essential to use this method for designing clinical trials in order to avoid wasting expensive resources by doing experiments that are too small to be able to detect a clinically important outcome. It is applicable both for quantitative and binary (e.g., dead/alive) outcomes. However, a power analysis is not so easy with more complex designs and in situations where a lot of characters are being measured. In this case the *Resource Equation* method, which depends on the law of diminishing returns, is useful.

6.2.5.2.4 Power Analysis

Power analysis depends on a mathematical relationship among the following six variables. For simplicity it will be assumed that there are only two treatment groups "Control" and "Treated":

- The size of effect to be detected. This is the magnitude of any difference between the treated and control groups, which is likely to be of clinical or scientific interest. A very small difference is not likely to be of much interest. A large one should certainly be detectable. The effect size is the dividing line between the two.
- The significance level. This is the probability of a false positive result, also known as a "Type I error" or "α error". It is usually set at $\alpha = 0.05$. Clearly false positive results are to be avoided, but setting it lower may mean that a larger sample size is needed, or the experiment may lack power.
- The sidedness of the test. A two-sided test should be used if it is not known whether the effect of the treatment will be to increase or decrease the treated group mean relative to the control. If the effect can only go one way, then a one-sided test should be used.
- The variability of the experimental material. The more variable the ExpUs are, the larger the required sample size required. The problem is that the experiment has not yet been done, so an estimate of the standard deviation (assuming a measurement character) must come from a previous experiment. This is the main weakness of the method, as estimates of the standard deviation can vary a lot between different experiments. The estimated sample size (or power) is also quite sensitive to differences in this figure.
- The power of the experiment. This is the probability that the experiment will be able to detect the specified effect and show it to be significant at the specified level, given the estimated variability. Somewhat arbitrarily, the power is usually set at 80–90%. Choice of the power really depends on the consequences of failing to detect a clinically or scientifically important effect. If this would have serious consequences, then a higher power should be specified, but the power can not be set at 100% or an infinitely large experiment would be needed.
- The sample size. This is the number of animals needed per group, based on the assumption that group sizes are equal.

In some cases the group size will be determined by the facilities available. In this case the same equations can be used to estimate the power of the experiment or the size of the effect likely to be detectable.

The equations linking these variables are somewhat complex, but fortunately most commercial software packages now offer power calculations for the simpler situations, and there are stand-alone

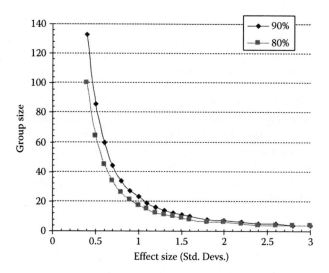

FIGURE 6.1 Group size required as a function of effect size in standard deviations, assuming a two-sided t-test, a 5% significance level and 80 or 90% power.

programs such as nQuery Advisor* (for more complex situations). There are also several websites[†] that will do the calculations. If the effect size is expressed in standard deviation units (this is the signal/noise ratio where the signal is the difference between the means and the noise is the standard deviation), then the required sample size can be read off from Figure 6.1. For example, if the heart weight of a rat is 800 mg with a standard deviation of 75 mg, then an experiment to detect a difference in heart weight between treated and control rats of one standard deviation (75 mg) would require about 18–23 rats per group whereas to detect an effect of 1.5 standard deviations (112 mg) would require only about 8–10 rats per group with the assumptions listed in the caption to Figure 6.1. Note from Figure 6.1 that very large sample sizes are required to detect effects of much less than one standard deviation. Note also that if many characters are being measured, then separate analyses are required for each one, or the most important one has to be chosen on which to power the experiment.

6.2.5.2.5 Resource Equation

This method, also known as the top-down method, depends on the law of diminishing returns. Adding one more animal to a small experiment will provide a useful amount of information. However, adding one more to a large experiment will have much less effect. The resource equation is suitable for more complex experiments with a quantitative outcome, which can be analysed by the ANOVA. The main problem in any experiment is to obtain a good estimate of the variability of the material, and in the ANOVA, the error mean square is an estimate of the within-group variance of the experimental material. The accuracy of this estimate depends on the size of the whole experiment, taking into account both the group size and the number of groups. As a general rule, an estimate of variance based on fewer than 10 degrees of freedom (the number of ExpUs that contribute to the estimated variance) will not be very accurate, increases above 20 degrees of freedom will not provide much extra information. For a completely randomised design:

$$E = \text{(total number of animals)} - \text{(total number of treatment groups)}.$$
E should be between about 10 and 20.

---

* http://www.statistical-solutions-software.com accessed on 15 August 2009.
† http://www.biomath.info/power/index.htm accessed on 15 August 2009.

However, there may be good reasons for E being slightly larger than this, such as in order to have equal numbers in each group. The E might also be larger in *in vitro* studies where ExpUs are often wells in a tissue culture dish, so may be very inexpensive.

As an example, suppose an experiment were to be planned with six treatment groups. With eight animals per group E = 48 − 6 = 42. This is far too many animals. With three animals per group, E would be 18 − 6 = 12. With four animals per group, E = 24 − 6 = 18. Either three or four animals per group would be within the suggested range.

6.2.5.3 Range of Applicability: The Factorial Experiment

Strictly, the results of an experiment only apply to the particular circumstances of that individual experiment. If only males were used, then the results may not apply to females. If it was done in one animal facility at a particular time it may not be repeatable at a different time or in another animal facility because environmental conditions may not be the same. However, no progress would be made if slight changes commonly resulted in unrepeatable results, so it is important to consider what variables are likely to have a major influence on the outcome, and try to factor these into the design of the experiment. This can be done using factorial experimental designs discussed below. Blocking will also increase the range of applicability to some extent because each block samples a slightly different environment or set of conditions.

6.2.5.4 Simplicity

Clearly, experiments should not be so complicated that mistakes are made either in the conduct of the experiment or in the subsequent statistical analysis. In all cases, protocols should be written before starting the experiment. Small pilot studies involving just a few subjects can be used to ensure that all the steps in the protocol are feasible, and to gather preliminary information such as whether dose or treatment levels appear to be about right. As far as possible, group sizes should be the same because this maximises power. This is particularly important for two-way experiments such as randomised block and factorial designs where unequal group sizes complicate statistical analysis and interpretation of the results. The main exception to this is where several treatment means are being compared with a single control group in a one-way design. In this case the control group may be made larger than the treated groups, depending on exactly how many groups are involved.

6.2.5.5 Amenable to Statistical Analysis

The aim of statistical analysis is twofold. First, the results need to be summarised, usually in terms of means and standard deviations, medians and inter-quartile ranges or counts and proportions. Second, in most cases a statistical analysis is used to determine the extent to which differences among groups are dependent on chance. If it is highly unlikely that differences could have arisen by chance, then it is assumed that the treatment differences are due to the effects of the treatment.

The statistical analysis should always be planned at the time that the experiment is designed, and experiments should be analysed and interpreted before the next one is planned so that any necessary adjustments can be made. Occasionally an experiment is so badly designed that no statistical analysis is possible. This mainly occurs when the ExpU has not been correctly identified. If control animals are housed in one cage and the treated animals in another one, it is possible to do a statistical analysis that will show whether the means of a particular character differ between cages. However, all this does is to show whether the two cages differ, not whether that difference is caused by the effect of the treatment. It is a unique property of the controlled experiment that it can detect causation rather than correlation.

6.2.6 Formal Designs

A number of formal experimental designs can be used to take account of the characteristics of the animals and the aims of the experiment.

6.2.6.1 The "Completely Randomised" or "Between Subject" Design

This is the most common design. Animals are assigned to treatments strictly at random, regardless of their characteristics, as shown in Table 6.1. The design assumes that the animals will be reasonably uniform, that the whole experiment will be done at the same time in the same location (or that it is not practical to take these factors into account in the design), and that there are no extraneous sources of variation that could be taken into account to improve the power of the experiment. There can be any number of treatment groups and animals per treatment group, and unequal group sizes pose no particular problem, although equal group sizes usually maximise over-all power. As suggested by the title, animals are assigned to groups at random and these are then maintained in the animal house in random order, and any measurements are also made in random order to avoid bias due to time trends.

6.2.6.2 The Randomised Block Design

A randomised block design is used to split the experiment up into a number of "mini-experiments" (Table 6.2). This may be because the experimental material has some sort of "natural structure", for example, when experimenting on neonatal animals litter differences may need to be taken into account. Blocks may be separated in time as, for example, when comparing genetically modified animals with their wild type litter-mates with litters only being available sporadically, or as

TABLE 6.1
Diagram of a Completely Randomised Between-Subject Design with Three Treatments Coded A, B, C, in Random Order

C	B	C	B	C
A	A	C	B	B
A	A	B	A	C

Note: The first animal caught would be assigned to treatment C, the next to B, and so on; the cages should be housed in the random order shown; and processed in the same order.

TABLE 6.2
Layout of a Randomised Block Design with Four Treatments in Five Blocks

Block 1	A	D	C	B
Block 2	B	C	A	D
Block 3	A	B	D	C
Block 4	D	A	B	C
Block 5	C	B	A	D

Note: Randomisation is done separately for each block.

above, when studying neonatal animals that are still being suckled by their mother. Alternatively, it may be useful to split an experiment in order to make the subjects more homogeneous such as when they vary in body weight more than is considered acceptable. In this case the heaver and lighter animals can be assigned to different blocks. Similarly, it may be worthwhile using a blocked design to take account of a suspected environmental variable such as shelf height in an animal room. Another reason for using a blocked design is to ensure reasonable repeatability. Blocks can be done at different times or even in different animal houses to check that the results are repeatable.

Typically, block size is the same as the number of treatments. So an experiment with four treatment groups will have a block size of four, although it can be some multiple of the number of treatments. However, if the natural structure of the animals suggests a block size of, say, four but there are more than four treatment groups, then an incomplete block design is necessary. Incomplete block designs can be complicated and are not considered here.

Randomisation of a block design is always done within the block. Thus if there are four treatments with a block size of four and there are to be six blocks, then six separate randomisations are necessary. A randomised block design is analysed using a two-way ANOVA "without interaction". A pooled standard deviation is estimated by taking the square root of the error mean square shown in the analysis of variance table. This can be reported with the treatment means and used for error bars, if necessary.

6.2.6.3 The Cross-Over, Repeated Measures or Within-Subject Design

Some experiments do not permanently alter the characteristics of the animals that are used. For example, an animal may be given an anaesthetic, and when fully recovered and rested will, presumably, not be significantly altered. So it is possible to compare different anaesthetics on the same animal in sequential order using a cross-over design. In this case the ExpU is the animal for a period of time. The order of the treatments needs to be randomised because animals may continue to grow or get fatter, or there may be time effects that need to be averaged out. However, generally this type of design is highly efficient because the variation within an animal over a period of time is usually much less than the variation between animals, so fewer animals are required. On the other hand it is a fairly rare design because there are relatively few treatments that do not permanently alter the animal.

The cross-over design is very like a randomised block design with the animals being regarded as the blocks. In Table 6.2, for example, if each animal is regarded as a "block" then the design could also apply to a cross-over experiment with four treatments. Differences between animals are usually of no interest. Indeed, the experiment could mix males and females with any sex difference being removed as differences between individuals, and outbred animals can be used because any genetic variation is removed as a block effect. The statistical analysis is identical to that of a randomised block design, again using a two-way ANOVA without interaction with a pooled standard deviation from the ANOVA table.

A before and after experiment in which something is measured in an animal before administering a treatment, and the same character is measured after the treatment is a special case of a repeated measures design. Although widely used, care needs to be taken in interpreting the results because there has been no randomisation of the ExpUs to before and after. If there were a time-related effect not related to the treatment, this would be confounded (i.e., mixed up) with any treatment effect.

The term "repeated measures" can be confusing. A study that involves taking measurements repeatedly over time, such as in a growth study, is not a repeated measures design unless distinctly different treatments have been applied during that period. The analysis of sequentially collected or longitudinal data usually involves some type of regression analysis and the fitting of curves, area under the curve, time to peak or alternatively a multivariate analysis such as a principle components analysis (Everit and Dunn 2001), is not discussed here.

TABLE 6.3
Layout of an Un-Randomised Latin Square Experiment

A	B	C	D	E
E	A	B	C	D
D	E	A	B	C
C	D	E	A	B
B	C	D	E	A

Note: Each row and each column has exactly one of each of the five treatments A–E. Randomisation is done by randomising first a whole column, then a whole row. This retains the balance of the design.

6.2.6.4 The Latin Square

A Latin square design resembles a randomised block design except that it can take account of two types of block-type variables at the same time. The number of animals needs to be the square of the number of treatments. So if there are six treatments there will need to be 36 animals altogether.

A randomised block design might split an experiment up into a number of "mini-experiments" over a period of time with, say, one block being done each week; the order in which the treatments are applied each week is randomised. This may mean that by chance more of one treatment group may be first. However, a Latin square design ensures that this is not the case and the experiment will be balanced both across days and within days (or whatever the nuisance variables are). Table 6.3 shows the layout of this type of design with four treatments. Note that each treatment occurs once in every row and every column, so that the experiment is balanced with respect to both these variables. A special method is used for randomisation of this type of design. In Table 6.3, the treatments are labelled A–D and are written as shown in the order A, B, C and D in the first row, then with the A offset one space in the second row, and so on. Randomisation is done by first randomising the rows and then the columns (or vice-versa). This retains the balanced properties of the design. The Latin square design is analysed by a three-way ANOVA without interaction, with the standard deviation estimated from the square root of the error mean square in the ANOVA table. Note that a single 2 × 2 Latin square experiment would be too small, but these can be replicated as many times as necessary to get an adequate sample size.

6.2.6.5 The Split-Plot Design

Split-plot designs have two sorts of ExpUs in one experiment. For example, suppose the aim were to compare a number of diet formulations, but within each cage half of the animals are to receive a vitamin injection and the other half the vehicle (Table 6.4). In this case, for comparing the effect of the vitamins, the individual animals are the ExpUs, but the cages are the ExpUs for comparing the diets. The approach is rather like a factorial design in that it will show whether there are statistically significant effects of the vitamins, the diets, and any diet × vitamin interaction (i.e., whether the response to the vitamin is the same for all diets). However, the statistical analysis is slightly complicated because it requires the use of a "nested" analysis of variance. This design is discussed in most textbooks on experimental design (Cochran and Cox 1957; Montgomery 1984). Investigators sometimes design this type of experiment without knowing that it is a split-plot design and then have difficulty analysing the results.

TABLE 6.4
Layout of a Split-Plot Design with Three Diet Treatments (A, B, C) Applied to Cages, and Two Vitamin Treatments x and y Given Individually to Animals Within a Cage, with four Blocks (1–4)

Block	Diet A		Diet B		Diet C	
1	x	y	x	y	x	y
2	x	y	x	y	x	y
3	x	y	x	y	x	y
4	x	y	x	y	x	y

Note: Cages should be assigned to diet treatment at random and within each cage individuals should be assigned at random to the vitamin or control treatment.

6.2.6.6 Sequential Designs

Sequential designs can be used in the situation where an individual animal (in an uncontrolled experiment) or a pair of animals (treated and control) can be treated and the result is obtained reasonably quickly. A decision can then be made on whether to treat the next individual or pair, or whether there is sufficient information to terminate the experiment and reach a statistically acceptable conclusion. The sample size in these designs varies, depending on the magnitude of the response to the treatment.

Sequential experiments are rare because the situations in which they can be used are rare. However, where they can be used they tend to be very efficient in using the smallest possible number of subjects. This is because experiments with a fixed sample size tend to over-estimate the number of ExpUs required in situations where there is a strong but unforeseen treatment response. In contrast, as soon as a significant difference is obtained (when comparing treated and control subjects), the sequential design experiment is terminated. However, although sequential designs are useful for hypothesis testing they are not as good as fixed sample size designs for estimating means and standard deviations (Johnstone 1998).

Both "up-and-down" and the "acute toxic class" are methods of assessing the acute toxicity of a chemical; they are sequential designs (Lipnick et al. 1995; Rispin et al. 2002). There is scope for more widespread use of these, although further discussion is beyond the scope of this chapter.

6.2.6.7 Factorial Designs

A factorial design is one in which two or more variables or "factors" vary at the same time. For example, one factor might be a treatment, with levels "control" and "treated", and another factor might be time with half the animals being measured at time one and the other half at time two (assuming it is not possible to measure the same animals at both times). The aim of such an experiment would be to determine whether the treatment and control means differ, whether the means at the two time periods differ, and whether the treatment effect differs between the two times. This is shown diagrammatically in Table 6.5. The statistical analysis compares the treated and control means averaged across time periods $(a + b) - (c + d)$, the means for each time averaged across treatments $(a + c) - (b + d)$, and the interaction between time and treatment $(a + d) - (b + c)$. There can be any number of factors and each can have any number of levels, although as the number of factors and levels increase, the number of means that have to be interpreted rises dramatically. The results are analysed using an "n"-way ANOVA with interaction, where "n" is the number of factors.

TABLE 6.5
Structure of a 2 × 2 Factorial Design

	Time 1	Time 2
Control	a	b
Treated	c	d

If the means a–d are based on, say, five or six animals in each group, then E will be 16–20, which according to the resource equation method, would be an appropriate sample size.
See text for more details.

Factorial experiments are common. In one survey of 48 experiments in two toxicology journals 17/48 or about a third of all experiments had a factorial structure; however, only four of the experiments were correctly analysed (Festing 1996), possibly because many investigators have not been taught how to use an ANOVA.

Factorial designs are a method of improving the generality of the results of an experiment. Strictly, a factorial design is not a "design"; it is an arrangement of treatments. Thus it is possible to have a factorial arrangement of treatments in a between-subjects design, a randomised block design, or even a Latin square design. However, when there are many factors and/or many levels of each, then they need separate consideration. Investigators often design experiments that, if properly done (i.e., with proper randomisation), could be factorial designs, but often they don't realise that when using a greater number of groups, the size of each group can be reduced, and also they fail to analyse them correctly.

A 2 (treatments) × 4 (mouse strains) factorial design was used to study the activity of a liver enzyme following treatment with the anti-oxidant BHA using a randomised block design and a total of 16 mice (Festing 2003). This found a large effect of the BHA treatment and a statistically significant, but small interaction due to the fact that one strain of mice responded slightly more than the other strains. Such designs are economical with animals and provide extra information at little or no cost. According to R. A. Fisher (1960):

> If the investigator ... confines his attention to any single factor we may infer either that he is the unfortunate victim of a doctrinaire theory as to how experimentation should proceed, or that the time, material or equipment at his disposal is too limited to allow him to give attention to more than one aspect of his problem. . . . Indeed in a wide class of cases (*by using factorial designs*) an experimental investigation, at the same time as it is made more comprehensive, may also be made more efficient if by more efficient we mean that more knowledge and a higher degree of precision are obtainable by the same number of observations. (italics added).

An example of a factorial design is given below in the section on statistical analysis.

In some cases factorial designs can also explore the effect of very many variables without using excessive numbers of animals, because high level interactions are usually quite small. Thus, an experiment with five variables such as treatment, time, prior treatment, strain, and sex each at two levels will involve 2^5 (= 32) treatment means and could be done using 64 animals. It could also be done with only 32 animals if the assumption is made that only main effects and two-way interactions are likely to be important. The high-order interactions would be pooled to provide an estimate of the variance. There is an extensive literature on factorial designs that are widely used in industrial experimentation.

6.3 STATISTICAL ANALYSIS

It is not possible in a chapter of this size to give detailed methods of statistical analysis of the experimental results. There are many textbooks that cover this in detail (see the Further Reading list at the end). The comments here are about the statistical analysis. This depends on the purpose of the experiment, the type of data, and the actual experimental design. Each case is unique and has to be considered on its individual merits. The aim here is to discuss briefly the reasons for doing a statistical analysis. This is firstly to summarise the results in a way that makes them easy to interpret, and secondly to quantify the importance of chance factors in determining those results.

6.3.1 EXAMINING THE DATA

The first step is always to screen the data for obvious mistakes. Serious outliers should be checked to ensure that they are not transcription errors. This can best be done using graphical methods that show the individual observations. However, outliers should only be deleted if they are clearly mistakes. The use of graphical methods will also help to show the distribution of the data, which will help to determine the type of statistical analysis.

With continuous rather than discrete (dead/alive, tumour/no tumour) data, so called parametric statistical methods can often be used. These include Student's t-test and the ANOVA, and requires the data to possess variation that is about the same in each treatment group, and the deviations from the group means (known as residuals) to have a normal (bell-shaped) distribution. In these cases, means and standard deviations are appropriate summary statistics. However, in some cases the data may be skewed by possessing a lot of fairly small numbers, but a few being five or 10 times as large. In such cases the residuals will have a non-normal distribution and the variation will often increase as the mean increases. This type of data can be summarised as medians and inter-quartile range, and graphically using box and whisker plots. Alternatively, it may be possible to transform the data to, say, a log scale. Each value, X, is replaced by $\log_{10}X$ (or $\log_{10}(X + 1)$ if there are genuine zero values as the logarithm of zero is undefined), and this is the data that is statistically analysed. This situation often arises when measuring the concentration of something in body fluids. There can not be a concentration of less than zero, there are often a lot of relatively low values and a few very high ones. The data may also consist of counts, a classical example being the number of cells in each square of a haemocytometer. If the mean is low, with quite a lot of zeroes or ones but there are a few higher values, the data may have a Poisson distribution and a square root transformation can be used to make the data suitable for a parametric analysis, with each count being replaced by its square root. Some data may be "censored". For example, litter size is left censored because litter sizes range from one upwards. A litter size of zero does not make sense. Survival is often right censored because the experiment may be stopped before all the animals have died. The analysis of survival needs specialised methods (Altman 1991).

If numerical data turns out to be unsuitable for a parametric analysis, then non-parametric tests will need to be used. These are less versatile than parametric methods and in some cases are not as powerful, so they produce more false negative results. Non-parametric methods are discussed briefly below.

Discrete data such as counts of dead/alive or small, medium and large usually summarised as percentages or proportions. It is important to indicate "n" for each percentage because this reflects the reliability of the figure. Obviously, 50% based on two animals is less reliable than 50% based on 100 animals.

6.3.2 PARAMETRIC STATISTICAL ANALYSIS

Once the data has been screened and, where necessary, transformed, then group means can be compared using either Student's t-test (but only if there are just two means to compare) or the ANOVA to assess whether or not any differences could be due just to chance sampling errors. If this is highly unlikely, then it is assumed that the differences are due to a treatment effect. Basically, the aim is to study the so called null hypothesis (that there are no differences between the group means) with the alternative hypothesis (that there are such differences). The statistical tests determine the probability that observed differences could be due to chance. If this probability (the p-value) is less than some pre-determined value, which is often set at 0.05, then the differences are said to be "statistically significant at the 0.05 probability level". In contrast, if the p-value is larger than 0.05 then this is interpreted as there being *no evidence* that the treatment is having an effect. Note that this is not the same as concluding that the treatment has *no* effect. It may have an effect that is undetectable by the experimental design and material that were actually used.

The ANOVA provides an over-all test of whether there are differences among two or more group means. When there are more than two means, further tests (usually so-called *post-hoc* comparisons) are required to determine which groups differ from each other. The ANOVA is much more versatile than the t-test because it can be used to analyse more complex experiments such as randomised block, Latin square, cross-over and factorial designs. Every biomedical scientist should understand the basic methods of the ANOVA. The actual calculations are done using a computer, so all that needs to be known is how to enter the data, generate the ANOVA and interpret the results. The ANOVA is explained in detail in most general statistical textbooks. An excellent introduction is given by Roberts and Russo (1999).

6.3.3 Non-Parametric Tests

Data which is not suitable for a statistical analysis using the t-test or ANOVA, even following a scale transformation, may often be analysed using one of several available non-parametric tests. In most cases the test requires that each observation is replaced by its rank order, and then determines whether the mean ranks differ between groups. In general these tests are not as powerful as parametric tests because they discard some data (the difference between an actual observation and its rank) and associated information. However, there may be little option but to use this type of test with some data sets.

The most widely used non-parametric method is the Mann–Whitney or Wilcoxon test (two names for the same test) for comparing the medians of two groups. This is the non-parametric equivalent of the t-test. The Kruskal–Wallis test is the non-parametric equivalent of the one-way ANOVA with several groups but only a single factor. In this case *post-hoc* comparisons usually use the Mann–Whitney test. The Friedman test is the non-parametric equivalent of the two-way ANOVA without interaction, used in the randomised block or repeated measures design. There is no non-parametric equivalent of the two-way ANOVA with interaction for a factorial design, although it has been suggested that an ANOVA on the ranked data could be used (Montgomery 1997).

6.3.4 Discrete Data

Data based on proportions or percentages, such as the percentage of animals with a tumour or some other attribute in different treatment groups, are usually analysed using a chi-squared test. This is based on the actual numbers in each group, not the percentages. In some circumstances other methods are used, for example, the percentage positive in two groups can be tested using a so-called normal approximation of the binomial distribution, provided none of the percentages are very low. In that case Fisher's exact test may be used. There are other methods and these are described in most statistical test books.

6.3.5 Interpreting and Reporting the Results

Although most papers are refereed by at least one person and read by the editor of a journal, it is surprising how often papers are badly presented. First, it is essential to give details of the animals used. Species, strain, genotype (with correct genetic nomenclature), sex, age/weight, origin, health status, housing methods, bedding, diet, cages, group sizes and refinements designed to minimise pain and distress should be described. The method of determining sample size and randomisation and whether or not blinding, with coded samples were used, should be stated.

Each experiment should be designated with a number or letter, and clearly described in the materials and methods section of the paper. The methods used for statistical analysis need to be described, with references if unusual methods are used. The results for each should also be clearly identified. Means, medians and percentages should be presented with some indication of "n" and a measure of variation such as a standard deviation, standard error or 95% confidence interval or inter-quartile range (depending on the type of data). A 95% confidence interval provides more information than a standard deviation or standard error. Similarly, figures should also give some indication of numbers. Ideally, data should be displayed in such a way that individual observations are shown, particularly for regression analyses when showing dose-response relationships. Whether error bars show standard deviations, standard errors or 95% confidence intervals should be clearly stated. Note that "n" needs to be known for every point with an error bar otherwise the bars can not be interpreted. All claims for differences between groups should be supported by an appropriate statistical analysis, to demonstrate that the results were not due to chance.

6.3.6 A Note on Statistical Software

As far as possible, investigators should use one of the commercially available statistical packages such as MINITAB, SPSS, SAS, S-PLUS, GraphPadPRISM and so on. These are generally reliable and most are menu-driven and so are relatively easy to use, although each has its particular idiosyncrasies so they all take some time to learn. General spreadsheets such as EXCEL are not recommended as they tend to give non-standard output and in some cases their reliability has been questioned. Some free software packages might be useful, but they should be treated with caution. The R statistical package is of particular interest as it is free and is widely used by professional statisticians, so it is certainly reliable. There are also several textbooks describing its use (see further reading section), but it is command driven and the commands are not easy to learn. However, the "R Commander" package provides a menu-driven interface that is a lot simpler to use, making it suitable for teaching, with each student being able to have his/her own copy on a laptop computer.

6.4 CONCLUSIONS

Well-designed experiments will ensure as far as possible that the minimum number of animals is used, consistent with achieving the scientific objectives of the experiment. Not only will this avoid pain and distress in the animals, but it will also save money, time, and scientific resources, so it is very much in the interests of each investigator to design their research strategy and individual experiments carefully. Moreover, careful attention to the design and analysis of the experiments will improve the quality of the experiments and increase the chances of publication in a high impact journal.

6.5 QUESTIONS UNRESOLVED

- Training of scientists
  Many scientists have little or no training in experimental design. Where they have had training in statistics, it often assumes that the data has already been generated and does not cover the design of experiments in any detail. They should know about: completely randomised, randomised block, cross-over and factorial designs, and should know how to analyse simple versions of these using a statistical package. This standard might be achieved by a 3–4 days short course.

- Statistical advice
  While research workers should be able to design and analyse their own relatively simple experiments, professional advice is essential for more complicated situations. This is often not available. Expert statistical advice should be available to all research scientists at nominal cost.
- Publication and refereeing
  Many papers are published with obvious statistical errors in experimental design, statistical analysis, and presentation of results. These should be corrected before publication, but editors have difficulty in finding well-qualified statistical referees. Statisticians are prepared to referee statistical papers by their colleagues, but there is no real incentive for them to spend their time refereeing papers by life-scientists. It might be necessary in this case to offer them some financial reward.
- Control of genetic variation
  A good experiment will, as far as possible, control all sources of variation so that the effect of any treatment can be more easily detected. Most scientists understand this. However, for some reason they often implicitly reject this rule for genetic variation, and use outbred genetically heterogeneous animals. This is particularly apparent in toxicity testing where it is probably leading to too many false negative results.

REFERENCES

Altman, D. G. 1991. *Practical statistics for medical research.* London, Glasgow, New York: Chapman and Hall.
Altman, D. G. 1982. Misuse of statistics is unethical. In *Statistics in practice*, ed. S. M. Gore, 1–2. London: British Medical Association.
Chia, R., F. Achilli, M. F. W. Festing, and E. M. C. Fisher. 2005. The origins and uses of mouse outbred stocks. *Nature Genetics* 37:1181–86.
Cochran, W. G., and C. M. Cox. 1957. *Experimental designs.* New York, London: John Wiley & Sons, Inc.
Cox, D. R. 1958. *Planning experiments.* New York: John Wiley and Sons.
Everit, B. S., and G. Dunn. 2001. *Applied multivariate data analysis.* London, New York: Arnold.
Festing, M. F. W. 1986. The case for isogenic strains in toxicological screening. *Archives of Toxicology*: 127–37.
Festing, M. F. W. 1996. Are animal experiments in toxicological research the "right" size? In *Statistics in toxicology*, ed. B. J. T. Morgan, 3–11. Oxford, UK: Clarendon Press.
Festing, M. F. W. 1997. Fat rats and carcinogenesis screening. *Nature* 388:321–22.
Festing, M. F. W. 1999. Warning: The use of heterogeneous mice may seriously damage your research. *Neurobiology of Aging* 20:237–44.
Festing, M. F. W. 2003. Principles: The need for better experimental design. *Trends in Pharmacological Sciences* 24:341–45.
Festing, M. F. W., P. Overend, R. G. Das, M. C. Borja, and M. Berdoy. 2002. *The design of animal experiments. Reducing the use of animals in research through better experimental design.* London: The Royal Society of Medicine Press Ltd.
Fisher, R. A. 1960. *The design of experiments.* New York: Hafner Publishing Company, Inc.
Garner, J. P. 2005. Stereotypies and other abnormal repetitive behaviors: Potential impact on validity, reliability, and replicability of scientific outcomes. *ILAR Journal/National Research Council, Institute of Laboratory Animal Resources* 46:106–17.
Johnstone, P. D. 1998. Should sequential design be used for comparative experiments on animals. *New Zealand Journal of Agricultural Research* 41:561–66.
Lipnick, R. L., J. A. Cotruvo, R. N. Hill, R. D. Bruce, K. A. Stitzel, A. P. Walker, I. Chu, et al. 1995. Comparison of the up-and-down, conventional LD50, and fixed-dose acute toxicity procedures. *Food and Chemical Toxicology* 33:223–31.
McCance, I. 1995. Assessment of statistical procedures used in papers in the Australian Veterinary Journal. *Australian Veterinary Journal* 72:322–28.
Mead, R. 1988. *The design of experiments.* Cambridge, New York: Cambridge University Press.
Montgomery, D. C. 1984. *Design and analysis of experiments.* New York: John Wiley & Sons, Inc.
Montgomery, D. C. 1997. *Design and analysis of experiments.* New York: Wiley.

Perel, P., I. Roberts, E. Sena, P. Wheble, C. Briscoe, P. Sandercock, M. Macleod, L. E. Mignini, P. Jayaram, and K. S. Khan. 2007. Comparison of treatment effects between animal experiments and clinical trials: Systematic review. *BMJ (Clinical Research Ed.)* 334:197.

Rispin, A., D. Farrar, E. Margosches, K. Gupta, K. Stizel, G. Carr, M. Greene, W. Mayer, and D. McCall. 2002. Alternative methods for the LD50 test: The up and down procedure for acute toxicity. *ILAR Journal* 43:233–43.

Roberts, M. J., and R. Russo. 1999. *A student's guide to the analysis of variance*. London: Routledge.

Russell, W. M. S., and R. L. Burch. 1959. *The principles of humane experimental technique*. Potters Bar, England: Special edition, Universities Federation for Animal Welfare.

FURTHER READING

Altman, D. G. 1982. Statistics in medical journals. *Statistics in Medicine* 1:59–71.

Altman, D. G., D. Machin, T. N. Bryant, and M. J. Gardiner. 2000. *Statistics with confidence*. London: BMJ Press.

Clarke, G. M. 1994. *Statistics and experimental design: An introduction for biologists and biochemists*. London, Melbourne, Aukland: Edward Arnold.

Dalgaard, P. 2003. *Introductory statistics with R*. New York, Berlin: Springer.

Duley, L., and B. Farrell. 2002. *Clinical trials*. London: BMJ Books.

Everitt, B. S., and T. Hothorn. 2006. *A handbook of statistical analysis using R*. London: Chapman & Hall.

Friedman, L. M., C. D. Furberg, and D. L. DeMets. 1996. *Fundamentals of clinical trials*. St. Louis, Baltimore, New York, London: Moseby.

Howell, D. C. 1999. *Fundamental statistics for the behavioral sciences*. PacificGrove, London, New York: Duxbury Press.

Kvanli, A. H. 1988. *Statistics: A computer integrated approach*. St. Paul, New York, Los Angeles, San Francisco: West Publishing Company.

Maxwell, S. E., and H. D. Delaney. 1989. *Designing experiments and analyzing data*. Belmont, California: Wadsworth Publishing Company.

McCleery, R. H., T. A. Watt, and T. Hart. 2007. *Introduction to statistics for biology*. London: Chapman & Hall.

Petrie, A., and P. Watson. 1999. *Statistics for veterinary and animal science*. Abingdon, Malden, USA, Winnipeg: Blackwell Science.

Snedecor, G. W., and W. G. Cochran. 1980. *Statistical methods*. Ames, Iowa: Iowa State University Press.

Sprent, P. 1993. *Applied nonparametric statistical methods*. London, Glasgow, New York: Chapman and Hall.

Verzani, J. 2005. *Using R for introductory statistics*. London: Chapman & Hall.

7 Animal Models: Selecting and Preparing Animals for a Study

Patrick Hardy, France and Sarah Wolfensohn, United Kingdom

CONTENTS

Objectives 152
Key Factors 152
7.1 Introduction 153
7.2 Study Objectives and Expectations, Introduction to Model Categories 154
7.3 Categories of Animal Model and Their Characteristics 155
    7.3.1 Physiological Models 155
        7.3.1.1 Outbred Animals 155
        7.3.1.2 Inbred Models 156
        7.3.1.3 F1 Hybrids 158
        7.3.1.4 Examples of Other Inbred-Derived Models or Definitions 158
    7.3.2 Human Disease Models 159
        7.3.2.1 Inbred Strains Used as Disease Models 159
        7.3.2.2 Polygenic Models Generated by Mono- or Bi-Directional Selection 159
        7.3.2.3 Monogenic Models: Spontaneous or Targeted Mutations, Additional Transgenesis 160
        7.3.2.4 Experimental Induction 162
        7.3.2.5 Combined Models 162
    7.3.3 Models of Animal Disease or in Animal Health Research 163
7.4 Review of Health Definitions and Categories 163
    7.4.1 Introduction 163
    7.4.2 Health Definitions and Standards 164
        7.4.2.1 Holoxenic or Conventional Animals 164
        7.4.2.2 Gnotoxenic Animals 164
        7.4.2.3 Agnotoxenic Animals: Heteroxenic/Specific Pathogen Free Health Standards 165
        7.4.2.4 Additional SPF Definitions 166
        7.4.2.5 Categories of Animal Micro-Flora 166
        7.4.2.6 Healthy Carriers 167
        7.4.2.7 Antibody-Free Animals 167
    7.4.3 Re-Derivation Techniques 167
        7.4.3.1 Aseptic Hysterectomy 167
        7.4.3.2 Aseptic Hysterotomy/Caesarean Section 168
        7.4.3.3 Embryo Transfer 168
        7.4.3.4 Aseptic Hysterectomy or Embryo Transfer Associated with other Procedures 168
        7.4.3.5 Genetic Issues When Re-Deriving a Colony 168
    7.4.4 Bioexclusion, Biocontainment and Health Monitoring 168
    7.4.5 Colony Termination and Recycling Policy 169

7.5 Animal Model Selection and Source 169
7.6 Environment Definition and Control 171
    7.6.1 Environmental Conditions 171
    7.6.2 Diet and Nutrition 171
    7.6.3 Transport Conditions 172
    7.6.4 Acclimatising Versus Quarantine Period 173
7.7 Conclusions 174
7.8 Questions Unresolved 174
References 175

OBJECTIVES

Selection of an animal model is one of the most important steps in the design of an experiment. This is true not only for studies involving living animals but also for replacement techniques such as those using *ex vivo* or *in vitro* cells, tissues and organs.

An animal model is defined as a complex "biological system", so the process of selection involves consideration of all its components, to ensure that its characteristics match the requirements of the experiment, and that they are adequately monitored over the duration of the entire study.

This chapter reviews the key steps of model selection:

- Defining study objectives and related expectations
- Reviewing the various animal model categories available and their key characteristics
- Selecting the most appropriate animal model and establishing controls for the required qualitative specifications, including health and environmental standards

KEY FACTORS

Considerations for model selection include:

- Selection of the most appropriate intrinsic animal components, including species characteristics; genetic standard; quality, sex and age.
- Relevance of the model to the study objective, including any scientific, technical, ethical or practical limitations.
- Appropriate "health standard" or "microbiological status" (i.e., ensuring the absence of pathogenic micro-organisms or parasites and the presence of an associated micro-flora), which is fully consistent with the desired characteristics of the animal model (including its specific and non-specific immunological competence) in relation to the research application, so as to guarantee the absence of interfering factors. Biosecurity safeguards and a health monitoring program, tailored to verify the specified health standard of animals concerned, must be put in place.
- The micro- and macro-environment, care, husbandry and handling practices, which must be appropriate both for the animal and the study requirements.
- Staff competency; that is, adequate education, training and experience of all personnel involved with the animal and maintaining their environment, including animal technicians and management, laboratory animal veterinarians, engineering and maintenance staff, site support and logistic services, quality assurance personnel and others.
- Close collaboration with the various suppliers (animal, consumables, equipments, services, etc.) or partners, in order to guarantee satisfactory quality of all animal model and study components.

- Last but not least, strict implementation of all regulations and good practices governing the responsible and ethical use of animals for scientific purposes (Russell and Burch 1959; European Commission 1986; ILAR/NRC 1996).

7.1 INTRODUCTION

A "model" has been defined as a "hypothetical description, often mathematical representation of a process, system or object developed to understand its behaviour or to make predictions. The representation always involves certain simplifications and assumptions" (Webster's 2009).

A more specific definition applicable to biomedical research is a "model organism is a species that is extensively studied to understand particular biological phenomena, with the expectation that discoveries made in the model organism will provide insight into the workings of other organisms. In particular, model organisms are widely used to explore potential causes of, and treatments for human disease when human experimentation would be unfeasible or unethical" (http://www.absoluteastronomy.com).

Animal models are also widely used to assess the efficacy and safety of new medicines or treatment strategies for human disease, before initiating clinical studies. The rationale for using animal models is based on the common phylogeny of living organisms and the conservation of physiological, metabolic, developmental, genetic and other characteristics over the course of evolution.

Using the term introduced by Russell and Burch (1959), the model should have high "discrimination"; in other words it should be sufficiently like humans or the target animal species in the expected field of application (anatomy, physiology, pathology, etc.) that the results can be reliably extrapolated to the condition and species of interest (Svendsen and Hau 1994). This holds true not only for studies involving living animals but also for *ex vivo* or *in vitro* models. As an example, distribution, metabolism and pharmacokinetics (DMPK) studies that make use of hepatocytes or microsomes extracted from them, must take account of the species and animal from which they are obtained (inbred/outbred, age, sex, etc.) in a similar way.

Because an animal model is a complex biological system it is necessary, during its selection, to consider each of its components, to define the standard required, then to make sure that each parameter is adequately monitored over the entire duration of the study.

The components are generally divided into three categories:

- The *intrinsic animal definition*: species, sex, age, genetic characteristics, etc.
- The *health definition*, which is environment-related and specifies the microbiological status of the animal, including the associated micro-flora, parasites, etc.
- The *living environment* (both micro- and macro-environment, at both the cage and room level), which includes parameters such as housing conditions (temperature, humidity, ventilation, lighting, noise level, etc.), the caging system (group or single housing, type of cage or enclosure, floor and/or bedding quality, type of environmental enrichment), feeding and nutrition (composition, consistency, quality and quantity of feed, type of water supply), care, nursing and handling practices, etc. (see Chapter 3).

The relevance, quality, and consistency of these components strongly influence the quality and the relevance of the study. Because they interact with each other and are relevant to the nature of the research, a comprehensive understanding and review of requirements is necessary before the specification for the animal model is established.

7.2 STUDY OBJECTIVES AND EXPECTATIONS, INTRODUCTION TO MODEL CATEGORIES

When selecting an animal model, the very first step is to define the objective of the study (i.e., the question to be addressed).

The easiest situation is when it is possible to use the target animal species as the study model, for example when investigating issues related to animal physiology or health. However, it is necessary to draw a distinction between:

- "Preclinical studies" or non-clinical studies that are conducted on animals that have been "purpose-bred" (or at least comply with genetic and health specifications), in a defined and controlled research environment. Depending on the objectives of the study, healthy or pathological models are used (a disease can be deliberately introduced to healthy animals to study pathogenesis or therapeutic interventions).
- "Clinical" or "field" studies, where animal models are no longer used (these clinical studies are conducted under real conditions, on pet or production animals, which are not classified as animal models).

In the case of research into human health problems, the findings are intended to inform understanding of human biology or pathology or to assess new therapeutic approaches including medicines, surgical techniques, biomaterials, diagnostics and so on.

Because the characteristics of animal species are never identical to those of humans, and at best are similar for some characteristics, it is critical to decide exactly what information is required from the animal model. We could say that one animal model addresses only one question or issue.

When the investigation is directed at complex questions, it may be necessary to use several animal models and different experimental approaches. A good example is the variety of studies and animal models used to assess drug safety: short- to long-term general toxicology and reproductive toxicology involving both rodent and non-rodent species; these include absorption, distribution, metabolism, excretion (ADME) and DMPK investigations, safety pharmacology assessment such as assessing cardiovascular, respiratory, central nervous system, immune system and side effects. All of these studies and models can be regarded as pieces of a complex puzzle, each of them contributing a part to the global picture.

It is possible to classify animal models into several different categories; these and their characteristics will be examined in more detail in the next section. To illustrate the basis of classification, several key model categories are reviewed here, each of them addressing a defined set of questions.

Physiological models: they are generally healthy animals (i.e., largely free of spontaneous pathology). The species is selected for its biological characteristics (Sharp and LaRegina 1998; Terril-Robb and Clemons 1998; Field and Sibold 1999; Hedrich and Bullock 2004; Fox et al. 2007), which have to be thoroughly researched and understood. Some species may present particular similarities with humans, for example in aspects of anatomy, physiology or metabolism, but also they may be used more generally to imitate humans.

As an example, the biology of rabbit models used in reproduction toxicology or rat models used in carcinogenicity studies is very well standardised, documented and understood, which ensures their use is relevant for detecting and analysing toxic effects. Two examples involve the use of rabbits:

- The teratogenic effect of thalidomide in the rabbit is particularly marked, and the toxic effect of the drug appears to be exerted mainly during the 7–12th day of pregnancy (Fabro et al. 1964).

- Several characteristics of the rabbit make it an excellent model for the study of lipoprotein metabolism and atherosclerosis. New Zealand White (NZW) rabbits have low plasma total cholesterol concentrations, high cholesteryl-ester transfer protein activity, low hepatic lipase (HL) activity and lack an analogue of human apolipoprotein apoA-II; this combination provides a unique system for assessing the effects of human transgenes on plasma lipoproteins and susceptibility to atherosclerosis (Brousseau and Hoeg 1999; Zhang et al. 2008).

Physiological models include:

- Outbred animals (all species)
- Inbred strains (rodents especially mice and rats, chicken)
- Other genetically defined models (F1 hybrids, recombinant strains, etc.)

Animal models of human diseases: classically, these have been used to mimic human pathology. However, no model can mimic the full characteristics of a particular human disease, so model selection should be based on the primary objective of the study, and be based on selection of an isomorphic or a homologous model. When assessing drug candidates in animal models, it is also necessary that they have a high predictive value (i.e., they more-or-less accurately predict its efficacy and safety in humans).

Animal models of human diseases may be based on:

- Inbred strains
- Polygenic models generated by mono- or bi-directional selection
- Monogenic models carrying spontaneous, chemically induced and/or targeted mutations, and which may have undergone additional transgenesis
- Experimentally induced diseases
- Combined models

7.3 CATEGORIES OF ANIMAL MODEL AND THEIR CHARACTERISTICS

7.3.1 Physiological Models

7.3.1.1 Outbred Animals

Outbred animals originate from a randomly mating population; that is, a large breeding pool of allelic forms (Hartl and Clark 1997; Hartl 2000). A population is defined by its allelic forms and their frequencies, and an individual within it is a random allelic sample. Each individual differs from other members of the population but originates from the same allelic pool. When adequately managed, animal breeds can be defined as closed populations. Populations are generally characterised by common phenotypic traits, but rather than characterising each individual, only the population can be described—in terms of its allelic forms and frequencies. Careful genetic management system is necessary to control genetic drift over time and to minimise the increase in the inbreeding coefficient (Falconer 1972; Green 1981; Foisil 1989; Hartl and Clark 1997). The breeding and genetic management practises of the supplier of the animal model should be carefully audited and reviewed.

All species used for experimental purposes are available as "outbred" or at least "limited inbred" colonies. Most are produced specifically for use in research; examples include mice (NMRI, CD1), rats (Wistar, CD), guinea-pigs (Dunkin-Hartley), hamsters (Syrian golden), gerbils (Mongolian), rabbits (NZW, Himalayan), hens (White Leghorns), cats, dogs (Beagles), mini-pigs (Göttingen, Yucatan), macaques (Cynomolgus) and vervets (Sharp and LaRegina 1998; Terril-Robb and Clemons 1998; Field and Sibold 1999; Fox et al. 2007).

For investigations into animal health, agronomical practices and other specific applications, it is possible to use farm animals such as pigs (Landrace, Large-White), sheep, goats, cattle, horses, hens, turkeys, Peking and Muscovy ducks. Care must be taken to comply with relevant animal protection laws, regulations and good practise guidelines that relate not only to laboratory animal science but also to agricultural practise.

Specialised breeders should be able to provide key population and historical data relating to their colonies and the characteristics of the animals in them, including information about growth curves, nutrition, husbandry and biological and experimental data.

Because of their genotypic and phenotypic variation, outbred stocks are said by some to more closely resemble human populations, at least in terms of their genetic variability. They are generally vigorous because heterozygocity masks the effect of detrimental recessive mutations, so that there is generally good resistance to environmental stresses or other influences. In regulatory drug safety, some outbred models have been used many years and benefit from impressive cumulated biological data, which is of course an important asset (see Chapter 6, Section 6.2.5.2.1).

It is important to bear in mind that, although colonies of outbred models such as Swiss mice or Sprague-Dawley or Wistar rats may have had a common historical origin, they can differ greatly, both genetically and phenotypically. As an example, the so-called albino Swiss mouse (Lynch 1969) originated from a small group of mice (two males and seven females), which was provided by the Centre Anticancéreux Romand de Lausanne (Switzerland) to the Rockefeller Institute for Medical Research, New York in 1926. These mice were not supplied with a name and were consequently referred to as Swiss mice. This name persisted although Switzerland was only the direct provenance of the stock, which had been obtained from Lausanne and, before that, from a breeding colony in Institut d'Hygiène in Strasbourg (France) in 1924. Prior to that, it appears that the line originally came from a non-inbred colony kept at the Pasteur Institute, Paris that originated with about 200 mice obtained from an unidentified dealer. The Rockefeller Institute distributed the mice widely and numerous further stocks were created from them, varying widely in the size of colonies (and the potential for "founder" or "bottleneck" effects) and the quality of genetic management of the colony, but also the intended or unintended introduction of alleles from other albino stocks. The history and relationships between the various Swiss breeding colonies, which have spread worldwide and are found in many research and breeding establishments, have not been documented with the exception of a few colonies originating directly from the Rockefeller Institute.

In consequence, it has to be accepted that there is considerable genetic and phenotypic variation between different Swiss stocks. When addressing scientific issues related to so-called Swiss stocks, it is important to rely exclusively on data available from the breeding establishment from which they were sourced, including the breeding and genetic management system employed, historical, zootechnical, biological and experimental data, and environmental conditions under which they were maintained, including diet and caging standards.

7.3.1.2 Inbred Models

Inbred models are available for only a few species—principally mice and rats, but some are also available for chickens and guinea-pigs. The inbreeding coefficient of an animal strain is the percentage of alleles that are fixed as homozygous. An inbred strain is defined as one having an inbreeding coefficient over 98.6%, which corresponds to 20 generations of brother × sister mating. Lines are managed so as to maintain or increase this coefficient, as well as to avoid genetic contamination and genetic drift into sub-strains. Inbred animals (Festing 1979, 1993) are homozygous and isogenic within the same strain (i.e., possess the same genotype). They are histocompatible and accept tissue grafts from each other without any rejection. This characteristic is of value in developing models of cancer based on syngenic grafts; that is, various types of tumours originating from inbred strains that are grafted onto histocompatible individuals.

The quality of genetic management is critical when selecting a supplier. Genetic contamination makes an inbred colony virtually unusable and genetic monitoring is essential to assure the integrity of the colony. If contamination is detected, the only option is to cull the colony and to restart from scratch with pure pedigreed inbred breeders.

Various sub-strains arise from the process of genetic drift, which in turn results from the accumulation of mutations that become fixed at random in the colony. Major divergences can lead to the development of sub-strains which may display different phenotypes or other characteristics.

The strict use of genetic nomenclature is critical if advantage is to be taken of clearly identified inbred strains and sub-strains (Staats 1985; Davisson et al. 1996; Eppig 2006). BALB/cByJ is a widely used sub-strain of the original BALB/cJ strain. C57BL/6J and C57BL/6N strains are two sub-strains that diverged in the early 1950s and must be now considered as two different and well-established inbred strains.

Inbreeding depression is reduced vigour, evidenced by features such as slower growth rate, poor reproductive performance and reduced life span, and is often associated with exacerbated sensitivity to stress, pathogens and environmental factors. For such animals, inadequate control of environmental conditions can lead to a dramatic increase in population variability (The Jackson Laboratory, The Staff of 1997).

Inbred strains are also characterised by a tendency to develop spontaneous pathology such as tumours, degenerative lesions, blindness, seizures, auto-immunity and haematopoietic disorders. They can be used as rodent human disease models, which will be addressed later. The characteristics of different inbred lines are published in numerous databases, for example the Jackson Laboratory Mice data base.

Examples of differences between inbred mouse strains:

C57BL/6J	DBA/2J
Resistant to audiogenic seizures	Susceptible to audiogenic seizures
High incidence of microphthalmia	Tendency to develop hereditary glaucoma
High susceptibility to diet-induced atherosclerosis	Low susceptibility to diet-induced atherosclerosis
Preference for alcohol and morphine	Extreme intolerance and avoidance of alcohol

Strain-related spontaneous pathology or other characteristics can have a marked influence on the outcome of a study as illustrated in two examples:

- Retinal degeneration caused by *Pde6b* < *rd1* > (a recessive mutation of *phosphodiesterase 6B* gene) causes blindness before the age of weaning in several inbred mouse strains, for example FVB/N, C3H/J, CBA/J and SJL/J (The Jackson Laboratory 2009).
- A chromosomal deletion in C57BL/6JOlaHsd results in ablation of the α-synuclein locus; this causes dysfunction of the gene coding for presynaptic telencephalic protein, and is implicated in the aetiology of a range of neurodegenerative diseases including Parkinsons and Alzheimer diseases (Specht and Schoepfer 2001).

Inbred strains are extremely diverse with respect to their sensitivity or resistance to stimuli or treatment and there may be strong inter-strain variation, as for example with the morphine analgesic tolerance (Kest et al. 2002), or in resistance to irradiation. This type of variation can be observed in metabolic or toxicological investigations and it is important to use a minimum of two inbred strains in parallel to identify how much of any effect is a consequence of strain genetics and how much is a general effect of the compound.

Similarly, because inbred strains are commonly used to provide a defined genetic background for mutations or transgenes (congenic models), it is essential to evaluate the phenotype induced by the locus in at least two different inbred backgrounds.

7.3.1.3 F1 Hybrids

F1 stands for "Filial 1", the first filial generation of offspring resulting from a cross-mating of distinctly different parental types. The term is commonly used to refer to cross-mating between two species (i.e., mules, which are F1 hybrids between a horse and donkey).

F1 hybrid rodents (mainly mice and rats) and poultry are produced by crossing animals of two different inbred strains. In this case, each strain provides one of the two allelic forms of each locus, for example, B6D2F1/J is the F1 produced by mating a C56BL/6J female (abbreviated in B6) and a DBA/2J male (abbreviated in D2). The D2B6F1 is obtained by mating a D2 female with a B6 male. The F1 hybrids produced in this way are isogenic: they are heterozygous at all loci and are genetically and phenotypically uniform.

F1 hybrids can be generated for as long as the parental strains exist. The F2 offspring produced by mating F1 mice are not isogenic anymore, but are a random mixture of alleles from both parental strains.

Features of F1 hybrids (Festing 1979) that make them particularly useful for laboratory animal science include the following:

- Genetic and phenotypic uniformity
- In comparison with inbred animals, they possess "hybrid vigour" or heterosis, which results in increased resistance to disease, stress or environmental variations, better reproduction performance (including embryo production), greater longevity, decreased spontaneous pathology, better tolerance of deleterious induced or spontaneous mutations, higher resistance to radiation
- More uniform responses in behavioural studies and assays for drugs and biologicals
- Histocompatibility for tissue transplants (tumours, ovaries, etc.) donated by other similar F1 hybrids or by either parental strain

In addition to being excellent animal models (both genetically identical and resistant to environmental challenges), they allow greater genotypic diversity to be established from a limited number of strains and are easy to breed. The principal disadvantage is that three breeding colonies must be maintained to produce one F1 hybrid (the two parental inbred strains and the F1 production colony).

7.3.1.4 Examples of Other Inbred-Derived Models or Definitions

Recombinant inbred strains (Bailey 1971; Festing 1979; Darvasi and Soller 1995) are generated by crossing two inbred strains to produce an F1 generation, followed by at least 20 generations of brother × sister mating. This results in a series of new inbred strains. Each recombinant inbred strain is unique, and individuals belonging to it carry the same percentage of alleles from each progenitor inbred strain. They are mainly used in genetic studies, for example mapping genes or identifying quantitative trait loci (QTLs; Demant and Hart 1986). Inbred strains that differ from each other at only one locus or one limited genetic region (small chromosomal segment) are called, respectively, *coisogenic* or *congenic* (Festing 1979).

- *Coisogenic* strains differ at a single locus as a consequence of a point mutation occurring in that strain; they may arise spontaneously (Festing 1979) or by targeted mutation of embryonic stem (ES) cells that have been crossed to and maintained on the inbred strain from which the ES cells originated (see Chapter 8).
- *Congenic* strains are created by mating two inbred strains, and repeatedly backcrossing the progeny to an inbred strain for a minimum of 10 generations, while selecting progeny that retain a character of interest (e.g., the major histocompatibility complex or a defect causing a type of immunodeficiency) at one selected locus (Festing 1979; Flaherty 1981). Most

genetically modified rodent models are used as congenics, in order to benefit from their well defined and stable inbred background, after careful selection of the most appropriate one. These are often generated using marker-assisted "speed backcrossing".
- *Consomic* strains (Nadeau et al. 2000) are inbred strains in which one chromosome has been replaced by the homologous chromosome of another inbred strain. They are created by marker-assisted backcrosses and are useful for investigating the association of a phenotype with a particular chromosome that differs in the donor and host.

7.3.2 Human Disease Models

Classically, animal models of human diseases are used to mimic a human pathology, with the intention to investigate the disease process or to develop a therapeutic strategy (diabetes type I or II, cancer, central degenerative disease, etc.). A suitable animal model may be readily available or one may have to be created. This category of model cannot mimic the full characteristics of the human disease concerned, so models are created and used which have high discrimination: that is, they model one or a few selected aspects of the human pathology, selected in accordance with the primary objective of the study.

- *Isomorphic* models replicate symptoms and clinical signs of the disease rather than the aetiology and physio-pathological processes that lead to the pathology. For example insulin-dependant *diabetus mellitus* may be induced in rats by selective chemical destruction of pancreatic beta cells, but the resulting lesions and clinical signs resemble those observed in humans, and the model can be used to assess new therapies.
- *Homologous* models on the other hand possess a similar aetiology or develop a similar physio-pathological process, although the symptoms and lesions may differ significantly. For example: insulin-dependant *diabetus mellitus* induced in rats or mice by viral infection that leads to auto-immune destruction of pancreatic beta cells similar to that which occurs in the human disease, even though the clinical signs differ from those seen in human patients. The model can be used to study the aetiology and development of the disease and to develop early diagnostic procedures and prophylactic treatments.

When assessing drug candidates in animal models, it is important that they have a high predictive value (i.e., reliably predict their efficacy in humans).

7.3.2.1 Inbred Strains Used as Disease Models

Most inbred strains of animals develop spontaneous pathology as they grow older. Therefore, in addition to being used as physiological models or because of their genetic background, they can also be valuable as disease models in fields such as oncology, degenerative or metabolic diseases or immunology.

7.3.2.2 Polygenic Models Generated by Mono- or Bi-Directional Selection

"Polygenic inheritance" refers to the inheritance of a quantitative trait or a phenotypic characteristic that varies according to the interactions between two or more genes and their environment. Quantitative trait loci (QTLs) can be briefly described as genes that can be molecularly identified and sequenced and are involved in specifying a quantitative trait.

Polygenic traits do not follow patterns of Mendelian inheritance and the resulting phenotype classically varies along a bell curve. Many pathologies with genetic components are polygenic, including cancer and diabetes.

Some oligogenic models have been created by combining different spontaneous or induced mutations and/or transgenes (see the paragraph below describing monogenic models).

Other polygenic models are based on the use of selective breeding or artificial selection, the process of breeding animals for particular genetic traits, resulting in the development of animal breeds for pet or farm animals or colonies in the case of laboratory rodents or birds (Okamoto and Aoki 1963). Control groups are commonly created alongside this process. One well-known example is the spontaneous hypertensive rat (SHR) created by Okamoto at the Kyoto School of Medicine in the 1960s. This originated as an outbred Wistar Kyoto (WKY) male, which had markedly elevated blood pressure and was mated to a female with slightly elevated blood pressure. Subsequently, brother × sister mating associated with continued selection for spontaneous hypertension resulted in the SHR rat (Okamoto and Aoki 1963). A SHR-SP (stroke prone) rat is a further development of the SHR, and not only has a higher blood pressure but also a strong tendency to die from stroke.

Polygenic models are also used in investigations into drug or alcohol addiction, growth, food intake, development of obesity and analgesic (anti-nociceptive) response (Anderson and McClearn 1981; Belknap et al. 1983).

In theory, polygenic models can be created with any species, but for practical reasons associated with the duration of the breeding cycle, they are generally limited to rodents. Research activity, scientific advances in the field and the need to address a range of public health challenges may stimulate the development of additional polygenic models, for example to address congenital malformations, cancer, obesity and diabetes mellitus, hypertension and schizophrenia amongst other conditions.

7.3.2.3 Monogenic Models: Spontaneous or Targeted Mutations, Additional Transgenesis

It is important to keep in mind that the term monogenic can be misleading. Strictly speaking, monogenic means that the model exploits a single main gene or gene of interest. It does not mean that the phenotype of the model is purely monogenic; numerous background genes, known as modifier genes, may vary the phenotype from total compensation to causing over-expression, acting by a variety of mechanisms. In the case of homozygous individuals, the same mutation can be lethal in one strain and viable on another or on a F1 hybrid. APP-SWE (Tg2576) transgenic mice with a C57BL/6 genetic background die before they are 21 weeks old, whereas mice on a mixed B6/SJL background show good survival. In oncology models, the genetic background may influence tumour location, time of onset and metastatic activity.

One locus or gene can be subject to several mutations, which need to be correctly identified (Maltais et al. 1997). By assessing the phenotypes of a mutation on different inbred backgrounds, it is possible to identify modifier loci, with potential application to better understanding of human diseases (Sundberg 1992).

Most monogenic models are rodents, especially mice, but spontaneous mutations exist in other species. MDX mice (X-linked muscular dystrophy) carry the Dmd^{mdx} allele and, like human patients who suffer from Duchenne muscular dystrophy (DMD) do not express dystrophin. Females homozygous and males hemizygous for the Dmd^{mdx} mutation are therefore used as an animal model of the disease, even though the course of the resultant myopathology is much less severe than in the human disease. A similar disease of dogs is Hereditary Dystrophic Myopathy, seen in Labrador Retrievers (HMLR). Another example is the myostatin mutation of Belgian Blue cattle (myostatin is a protein that counteracts muscle growth): its truncated allele causes accelerated muscle growth.

The main ways in which genetic variation arises are:

- Spontaneous mutations
- Chemically induced mutation, by ENU (N-ethyl-N-nitrosourea)
- Genetic engineering
    i. Gene targeting in mice ("knock-out" or "knock-in" mice)
    ii. Gene over-expression

Genetically engineered mice created by random integration and over-expression of genes are technically simple and quick to generate and relatively inexpensive (see Chapter 8). However, gene integration is random and the method is suitable only for gene over-expression, there is potential for interference by equivalent endogenous genes and a risk of "insertion mutation". Furthermore, because the insertion of one transgene leads to the production of several founder individuals and their progeny lines, they all need to be genotyped, then phenotyped. In theory, random gene integration and over-expression can be carried out in most species, but in practise the technique is generally limited to mouse, rat, or rabbits models. Rabbits are used for bioprotein production and in the preparation of some type of models.

Genetically engineered mice created by gene targeting (a more complex technique) benefit from a single targeted integration locus and the presence of regulatory sequences. Most genetically engineered models are not created on an easily exploited genetic background and must be backcrossed onto an inbred genetic background (Silver 1995; Silva et al. 1997; Linder 2001). Marker-assisted backcrossing techniques have the advantage that they make possible much quicker generation of such congenic lines (Markel et al. 1997; Wakeland et al. 1997).

While most models exploit the insertion or deletion of one gene of interest, for example to mimic genetic diseases, other models are generated by combining two or three mutations and/or transgenes as models of oligogenic diseases, and are of value for analysis or modelling of more complex pathologies. For more information about the creation of genetically modified rodents, see Chapter 8. Non-genetically engineered models may be based on protein inactivation by antibody or on functional KO, which relies on inducing intolerance towards an endogenous protein.

Critical to the long-term standardisation and control of phenotypic expression in models is the adoption of a homogenous, well defined and stable genetic background (creation of congenic models) maintained in accordance with a relevant health standard and within a strictly controlled environment (Hardy 1998). These features are also key components of model definition, because they strongly influence expression of the gene(s) of interest, as illustrated in the following examples of spontaneous or targeted mutations in mice:

$Lepr^{db}$ mutation (leptin receptor), homozygous
    On C57BL/6: obesity, no or transient diabetes
    On C57BL/Ks: obesity, overt diabetes, increased sensitivity to atheromatosis

Fec^{m1Pas} (ferrochelatase deficiency), homozygous (Boulechfar et al. 1993)
    On BALB/c: anaemia, jaundice, hepato-splenomegaly, protoporphyrin crystals in the liver (model of human erythropoeitic protoporphyria)
    On SJL/J: mild phenotype
    On C57BL/6: almost normal phenotype

$Cftr^{tm1Hgu}$ (cystic fibrosis transmembrane receptor)
    A first modifier locus was identified (Rozmahel et al. 1996), modulating the severity of disease in a cystic fibrosis mouse model caused by deficiency of a transmembrane conductance regulator by a secondary genetic factor and affecting the intestinal phenotype.

If an inbred genetic background is used to maintain a mutation or transgene (e.g., congenic lines) the wild inbred strain makes an ideal control, at least for the first few generations. The adoption of a standardised and well-documented genetic background is recommended also by the Banbury Conference Guidelines (Silva et al. 1997):

- Published reports must include a detailed description of the mice studied that is sufficient to allow replication of the study.
- The genetic background chosen for studies should not be so complex as to preclude replication.
- Use of common or standardised genetic backgrounds facilitates comparison of experimental results.

7.3.2.4 Experimental Induction

This term refers to the creation of a model using non-genetic techniques.

Advantages

The approach allows:

- Creation of models when genetic approaches cannot be used
- Creation of combined/complex models
- Facilitating the expression of a genetically determined pathology for example NaCl in drinking water of SHR-SP rats or tumour induction
- Quicker and less expensive to create than genetically modified models

Drawbacks

- Generally time-consuming because the induction process has to be conducted on each individual
- A period of recovery is necessary post-induction (especially with surgical procedures)
- Models may be subject to interference or complications, especially when surgical preparation is involved, which may result in a variable percentage of non-usable animals
- There is usually considerable inter-individual variation
- Success depends on personnel skill, training and practise (especially of surgical procedures) and sometimes there is a requirement for specific equipment or techniques such as irradiation or surgery based on imaging.

Categories of experimental induction and some examples:

- Chemical induction: Examples of this are dietary deficiency, feeding high fat diets, use of streptozotocin to destroy pancreatic beta cells, corticoid or antimitotic-induced immunodeficiency, administration of cytotoxic, mutagenic or carcinogenic agents, administration of irritants to induce oedema, local inflammation and arthritis.
- Physical induction: γ radiation is used to prepare models for studies into immunology and oncology, hyper or hypothermia may be used for metabolic studies, hypercapnia or hypoxamia may be induced for CNS studies, or light cycle and intensity may be varied to modify the diurnal rhythm.
- Mechanical or surgical interventions: For example to create cardiac valve lesions, obstruction (by tying the renal artery to induce renal hypertension), hypophysectomy for studies in endocrinology, the use of cerebral ischaemia and stereotaxic targeted lesion in CNS studies, knee dislocation or section of ligaments to study arthrosis and induction of micro-embolic lesions.
- Biological induction: For example experimental infection of animals, cell or tumour grafting (including syngenic or xenogenic grafts, orthotopic or ectopic grafts), use of gnotoxenic animals to investigate gut metabolism or local immune reactivity, treatment with anti-lymphocyte serum or implantation of xenografts such as haematopoietic cells in immuno-deficient strains for studies into tissue immune processes.

7.3.2.5 Combined Models

Combined models are based on two or several different models, either created by similar or different approaches. They are used to mimic multifactorial or complex pathologies. Examples include:

- Chimeric systems:
    i. For example xenografts of tumours originating from humans (solid tumours or leukaemia cells) inserted into immuno-deficient outbred or inbred mice with T-cell

deficiency (*nude* mutation: *Hfh11^{nu}*) or combined T- and B-cell deficiency (*scid* mutation: *Prkdc^{scid}*).
   ii. A similar model for tumour implantation, but with greater immuno-suppression, can be generated by a using a facilitating dose of irradiation that inhibits also non-specific immunity such as NK cells (e.g., 350 Rad/3.5 Gy for a BALB/c inbred background or 500 Rad/5 Gy for an outbred NU-*nude* mouse).
   iii. Xenografts of human haematopoietic cells, for example human foetal hepatocytes, cord cells or blood lymphocytes can be prepared in immuno-deficient strains of mice, which can then be infected with the HIV virus.
- ApoE knock-out mice may be maintained on a high fat diet (Bourdillon et al. 2000).
- SHR-SP rats may be supplied with drinking water supplemented with a small amount of sodium chloride.

7.3.3 Models of Animal Disease or in Animal Health Research

Logically, animal models are also useful for investigations into animal diseases, either to elucidate aspects of animal pathology or to develop new treatments or vaccines to improve the health and well-being of both companion and production animals.

In the field of R&D into veterinary drug and vaccine development, three categories of studies can be distinguished:

- Clinical studies, which are outside the scope of this chapter. These do not use animal models but rather the target species, maintained in realistic field conditions such as shelters or breeding colonies of pet animals, household pets, commercial breeding colonies, farm or ranching animals, animals in the wild, collaborative studies with veterinary hospitals or veterinary school clinics.
- Preclinical studies in the target species. Even if the target species is used, the animals are sourced from specialised breeders or, in the case of farm animals, must conform to strict study specifications such as health and/or genetic status, antibody-free status, feeding regimen, age and/or body weight. Usually, the study is conducted in research animal facilities, and complies with the regulatory and quality standards applied to laboratory animals under strictly controlled health and environmental conditions.
- Preclinical studies in species other than the target species. In this case, the situation and rationale are very similar to the use of models for human health R&D or biomedical research.

7.4 REVIEW OF HEALTH DEFINITIONS AND CATEGORIES

7.4.1 Introduction

The interaction between environmental microbiological agents and animal health can be either beneficial or deleterious. The ability to breed germfree rats was reported in 1946 (Reyniers, Trexler, and Ervin 1946) using biocontainment housing systems and sterilisation systems. The use of aseptic hysterectomy or hysterotomy to generate specified pathogen-free (SPF) animals and the breeding and maintenance of gnotobiotic rat and mouse colonies became a common practice in the late 1950s.

Techniques developed in gnotobiology for re-derivation of contaminated lines or colonies to produce gnotoxenic individuals paved the way for the creation, breeding, and experimental use of SPF animals. SPF status guarantees the absence of a defined list of species-specific pathogenic agents ("bioexclusion list"). As advances were made in developing virology, serology, and molecular biology techniques, the definitions SPF was progressively extended to signify the absence of zoonotic, pathogenic, and research interfering agents (Hansen et al. 2000; Nicklas et al. 2002; Clifford and Watson 2008).

Use of immuno-compromised research models began when the *nude* mouse became available, and expanded greatly with the availability of models lacking additional elements of their immune system and prompted extension of SPF exclusion lists to include opportunistic organisms, corresponding to current "Specific and Opportunistic Pathogen Free" standards (Perrot 1976; Shomer et al. 1998).

In light of the growth of research applications in immuno-modulated mice, the next step could be an extension of the exclusion list to some murine retroviruses, transmissible spongiform encephalopathy agents or any other micro-organisms known or suspected to interfere with research generally or under certain experimental conditions.

7.4.2 Health Definitions and Standards

Classically, animal models are divided into four main categories:

- Holoxenic or conventional
- Gnotoxenic
- Agnotoxenic: heteroxenic/specific pathogen free
- Agnotoxenic: heteroxenic, extended specific pathogen free definitions

7.4.2.1 Holoxenic or Conventional Animals

This category includes rodents associated with an uncontrolled, undefined, or unknown micro-flora, such as that observed in wild rodents within their natural environment.

For reasons of good ethical (Russell and Burch 1959), scientific and quality assurance principles, the regulations governing the use of animals for scientific purposes (European Commission 1986) require close veterinary supervision of experimental animals to guarantee the absence of lesions or clinical signs of disease. However, in order to exclude healthy carriers, which may harbour pathogenic or zoonotic agents capable of interfering with research, a suitable health monitoring programme should be designed and implemented, this is especially important in the case of colonies of non-human primates that may carry pathogens of public health importance, such as Herpes B virus, tuberculosis, and pathogenic enterobacteria. Vaccination and de-worming programmes are common in colonies of conventional animals such as dogs, cats, pigs and non-human primates.

7.4.2.2 Gnotoxenic Animals

The term "gnotoxenic" or "gnotobiotic" describes animals living in the absence of detectable micro-organisms (axenic or germfree) or associated with a well-defined micro-flora (Trexler 1987). Species such as ruminants, guinea-pigs and rabbits cannot live in a germ-free environment whereas other species including mice, rats, chickens and pigs can adapt easily.

The establishment of a gnotoxenic colony requires the creation of an axenic (or germ-free) colony, maintained under strict bioexclusion conditions within a sterile environment. A selected bacterial micro-flora may be introduced to create a polyxenic or oligoxenic colony (depending on the number of associated microbial species). Maintaining a gnotoxenic colony requires the use of isolators, with very strict management of biosecurity to ensure exclusion of microbial contamination; this involves the provision of sterile water, food, and other supplies and high-efficiency air filtration (Heine 1978; Obernier and Baldwin 2006).

For practical and economical reasons, the vast majority of animals used in research are not gnotoxenic; that is, there is no exclusive and exhaustive definition of their resident micro-flora; such animals are better defined as agnotoxenic or specified pathogen free.

7.4.2.3 Agnotoxenic Animals: Heteroxenic/Specific Pathogen Free Health Standards

Heteroxenic animals are originally obtained by aseptic hysterectomy or embryo transfer, followed by controlled introduction of micro-flora (gnotoxenic step); they are then maintained under barrier-conditions where they are exposed to environmental and human-borne micro-organisms, hence the "heteroxenic" definition (Schaedler, Dubs, and Costello 1965; Perrot 1976).

For practical and economical reasons, the great majority of animals used in research are not gnotoxenic (i.e., with a positive and exhaustive definition of their resident micro-flora), but are defined as agnotoxenic and specific pathogen free (SPF). After transferring a defined micro-flora, they are maintained in barrier-rooms where they are progressively exposed to environmental and human-borne micro-organisms, hence the term "heteroxenic" (Dabard et al. 1977).

Specific pathogen free animals are defined according to an exclusion list (i.e., a definition), which details all organisms that have been excluded, including bacteria, viruses and other parasites (both unicellular and multicellular; Perrot 1976; van Zwieten et al. 1980; Ward et al. 1994; Hansen et al. 2000; Nicklas et al. 2002). For more specific applications, ecotropic retroviruses, and non-conventional transmissible agents may also be excluded.

Each breeder or user should establish, validate and assure a more or less restrictive definition of SPF, which matches the specific expectations and research activities for which the animals are to be used and that clearly references the methods used to control exclusion and to monitor the microbiological status of the animals (Nicklas et al. 2002). A SPF exclusion list should include at least the primary pathogens for the species concerned, zoonotic and other major interfering agents.

Originally the term *Specific* Pathogen Free was used to refer to the screening and exclusion of *species*-specific pathogens. Because of progressive extension of the definition to include an increasing number of agents that are not specific pathogens, but rather may interfere with the interpretation of scientific results (Baker 1988; McKisic et al. 1995; Jacoby et al. 1996; Ueno et al. 1997; Nicklas et al. 1999; Clifford and Watson 2008), it is common to refer to *Specified* Pathogen Free colonies, emphasising the need for the term to be linked to an accurately defined exclusion list.

It is important to appreciate the difference between the *screening* list and the *exclusion* list. The former can be purely informative and may include the monitoring of resident micro-flora as an indication of the efficiency of the bioexclusion precautions. The exclusion list is of more immediate scientific importance because if a positive result is found, this may invalidate experimental results and perhaps trigger a decision to replace the colony, with ethical, practical and economical consequences (Nicklas et al. 1999, 2002).

Some commercial breeders have registered "trade marks" such as VAF for Virus Antibody Free" that supplement the term SPF by indicating the criteria on which the claim of SPF status is based; in the example given, it is the absence of viruses as detected by serological methods. Similarly the term Murine Pathogen Free (MPF) refers to animals that are free of micro-organisms known to be pathogenic under certain study conditions. Whatever wording is used, the breeder or supplier must provide:

- the list of agents for which screening is carried out, the frequency of testing, and sampling methods;
- the exclusion list, with a predefined policy for each agent that may be identified. For example, policies may include immediate removal of the colony if the contaminating agent is pathogenic or may interfere substantially with scientific use; delayed or planned closure of the colony for agents that interfere only in a minor way, for example if there are very few experimental projects or research activities that would be affected; or no action if the micro-organisms are commensal rather than harmful.

7.4.2.4 Additional SPF Definitions

The introduction of immuno-compromised or debilitated rodents made it necessary to develop more efficient bioexclusion techniques, including filter-top cages, individually ventilated cages (IVCs), or isolators and to extend the SPF exclusion list to incorporate opportunistic agents (mainly environmental bacteria or agents of human origin), which have the potential to cause local inflammatory responses such as abscesses or conjunctivitis and systemic illnesses including septicaemia, wasting syndrome or respiratory infections or to shorten life span or breeding life.

In other cases precautions are needed to avoid interference with experimental findings in the case of severely debilitated or immuno-deficient strains (Perrot 1978) or to investigate factors that influence microbial interactions with genetically modified mice (Mahler et al. 2002).

This has led to the introduction of Specific and Opportunistic Pathogen Free or Restricted Flora definitions. Any opportunistic agent considered to be a synergistic or causative agent in lesions found in immuno-compromised strains should be considered for inclusion in a SOPF exclusion list (Saito et al. 1978).

7.4.2.5 Categories of Animal Micro-Flora

The micro-flora associated with higher organisms is normally very complex, highly variable and incredibly rich (Woodroffe and Shaw 1974). The commensal micro-flora is influenced both by the environment and species. When defining or controlling health standards (exclusion and screening lists) and monitoring programmes, it is essential to be aware of the various categories of microorganisms that may be present.

7.4.2.5.1 Commensal or Resident Micro-Flora

Normal, stable micro-flora associated with the host organism over the long-term. The commensal micro-flora includes:

- *Synergistic micro-organisms,* which have a demonstrated benefit to the host (e.g., facilitating digestion).
- *Barrier micro-flora* such as gastro-intestinal and skin bacteria that are ecologically adapted to the host organism and prevent colonisation by other micro-organisms including pathogens and opportunists.
- *Opportunistic micro-flora* (commensal micro-organisms, not included on the SPF exclusion list but able to create either pathology or interfere with experiments under special conditions (Perrot 1978) such as immunodeficiency, stress, poor environmental conditions, disease models or ageing. It is the responsibility of the breeder and the investigator to establish whether they are likely to be relevant and if so to add them to the exclusion or the screening list. Agents interfering with research applications or playing a synergistic role when associated to other viruses or other bacteria should be addressed in a similar way. Both concepts of *interfering* and *opportunistic* agent are relative concepts and must be considered in the context of the research environment and sometimes there is documented evidence in peer-reviewed literature (Shomer et al. 1998).

7.4.2.5.2 Transit Micro-Flora

Some environmental or human-borne micro-organisms can be detected only briefly before they disappear or reach a non-detectable level. The revised version of the FELASA recommendations duly acknowledged this situation in accepting that reporting of these agents can be discontinued if they have not been detected for 18 months or six successive quarterly screenings (Nicklas et al. 2002).

7.4.2.5.3 Pathogenic Agents

All viruses, bacteria, fungi, protozoa, metazoans, endo and ectoparasites known to be major or minor pathogens should be listed in the Specific Pathogen Free exclusion list. It is also recommended that organisms that have been well-documented as interfering with experimental results in the principal research fields should also be included.

7.4.2.5.4 *Interfering Agents*

Any parasite or micro-organism that has the potential to interfere with the sensitivity or characteristics of the animal model, resulting in experimental bias and study invalidation. Interfering agents are specific to particular research fields.

7.4.2.5.5 *Zoonotic Agents*

Their absence should always be guaranteed by strictly implemented bioexclusion and health monitoring programmes. Any suspicion of their presence should be reported and addressed immediately.

7.4.2.6 Healthy Carriers

Pathogenic, parasitic, or zoonotic agents can also be found in *healthy carrier animals* (i.e., individuals not displaying any clinical signs or lesions). The risk of rodents or rabbits carrying and transmitting pathogens to humans is very low, especially for barrier-bred animals. An example of a murine zoonotic agent is the Lymphocytic Choriomenigitis Virus (LCMV), an Arenavirus causing severe flu-like symptoms in man but asymptomatic in mice. A health monitoring programme based on serology is the best way to guarantee the absence of carriers.

7.4.2.7 Antibody-Free Animals

Scientific protocols may require the use of antibody-free animals; this is usually the case for research and development of veterinary vaccines. In this case, serological screening is not only a means of monitoring the absence of the micro-organisms in the breeding colony, but also a prerequisite for some categories of vaccine efficiency and safety studies.

7.4.3 RE-DERIVATION TECHNIQUES

When an animal colony is confirmed to be contaminated (i.e., not conforming to the exclusion criteria), there are several approaches to eliminating the contaminant or re-deriving the colony. The most appropriate measure depends on the number and the nature of the agent(s) to be eliminated and on the expected use of the animals.

The term "re-derivation" is generally used for the most efficient and sophisticated techniques involving aseptic hysterectomy or embryo transfer, especially for rodent colonies. These are based on a very simple (but not absolute) principle: the sterility of the reproductive tract during pregnancy until the onset of parturition and opening of cervix. Failure may occur if there is false vertical transmission, resulting from contamination during the process (Hill and Stalley 1991) or real vertical transmission when the pathogen infects the offspring *in utero*, before or after embryo implantation. Other techniques are also usable. Generally they are less reliable and efficient.

7.4.3.1 Aseptic Hysterectomy

Technical details of this method are available in several publications (Heine 1978). It is the classical method of eliminating infections in rodents and it can be summarised as follows. Future foster mothers originating from a gnotoxenic or SPF colony are maintained in an isolator or in IVCs; they are time-mated. About 24 hours later, contaminated donor females are also mated. Aseptic hysterectomy is carried out on donor females a few hours before their anticipated littering time and shortly after the foster females have littered normally. The pre-parturient donor females are killed humanely before the cervix opens prior to parturition and the entire gravid uterus is removed and transferred aseptically into a sterile isolator through a germicidal bath. Then, the uterus is incised and the pups are removed, warmed-up and substituted for the foster mother's litter. A health screen of the re-derived litters is conducted at least once before success of the process is assured.

Depending on the breeding objective and genetic considerations (e.g., outbred versus inbred) the total number of re-derived females may vary from a few to over 100 pairs.

7.4.3.2 Aseptic Hysterotomy/Caesarean Section

In principle this technique is similar to aseptic hysterectomy. Rather than removing the entire uterus and euthanising the donor female, the pups are removed from the uterus by caesarean section and transferred aseptically to a sterile housing isolator. The pups are usually hand-fed on artificial milk until weaning age; they are kept isolated until reliable health monitoring results are available. A similar technique is used to obtain colostrum-deprived and antibody-free ruminants.

7.4.3.3 Embryo Transfer

Technical details are available in specialised handbooks (Hogan et al. 1994; Suckow, Danneman, and Brayton 2001) and publications (Okamoto and Matsumoto 1999; see also Chapter 9). The embryo-transfer technique varies with different species. In mice, contaminated donor females are time mated; at the same time SPF recipient females are mated to vasectomised males to become pseudo-pregnant. Technical options include the use of super-ovulation and/or *in vitro* fertilisation. Donor females are anaesthetised in order to collect two-cell-embryos that are washed and morphologically screened before implantation into the oviduct of recipient females. After transfer into an isolator and weaning, the recipient females undergo health screening to assess the success of the re-derivation, specifically for any infective micro-organisms identified beforehand. One advantage of embryo transfer in comparison with hysterectomy or hysterotomy is that it avoids the possibility of post-implantation vertical transmission of infections (Reetz et al. 1988).

7.4.3.4 Aseptic Hysterectomy or Embryo Transfer Associated with other Procedures

To eliminate a vertically transmitted infectious agent such as LCMV in mice or to avoid intra-uterine contamination with organisms such as *Mycoplasma pulmonis* (Hill and Stalley 1991), it is possible to combine classical re-derivation techniques with isolation and serological screening and/or also washing the embryos in sterile media and/or treatment with an antimicrobial compound, depending on the nature of the contaminant. To be reliable, screening should be conducted several times before success is regarded as proven. Particular difficulty arises with elimination of vertically transmitted endogenous ecotropic retroviruses such as MuLV in mouse (Richoux et al. 1989; Hesse et al. 1999).

Occasionally other techniques can be used in very specific situations. They are not detailed in this chapter, but are described in specialised publications (Hedrich and Bullock 2004). They are based on isolation of the animal colony, combined with a process aiming to eliminate the contaminant by seroconversion/burning-out (Homberger 1997), immediate post-natal transfer to a clean foster mother (Truett, Walker, and Baker 2000) or medical treatment (Huerkamp 1993; Zenner 1998). Whatever the process used, a reliable health screening technique is required to provide assurance that the colony is no longer infected.

7.4.3.5 Genetic Issues When Re-Deriving a Colony

The number of pairs and their origins that are used to re-derive pups for breeding is very important in the case of outbred stocks. It is essential to avoid a genetic bottleneck (Rapp and Burrow 1979). With rodents, progeny from at least 80 pairs should be used to found the next generation, with each pair contributing equally to it. For larger species, a case by case genetic assessment has to be conducted when designing the re-derivation protocol.

7.4.4 BIOEXCLUSION, BIOCONTAINMENT AND HEALTH MONITORING

Maintenance of a defined health standard requires a properly implemented and comprehensive bioexclusion programme, aiming at eliminating contamination of the animal colonies. It requires careful facility design, attention to finishes, housing, caging and handling equipment and rigorous application of procedures associated with education and training of personnel. A health monitoring programme should be designed, kept under review and duly implemented (Nicklas et al. 2002; Reuter and Dysko 2003). Microbiologically undefined or contaminated animals should be kept under biocontainment conditions in order to prevent spread of undesirable or hazardous infective agents to cleaner animals or to humans (see also Chapter 2).

7.4.5 COLONY TERMINATION AND RECYCLING POLICY

There are occasions when it is necessary to terminate, or to re-derive a colony; examples are:

- when major contamination has occurred, for example by an agent on the exclusion list defined when the colony was established;
- less urgently, if opportunistic contamination(s) is found in association with pathological lesions;
- if genetic contamination of inbred animals is found to have occurred;
- in the case of rodents particularly, when different stocks and strains have to be re-organised and re-allocated to different barrier-areas.

7.5 ANIMAL MODEL SELECTION AND SOURCE

The first stage in designing an investigation is to establish whether the scientific objectives can be achieved using only replacement strategies such as *in vitro* observations on cells in culture, or by using invertebrates such as the insect *Drosophila* or the nematode *Caenorhabditis elegans,* which have proven especially useful in genetics and developmental biology. The next step critical to a successful scientific outcome is selection of an appropriate animal model. Major factors influencing the decision include identifying the species that is least likely to suffer, and requirements for housing and husbandry, consistent with obtaining satisfactory scientific results whilst bearing in mind that the model should be the most appropriate to one address the scientific issue or question.

High quality laboratory animal science requires the use of high quality and high health status animal models that have been purpose-bred, and this is made explicit in the current proposal for revision of the Directive 86/609/EC (and its expected revised version). The revised Appendix A of the European Convention ETS 123 (2006) and the recommendations prepared by FELASA (2007) also require that regular health monitoring programs be implemented in breeding establishments, to ensure that laboratory animals remain disease free. Many commercial breeders are now able to deliver high quality, purpose-bred animals.

The number of animals required depends on the number and size of experimental groups, and also the expected variability of the experimental data. Moreover it is important to ensure that all necessary data is collected to enable robust statistical analysis (see Chapter 6); the need for groups of animals of similar genotype, sex, weight and age may constrain some of the options for the model selection.

Occasionally, the choice of model is outside the control of the investigator. This is the case in regulatory studies such as vaccine quality control conducted in accordance with Pharmacopeia or other requirements. In the case of regulatory preclinical drug safety assessment, the selection of animal model is often purposefully very restricted in order to take advantage of accumulated historical data of control groups, comparisons over time and particularly well-controlled models (White and Lee 1998).

One common pitfall when selecting an animal model is to confidently or blindly adopt a reference publication, and to rely on the assumed expertise of the authors. Critical review of the literature is of extreme importance when developing a new protocol, and must be accompanied by very careful evaluation of options.

After addressing the questions "What is my research objective, what is the issue, and the question to be addressed?", and comprehensively reviewing available animal models, several options can be selected from the categories of available species and models.

When making a selection between several options, it is often helpful to consider a series of other questions, requirements and constraints:

- Ethical issues, implementation of the three Rs
- Availability of the model (number of animals, homogenous groups, delivery, etc.)
- Recurrent use of the model over a long period of time (long-term availability) versus requirements for a greater number of animals to be procured over to a short period
- Quality of health and genetic standard
- Quality assurance (good laboratory practises) or regulatory requirements
- Species size and behaviour (e.g., when stringent biocontainment or bioexclusion is a key experimental requirement)
- Intrinsic limitations of the model (life span, weight, anatomical limitation, organ size, blood volume, etc.)
- Qualitative or capacity limitations of the animal facility, ability to ensure adequate quality of care, social housing and environmental needs
- Availability of technical expertise on site
- Budgetary limitations

Leader and Padgett (1980) listed the following criteria of a good animal model system:

1. It should accurately reproduce the disease or lesion under study.
2. It should be available to multiple investigators.
3. It should be exportable.
4. If genetic, it should be in a polytocous species.
5. It should be large enough for multiple biopsies of samples.
6. It should fit into available animal facilities of most laboratories.
7. It should be easily handled by most investigators.
8. It should be available in multiple species.
9. It should survive long enough to be usable.

Ultimately, selection of the most appropriate model may be strongly influenced by the nature of the research (e.g., fundamental research or drug discovery), because the requirements or constraints are specific to each research need. Despite this, more than 85% of scientific procedures are currently carried out on rodents.

The sourcing of laboratory animals is regulated by EU Directive 86/609/EEC (and the expected revised version). This Directive recognises the need for high-quality laboratory animals and ensuring that, whenever possible, animals used have been specifically bred for the purpose. The code of practice most widely adopted in Europe (FELASA 2007) also requires that regular health monitoring programs be implemented in breeding establishments, to ensure that laboratory animals remain disease free. There are now many commercial breeders who can deliver high-quality, purpose-bred animals. In the case of other species, such as livestock, purpose-bred animals may not be available but similar expectations, tailored to the research application, should be defined and applied.

It is highly recommended to ensure that the animal supplier complies not only with legal regulations in force, but also with best practice in areas such as quality of management and organisation, adequacy of resources (quantitative and qualitative), implementation of animal welfare standards, ethical principles and quality of procedures and practises.

This can be achieved by visiting/auditing the facilities and/or by relying on accreditation schemes (AAALAC International). Because housing conditions and care in the breeding facility may influence characteristics of the animal model and the outcome of the study, it is important for the investigator to have an appreciation of these also. In some cases, it is necessary to develop a close partnership with the supplier, which might include discussing and setting particular specifications.

7.6 ENVIRONMENT DEFINITION AND CONTROL

7.6.1 ENVIRONMENTAL CONDITIONS

The material and methods section of funding applications, publications and reports require a comprehensive definition and justification of the animal model. In addition to genetic and health definitions, this should include all aspects of the model that have a bearing on the experimental outcomes, including housing and caging conditions (the macro- and micro-environment, respectively, at the cage or room level), key parameters such as ventilation, temperature, humidity, lighting cycle and intensity, sound intensity (Gamble 1982), the system of caging or enclosure, the level and method of bioexclusion or biocontainment, social interactions (group or single housing; Bruce 1970; Whitten 1973), feeding and nutrition (NCR 1995; GV-SOLAS 2002), water quality and watering system, floor and/or bedding type and quality, enrichment objects, frequency and method of cage changing and cleaning, daily care and nursing and handling routines.

For each environmental parameter, the qualitative and quantitative requirements must be adapted to the study requirements (Jennings et al. 1998) as well as the needs of the animals. Noise, light level (Greenman et al. 1982), odours including pheromones (Bruce 1970), handling, and all potential interfering or stress factors must be adequately controlled, monitored, and recorded. These are considered at greater length in Chapter 3. In order to ensure the quality of studies and experimental data, and to meet our ethical obligations under the Three Rs (Russell and Burch 1959), animal facilities should be designed to meet not only regulatory requirements but also species- and animal model-specific needs (European Commission 1986; ILAR/NRC 1996; The Jackson Laboratory, The Staff of 1997; Clough 1999; CCAC 2003; Council of Europe 2006; FELASA 2007; see also Chapter 2).

Specifications for environmental enrichment strategies should be carefully drawn up, because the quality of the environment can influence learning, memory and neuronal synaptic capacity and can mask or compensate for deficiencies inherent in the model (see Chapter 4). Enrichment is now common practice in regulatory safety studies, but here too must be carefully managed (Dean 1999). Hygiene procedures should be appropriate for the animal model and the type of study, and adhered to at all times (Wolfensohn and Lloyd 2003).

Environmental conditions in animal holding and breeding rooms can markedly influence the characteristics of animal models and the principal investigator should maintain close liaison with animal facility management as well as with care staff, breeders and suppliers, and should regularly visit the animals.

7.6.2 DIET AND NUTRITION

Diet is an important variable in animal studies (NCR 1995; GV-SOLAS 2002). Both caloric intake and nutrient composition can influence adult body weight, survival and the incidence of spontaneous pathology. In rodents, especially in rats on long-term studies, *ad libitum* feeding combined with single housing is generally associated with decreased activity, excessive caloric intake, excessive body weight, abnormally high incidence of spontaneous diseases (nephropathies and other degenerative lesions, tumours, etc.) and reduced life span. In species such as dogs, pigs and primates, the food intake of which is normally restricted, social interaction and dominance behaviour may impact negatively on access to food.

Different species have very different needs. For example, guinea-pigs and primates have lost the ability to synthesise vitamin C and are very sensitive to its deficiency. New world primates also are susceptible to vitamin D and E deficiency.

Purified ingredient diets are generally based on open formulae and make use of purified and chemically defined ingredients such as casein or even amino acids, starch or defined mono- or disaccharides, dextrin, oils, trace elements, vitamins. They are expensive but well defined and highly reproducible, and offer clear advantages in some types of studies where it is important to minimise variation over time and to avoid exposure to contaminants. However, such diets are produced mostly for rodents, and their use is almost entirely restricted to nutritional studies.

Chow diets have been used since the beginning of animal experimentation; formulations are available for breeding, maintenance and long-term studies. They are cereal-based (wheat, oats, corn) supplemented by plant (alfalfa/soybean meal) or animal (fishmeal, milk) proteins, fats/oils and vitamin/mineral premix to ensure nutritional adequacy. Plant ingredients are chemically complex and contain carbohydrates, proteins, lipids, vitamins and minerals as well as non-nutritive components (fibres). Chow diet must be formulated to meet laboratory animal specifications, avoiding the least-cost and productivity-based approaches used for agricultural animals. Formulae should be appropriate for the species, age and physiological condition (breeding, lactation, ageing) of the animals as well as any other study-related requirements. Vitamin levels should take account of the maximum shelf life and the stability of the formulation. Autoclavable diets should contain an excess of vitamins to compensate loss due to the heating process.

Chow diets are generally based on a closed formula, meaning that the exact amount or concentration of each ingredient is kept confidential by the manufacturer. Subject to agreeing confidentiality terms and restrictions on sharing the information, the formula and specifications should be available for assessment. An open formula lists the percentage or amount per ton of all ingredients in the complete feed, premixes or supplements. In practice, a target or a range (e.g., a percentage of protein) is set for each nutrient (carbohydrate, protein, fat, ash and fibre) and the specification is met by combining them in a constant way. This approach must be supplemented by careful selection of raw materials and suppliers and specification of ingredients.

Pellet hardness influences food intake—harder pellets may be suitable for some rodent stocks or strains but could be inappropriate for others that require softer or more friable pellets. The degree of hardness is an important factor in the ability of rodents to thrive.

Since the introduction of GLP, especially for rodent diets and other supplies used during toxicology studies, it has been necessary for producers to supply a certificate of analysis with each batch produced in order to guarantee the absence of pesticides, heavy metals or other contaminants resulting from agricultural or industrial practises or pollution (e.g., pesticides, heavy metals, phyto-oestrogens, or mycotoxins). The range of contaminants likely to be present depends on the type of ingredients used, their origin, transport and storage conditions.

7.6.3 Transport Conditions

Research animals may come from a wide variety of sources, including from within the same facility as that in which the experiments are being carried out, from another animal facility within the same organisation (for example, another animal unit of the same university), from colleagues or collaborators at another research institute working in the same field, or from a commercial supplier. Some of these sources may be located in another country that may or may not be within the EU.

Selection of a supplier must also take account of the need to transport animals to the experimental facility and audits of suppliers must take account of facilities for transport, the duration of journeys, compliance with regulations, arrangements for maintaining the health status, transport containers and quality of environmental conditions (Swallow et al. 2005; see also Chapter 3).

Council Directive 95/29/EC (European Commission 1995) amending Directive 91/628/EEC and European Regulation No 1/2005 (European Commission 2005) addresses the protection of animals during transport. It requires transporters to be registered and the use of vehicles appropriately equipped with ventilation and temperature control. All animals should be inspected before shipment and certified as fit to travel. Air transport is subject to specific regulations (IATA 2009) and additional controls apply in the case of endangered species (CITES 1973).

Council Directive 92/65/EC of 13 July 1992 (European Commission 1992) also lays down animal health considerations relating to trade in, and imports to the European Community of animals, semen, ova and embryos or animals originating from EU countries. European Regulation 1282/2002 amends Annexes to Directive 92/65/EC and stipulates requirements for health certificates.

Arrangements for the transport of animals should ensure that their well-being is not jeopardised, and that they arrive at their destination in good health. Particular attention should be paid to their health and welfare prior to shipment, in order to ensure that they are fit to travel and that there are no sub-clinical infections that could cause illness during or shortly after transport. Containers should be designed to provide comfort and minimise stress to the animals. They must be leak proof and escape proof, be so constructed that animals cannot damage themselves during transport and may be either single-use or capable of disinfection. Sufficient space should be provided for the animals to not feel any discomfort, taking into consideration the conditions that will prevail during transport. Animals to be transported together should be of the same age and sex and from the same source. Species must not be mixed in containers, and animals should not be transported in the same vehicle as their natural predators. It is important that there is adequate ventilation within the container; vents should be placed on opposite sides of the container and located so that they cannot be occluded. Temperature should be maintained as close as possible to the thermo-neutral zone for the animal. Litter materials and bedding should provide comfort and be able to absorb moisture as well as insulating the animal against temperature changes and vibration. Access to food and water should be adapted to each situation (species, age, duration of transport). Except for journeys of very short duration, water should be available throughout the period of transport, and provided in leak proof containers, or as mash, gel, fruit or vegetables, taking account of the risk of microbiological contamination. Dedicated vehicles and staff with suitable experience and training should be used for transportation. Vehicles should be insulated and air conditioned, with a ventilation system that is powered independently of the main engine. Alarms should warn of failures of the heating and ventilation system. It must be possible to clean and disinfect the vehicle thoroughly, and a back-up system should be available (and tested periodically) in case the vehicle breaks down.

Adequately designed and fitted rooms or airlocks should be available for all steps of animal reception. Transport should be arranged in advance so that animals can be unboxed and placed in fresh, ready-prepared cages as soon as they arrive; immediately on arrival all animals should be examined by a competent person and any concerns addressed by an appropriate person.

In the same way that European Directives 98/81/EC (European Commission 1998) and 2001/18/EC (European Commission 2001) govern the safe use of GMOs, respectively, for contained use and for deliberate release; United Nations recommendations on the transport of dangerous goods (2007) define conditions for packaging and transport of "dangerous goods". Class 6.2 ("Infectious Substances") deals with substances containing viable micro-organisms that are known or reasonably believed to cause disease in animals or humans. Genetically modified micro-organisms and organisms, biological products, diagnostic specimens and infected live animals shall be assigned to this class if they meet the conditions for Class 6.2. Class 9 ("Miscellaneous Dangerous Substances and Articles") covers substances and articles that, during carriage, present a danger not covered by the headings of other classes. Included in this class are genetically modified micro-organisms or organisms, that are not dangerous for animals or humans, but which could modify animals, plants, microbiological substances, and ecosystems in a way that does not occur naturally.

7.6.4 ACCLIMATISING VERSUS QUARANTINE PERIOD

As soon as possible after arrival at their destination, animals should be removed from their containers, inspected by competent staff and, if necessary, receive skilled special or veterinary care.

Quarantine may be required if the health status of the animals received is known or suspected to pose a risk of contaminating other animals, if their health status is unknown or if the accompanying health records are either incomplete or outdated. Quarantine requires complete isolation of the consignment for several weeks, during which the health of the animals is monitored and a screen is performed, before a decision is made about whether or not to introduce them to the experimental facilities (see also Chapter 2).

If the animals comply with the health quality requirements, they should then undergo an acclimatising or acclimation period. The purpose of this is to allow the animals to recover from the stress of transport and to become adjusted to the changed environment (housing and caging conditions, social group, watering system and food, staff, and so on). During this period, efforts should be made to minimise the impact of the new environment by initially retaining the same social groups, litter material or food, and only gradually introducing the materials that will be used during the study.

Acclimatisation usually requires less time than quarantine (from a few days for rodents to a few weeks for non-human primates). Its duration should be adjusted to take account of the amount of environment change and the species' sensitivity.

7.7 CONCLUSIONS

In the words of Michael Festing, "Animal experiments should be done using high quality animals which are free of clinical or subclinical disease, 'genetically defined' (i.e. isogenic, mutant or transgenic), and maintained on a nutritionally adequate diet and in a good environment" (Festing et al. 2002, p. 17). This chapter has examined the wide range of animal models available and discussed the importance of ensuring that they do not harbour clinical or sub-clinical disease. Guidance is also provided on the selection, sourcing and acquisition of models.

7.8 QUESTIONS UNRESOLVED

- There is an urgent need to develop non-sentient models so as to minimise the number of animals needed.
- The biomedical research community benefits from major advances in laboratory animal science and welfare made over the last 50 years, including our understanding of, and ability to control the genetic, health and welfare characteristics of animal models. There is a need to continuously build on this acquired knowledge by reviewing quality standards in light of scientific advances and to ensure the highest welfare standards during care and use.
- New animal models must be developed (or existing ones refined) to meet the needs of future research programmes, assuring their availability, adhering to high quality standards and establishing the most appropriate way to care for and use them.
- A clearer understanding is needed of the subtle and complex interactions between gut micro-flora and the development, maturation and reactivity of the animal's immune system. This would contribute to better controlled and more relevant models for studying immunity, allergy and nutrition, and so on.
- More information is required about QTLs in relation to polygenic or multigenic diseases, to allow models to be developed for therapeutic strategies to treat conditions such as cancer and metabolic and neurodegenerative diseases.
- Better models are needed for investigating infectious or parasitic diseases and to sustain investigations into emerging or re-emerging diseases in both human and veterinary health.

REFERENCES

Anderson, S. M., and G. E. McClearn. 1981. Ethanol consumption: Selective breeding in mice. *Behavior Genetics* 11:291–301.
Bailey, D. W. 1971. Recombinant-inbred strains. An aid to finding identity, linkage, and function of histocompatibility and other genes. *Transplantation* 11:325–27.
Baker, D. G. 1988. Natural pathogens of laboratory rats, mice, and rabbits and their effects on research. *Clinical Microbiology Reviews* 11:231–66.
Belknap, J. K., N. R. Haltli, D. M. Goebel, and M. Lame. 1983. Selective breeding for high and low levels of opiate-induced analgesia in mice. *Behavior Genetics* 13:383–96.
Boulechfar, S., J. Lamoril, X. Montagutelli, J. L. Guenet, J. C. Deybach, Y. Nordmann, H. Dailey, B. Grandchamp, and H. de Verneuil. 1993. Ferrochelatase structural mutant (Fechm1Pas) in the house mouse. *Genomics* 16:645–48.
Bourdillon, M. C., R. N. Poston, C. Covacho, E. Chignier, G. Bricca, and J. L. McGregor. 2000. ICAM-1 deficiency reduces atherosclerotic lesions in double-knockout mice (ApoE(-/-)/ICAM-1(-/-)) fed a fat or a chow diet. *Arteriosclerosis, Thrombosis, and Vascular Biology* 20:2630–35.
Brousseau, M. E., and J. M. Hoeg. 1999. Transgenic rabbits as models for atherosclerosis research. *Journal of Lipid Research* 40:365–75.
Bruce, H. M. 1970. Pheromones. *British Medical Bulletin* 26:10–13.
CCAC. 2003, March 3. *CCAC (Canadian Council on Animal Care) guidelines on laboratory animal facilities, characteristics, design and development.* Available from http://www.ccac.ca/en/CCAC_Main.htm, Accessed 25 June 2010.
CITES. 1973. *Convention on international trade in endangered species of wild fauna and flora.* Washington, DC.
Clifford, C. B., and J. Watson. 2008. Old enemies, still with us after all these years. *ILAR Journal* 49:291–302.
Clough, G. 1999. The animal house; design, equipment and environmental control. In *The UFAW handbook on the care and management of laboratory animals*, 97–135, 7th ed., Vol. 1. Oxford, UK: Blackwell Science.
Council of Europe. 2006. *Appendix of the European convention for the protection of vertebrate animals used for experimental and other scientific purposes (ETS123). Guidelines for accommodation and care of animals (article 5 of the convention) approved by the multilateral consultation.* Trans. Council of Europe. Vol. Cons 123 (2006) 3. Strasbourg.
Dabard, J., R. P. Dechambre, R. Ducluzeau, C. Gosse, J. C. Guillon, A. Perrot, P. Raibaud, M. Sabourdy, E. Sacquet, and C. Tancrède. 1977. Axenia-gnotoxenia-holoxenia-heteroxenia: Definitions and comments. *Sciences et Techniques de l'Animal de Laboratoire* 2:9–11.
Darvasi, A., and M. Soller. 1995. Advanced intercross lines, an experimental population for fine genetic mapping. *Genetics* 141:1199–207.
Davisson, M. T., chair. 1996. Committee on standardized genetic nomenclature for mice, rules for nomenclature of inbred strains. In *Genetic variants and strains of the laboratory mouse*, eds. M. F. Lyon, S. Rastan, and S. D. M. Brown, 1532–36, 3rd ed.. Oxford: Oxford University Press.
Dean, S. W. 1999. Environmental enrichment of laboratory animals used in regulatory toxicology studies. *Laboratory Animals* 33:309–27.
Demant, P., and A. A. Hart. 1986. Recombinant congenic strains—A new tool for analyzing genetic traits determined by more than one gene. *Immunogenetics* 24:416–22.
Eppig, J. T. 2006. Mouse strain and genetic nomenclature: An abbreviated guide. In *The mouse in biomedical research*, eds. J. Fox, S. Barthold, M. Davisson, C. Newcomer, F. Quimby, and A. Smith, 79–98, 2nd ed., Vol. 1. New York: Academic Press.
European Commission. 1986. *European directive 86/609/EC on protection animals used for experimental and other scientific purposes.*
European Commission. 1992. *European directive 92/65/EC laying down animal health requirements governing trade in and imports into the community of animals, semen, ova and embryos not subject to animal health requirements laid down in specific community rules referred to in annex A (I) to directive 90/425/EEC* Available from http://eur-lex.europa.eu/fr/index.htm, Accessed 25 June 2010.
European Commission. 1995. *European directive 95/29/EC amending directive 91/628/EEC concerning the protection of animals during transport.*
European Commission. 1998. *European directive 98/81/EC amending directive 90/219/EEC on the contained use of genetically modified micro-organisms.*

European Commission. 2001. *European directive 2001/18/EC on the deliberate release into the environment of genetically modified organisms and repealing council directive 90/220/EEC.*
European Commission. 2005. *European regulations no 1/2005—European directive 91/628/EC on the protection of animals during transport and amending directives 90/425/EEC and 91/496/EEC, amended by council directive 95/29/EC of 29 June 1995 repealed and replaced by regulation no 1/2005.*
Fabro, S., R. L. Schumacher, L. Smith, and R. T. Williams. 1964. Identification of thalidomide in rabbit blastocyst. *Nature* 201:1125–26.
Falconer, D. S. 1972. Genetic aspects of breeding methods. In *The UFAW handbook on the care and management of laboratory animals,* 5–25, 4th ed. Edinburgh: Churchill Livingstone.
FELASA. 2007. *Euroguide on the accommodation and care of animals used for experimental and other scientific purposes.* London: RSM Press.
Festing, M. F. W. 1979. *Inbred strains in biomedical research.* Oxford, New York: MacMillan Press. London: University Press.
Festing, M. F. W. 1993. Origins and characteristics of inbred strains of mice, 11th listing. *Mouse Genome* 91:393–550.
Festing, M. F., W., P. Overend, R. Gaines Das, M. Cortina Borja, and M. Berdoy. 2002. *The Design of Animal Experiments.* Laboratory Animals Ltd.: London.
Field, K., and A. L. Sibold. 1999. *The laboratory hamster and gerbil.* Boca Raton, FL: CRC Press.
Flaherty, L. 1981. Congenic strains. In *The mouse in biomedical research,* eds. H. L. Foster, J. D. Small, and J. G. Fox, 215–22, Vol. 1. New York: Academic Press.
Foisil, L. 1989. Identification genetique de la souche non consanguine de rats OFA Spraque-Dawley et strategie d'elevage pour le maintien de la variabilite genetique. Bourges, France: Faculte des sciences d'Orleans.
Fox, J. G., S. W. Barthold, M. T. Davisson, C. E. Newcomer, F. W. Quimby, and A. L. Smith, eds. 2007. *The mouse in biomedical research.* Vol. 1 History, Wild Mice and Genetics. New York: Academic Press.
Gamble, M. R. 1982. Sound and its significance for laboratory animals. *Biological Reviews of the Cambridge Philosophical Society* 57:395–421.
Green, E. L. 1981. Breeding systems. In *Mouse in biomedical research,* eds. H. L. Foster, J. D. Small, and J. G. Fox, 91–104,. Vol. 1. San Diego, CA: Academic Press.
Greenman, D. L., P. Bryant, R. L. Kodell, and W. Sheldon. 1982. Influence of cage shelf level on retinal atrophy in mice. *Laboratory Animal Science* 32:353–56.
GV-SOLAS. 2002. *German laboratory animal science association—Guidelines for the quality-assured production of laboratory animal diets.* [Cited October 28, 2009]. Available from http://www.gv-solas.de/auss/ern/leitlinien_futter_e.pdf
Hansen, A. K., S. Velschow, O. Clausen, H. Amtoft-Neubauer, K. Kristensen, and P. H. Jorgensen. 2000. New infections to be considered in health monitoring of laboratory rodents. *Scandinavian Journal of Laboratory Animal Science* 27:65–84.
Hardy, P. 1998. OECD proceedings. Novel systems for the study of human disease. From basic research to applications. In *Transgenic models in drug discovery: An industrial perspective,* 359–64. New York: John Wiley & Sons, Ltd.
Hartl, D. L. 2000. *A primer of population genetics,* 3rd ed. Sunderland, MA: Sinauer Associates.
Hartl, D. L., and A. C. Clark. 1997. *Principles of population genetics,.* 3rd ed. Sunderland, MA: Sinauer Associates.
Hedrich, H. J., and G. R. Bullock. 2004. The laboratory mouse. *The handbook of experimental animal series,* ed. H. Hedrich. New York: Elsevier Academic Press.
Heine, W. 1978. Operating procedures, equipment and housing facilities to maintain small laboratory animals under SPF conditions in large breeding colonies and in experiments. *Journal of the South African Veterinary Association* 49:171–74.
Hesse, I., A. Luz, B. Kohleisen, V. Erfle, and J. Schmidt. 1999. Prenatal transmission and pathogenicity of endogenous ecotropic murine leukemia virus akv. *Laboratory Animal Science* 49:488–95.
Hill, A. C., and G. P. Stalley. 1991. Mycoplasma pulmonis infection with regard to embryo freezing and hysterectomy derivation. *Laboratory Animal Science* 41:563–66.
Hogan, B., R. Beddington, F. Costantini, and E. Lacy, eds. 1994. *Manipulating the mouse embryo, A laboratory manual,* 2nd ed. New York: Cold Spring Harbor Laboratory Press.
Homberger, F. R. 1997. *Enterotropic mouse hepatitis virus.* Habilitationsschrift: University of Zurich.
Huerkamp, M. J. 1993. Ivermectin eradication of pinworms from rats kept in ventilated cages. *Laboratory Animal Science* 43:86–90.
IATA. 2009. *Live animal regulations.* Available from http://www.iata.org/ps/publications/Pages/live-animals.aspx, Accessed 25 June 2010.
ILAR/NRC. 1996. *Guide for the care and use of laboratory animals.* Washington, DC: National Academy Press.

The Jackson Laboratory (The Staff of). 1997. *General husbandry, "Handbook on genetically standardized JAX mice"*, eds. R. R. Fox and B. A. Witham, 107–29, 5th ed. Bar Harbour, MA: The Jackson Laboratory.

The Jackson Laboratory. 2009. *The Jackson laboratory mice database*. [Cited October 27, 2009]. Available from http://www.jax.org, Accessed 25 June 2010.

Jacoby, R. O., L. J. Ball-Goodrich, D. G. Besselsen, M. D. McKisic, L. K. Riley, and A. L. Smith. 1996. Rodent parvovirus infections. *Laboratory Animal Science* 46:370–80.

Jennings, M., G. R. Batchelor, P. F. Brain, A. Dick, H. Elliott, R. J. Francis, R. C. Hubrecht, et al. 1998. Refining rodent husbandry: The mouse. Report of the rodent refinement working party. *Laboratory Animals* 32:233–59.

Kest, B., E. Hopkins, C. A. Palmese, M. Adler, and J. S. Mogil. 2002. Genetic variation in morphine analgesic tolerance: A survey of 11 inbred mouse strains. *Pharmacology, Biochemistry, and Behavior* 73:821–28.

Leader, R. W., and G. A. Padgett. 1980. The genesis and validation of animal models. *The American Journal of Pathology* 101:S11–S16.

Linder, C. C. 2001. The influence of genetic background on spontaneous and genetically engineered mouse models of complex diseases. *Laboratory Animals* (NY) 30:34–39.

Lynch, C. J. 1969. The so-called Swiss mouse. *Laboratory Animal Care* 19:214–20.

Mahler, M., C. Janke, S. Wagner, and H. J. Hedrich. 2002. Differential susceptibility of inbred mouse strains to *Helicobacter pylori* infection. *Scandinavian Journal of Gastroenterology* 37:267–78.

Maltais, L. J., J. A. Blake, J. T. Eppig, and M. T. Davisson. 1997. Rules and guidelines for mouse gene nomenclature: A condensed version. International committee on standardized genetic nomenclature for mice. *Genomics* 45:471–76.

Markel, P., P. Shu, C. Ebeling, G. A. Carlson, D. L. Nagle, J. S. Smutko, and K. J. Moore. 1997. Theoretical and empirical issues for marker-assisted breeding of congenic mouse strains. *Nature Genetics* 17:280–84.

McKisic, M. D., F. X. Paturzo, D. J. Gaertner, R. O. Jacoby, and A. L. Smith. 1995. A nonlethal rat parvovirus infection suppresses rat T lymphocyte effector functions. *Journal of Immunology* (Baltimore, MD: 1950) 155:3979–86.

Nadeau, J. H., J. B. Singer, A. Matin, and E. S. Lander. 2000. Analysing complex genetic traits with chromosome substitution strains. *Nature Genetics* 24:221–25.

NCR. 1995. *Nutrient requirements of laboratory animals*. Washington, DC: National Academy Press.

Nicklas, W., F. R. Homberger, B. Illgen-Wilcke, K. Jacobi, V. Kraft, I. Kunstyr, M. Mahler, H. Meyer, G. Pohlmeyer-Esch, and GVSOLAS Working Grp Hyg. 1999. Implications of infectious agents on results of animal experiments—Report of the working group on hygiene of the gesellschaft fur versuchstierkunde—Society for laboratory animal science (GV-SOLAS). *Laboratory Animals* 33:39–87.

Nicklas, W., P. Baneux, R. Boot, T. Decelle, A. A. Deeny, M. Fumanelli, and B. Illgen-Wilcke. 2002. Recommendations for the health monitoring of rodent and rabbit colonies in breeding and experimental units. *Laboratory Animals* 36:20–42.

Obernier, J. A., and R. L. Baldwin. 2006. Establishing an appropriate period of acclimatization following transportation of laboratory animals. *ILAR Journal* 47:364–69.

Okamoto, K., and K. Aoki. 1963. Development of a strain of spontaneously hypertensive rats. *Japanese Circulation Journal* 27:282–93.

Okamoto, M., and T. Matsumoto. 1999. Production of germfree mice by embryo transfer. *Experimental Animals* 48:59–62.

Perrot, A. 1976. Evolution of the digestive microflora in a unit of specified-pathogen-free mice: Efficiency of the barrier. *Laboratory Animals* 10:143–56.

Perrot, A. 1978. Evolution of digestive microflora of SPF rats during a 5 month experiment and a 7 month experiment in a barrier unit. *Sciences Et Techniques De L'Animal De Laboratoire* 3:49–56.

Reetz, I. C., M. Wullenweber-Schmidt, V. Kraft, and H. J. Hedrich. 1988. Rederivation of inbred strains of mice by means of embryo transfer. *Laboratory Animal Science* 38:696–701.

Reuter, J. D., and R. C. Dysko. 2003. Quality assurance/surveillance monitoring programs for rodent colonies. In *Laboratory animal medicine and management*, eds. J. D. Reuter and M. A. Suchow. Available from http://www.ivis.org/advances/Reuter/reuter3/chapter.asp?LA=1

Reyniers, J. A., P. C. Trexler, and R. F. Ervin. 1946. *Rearing germfree albino rats*. Lobund Reports. Notre Dame, IN: University of Notre Dame.

Richoux, V., J. J. Panthier, A. M. Salmon, and H. Condamine. 1989. Acquisition of endogenous ecotropic MuLV can occur before the late one-cell stage in the genital tract of SWR/J-RF/J hybrid females. *The Journal of Experimental Zoology* 252:96–100.

Rozmahel, R., M. Wilschanski, A. Matin, S. Plyte, M. Oliver, W. Auerbach, A. Moore, et al. 1996. Modulation of disease severity in cystic fibrosis transmembrane conductance regulator deficient mice by a secondary genetic factor. *Nature Genetics* 12:280–87.

Russell, W. M. S., and R. L. Burch. 1959. *The principles of humane experimental technique.* Potters Bar, England: Special edition, Universities Federation for Animal Welfare.

Saito, M., M. Nakagawa, K. Kinoshita, and K. Imaizumi. 1978. Etiological studies on natural outbreaks of pneumonia in mice. *The Japanese Journal of Veterinary Science* 40:283–90.

Schaedler, R. W., R. Dubs, and R. Costello. 1965. Association of germfree mice with bacteria isolated from normal mice. *The Journal of Experimental Medicine* 122:77–82.

Sharp, P., and M. LaRegina. 1998. *The laboratory rat.* Boca Raton, FL: CRC Press.

Shomer, N. H., C. A. Dangler, R. P. Marini, and J. G. Fox. 1998. *Helicobacter bilis/Helicobacter rodentium* co-infection associated with diarrhea in a colony of SCID mice. *Laboratory Animal Science* 48:455–59.

Silva, A. J., E. M. Simpson, J. S. Takahashi, H. P. Lipp, S. Nakanishi, J. M. Wehner, K. P. Giese, et al. 1997. Mutant mice and neuroscience: Recommendations concerning genetic background. *Neuron* 19:755–59.

Silver, L. M. 1995. *Mouse genetics: Concepts and applications.* Oxford: Oxford University Press.

Specht, C. G., and R. Schoepfer. 2001. Deletion of the alpha-synuclein locus in a subpopulation of C57BL/6J inbred mice. *BMC Neuroscience* 2:11.

Staats, J. 1985. Standardized nomenclature for inbred strains of mice: Eighth listing. *Cancer Research* 45:945–77.

Suckow, M. A., P. Danneman, and C. Brayton. 2001. *The laboratory mouse.* Boca Raton, FL: CRC Press.

Sundberg, J. P. 1992. Conceptual evaluation of animal models as tools for the study of diseases in other species. *Laboratory Animals* 21:48–51.

Svendsen, P., and J. Hau, eds. 1994. *Handbook of laboratory animal science: Volume 1. Selection and handling of animals in biomedical research.* Boca Raton, FL: CRC Press.

Swallow, J., D. Anderson, A. C. Buckwell, T. Harris, P. Hawkins, J. Kirkwood, M. Lomas, et al. 2005. Guidance on the transport of laboratory animals. *Laboratory Animals* 39:1–39.

Terril-Robb, L., and D. Clemons. 1998. *The laboratory guinea pig.* Boca Raton, FL: CRC Press.

Trexler, P. C. 1987. *The UFAW Handbook on the care and management of laboratory animals,* ed. T. Poole, 85–98, 6th ed. Harlow, UK: Longman Scientific & Technical.

Truett, G. E., J. A. Walker, and D. G. Baker. 2000. Eradication of infection with *Helicobacter spp.* by use of neonatal transfer. *Comparative Medicine* 50:444–51.

Ueno, Y., F. Sugiyama, Y. Sugiyama, K. Ohsawa, H. Sato, and K. Yagami. 1997. Epidemiological characterization of newly recognized rat parvovirus, "rat orphan parvovirus". *Journal of Veterinary Medical Science* 59:265–69.

van Zwieten, M. J., H. A. Solleveld, J. R. Lindsey, F. G. de Groot, C. Zurcher, and C. F. Hollander. 1980. Respiratory disease in rats associated with a filamentous bacterium: A preliminary report. *Laboratory Animal Science* 30:215–21.

Wakeland, E., L. Morel, K. Achey, M. Yui, and J. Longmate. 1997. Speed congenics: A classic technique in the fast lane (relatively speaking). *Immunology Today* 18:472–77.

Ward, J. M., J. G. Fox, M. R. Anver, D. C. Haines, C. V. George, M. J. Collins, Jr., P. L. Gorelick, K. Nagashima, M. A. Gonda, and R. V. Gilden. 1994. Chronic active hepatitis and associated liver tumors in mice caused by a persistent bacterial infection with a novel helicobacter species. *Journal of the National Cancer Institute* 86:1222–27.

Webster's. 2009. *Webster's online dictionary.* [Cited October 28 2009]. Available from http://www.websters-online-dictionary.org

White, W. J., and C. S. Lee. 1998. The development and maintenance of the crl:CD(SD)IGSBR rat breeding system. In *Biological reference data on CD(SD) IGS rats, CD(SD)IGS study group,* eds. M. Toshiaki and H. Inoue. Yokohama, Japan.

Wolfensohn, S., and M. Lloyd. 2003. *Handbook of laboratory animal management and welfare,* 3rd ed. Oxford, UK: Blackwell Publishing.

Woodroffe, R. C. S., and D. A. Shaw. 1974. *The normal flora of man.* London/New York: Academic Press.

Zenner, L. 1998. Effective eradication of pinworms (*Syphacia muris, Syphacia obvelata* and *Aspiculuris tetraptera*) from a rodent breeding colony by oral anthelmintic therapy. *Laboratory Animals* 32:337–42.

Zhang, X., D. L. Chinkes, A. Aarland, D. N. Herndon, and R. R. Wolfe. 2008. Nutrition and disease: Lipid metabolism in diet-induced obese rabbits is similar to that of obese humans. *The Journal of Nutrition* 138:515–18.

8 Creation of Genetically Modified Animals

Belen Pintado, Spain and Marian van Roon, The Netherlands

CONTENTS

Objectives.. 179
Key Factors 180
8.1 Getting Started: Search for Information... 180
8.2 Genetically Modified Models to Answer Biological Questions: Types of Modifications 181
    8.2.1 Overview.. 181
    8.2.2 Additive Transgenesis.. 182
        8.2.2.1 Pronuclear Microinjection 183
        8.2.2.2 Vector Mediated Additive Transgenesis 185
        8.2.2.3 Additive Transgenesis Mediated by ES Cells 187
        8.2.2.4 Nuclear Transfer (Cloning) 187
        8.2.2.5 Use of Transposons 187
    8.2.3 Gene Targeting 188
        8.2.3.1 Gene Transfer in ES Cells... 189
        8.2.3.2 Nuclear Transfer.. 190
    8.2.4 Inducible Transgenesis... 190
    8.2.5 Conditional Mutagenesis 192
    8.2.6 Other Technologies Related to GMOs 194
        8.2.6.1 ENU Mutagenesis 194
        8.2.6.2 Interfering RNA 194
        8.2.6.3 Gene Trapping... 195
8.3 Construct Design 195
    8.3.1 Design of Transgenic Constructs for Gene Transfer in Zygotes............ 195
    8.3.2 Design of Gene Targeting Constructs for Gene Transfer in ES Cells 196
8.4 Influence of Genetic Background 197
8.5 Transgenic Identification: Genotyping 199
8.6 Conclusions.. 200
8.7 Questions Unresolved 200
References... 201

OBJECTIVES

The ability to genetically modify animals and their appearance in the scientific toolbox in the early 1980s has added substantially to the repertoire of research models ever since. The exponential growth in the use of such models over the past 30 years has resulted in substantial changes to animal facilities, not only as a result of the preponderance of certain species, principally the mouse, but also the implementation of new technologies for their care, evaluation and use.

    The generation of a genetically modified animal is always a long-term project. It demands substantial logistic investment including technical expertise, specialised infrastructure and a significant number of animals because of the relative inefficiency of the process. For this reason it is of utmost importance to ensure that the experimental approach ensures maximum efficiency so as to optimise the results. This chapter points out of those key factors that always have to be considered and on which the success of the project depends. We will also review different methods of permanently changing the genome, their applications and limitations. Finally we will review certain closely related techniques such as genotyping, which are integral parts of the whole process. Our aim is to provide the reader with basic information that will help to select the best experimental approach and to design an efficient and effective experiment when creating such models.

KEY FACTORS

The correct approach to the creation of a genetically modified animal model requires consideration of several key questions before starting the process:

- Has this model, or a similar one, been generated already?
- What kind of genetic modification would best address the biological question being asked?
- What is the likely impact of the alteration on the well-being of the animal and can adverse effects be ameliorated?
- If the host animal's genetic background might modulate the function of the exogenous genetic material, which strain of host animal would minimise this effect?
- What steps can be taken to minimise the number of animals required at each stage?
- What is the most reliable and efficient method of assessing the genotype of the model being created?

8.1 GETTING STARTED: SEARCH FOR INFORMATION

Before creating a new genetically modified (GM) line, a search of lines must be made available in international repositories or in the scientific literature. There are three reasons for this. Firstly, using available models avoids unnecessary use of animals in research, which is fundamental to the principle of the Three Rs. Secondly, even in the best scenarios, such experiments are expensive and it takes several months before the founders are available, even when there are no complications. Moreover, expression of the modified genotype cannot be predicted with certainty, so that use of a transgenic line, which has been characterised already, always saves time and effort. The third reason is that when laboratory animals are involved, ethical evaluation of the procedure requires assessment of its impact on animal welfare, and authorisation would be subject to cost-benefit assessment and reassurance that such genetic manipulation is justified. Moreover, EU countries also require evaluation of the potential risks to human health and the environment of any experiment involving the use of genetically modified organisms (GMOs).

The best starting point of any search of this nature is the Mouse Genome Informatics (MGI) database, which is accessible free through the web at: http://www.informatics.jax.org. This database is the result of a multi-national collaboration that started during the 1990s. It provides, within a unified information about genetics, genomics, phenotype and biology of hundreds of mutant strains of the laboratory mouse, drawn from many different sources. Links are provided to related fields when complex questions are posted to the web place. In addition when a potential match is found, the database also indicates whether the mutation is at cell line level or whether a mouse line has already been generated, and if so, in which international repository it is held. The potential of this approach to support research requirements is unquestionable and its use before, during and after the generation of experiments involving GM mouse models is most important. Unfortunately, similar databases dealing with other species are not available.

At the time of writing, about 20% of the mouse's genes have been modified experimentally. Sequencing of the human and the murine genome, recently completed, will further enhance demand for mutants as primary tools for elucidating functional roles of newly discovered genes. In view of this, international consortia (EUCOMM, NORCOMM, and KOMP) propose to develop a genetically engineered model organism for each single mouse gene within the next few years and results can be found on their websites: http://www.eucomm.org; http://www.norcomm.org; http://www.komp.org; http://www.knockoutmouse.org; http://www.ensembl.org; and http://www.sanger.ac.uk/htgt/welcome

For maintaining databases such as these, the use of an effective and standardised nomenclature system is a *conditio sine qua non* to ensure that information on the growing number of animal models carrying spontaneous or induced mutations is meaningfully recorded and accessible. A rigorous, consistent identification system is essential for several reasons; firstly, for accurate communication of the precise mutant the investigators are working with and to facilitate searches for information about what to expect in terms of phenotype. Secondly, to enable efficient communication of experimental findings. Thirdly, to facilitate acquisition of a desired mutant line and retrieval from breeding colonies or repositories, as well as searches for important information about its current availability, geno- and phenotypic characteristics, special needs for housing, husbandry, handling, and experimental use of the mutants and so on.

To achieve these aims, rules for nomenclature of rodent strains have been established by international committees and published on the World Wide Web (http://research.jax.org/resources/index.html). Since 2003, guidelines for mice and rats for the nomenclature of genes, alleles and mutations have been unified and updated annually by the two committees responsible. The importance of consistently applying these rules of nomenclature has been emphasised in Chapter 9 (9.2.4.1). Not only does the correct use of nomenclature supply information about the origin of transgenic sequences, the involvement of reporter genes and the animal's genetic background but also it ensures that all this information is documented in a unequivocal and understandable language and remains closely associated with the animals concerned.

8.2 GENETICALLY MODIFIED MODELS TO ANSWER BIOLOGICAL QUESTIONS: TYPES OF MODIFICATIONS

8.2.1 OVERVIEW

Before the development of methods for controlled genetic modification of animals, studies were restricted to animal species whose biological characteristics sufficiently resembled those of the species of interest—frequently man. The amount and quality of information obtained was clearly limited by the extent to which the features of interest were replicated by the model.

Identification of spontaneous mutations in animals, principally laboratory animals, which led to conditions that strongly resembled human diseases resulted in development of a method called "Forward Genetics". In this approach, a phenotypic abnormality is identified, then studies are carried out that lead step by step to identification of the affected gene and eventually to its location in the genome.

Transgenesis can be defined as the permanent transfer of genetic material, obtained using recombinant DNA techniques, into the genome of an organism, which can transmit the new information to its progeny. The great advantage of transgenesis is that it allows studies to be made in the converse direction, from the gene to the phenotype; for this reason such studies involving transgenic animals are described by the term "Reverse Genetics". In this second approach, the DNA sequence is inserted into the genome of a host animal to produce a phenotype that has been predicted beforehand and can be studied afterwards. In addition to this, transgenesis also makes it possible to study the action of a protein from a different species—such as human—when it has been integrated into a whole living organism, which retains the characteristic interactions of all organs and tissues instead of the bi-dimensional insight that *in-vitro* models provide. When addressing complex physiological processes such as development, behaviour, immune responsiveness or processes that require interaction between several tissues or organs, the experimental approach needs to be more complex than an *in-vitro* model. This is true also for many pathological situations, such as cancer. This requirement explains the exponential increase since the mid-1980s in the use of models based on GMOs.

The term "genetically modified animal" originally meant an animal in which the genome has been changed by introduction of an exogenous sequence of DNA using recombinant DNA technologies. This definition excludes alterations to the genome that are either spontaneous or have been induced with mutagenic agents such as N-ethyl nitrosourea (ENU). However, interest in these latter is increasing, a fact reflected by the number of consortia dedicated to the identification and production of such mutants. Many within the scientific community now include such induced mutations in a broader definition of "Genetically Modified Animals", which takes no account of the way the genome has been altered, and it is this wider definition that will be used in this chapter.

Genetically modified animals are generated by accessing the germline via gametes, zygotes or embryonic stem (ES) cells. Each of these approaches requires specific gene transfer techniques and applications. Changes in methodology over time reflect the need to provide precise and definitive answers to a greater number of increasingly complex questions. The first GMOs were created by inserting a fragment of foreign DNA into the genome at random, a process known as additive transgenesis. Using this approach it quickly became clear that the site within the genome at which the insertion was located had a crucial effect on its expression. Subsequently, generation of the first knock out and knock in mice through the use of ES cells established an alternative way of creating transgenic animals, and provided a method of targeted mutagenesis, in which the genetic modification is restricted to a specific sequence of the genome through homologous recombination. Even though this approach resulted in more predictable expression, important limitations became apparent, leading to the development of new ways of regulating the transgene. Two of these are of the utmost importance: inducible transgenesis and conditional mutagenesis, both of which are major advances over their original technologies. Inducible transgenesis is an improvement on additive transgenesis because it allows functionality of the transgene to be controlled by use of a genetic switch that allows expression to be either started or stopped at will. Conditional mutagenesis, on the other hand, allows the silencing of a gene located in a specific tissue or organ or at a certain developmental stage, thereby making it possible to knock out crucial genes in specific tissues or stages of development that would otherwise result in a lethal phenotype.

Although most transgenic models have involved modifying the genome of mice, genetic modifications have been achieved in many different species including zebrafish, poultry, rat, rabbit, pig, sheep, goat and cattle. Some of the techniques developed can be applied universally, whereas others are applicable only to certain species. The degree of success varies considerably. We will attempt to summarise these gene transfer techniques, their advantages and disadvantages and caveats associated with their use. A summary can be found in Table 8.1. However, there are publications and handbooks that provide a much more detailed and deeper insight and these should be consulted by anyone who intends to undertake any such experiments (Clarke 2002; Pinkert 2002; Nagy et al. 2003; Glaser, Anastassiadis, and Stewart 2005; Pease and Lois 2006).

8.2.2 Additive Transgenesis

Additive transgenesis is usually designed to:

- Investigate the function of a particular gene by overexpressing it using a constitutive promoter.
- Investigate the regulatory elements of a certain gene, usually by placing these putative regulatory elements in front of a reporter gene, which allows the time and place of expression to be monitored by the reporter.

- Investigate the phenotype of a specific mutation by overexpressing that mutation in the form of a dominant negative mutant. In this case a mutated version of the gene of interest (GOI) is overexpressed by using a constitutive promoter.

All methods of additive transgenesis involve the incorporation of new genetic information into the genome in a random manner and are exerted directly on the embryo or its precursors, the gametes. The classical method of pronuclear microinjection has been complemented with various other approaches, the success rates of which vary; we will also summarise these.

8.2.2.1 Pronuclear Microinjection

This technique is summarised in Figure 8.1. It involves use of a micromanipulator to directly inject genetic material into the pronuclei of fertilised oocytes before syngamy has been completed. The mechanism of integration is still poorly understood. Pronuclear microinjection is used principally

TABLE 8.1
Summary of Gene Transfer Techniques, Advantages, Disadvantages

Technique	Advantages	Disadvantages
	Gene transfer in gametes	
Sperm mediated gene transfer (SMGT)	• Useful for all species forming sperm	• Not reproducible
Transgenesis by ICSI	• Efficient with large constructs	• Questionable genetic integrity • Technically difficult
	Gene transfer in zygotes	
Pronuclear injection	• Moderate efficiency • DNA introduced is not species specific	• Random integration into the genome • Low efficiency with large constructs • Disruption of endogenous sequences • Insertion site affects expression
Nuclear transfer (cloning)	• Applicable to many species	• Low efficiency • Abnormalities in offspring are common
Use of lentiviral vectors	• High efficiency • Avoids the need for microinjection equipment	• Multiple integrations, need for segregation • Safety concerns • Limited transgene size • Production of viral vectors with sufficient titre to allow efficient insertion
Use of transposons	• Single copy integration • Easily traceable	• Limited transgene size • Bias towards insertion in certain regions
	Gene transfer in ES cells	
Homologous recombination	• *In vivo* and *in vitro* studies possible • Very specific modification	• Only available for species with ES cells (mice and rat) • Need for isogenic DNA in the arms of homology
Site specific recombination	• Controlled copy number	• Controlled copy number
Conditional mutagenesis	• Conditional and inducible	• Complex colony management
RNA mediated interference	• Post-transcriptional, no need for homozygosity • Combination with lentiviral vectors	• There may be immune response side effects
Use of lentiviral vectors	• High efficiency • Low costs • Technically not difficult	• Multiple integrations, need for segregation • Limited transgene size

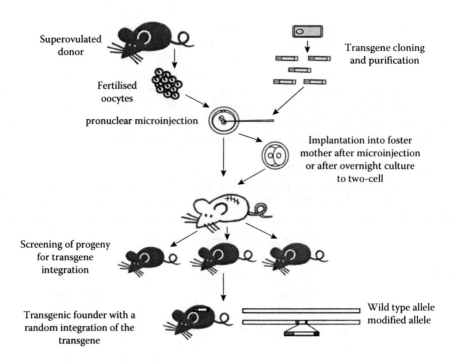

FIGURE 8.1 Pronuclear microinjection. Designed and manufactured transgenes are purified and injected intro pronuclei of embryos obtained from superovulated donors. Injected embryos are transferred into pseudopregnant recipients and resulting offspring are analysed to detect whether the transgene is present.

to obtain "gain-of-function" mutants. An additional gene is added and the modified animal can be studied by reverse genetic screening to determine its phenotype. The major disadvantage of the method is that the DNA injected integrates into the genome of the host animal at random. One or multiple copies (mostly head-to-tail concatemers) of the transgene are inserted into the genome and this can interfere with the expression of an endogenous gene thereby leading to a phenotype, which is unrelated to the transgene. The gene may also be inserted into an inactive region of the genome, so that it is silenced, failing to produce a phenotype. Conversely, a gene can integrate into a strongly regulating sequence of the genome and become associated with strong regulatory signals that are not related to the integrated DNA. To distinguish effects resulting from the integrated DNA from those related to the site of integration, at least two independent founder animals should be assessed. If they have the same phenotype, then this can be attributed to the integrated DNA, whereas if they differ in phenotype, the effects could be attributed to the site of integration. In most cases, pronuclear integration takes place at one site in the genome but it is still necessary to screen for multiple sites of integration. Pronuclear microinjection has been successfully carried out in the mouse, rat, rabbit, sheep and pig, but so far has proved very difficult in the cow and impossible in birds (Pease and Lois 2006).

The efficiency of this method in achieving transgenesis depends on a variety of factors, many of which were established several years ago (Brinster et al. 1985). The choice of injection buffer plays a crucial role. Success rates greatly improved when linearised DNA was used. The optimal size of the DNA construct is between 7 and 8 kb, although smaller fragments, and larger ones have been used, including yeast artificial chromosomes (YACs), bacterial artificial chromosomes (BACs) and P1-derived artificial chromosomes (PACs), although with a markedly lower integration efficiency. Genomic DNA should be favoured over cDNA in designing the constructs and the addition of a synthetic intron improves expression. Conversely, the presence of prokaryotic DNA sequences in the transgene lower the expression efficiency. Embryo quality, survival after microinjection and the host background strain also influence the overall efficiency. Success rates vary from about 10%–15% of animals born in mice, down to 1 or 1.5% in livestock species.

Pronuclear microinjection is performed in the rat in a very similar way to that done in the mouse (Charreau et al. 1996), although several reproductive aspects limit success. The first constraint is the difficulty in inducing superovulation. In a mouse, the technique consists of two intraperitoneal injections of gonadotrophins, eCG (equine gonadotrophin) and hCG (human gonadotrophin), 48 hours apart; however in the rat a more complicated regime is necessary because the response to this protocol is somewhat unpredictable. Recently published superovulation protocols appear to have overcome these problems (Filipiak and Saunders 2008). Regardless of the method used for superovulation, as in mice the results are strongly dependant on age or strain (Charreau et al. 1996).

The second reason for the lack of progress with transgenesis by microinjection in rats has been the lack of a suitable embryo culture medium, even though significant improvements have been made (Zhou et al. 2003). The advantages of the rat as a model of several specific diseases has been reviewed extensively (Iannaccone and Galat 2002) and this justifies the great effort being made to overcome these technical limitations and to develop efficient alternative systems that will be described later.

8.2.2.2 Vector Mediated Additive Transgenesis

In microinjection, a foreign DNA sequence is delivered with the aid of a microneedle directly into the pronucleus of the host cell. However, DNA transport can also be achieved by using vectors. In some cases those vectors interact with the DNA they carry, in other cases they don't, merely acting as physical support. Viral vectors are an example of the first situation, intracytoplasmic sperm injection (ICSI) mediated transgenesis an example of the second.

8.2.2.2.1 Viral Vectors

Several viruses have developed a survival strategy that is based on their incorporation into the genome of the host cell and the use of that cell's metabolic machinery to replicate viral proteins and other components necessary for them to complete their life-cycle. Making use of this ability pioneer experimenters (Jaenisch and Mintz, 1974) were able to achieve transformation of embryos at early developmental stages (2–8 cells). All of those first viral vectors, known as oncoretroviral vectors, belonged to the retrovirus family; they were characterised for being only able to infect replicating cells and also for being always related to acute neoplastic pathologies such as Rous sarcoma or Mouse mammary tumour. Even though this was a very effective way of transforming cells, expression of transgene was very often silenced and there were other limitations, so this strategy was abandoned for several years. However its effectiveness for creating transgenic chickens (Ronfort, Legras, and Verdier 1997) and the efficient insertion of foreign DNA into oocytes of the cow and monkey (Chan et al. 1998, 2001) has led to a renewal of interest in the method.

Within the retrovirus family, one gene gained unfortunate celebrity because one of its components has been identified as the agent that induces AIDS. This RNA virus can colonise the host genome of cells in either replicative or non-replicative stages. It causes diseases that have a very long time-course, a characteristic that gives them the name, lentiviruses. Lentiviral vectors proved to be highly efficient in transforming the mouse genome after injection into one cell embryos or when co-cultivated with denuded mouse oocytes (Lois et al. 2002). The lentiviral gene transfer technique works in all mammals, is highly efficient, inexpensive and is no more technically demanding than pronuclear injection; moreover, the technique of co-cultivation obviates the need for microinjection equipment and skills. However the most restrictive step in this methodology is the production of the viral vectors at a sufficiently high titre to allow efficient insertion (Wall 2002). The manipulation of denuded embryos is also demanding, and birth rates after transfer into the uterus are lower. The DNA insert should be ≤10 kb and insertion of cDNA is favoured over genomic DNA to circumvent the risk that the viral splicing machinery might modify the construct.

Another disadvantage of using lentiviral vectors is that copies of the virus tend to be integrated at multiple loci in the genome so that it is necessary to segregate them by backcrossing the initial founder until progeny bearing only one copy of the transgene is obtained. During this process selection for good expression is an advantage. In addition there is a remote possibility that recombination may take place between the viral vector and proviral particles already present in the genome. For this reason use of lentiviruses is often restricted to facilities with higher biological security, which considerably impedes their use.

8.2.2.2.2 The Male Gamete as Vector

In addition to using lentiviruses to transform spermatogoniae, it is possible to achieve germline gene transfer by injecting plasmid DNA-liposome complexes into the seminiferous tubules of male rats that are then mated with wild-type females. Other approaches include *in vivo* electroporation of DNA into the mouse testis and the transduction of spermatogenic germ cells with lentiviral vectors, which has proved to be a relatively efficient process. These germ cells are subsequently transferred to the testicles and give rise to genetically altered spermatozoa and transgenic offspring. Those methods rely on transforming spermatozoa precursor cells and in general have proved to be less effective ways of inducing germline integration of foreign DNA. On the contrary, other methods using the male gamete as vector offer advantages over pronuclear microinjection

8.2.2.2.2.1 Sperm Mediated DNA Transfer Sperm mediated gene transfer (SMGT) was the first method used to generate a transgenic animal and reported in the scientific literature (Brackett et al. 1971). Unfortunately at that time the authors lacked the tools necessary to demonstrate their observations. The strategy was subsequently reintroduced (Lavitrano et al. 1989) but led to a great deal of controversy because many laboratories were unable to repeat the procedure. Recently the technique has been successfully demonstrated again, with ejaculated pig sperm (Lavitrano et al. 2002, 2006); however the approach remains controversial. In order to improve the efficiency and integrity of incorporation of DNA, several techniques have been developed to assist the internalisation of the DNA, for instance electroporation, use of restriction enzyme mediated integration (REMI) or employing antibodies bound to the DNA of interest and which recognise a target protein on the sperm's surface (Houdebine 2003).

8.2.2.2.2.2 Transgenesis by ICSI The efficiency of all the above methods relies to a large extent on the quality and viability of the sperm, which can be influenced also by the way it is handled. ICSI has been used to circumvent problems of penetration/fertilisation by the sperm. ICSI can also be used to co-inject sperm and DNA into the oocyte, as a way of achieving transgenesis, (Perry et al. 1999) with similar efficiency to pronuclear microinjection; this method has proved successful not only in mice but also in rats (Hirabayashi et al. 2008). This technique has been also reported to produce YAC transgenics (Moreira et al. 2004) with an efficiency 10 times greater than that usually obtained by standard microinjection. Even more interestingly, all transgenic animals produced were able to transmit the transgene to their progeny, thereby avoiding one of the problems associated with microinjection, following which more than 60% of founders may be mosaic (Whitelaw et al. 1993).

One major concern with ICSI is the genetic integrity of the offspring. This method of assisted reproductive technology circumvents the forces of natural selection that are based on survival of the fittest and for that reason can result in survival and germline transmission of genetic material that in natural circumstances would not have led to progeny. Finally, the technique of ICSI is relatively easy in humans but in mice a piezo drill or laser needs to be used to penetrate the oocyte if it is to be efficient. In other species, such as cattle, ICSI is even more complicated because release of the sperm head into the oocyte is not enough to activate the female gamete (Malcuit et al. 2006).

8.2.2.3 Additive Transgenesis Mediated by ES Cells

The principal use of ES cells (see Section 8.2.3), is through homologous recombination to incorporate new genetic information into a specific locus on the genome, aimed at inactivating a gene sequence or replacing an existing one. However a substantial number of non-homologous recombination events can lead to incorporation of exogenous information in a random pattern and this can be used as an alternative approach to additive transgenesis. By co-electroporating a transgene and a selection gene based on resistance it is possible to isolate ES cell colonies that carry the resistance trait and a significant percentage of these would have also integrated the transgene, in a random pattern. When these cells have been identified, they can be injected into blastocysts or aggregated into morulae to produce chimeric animals.

One refinement to this strategy allows insertion of a single copy of the construct at a specific point on the genome (Bronson et al. 1996). By directing homologous recombination onto murine hypoxanthine phosphoribosyltransferase (*Hprt*) it has proved possible to achieve a highly efficient, directly selectable, recombination event. There is an additional advantage: different constructs can be compared without the complication of different copy numbers or the influences arising from the integration site. This system also has proven efficient for large constructs like BAC (Heaney and Bronson 2006). However this approach has the drawback that *Hprt* is X-linked and is subject to X inactivation.

8.2.2.4 Nuclear Transfer (Cloning)

Nuclear transfer (NT; Hochedlinger and Jaenisch 2002a, 2002b, 2003), nuclear transplantation or nuclear cloning involves replacement of the nuclear genetic material of an oocyte with that of an exogenous diploid cell. The resulting cell is then activated to initiate the first cell cycle. The genome of the resulting embryo is an exact copy of that of the exogenous cell. This technique was developed 50 years ago in frogs and can be performed in many mammals including goats, mice, pigs, cows, rabbits, cats, dogs, rats, sheep and in zebrafish. The efficiency of the method is low, possibly because it relies on the existence of the nucleus of a stem cell, which may occur in an otherwise differentiated population. There is an urgent need for further research into this question, in which reprogramming is a central issue. The use of cloning has been an invaluable tool for genetic manipulation of livestock because it allows substantial shortening of the experimental procedure. Inclusion of the transgene into a fibroblast cell line by means of electroporation with a reporter gene allows transformed cells to be easily selected, thereby ensuring that all nuclei used for cloning carry the desired information. The greatest advantage is that all animals born following transfer of reconstructed embryos are transgenic, a particular benefit with cattle and monotocous species with a long gestation period. In multiparous species such as the mouse—or even pig—this may be less important. The major limitations are that relatively few offspring are produced and there is a high occurrence of abnormalities in them that are unrelated to the donor cell and include shortened telomeres, developmental abnormalities, overgrowth, premature death and neonatal cardio-respiratory failure.

8.2.2.5 Use of Transposons

Transposable elements or transposons are endogenous genetic elements that are able to change their location within the genome. They insert in or near genes, but can be excised from one locus and reinserted into another under the action of specific enzymes called transposases. Transposons have been of great value in studying the genetics of lower organisms, and interest in their role in mammals has recently been revived. The most widely used ones are Sleeping Beauty (Dupuy et al. 2005), piggyBac (Ding et al. 2005) and Minos (Collier et al. 2005). All three of them are active in the mouse germline and can be used as tools for transgenesis or inserting mutagens in forward genetic studies. Sleeping Beauty has a somewhat lower transposition frequency and shows an integration bias towards genes and their upstream regulatory sequences, whereas piggyBac has a somewhat higher transposition frequency and so can span larger genomic distances, and shows a bias for integration into transcription units without chromosomal region preferences.

Initial integration of the transposon is stable and it is transmitted through the germline. Excision and reinsertion only takes place on addition and activation of the transposase. The advantages of transgenesis using transposons is the fact that only one copy is integrated, that individual integrations can be traced easily by use of inverse PCR, and that the transposon element allows expression of one or more transgenes. The co-expression of two transgenes within one transposon has the additional advantage that it is easy to use a reporter gene (RFP, GFP) for tracing purposes together with the functional transgene. Transposons can carry transgenes up to 9 kb without reduction in transposition frequency and up to 14 kb with some loss of transposition frequency. Transposons are introduced into the genome by pronuclear injection, and they can be co-injected with the transposase; it is possible to make the transposase tissue specific, inducible or conditional.

Transposons can also be used as insertional mutagens, but the mutagenesis frequency is much lower than that induced by ENU and there is a clustering of mutations in the proximity of the initial transposon insertion.

8.2.3 Gene Targeting

This technique allows delivery of the genetic information to a specific location in the genome and is performed by homologous recombination. Under ideal conditions, homologous recombination occurs in one of 1000 cells and this high inefficiency is the reason why it is not performed directly on embryos, but through an intermediate step (Figure 8.2).

This approach introduces the genetic modification into a cell line by a method that allows cells that have been modified through homologous recombination to be differentiated and cultured selectively. The approach most widely used has utilised ES cell lines; however, the availability of NT (cloning) of GM somatic cells (usually fibroblasts) offers an alternative for those species in which there are no available ES cell lines.

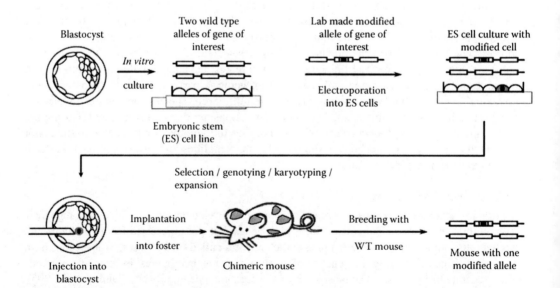

FIGURE 8.2 Gene transfer by microinjection of modified ES cells into blastocysts: ES cell lines are electroporated with the targeting vector. Homologous recombination occurs in some cells. After selection and karyotyping, several cells are injected into embryos to produce chimeric animals carrying the allele of interest in a proportion of their cells. The progeny is analysed to detect germline transmission.

Gene targeting makes it possible to interrupt the normal function of a specific gene (knock out), or to replace one functional gene by another one (knock in). Knock out mice have been as central to reverse genetic investigations as spontaneous mutation has been for forward genetics. In many cases knock out mice make it possible to study the phenotypic consequence of silencing a specific gene. On occasions, the result is completely unexpected because other genes may compensate for the loss of gene function (gene redundancy), or it may be that the resulting phenotype is lethal. Conditional mutagenesis offers a way of avoiding the latter limitation and will be addressed later.

Although most knock out mice have been obtained through homologous recombination, there are other ways of silencing a gene, such as ENU mutagenesis or insertional mutagenesis resulting from microinjection, when the transgene inserts at a sensible part of a gene and interrupts its normal function. Those gene disruptions are fortuitous, but a recent description of a knock out procedure based on the use of zinc finger nucleases indicates that it is possible to create a knock out in rats targeted at a specific endogenous gene, thereby circumventing the need for a stem cell line (Geurts et al. 2009).

Knock in mice, on the other hand, are also a very interesting research model. This technique is based on substitution of a gene by a mutant that is sometimes available in nature, but may be just a theoretical mutation of scientific interest. This technique also makes it possible to place unrelated genes under the control of the specific promoter of an endogenous gene, or if the complete gene sequence is substituted, to study the chromatin environment as a way of identifying the boundary domains that inhibit or enhance gene expression.

8.2.3.1 Gene Transfer in ES Cells

Most of the methods for generating GM animals that have been discussed so far apply to many species. However, the technique of gene transfer in ES cells is, at present, applicable only to the mouse. Human ES cells are available, but for obvious reasons cannot be used for gene transfer. The recent development of rat ES cell lines that contribute to the germline (Li et al. 2008) shows that, for the first time, it may be possible also to produce knock out or knock in rats through homologous recombination. The ES cells are characterised by their potential to be cultured and modified without losing their pluripotency, or their ability to form a fully differentiated animal again. To maintain this potency, cells must be cultured under conditions that keep them undifferentiated, for instance on feeder layers of mouse embryonic fibroblasts, in the presence of leukaemia inhibitory factor (LIF) and on media and sera that have been selected for their ability to keep the ES cells undifferentiated.

ES cells with high germline potential are being isolated from an increasing number of mouse backgrounds, 129, C57BL/6, BALB/c, FVB, CH3, DBA and it is likely that more lines will follow. ES cell lines derived from an F1 cross of mice (e.g., C57BL/6 × 129) have the ability to form the embryo completely by themselves (Nagy et al. 1990) and give rise to live animals when aggregated with a tetraploid embryo in the four-cell stage (a procedure that is relatively easy to perform and doesn't require special equipment) or injected in a tetraploid blastocyst (technically more demanding and needing a Piezo or laser-assisted injection microscope). The tetraploid host forms the extra-embryonic tissues while the diploid cells give rise to the embryo proper. It is not known when the tetraploid cells die during development, leaving the diploid cells alone in the embryo, although it is possible that some develop to the somite-setting stage before they die. The potential of the diploid ES cells to form a complete embryo and become a wholly ES cell-derived adult mouse is only seen in F1 ES cells and never in inbred cell lines. The tetraploid aggregation technique has greatly advanced ES cell technology and bypasses the need to generate chimeras. Set against this, however, is the disadvantage that the genetic background of animals derived in this way is no longer pure.

Recently it has proved possible to produce F0 generation mice from gene-targeted ES cells by laser-assisted injection of either inbred or hybrid ES cells into eight-cell stage embryos; this has resulted in 100% germline transmission (Poueymirou et al. 2007). However, the use of a laser is not essential, because similar results have been published already using a standard bevelled needle

(Tokunaga and Tsunoda 1992). The critical factor is the embryo stage at which the injection is performed—eight-cell embryos instead of compacted morulae. Other technical factors, such as the potential of the ES cells and the culture system used, determine the extent to which the ES cells contribute to the embryo (Ramirez et al. 2009).

8.2.3.2 Nuclear Transfer

Despite tremendous efforts to generate ES lines in other species none has been successful except for a recent approach in the rat (Li et al. 2008). The potential benefits of producing *knock outs* and *knock ins* in species other than mice have fuelled studies in which homologous recombination has been attempted in fibroblast cultures. The advantage of this approach is that it is possible to obtain progeny from the nuclei of differentiated cells by NT, so that if a homologous recombination is achieved in such cell cultures, those cell nuclei can be used to produce a GM animal carrying the targeted mutation. Such a method has been attempted with sheep; unfortunately the severe phenotype caused by the cloning procedure led to death of the first knock out lambs within hours of their birth (Denning et al. 2001). However, this failure was not attributed to inactivation of the gene but rather to the NT methodology, which led to major changes in the epigenetic expression of crucial genes. A similar approach has been used to produce cattle lacking the bovine PrP gene, and which are consequently immune to transmissible bovine spongiform encephalopathy (Richt et al. 2007). In spite of poor perinatal viability of the cloned animals the health and productive performance of those that survived to adulthood and their offspring did not differ significantly from conventionally bred cattle (Watanabe and Nagai 2008). In addition, some of the abnormal phenotypes observed in cloned animals are not transmitted to their offspring (Tamashiro et al. 2002).

8.2.4 INDUCIBLE TRANSGENESIS

Genetic modifications can be classified according to the nature of the integration; it is possible to distinguish between additive transgenesis, in which a new DNA fragment (a transgene) is added to the genome and gene targeting in which a fragment of the host genome is removed and replaced by the exogenous construct. However, if the reference point is expression, a distinction can be made between those constructs whose expression is controlled only by the promoter, and those constructs that are subjected to external control of expression. This latter strategy arose from the need to refine the pattern of expression and resulted in the development of inducible transgenesis. The first approach in this direction was based on the metallothionein promoter, first described in 1989 (Miller et al. 1989) as a way of controlling the expression of growth hormone. However this was soon replaced by a system developed by Bujard and Gossen in the early 1990s (Gossen and Bujard 1992; Gossen et al. 1995), which was based on the resistance of some bacteria to tetracycline. In natural conditions the sequence responsible, known as the tet operator is blocked by a protein that binds to the tet promoter thereby preventing its transcription. This tet repressor protein has a high affinity for tetracycline, to which it binds in preference; by leaving the tet operator free it allows expression of the resistance gene.

This system has been used as a way of controlling expression of the transgene by use of tetracycline or its derivate doxycycline. In the tet-off system (Figure 8.3) addition of doxycycline stops the expression of the GOI meanwhile in the tet-on system (Figure 8.4), addition of doxycycline activates expression.

Traditionally this strategy is achieved by crossing two transgenic lines, the first one carrying the GOI, under control of the tetracycline inducible promoter, the second line bearing the inductor (activator or repressor) under a promoter specific to the tissue, organ or developmental stage at which the gene is to be studied (Blau and Rossi 1999; Yamamoto, Hen, and Dauer 2001). The phenotype only becomes evident in the progeny resulting from crossing both lines - in double transgenic animals. The level of control is not always high. There are descriptions of low-level basal expression in tet-off

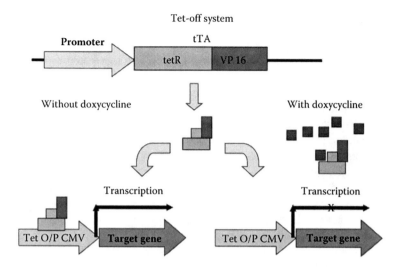

FIGURE 8.3 The tet-off system is based on the ability of tetracycline transactivator (tTA) to bind a specific DNA sequence, the tet operator (Tet O). tTA is the product of fusing the tetracycline repressor (TetR) from *E.coli* and a viral protein VP16. This resulting protein binds to the tet operator and activates the promoter coupled to this sequence so inducing gene expression. When tetracycline or its derivate doxycycline is present, tTA binds to this compound instead of Tet O, so the gene is not expressed (i.e., tetracycline turns the expression off). (From Pintado, B. and Gutiérrez-Adán, A., Modificación genética animal Mutagénesis tradicional y transgénesis. El mantenimiento de las mutaciones. Cienciay Tecnología del Animal de Laboratorio. Publ. 2008. With permission.)

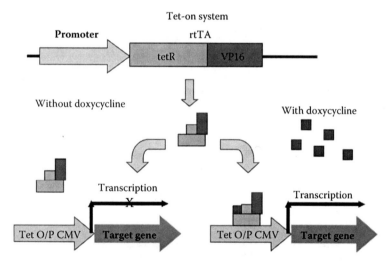

FIGURE 8.4 In tet-on inducible transgenesis, expression of a gene is induced by the presence of tetracycline or its derivate doxycycline. A modified tetracycline transactivator (rtTA) is only capable of binding the tet operator and thus inducing transcription when bound to tetracycline. (From Pintado, B. and Gutiérrez-Adán, A., Modificación genética animal Mutagénesis tradicional y transgénesis. El mantenimiento de las mutaciones. Cienciay Tecnología del Animal de Laboratorio. Publ. 2008. With permission.)

systems, lack of activation in tet-on systems, and influences of genetic background on the appearance of these effects (Zhu et al. 2001).

Finally, the level of expression of the inducible line varies depending on the inductor line with which it has been backcrossed and because expression only becomes evident in second generation animals carrying both transgenes; this means that experimental periods may be quite long.

8.2.5 Conditional Mutagenesis

Analysis of loss-of-function mutants derived from homologous recombination in ES cells reveals three kinds of phenotype. First, there is the expected phenotype based on the predicted function of the GOI. Second, there may be no phenotype at all if the function of the GOI is so important to the mouse that there is a backup system that takes over the function of the inactivated gene. In this case we speak of gene redundancy: another gene is compensating for the GOI. Tracking the stand-in is an interesting but difficult task, even if there is evidence about the gene family to which it belongs. However the assumption that there is no phenotype at all should be considered cautiously because it may be the case that the phenotype has not yet been identified (Barbaric, Miller, and Dear 2007). Third, a lethal phenotype may exist if the GOI plays an important (additional) role in early development and the task cannot be taken over by another gene. In this latter case it may be of interest to study the effect of inactivating the gene later in development or to confine the investigation to one particular organ. In other words there is a requirement to confine expression to a specific time and/or place. The answer is conditional mutagenesis.

Conditional mutagenesis can be achieved by using site-specific recombinases. These are enzymes that provoke crossing over between specific sequences in the genome (Figure 8.5) thereby deleting or inverting the genomic sequences in these segments. The discovery and use of site-specific recombinases have opened up ES cell technology to many new experimental approaches. The inclusion of cross-over loci in the targeting construct makes it possible to manipulate the GOI in a more subtle and conditional way than simply introducing a stop signal. Hundreds of lines have been created using this strategy and most of these have been compiled in a database called Cre-X-mice that lists not only floxed genes, but also lines expressing Cre-recombinase under a range of different promoters. This database is available on the internet.*

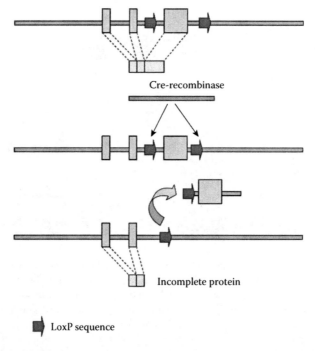

FIGURE 8.5 Cre-recombinase mediated deletion of a gene. LoxP sites are inserted so as to flank a DNA region crucial for gene expression (Floxed gene). In the absence of Cre-recombinase, transcription and translation occurs normally. When present, Cre-recombinase catalyzes recombination of the two loxP sites oriented in the same direction so that the DNA fragment between them is deleted.

---

* http://nagy.mshri.on.ca/cre/

The best known recombinases and cross-over loci are the Cre-loxP system (Gu, Zou, and Rajewsky 1993; van der Weyden, Adams, and Bradley 2002; Branda and Dymecki 2004; Schnutgen, von Stewart, and Anastassiadis 2006), isolated from bacteriophage λ and the FLP-FRT system (Rodriguez et al. 2000; Schaft et al. 2001), isolated from yeast but which is less efficient than the Cre-recombinase. When a gene is placed between two loxP sites ("floxing a gene") that have the same orientation, it is possible to remove or inactivate that specific gene where and when the Cre-enzyme is active. By crossing two transgenic lines, one bearing the Cre-enzyme and the other with floxed GOI it is possible to obtain double transgenic offspring in which the modification is conditioned (Figure 8.6). This conditional recombination can also be made inducible by adding an estrogen receptor to the Cre-enzyme in a so called Cre-MER (Metzger et al. 1995) construct after which Cre action only occurs after addition of hormones.

The orientation of the cross-over loci relative to each other determines whether the effect will be excision (same orientation) or inversion (opposite orientation). Intramolecular cross-over is favoured over intermolecular crossing over, although crossing over can take place over very long distances. The introduction of loxP sites close to centromeres of chromosomes and use of Cre-recombinase can even result in, or mimic chromosomal translocations. Introduction of a single cross-over locus into a transgene construct makes it possible to use recombinase to recombine all loci of cross-over into a concatamer, leaving just one behind and thereby introducing only a single copy of the transgene into the genome. A major problem with use of site-specific recombination is the existence of cryptic endogenous loci of cross-over, although the benefits of the technique outweigh the hazard that this presents.

The Cre-loxP construct is used widely to achieve conditional mutagenesis, whereas the FLP-FRT system is used were the efficiency of recombination is not critical (for instance removing marker genes in targeted ES cells). However, there is a need for additional recombinases for use in

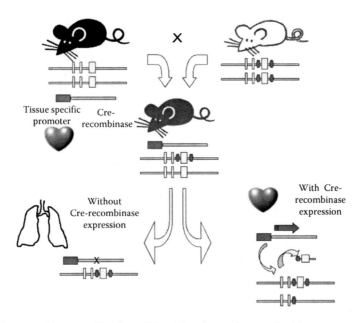

FIGURE 8.6 Cre-recombinase mediated conditional knock out of a gene. In this system two mouse lines are used, one of which carries a floxed gene (gene flanked by loxP regions in the same orientation), which is to be removed in a particular tissue. The second line provides expression of Cre-recombinase under the control of a promoter specific for the tissue or cell line where the knock out of the gene is desired. The animal resulting from crossing both lines experiences deletion of the gene in the specific tissue where Cre-recombinase is expressed, but not elsewhere.

recombinase mediated cassette exchange (RCME) in which one version of a targeted mutation is exchanged for another at the same locus. New players in the field that are attracting interest include Dre-rox (Sauer and McDermott 2004) and φC31 (Belteki et al. 2003), a recombinase derived from Streptomyces, which seems to favour intermolecular recombination and is irreversible.

8.2.6 OTHER TECHNOLOGIES RELATED TO GMOs

8.2.6.1 ENU Mutagenesis

N-ethyl-N-nitrosourea (ENU) is an alkylating agent that has the ability to transfer its ethyl group to nucleobases in DNA and by this means induce a mismatch of base pairs; unless this is repaired the result is a point mutation (Russell et al. 1979; Justice et al. 1999). The agent has a specific affinity for spermatogonial stem cells and induces one new mutation for every 700 gametes. As a consequence of point mutations, patterns of expression vary from complete silencing of the gene to overexpression in some cases (Yu et al. 2005). The enormous potential of ENU mutagenesis is its ability to create an alleleic series of a single gene that may include nullimorphs with no expression, hypermorphs with overexpression, hypomorphs with down-regulated expression and even antimorphs or neomorphs. ENU, like other mutagenic agents, induces changes throughout the whole genome, and the process of identifying changes is laborious and time consuming. For this reason several international consortia have been formed that undertake the creation of ENU mutants, identification of the resulting phenotypes and publication for the use of the scientific community.

ENU makes it possible to examine genes associated with phenotypes without any prior assumptions about the nature of those genes, which is a perfect example of Forward Genetics; also has made it possible to combine Forward and Reverse Genetics into a single process by using so-called gene driven ENU mutagenesis. In this method, recombinant DNA techniques are used to identify which of thousands of different ENU mutations specifically affect a certain DNA region. When these have been identified it is possible to generate lines of those animals or their frozen sperm and to perform a comparative study. Since the initial description of this approach (Coghill et al. 2002), it is being increasingly used for several reasons: as part of the initiative to produce mutations in every single gene of the mouse genome, it can be used to produce different expression levels of the same gene, and thirdly, it can be applied to any species and has been used successfully in rats and fish. It was this approach that made it possible to create a functional knock out in the rat (Zan et al. 2003), an outcome that had proved impossible by gene targeting.

8.2.6.2 Interfering RNA

RNAi provides a mechanism for post-transcriptional modification by degrading mRNA into short homologous RNA sequences. RNAi is a mechanism of gene regulation native to almost all eukaryotes. RNA mediated interference can be used to study the result of loss-of-function but is never complete and most of the time not permanent. In contrast to the terms knock out and knock in, the effects of RNAi are described as knock down. Initial attempts to manipulate RNAi were based on long double strand RNA (ds-RNA) molecules (500-1000 nt), but these long RNA molecules also activated the PKR signal pathway resulting in non-specific immunity and an interferon response. Although in zygotes, ES cells and EC cells, these long dsRNA molecules can give rise to specific silencing, in other organisms the use of small (≤ 21 bp) interfering RNAs (siRNAs), or short hairpin RNA (shRNAs) is preferred because these better represent the natural processing pathway. The effectiveness of this method has been demonstrated not only in cell cultures, but also in living animals (Gallozzi et al. 2008).

The benefits of using RNAi are the speed with which gene modification is achieved, and the fact that the knock down phenotype is not an all-or-nothing change but one that can be regulated and disappears with cell division. Another big advantage is the fact that RNAi mediated effects can be studied without the need for the mutation to be homozygous. Introduction of siRNA affects both alleles at the post-transcriptional level.

It is possible to deliver naked RNAs locally to the eye, skin and lungs, but the full potential of RNA mediated interference is realised when it is combined with (lenti)viral vectors. Lentiviral infection with shRNA expressing vectors can result in persistent production of RNAi molecules and infect even non-dividing cells. There is considerable interest in the prospect of combining human NT-ES cells, RNAi and lentiviral vectors in development of therapeutic ES cell cloning techniques. ES cells of human or mouse origin, derived from NT, can be differentiated *in vitro* after being modified with RNAi and provide a means of investigating the processes of differentiation, remaining pluripotent, becoming malignant and regulation by epigenetic or genetic factors. As soon as issues concerning safety (for example how to establish with certainty that only stem cells are present), immune response and efficiency (derivation of all cell types, *in vivo* delivery to all cell types and target specificity) have been resolved there will be great implications for gene therapy.

8.2.6.3 Gene Trapping

This technique provides an alternative method of random based gene targeting to ENU mutagenesis. It involves the application of recombinant DNA technology to ES cells. In summary the method entails inclusion of so-called trap genes into the genome of ES cells. Clones of these modified cells are then generated and may be used eventually to produce mice. Gene trapping vectors consist of a promoterless reporter gene preceded by a splicing region. This vector, once included randomly into the genome of the ES cell, generates a transcript that is a combination of the endogenous gene and the reporter gene; this transcript makes it easy to identify the trapped gene and its control regions. In many cases the insertion interrupts an endogenous gene and the cDNA sequence of the trapped gene can be determined by rapid amplification of cDNA 5 ends (5RACE; (Hicks et al. 1997). Several gene trap vectors have been described, and the more efficient of these not only encode a reporter gene, but also a selection marker, thus allowing the cells in which an active gene has been "trapped" to be selected. The main limitation is that gene trapping is appropriate only for studying genes that are functional *in vitro* because it requires their expression.

8.3 CONSTRUCT DESIGN

Because there are so many ways of introducing genetic material into the germline and a wide range of research questions to be solved, there are many approaches for designing the construct and making modifications, dependant on the skill and creativity of the researcher. This section will review some general aspects of designing constructs. More comprehensive overviews are available in several publications (Clarke 2002; Kwan 2002; Houdebine 2003; Nagy et al. 2003; Pease and Lois 2006) and can be found on various websites.*

8.3.1 Design of Transgenic Constructs for Gene Transfer in Zygotes

When designing a transgene for pronuclear microinjection, the following portions should be included: a promoter, the protein coding sequence and a termination signal. Depending on the kind of DNA used to code for the protein to be expressed, the construct may need additional components. If it is genomic DNA that includes introns and exons, expression will be more or less guaranteed. However if the DNA that encodes the protein is cloned and lacks introns, expression is likely to fail. To avoid this problem, it is important to include a spliced donor and acceptor sequences flanking an intron positioned at 3′ of the cDNA before the polyadenylation signal. The cloning design should also include a strategy to facilitate detection of the transgene when it has been integrated in the genome so that it is clearly discernible from the host's endogenous genomic DNA.

---

* See, for example, http://www.med.umich.edu/tamc or http://www.research.uci.edu/tmf/constructDesign.htm

In addition to this general structure, the following aspects should also be considered:

- Avoid prokaryotic and plasmid vector sequences that can severely compromise gene expression.
- Choose a cloning vector of a different size to the insert to avoid contamination during purification.
- Give preference to linear fragments and linearise the BACs to avoid disruption of double stranded DNA and damage to the transgene.
- Use genomic DNA. If cDNA is used, an artificial intron should be added 3' of the cDNA to stimulate mRNA transport out of the nucleus and splicing.
- The length of the construct is limited only by cloning and handling. It is preferable to introduce large fragments (BAC, YAC and PAC) because they are more likely to retain elements regulating expression—enhancer, promoter and insulator sequences. However, handling larger DNA molecules demands more skill than plasmidic DNA. In addition, polyamines should be used in the preparation and injection buffer to maintain the integrity of the construct and to ensure integration of the complete fragment. The overall efficiency of microinjection is appreciably reduced with large constructs.
- Make sure that expression of the transgene is discernible from that of the endogenous gene, for example by creating a fusion protein (with AP or GFP) or link expression of the transgene and the reporter by placing an internal ribosome entry site (IRES) sequence in between.
- Reporter genes can be used to identify regulatory sequences and their expression pattern (lacZ, CAT chloramphenicol acetyltransferase, alkaline phosphatase AP, luciferase and GFP); however, some of them of prokaryotic origin may cause variegate expression and even silencing of the transgene.
- The choice of promoter depends on the aim of research, but examples are:
    i. Strong constitutive ubiquitous promoters like CMV, PGK, ß-actin and the promoter of the Rosa 26 locus. However, the expression of these promoters remains dependent on and can be influenced by the site of integration.
    ii. Tissue specific promoters
    iii. Inducible promoters
    iv. Developmentally regulated promoters
- The construct should be carefully purified. DNA that is contaminated by traces of resins or chemicals used during purification may be difficult to inject and may decrease survival of the embryo.

8.3.2 Design of Gene Targeting Constructs for Gene Transfer in ES Cells

Gene targeting constructs are designed to undergo homologous recombination into a specific locus thereby replacing the endogenous sequences, with the aim of disrupting the function of the target gene (knock out) or mutating it (knock in). Because of the low frequency of homologous recombination compared with random integration, homologous recombination has to be performed *in vitro* in mouse ES cells and modified cells have to be selected for injection into the germline of mice.

The simplest targeting construct consists of two arms homologous to the GOI with a positive selection cassette in between to facilitate selection of homologous recombinants from non-integrants. A second, negative selection marker can be introduced at the 3' end of the targeting construct to discriminate random integrants from homologous recombinants.

- Selection cassette
  In most cases the positive selection marker should be driven by its own promoter and pA site: Pgk-1 is the gene promoter of choice in ES cells and bGHpA is a good choice as pA

signal. If the GOI is known to be expressed in ES cells, a promoterless construct can be used to select for homologous recombinants, because only homologous recombination will lead to an in-frame recombination of the GOI and the selection marker enables positive selection.
- Positive selectable markers
  - neo; neomycin phosphotransferase or aminoglycoside 3′-phosphotransferase gene; is the most generally applied selection marker of *E.coli* origin. The selective drug is G418, an aminoglycosidic antibiotic that inhibits protein synthesis.
  - puro; puromycin resistance gene; puromycin is an aminonucleoside antibiotic from *Streptomyces alboniger*. It specifically inhibits peptidyl transfer on ribosomes, thereby inhibiting the growth of various insect and animal cells, including ES cells.
  - hygro; hygromycin resistance gene; Hygromycin B is an aminoglycosidic antibiotic produced by *Streptomyces hygroscopicus* and used for the selection and maintenance of prokaryotic and eukaryotic cells transfected with the hygromycin-resistance gene of *E.coli* origin. It kills cells by inhibiting protein synthesis.
- Negative selectable markers are used when the loss of the gene is the event to be selected. Examples are
  - HSV-tk; herpes simplex virus-1 thymidine kinase gene; this gene kills cells in the presence of ganciclovir because ganciclovir can be converted to the triphosphate form that stops DNA chain elongation. Expression of HSV-tk in germ cells is known to cause sterility (Wilkie et al. 1991) and therefore ES cells containing the HSV-tk gene are usually not germline competent; this marker should be deleted before making a mouse containing it. However, a truncated HSV-tk gene is germline compatible (Salomon et al. 1995).
  - dt; diphtheria toxin; diphtheria toxin from *Corynebacterium diphteriae* is so poisonous to cells that a single molecule may be sufficient to kill it. It inactivates elongation factor-2 and inhibits chain elongation in protein synthesis.
- Positive/negative selectable marker
  - hprt; hypoxanthine phosphoribosyltransferase; this enzyme can be used as a positive selectable marker in hprt-deficient cells but can also be used as a negative selectable marker in hprt-containing cells. The hprt-encoded enzyme converts free purine bases into the corresponding nucleoside phosphate and thus makes them available for the synthesis of nucleic acids. Hprt-deficient cells lack this capacity and therefore cannot survive hypoxanthine, aminopterin, thymidine (HAT) medium selection. If an hprt minigene is introduced into hprt-deficient cells the capacity is restored and allows positive selection. Hprt-containing cells are killed by the presence of 6-thioguanine in the medium, so it can be used as a negative selecting agent.
- Arms of homologous sequences

The frequency of homologous recombination depends on the degree and length of homology between the targeting construct and the endogenous sequence of the GOI. For efficient targeting frequencies, the cloned DNA should be of isogenic origin (te Riele et al. 1992) and be at least 6–8 kb (the longer the better), and not shorter than 2 Kb at either end. Large regions of repetitive sequences should be avoided. The length of the short arm of homology depends also on the method used for initial screening of targeting events; if PCR is to be used it should be around 2.5 Kb.

8.4 INFLUENCE OF GENETIC BACKGROUND

The sum of the hereditary information stored in the chromosomes of an organism is referred to as its genome. The genotype of an individual is its genetic make-up, which can be defined as the combination of alleles throughout its genome, although it is most often used to refer to the combination of alleles carried at a specific locus. The terms *genetic background* and *genotype* are frequently used interchangeably in discussing their impact on the phenotype of a mutant strain. After it was realised that genetic background can influence the phenotype of a mutation, whether occurring spontaneously or induced by transgenesis, the influence of genotype has become an important topic of scientific interest. This variability in phenotype, sometimes referred to as its penetrance, can be attributed to the presence of independent modifier genes with allelic variations; the only way to determine whether or not a phenotype is affected by such modifiers is to investigate the mutation on different genetic backgrounds (Linder 2001; Nadeau 2001). This is why it is essential to specifically address this point when designing the experiment, why it is considered to be a key factor in the creation of a GM animal and why correct nomenclature is so important.

There is no universal ideal genetic background, onto which it would always be appropriate to introduce and study new mutations. When generating a new transgenic line there may be technical reasons for selecting a particular genetic background. For example, it is important to consider the characteristics of different strains or substrains of hosts. Some inbred strains of mice and rats are blind; some become deaf early in life, whilst others have poor learning abilities, and so on. Animals of inbred strains and even of sub-strains within these (e.g., C57 Bl6/J or C57BL6/N) may exhibit extreme phenotypic characteristics that render them inappropriate models for particular studies, or may unnecessarily impair the welfare of a newly developed mutant. Although C57BL/6 is widely used for mouse transgenic studies, it is important to recognise that inbred strains are narrow windows that may make extrapolations to the wide population unreliable (Montagutelli 2000), and that although levels of expression are more consistent between animals in the same experiment, these levels may differ markedly in different strains.

Difficulties in culturing embryos of many inbred strains have led to the general use of F1 hybrids that respond well to superovulation treatments and are robust enough to withstand the trauma of microinjection (Table 8.2).

However the use of F1 hybrids results in a mixed genetic background following onward breeding, and consequently an inbred strain (FVB/N) has been developed that has easy detectable pronuclei

TABLE 8.2
Comparative Study of Transgenic Production Efficiency with Two Different Genetic Backgrounds

	C57BL/6	B6CBAF1
Microinjection sessions	23	28
Number of superovulated females	174	151
Percentage mating	72,41	90,1
Percentage embryos surviving injection	76,8	84,2
Mean number of injected embryos per donor	10,77	17
Number of transferred embryos	1002	1825
Percentage pregnancy	67,44	62
Mean born animals per pregnant recipient	2,93	4,28
Positives/analyzed animals (%)	17/80 (21.25)	46/197 (23.35)
Efficiency according to the total number of injected embryos	1.25 %	1.99 %

Source: Pintado, B. and Gutiérrez-Adàn, A., Modificación genética animal Mutagénesis tradicional y transgénesis. El mantenimiento de las mutaciones. Ciencia y Tecnología del Animal de Laboratorio. Publ. 2008. With permission.

and tolerates oocyte manipulation well (Taketo et al. 1991). This is seen as an improvement on existing inbred lines such as C57BL/6, which have lower efficiency.

In the case of targeted mutagenesis, most of the original stem cell lines were derived from the 129 strain, although this has subsequently been shown to consist of a large number of different substrains, in part because of genetic contamination. ES lines from other genetic backgrounds lacked their efficiency and for that reason were less popular. Today it is technically possible to microinject the construct directly into several inbred strains or to use stem cell lines that are genetically defined, although the strategy adopted should be determined by the research objectives and the breeding scheme that will be followed when the line is established (see Chapter 9). Speed congenics involves selectively breeding animals of a mixed background towards the desired background by selecting breeding stock with the aid of satellite markers for a specific, favoured background; this can be used to generate animals on the favoured/desired background more quickly than by conventional breeding (Weil, Brown, and Serachitopol 1997).

8.5 TRANSGENIC IDENTIFICATION: GENOTYPING

The initial genotyping is the most important step in establishing a GM line. The method used for screening should be tested and validated to be able to identify one single copy of the modified DNA against the genomic background and it must be planned in advance, while the transgene is being designed. At least one unique restriction site has to be present in the transgene and the sequence should include targets for primers that allow a distinction to be made between the transgene and an endogenous sequence if such a coincidence exists.

The following molecular techniques are available:

- Southern blotting
- Dot blotting
- PCR
- Quantitative PCR

Sample collection and individual identification can be done simultaneously when tissue biopsies are used for DNA isolation. Tissue can be taken by ear punching, but tail snipping, toe clipping and blood sampling are sometimes used, but are considered more invasive. Each method has its optimal stage of tissue sampling and limitations but it is important to make a decision for a method based on welfare evidence and scientific arguments. Non-invasive DNA sampling techniques are based on saliva and stool isolation, but these methods have to be accompanied by identification techniques, which usually are invasive.

The initial identification of founders demands careful screening. Southern blotting has been the method of choice for determining transgene copy number and it is still the method to discern when there is more than one integration site. In addition it is not prone to provide false positive results. However, it is time consuming and although there are non-radioactive alternatives, most laboratories still use radioactive probes to perform it, with concomitant issues relating to health and safety. It is also possible that some founders may be strong mosaics and for that reason provide a low signal, that may be overlooked with this method.

PCR-based genotyping requires smaller quantities of DNA, allowing the use of non-invasive methods for collecting samples and results are obtained more quickly. However it has to be used very carefully for detecting founders. Many purification procedures used to extract DNA from tissues for genotyping are based on fast protocols that rarely provide clean samples. These conditions complicate the identification of a founder and it is always good practise to verify that an endogenous gene, present as a single copy, can be identified in the very same sample. Identification of a transgene inserted in genomic DNA may require different PCR conditions than those optimised when cloning the construct; on one hand this results in false negatives, on the other, PCR is also prone to false positives unless the methodology has been correctly validated. Finally, it is always risky to assert a gene deletion in the absence of amplification, because failure of amplification for any reason will lead eventually to the identification of a false KO.

Quantitative PCR can be considered a refinement of standard PCR for genotyping, because it allows determination of the number of copies that have been incorporated into the genome (Mitrecic et al. 2005; Chandler et al. 2007). However a much more important aspect of this technique is that it allows differentiation of heterozygotes from homozygotes (Wishart et al. 2007) and this is of obvious value when establishing a transgenic line that has to be bred to homozygosity. This is easily resolved in KO mice, but demands the use of more animals when additive transgenesis is involved, because it is necessary to perform a homozygosity test in order to be sure that the transgene is present in both chromosomes. Quantitative PCR may also discern between single and multiple integrations, but there are technical limitations, including the need for amplicons to be as small as reasonably possible, 70–150 bp for designs using hybridization probes and less than 300 bp for SYBR Green assays (Ginzinger 2002); this makes it more difficult to design primers, because a very specific region has to be targeted

8.6 CONCLUSIONS

The development of any transgenic animal model is a technically demanding process that requires expertise in several fields and a substantial investment of resources. In addition, many factors are involved and there will always be a small percentage of instances in which chance plays a crucial role. Our knowledge of the structure and function of genetic material is increasing day-by-day, but we are still far from being able to control all of the factors involved. Transgene design has evolved from the relatively simplistic view of a construct in the 1980s to a complex process in which other factors play a role and allow up- and down-regulation of gene function. However, in spite of tremendous technical advances during the last 30 years, manipulation of the animal genome is still something of an art and it is important to keep an open, enquiring mind, because sometimes failures may give insights into unexpected biological processes. Even with such limitations, the generation of GM animal models has been and will continue for many years to be an invaluable tool for research.

8.7 QUESTIONS UNRESOLVED

It is difficult to foretell what new developments will arise in the near future but some specific questions need to be addressed. Amongst others, these include the need to generate stem cell lines in species other than mice and rats for preparing knock out and knock in models. The role that induced pluripotent stem (iPS) cells will play in this field is still unclear but NT has proven to be a technical alternative in farm animals. However, efficiency is still extremely low. On the other hand, there are serious concerns regarding the welfare of cloned animals highlighting the need to investigate cell reprogramming in order to identify the genetic and epigenetic factors that underlie such alterations.

The discovery of different sequences that modulate expression or the need to develop complex models of inherited diseases, has promoted the need for larger constructs such as BACs and YACs. With these, the efficiency of microinjection decreases considerably, leaving room for technical improvements besides SMGT, which is based on ICSI (a challenging technique in mice as well as in other mammals). Attention should also be given to other upcoming animal models such as zebrafish or medaka where the efficiency of transgenesis has yet to attain a level comparable to some mammalian species.

Last, but no less important, we still need robust methods for genotyping animals, using minimal quantities of tissue but which are sufficiently accurate to allow correct identification of positives and even to discriminate between hemizygotic and homozygotic animals.

REFERENCES

Barbaric, I., G. Miller, and T. N. Dear. 2007. Appearances can be deceiving: Phenotypes of knockout mice. *Briefings in Functional Genomics & Proteomics* 6:91–103.

Belteki, G., M. Gertsenstein, D. W. Ow, and A. Nagy. 2003. Site-specific cassette exchange and germline transmission with mouse ES cells expressing phiC31 integrase. *Nature Biotechnology* 21:321–24.

Blau, H. M. and F. M. Rossi. 1999. Tet B or not tet B: Advances in tetracycline-inducible gene expression. *Proceedings of the National Academy of Sciences of the United States of America* 96:797–99.

Brackett, B. G., W. Baranska, W. Sawicki, and H. Koprowski. 1971. Uptake of heterologous genome by mammalian spermatozoa and its transfer to ova through fertilization. *Proceedings of the National Academy of Sciences of the United States of America* 68:353–57.

Branda, C. S. and S. M. Dymecki. 2004. Talking about a revolution: The impact of site-specific recombinases on genetic analyses in mice. *Developmental Cell* 6:7–28.

Brinster, R. L., H. Y. Chen, M. E. Trumbauer, M. K. Yagle, and R. D. Palmiter. 1985. Factors affecting the efficiency of introducing foreign DNA into mice by microinjecting eggs. *Proceedings of the National Academy of Sciences of the United States of America* 82:4438–42.

Bronson, S. K., E. G. Plaehn, K. D. Kluckman, J. R. Hagaman, N. Maeda, and O. Smithies. 1996. Single-copy transgenic mice with chosen-site integration. *Proceedings of the National Academy of Sciences of the United States of America* 93:9067–72.

Chan, A. W., E. J. Homan, L. U. Ballou, J. C. Burns, and R. D. Bremel. 1998. Transgenic cattle produced by reverse-transcribed gene transfer in oocytes. *Proceedings of the National Academy of Sciences of the United States of America* 95:14028–33.

Chan, A. W., K. Y. Chong, C. Martinovich, C. Simerly, and G. Schatten. 2001. Transgenic monkeys produced by retroviral gene transfer into mature oocytes. *Science* 291:309–12.

Chandler, K. J., R. L. Chandler, E. M. Broeckelmann, Y. Hou, E. M. Southard-Smith, and D. P. Mortlock. 2007. Relevance of BAC transgene copy number in mice: Transgene copy number variation across multiple transgenic lines and correlations with transgene integrity and expression. *Mammalian Genome: Official Journal of the International Mammalian Genome Society* 18:693–708.

Charreau, B., L. Tesson, J. P. Soulillou, C. Pourcel, and I. Anegon. 1996. Transgenesis in rats: Technical aspects and models. *Transgenic Research* 5:223–34.

Clarke, A. R. 2002. *Methods in molecular biology; transgenesis techniques, principles and protocols*. New York: Humana Press Inc.

Coghill, E. L., A. Hugill, N. Parkinson, C. Davison, P. Glenister, S. Clements, J. Hunter, R. D. Cox, and S. D. Brown. 2002. A gene-driven approach to the identification of ENU mutants in the mouse. *Nature Genetics* 30:255–56.

Collier, L. S., C. M. Carlson, S. Ravimohan, A. J. Dupuy, and D. A. Largaespada. 2005. Cancer gene discovery in solid tumours using transposon-based somatic mutagenesis in the mouse. *Nature* 436:272–76.

Denning, C., S. Burl, A. Ainslie, J. Bracken, A. Dinnyes, J. Fletcher, T. King, et al. 2001. Deletion of the alpha(1,3)galactosyl transferase (GGTA1) gene and the prion protein (PrP) gene in sheep. *Nature Biotechnology* 19:559–62.

Ding, S., X. Wu, G. Li, M. Han, Y. Zhuang, and T. Xu. 2005. Efficient transposition of the piggyBac (PB) transposon in mammalian cells and mice. *Cell* 122:473–83.

Dupuy, A. J., K. Akagi, D. A. Largaespada, N. G. Copeland, and N. A. Jenkins. 2005. Mammalian mutagenesis using a highly mobile somatic sleeping beauty transposon system. *Nature* 436:221–26.

Filipiak, W. and T. Saunders. 2008. Advances in rat transgenesis. *Transgenic Research* 17:1007.

Gallozzi, M., J. Chapuis, F. Le Provost, A. Le Dur, C. Morgenthaler, C. Peyre, N. Daniel-Carlier, et al. 2008. Prnp knockdown in transgenic mice using RNA interference. *Transgenic Research* 17:783–91.

Geurts, A. M., G. J. Cost, Y. Freyvert, B. Zeitler, J. C. Miller, V. M. Choi, S. S. Jenkins, et al. 2009. Knockout rats via embryo microinjection of zinc-finger nucleases. *Science* 325:433.

Ginzinger, D. G. 2002. Gene quantification using real-time quantitative PCR: An emerging technology hits the mainstream. *Experimental Hematology* 30:503–12.

Glaser, S., K. Anastassiadis, and A. F. Stewart. 2005. Current issues in mouse genome engineering. *Nature Genetics* 37:1187–93.

Gossen, M. and H. Bujard. 1992. Tight control of gene expression in mammalian cells by tetracycline-responsive promoters. *Proceedings of the National Academy of Sciences of the United States of America* 89:5547–51.

Gossen, M., S. Freundlieb, G. Bender, G. Muller, W. Hillen, and H. Bujard. 1995. Transcriptional activation by tetracyclines in mammalian cells. *Science* 268:1766–69.

Gu, H., Y. R. Zou, and K. Rajewsky. 1993. Independent control of immunoglobulin switch recombination at individual switch regions evidenced through cre-loxP-mediated gene targeting. *Cell* 73:1155–64.

Heaney, J. D., and S. K. Bronson. 2006. Artificial chromosome-based transgenes in the study of genome function. *Mammalian Genome* 17:791–807.

Hicks, G. G., E. G. Shi, X. M. Li, C. H. Li, M. Pawlak, and H. E. Ruley. 1997. Functional genomics in mice by tagged sequence mutagenesis. *Nature Genetics* 16:338–44.

Hirabayashi, M., M. Kato, K. Amemiya, and S. Hochi. 2008. Direct comparison between ICSI-mediated DNA transfer and pronuclear DNA microinjection for producing transgenic rats. *Experimental Animals* 57:145–48.

Hochedlinger, K., and R. Jaenisch. 2002a. Monoclonal mice generated by nuclear transfer from mature B and T donor cells. *Nature* 415:1035–38.

Hochedlinger, K., and R. Jaenisch. 2002b. Nuclear transplantation: Lessons from frogs and mice. *Current Opinion in Cell Biology* 14:741–48.

Hochedlinger, K., and R. Jaenisch. 2003. Nuclear transplantation, embryonic stem cells, and the potential for cell therapy. *The New England Journal of Medicine* 349:275–86.

Houdebine, L.-M. 2003. *Animal transgenesis and cloning*. New York: Wiley.

Iannaccone, P., and V. Galat. 2002. Production of transgenic rats. In *Transgenic animal technology*, ed. Carl A. Pinkert, 235–50, 2nd ed. San Diego, CA: Academic Press.

Jaenisch, R. and B. Mintz. 1974. Simian virus 40 DNA sequences in DNA of healthy adult mice derived from preimplantation blastocysts injected with viral DNA. *Proceedings of the National Academy of Sciences of the United States of America* 71:1250–54.

Justice, M. J., J. K. Noveroske, J. S. Weber, B. Zheng, and A. Bradley. 1999. Mouse ENU mutagenesis. *Human Molecular Genetics* 8:1955–63.

Kwan, K. M. 2002. Conditional alleles in mice: Practical considerations for tissue-specific knockouts. *Genesis* 32:49–62.

Lavitrano, M., A. Camaioni, V. Fazio, S. Dolci, G. Farace, and C. Spadafora. 1989. Sperm cells as vectors for introducing foreign DNA into eggs: Genetic transformation of mice. *Cell* 57:717–23.

Lavitrano, M., M. Busnelli, M. G. Cerrito, R. Giovannoni, S. Manzini, and A. Vargiolu. 2006. Sperm-mediated gene transfer. *Reproduction, Fertility, and Development* 18:19–23.

Lavitrano, M., M. L. Bacci, M. Forni, D. Lazzereschi, C. Di Stefano, D. Fioretti, P. Giancotti, et al. 2002. Efficient production by sperm-mediated gene transfer of human decay accelerating factor (hDAF) transgenic pigs for xenotransplantation. *Proceedings of the National Academy of Sciences of the United States of America* 99:14230–35.

Li, P., C. Tong, R. Mehrian-Shai, L. Jia, N. Wu, Y. Yan, R. E. Maxson, et al. 2008. Germline competent embryonic stem cells derived from rat blastocysts. *Cell* 135:1299–1310.

Linder, C. C. 2001. The influence of genetic background on spontaneous and genetically engineered mouse models of complex diseases. *Laboratory Animal* 30:34–39.

Lois, C., E. J. Hong, S. Pease, E. J. Brown, and D. Baltimore. 2002. Germline transmission and tissue-specific expression of transgenes delivered by lentiviral vectors. *Science* 295:868–72.

Malcuit, C., M. Maserati, Y. Takahashi, R. Page, and R. A. Fissore. 2006. Intracytoplasmic sperm injection in the bovine induces abnormal [Ca2+]i responses and oocyte activation. *Reproduction, Fertility, and Development* 18:39–51.

Metzger, D., J. Clifford, H. Chiba, and P. Chambon. 1995. Conditional site-specific recombination in mammalian cells using a ligand-dependent chimeric cre recombinase. *Proceedings of the National Academy of Sciences of the United States of America* 92:6991–95.

Miller, K. F., D. J. Bolt, V. G. Pursel, R. E. Hammer, C. A. Pinkert, R. D. Palmiter, and R. L. Brinster. 1989. Expression of human or bovine growth hormone gene with a mouse metallothionein-1 promoter in transgenic swine alters the secretion of porcine growth hormone and insulin-like growth factor-I. *The Journal of Endocrinology* 120:481–88.

Mitrecic, D., M. Huzak, M. Curlin, and S. Gajovic. 2005. An improved method for determination of gene copy numbers in transgenic mice by serial dilution curves obtained by real-time quantitative PCR assay. *Journal of Biochemical and Biophysical Methods* 64:83–98.

Montagutelli, X. 2000. Effect of the genetic background on the phenotype of mouse mutations. *Journal of the American Society of Nephrology* 11 Suppl 16: S101–S105.

Moreira, P. N., P. Giraldo, P. Cozar, J. Pozueta, A. Jimenez, L. Montoliu, and A. Gutierrez-Adan. 2004. Efficient generation of transgenic mice with intact yeast artificial chromosomes by intracytoplasmic sperm injection. *Biology of Reproduction* 71:1943–47.

Nadeau, J. H. 2001. Modifier genes in mice and humans. *Nature Reviews Genetics* 2:165–74.

Nagy, A., E. Gocza, E. M. Diaz, V. R. Prideaux, E. Ivanyi, M. Markkula, and J. Rossant. 1990. Embryonic stem cells alone are able to support fetal development in the mouse. *Development* 110:815–21.

Nagy, A., M. Gertsenstein, K. Vintersten, and R. Behringer. 2003. *Manipulating the mouse embryo. A laboratory manual,* 3rd ed. Cold Spring Harbor: Cold Spring Harbor Laboratory Press.

Pease, S., and C. Lois. 2006. *Mammalian and avian transgenesis—New approaches. Principles and practice.* Berlin, Heidelberg: Springer-Verlag.

Perry, A. C., T. Wakayama, H. Kishikawa, T. Kasai, M. Okabe, Y. Toyoda, and R. Yanagimachi. 1999. Mammalian transgenesis by intracytoplasmic sperm injection. *Science* 284:1180–83.

Pinkert, C. 2002. *Transgenic animal technology. A laboratory handbook,* 2nd ed. San Diego, CA: Academic Press.

Poueymirou, W. T., W. Auerbach, D. Frendewey, J. F. Hickey, J. M. Escaravage, L. Esau, A. T. Dore, et al. 2007. F0 generation mice fully derived from gene-targeted embryonic stem cells allowing immediate phenotypic analyses. *Nature Biotechnology* 25:91–99.

Ramírez, M. A., R. Fernández-González, M. Pérez-Crespo, E. Pericuesta, and A. Gutiérrez-Adán. 2009. Effect of stem cell activation, culture media of manipulated embryos, and site of embryo transfer in the production of F0 embryonic stem cell mice. *Biology of Reproduction* 80:1216–22.

Richt, J. A., P. Kasinathan, A. N. Hamir et al. 2007. Production of cattle lacking prion protein. *Nature Biotechnology* 25:132–38.

Rodriguez, C. I., F. Buchholz, J. Galloway, R. Sequerra, J. Kasper, R. Ayala, A. F. Stewart, and S. M. Dymecki. 2000. High-efficiency deleter mice show that FLPe is an alternative to cre-loxP. *Nature Genetics* 25:139–40.

Ronfort, C. M., C. Legras, and G. Verdier. 1997. The use of retroviral vectors for gene transfer into bird embryos. In *Transgenic animals generation and use,* ed. L. M. Houdebine, 83–95. Amsterdam: Harwood.

Russell, W. L., E. M. Kelly, P. R. Hunsicker, J. W. Bangham, S. C. Maddux, and E. L. Phipps. 1979. Specific-locus test shows ethylnitrosourea to be the most potent mutagen in the mouse. *Proceedings of the National Academy of Sciences of the United States of America* 76:5818–19.

Salomon, B. S. Maury, L. Loubiere, M. Caruso, R. Onclercq, and D. Klatzmann. 1995. A truncated herpes simplex virus thymidine kinase phosphorylates thymidine and nucleoside analogs and does not cause sterility in transgenic mice. *Molecular and Cellular Biology* 15:5322–28.

Sauer, B., and J. McDermott. 2004. DNA recombination with a heterospecific cre homolog identified from comparison of the pac-c1 regions of P1-related phages. *Nucleic Acids Research* 32:6086–95.

Schaft, J., R. H. F. van der Shery-Padan, P. Gruss, and A. F. Stewart. 2001. Efficient FLP recombination in mouse ES cells and oocytes. *Genesis* 31:6–10.

Schnutgen, F., A. F. Stewart, H. Von Melchner, and K. Anastassiadis. 2006. Engineering embryonic stem cells with recombinase systems. *Methods in Enzymology* 420:100–36.

Taketo, M., A. C. Schroeder, L. E. Mobraaten, K. B. Gunning, G. Hanten, R. R. Fox, T. H. Roderick, C. L. Stewart, F. Lilly, and C. T. Hansen. 1991. FVB/N: An inbred mouse strain preferable for transgenic analyses. *Proceedings of the National Academy of Sciences of the United States of America* 88:2065–69.

Tamashiro, K. L., T. Wakayama, H. Akutsu, Y. Yamazaki, J. L. Lachey, M. D. Wortman, R. J. Seeley, et al. 2002. Cloned mice have an obese phenotype not transmitted to their offspring. *Nature Medicine* 8:262–67.

te Riele, H., E. R. Maandag, and A. Berns. 1992. Highly efficient gene targeting in embryonic stem cells through homologous recombination with isogenic DNA constructs. *Proceedings of the National Academy of Sciences of the United States of America* 89:5128–32.

Tokunaga, T., and Y. Tsunoda. 1992. Efficacious production of viable germ-line chimeras between embryonic stem (ES) cells and 8-cell stage embryos. *Development, Growth & Differentiation* 34:561, 566.

van der Weyden, L., D. J. Adams, and A. Bradley. 2002. Tools for targeted manipulation of the mouse genome. *Physiological Genomics* 11:133–64.

Wall, R. J. 2002. New gene transfer methods. *Theriogenology* 57:189–201.

Watanabe, S. and T. Nagai. 2008. Health Status and Reproductive Performance of Somatic Cell Cloned Cattle and their Offspring Produced in Japan. *Journal of Reproduction and Development* 54:6–17.

Weil, M. M., B. W. Brown, and D. M. Serachitopol. 1997. Genotype selection to rapidly breed congenic strains. *Genetics* 146:1061–69.

Whitelaw, C. B., A. J. Springbett, J. Webster, and J. Clark. 1993. The majority of G0 transgenic mice are derived from mosaic embryos. *Transgenic Research* 2:29–32.

Wilkie, T. M., R. E. Braun, W. J. Ehrman, R. D. Palmiter, and R. E. Hammer. 1991. Germ-line intrachromosomal recombination restores fertility in transgenic MyK-103 male mice. *Genes Developement* 5:38–48.

Wishart, T. M., S. H. Macdonald, P. E. Chen, M. J. Shipston, M. P. Coleman, T. H. Gillingwater, and R. R. Ribchester. 2007. Design of a novel quantitative PCR (QPCR)-based protocol for genotyping mice carrying the neuroprotective Wallerian degeneration slow (Wlds) gene. *Molecular Neurodegeneration* 2:21.

Yamamoto, A., R. Hen, and W. T. Dauer. 2001. The ons and offs of inducible transgenic technology: a review. *Neurobiology of Disease* 8:923–32.

Yu, P., R. Constien, N. Dear et al. 2005. Autoimmunity and inflammation due to a gain-of-function mutation in phospholipase C gamma 2 that specifically increases external Ca^{2+} entry. *Immunity* 22:451–65.

Zan, Y., J. D. Haag, K. S. Chen et al. 2003. Production of knockout rats using ENU mutagenesis and a yeast-based screening assay. *Nature Biotechnology* 21:645–51.

Zhou, Y., V. Galat, R. Garton, G. Taborn, K. Niwa, and P. Iannaccone, 2003. Two-phase chemically defined culture system for preimplantation rat embryos. *Genesis* 36:129–33.

Zhu, Z., B. Ma, R. J. Homer, T. Zheng, and J. A. Elias. 2001. Use of the tetracycline-controlled transcriptional silencer (tTS) to eliminate transgene leak in inducible overexpression transgenic mice. *The Journal of Biological Chemistry* 276:25222–29.

9 Management of Genetically Modified Rodents

Jan-Bas Prins, The Netherlands

CONTENTS

Objectives 205
Key Factors 206
9.1 Breeding 206
     9.1.1 Breeding Schemes 206
          9.1.1.1 Congenic Breeding 206
          9.1.1.2 Pair Mating 207
          9.1.1.3 Trio Mating 207
          9.1.1.4 Harem Mating 207
     9.1.2 Breeding Records 208
          9.1.2.1 Records of Individual Animals 208
          9.1.2.2 Lineage Records 208
     9.1.3 Genotyping 209
          9.1.3.1 Techniques 209
          9.1.3.2 DNA Markers 209
     9.1.4 Phenotyping 209
          9.1.4.1 Phenotyping Strategy 210
          9.1.4.2 Phenotyping Protocols 213
9.2 Management 215
     9.2.1 Husbandry, Animal Care and Welfare 215
     9.2.2 Housing and Transport 215
     9.2.3 Health Monitoring 215
     9.2.4 Preservation and Recovery 216
          9.2.4.1 General Aspects 216
          9.2.4.2 Public and Private Repositories 217
          9.2.4.3 Cryopreservation of GM Strains 217
          9.2.4.4 Genome Resource Banking Management 222
9.3 Conclusions 222
9.4 Questions Unresolved 222
References 222

OBJECTIVES

The technology of generating genetically modified (GM) rodents is well-established, although specialised laboratories are needed to apply the technology successfully (see Chapter 8). Most often these "transgenic facilities" are integrated into facilities that accommodate animal research. Technical advances and availability of detailed genomic information of the mouse and rat are contributory factors to the increasing demand for GM animals.

When a founder animal has been produced, its genomic content must be propagated to produce sufficient numbers of offspring to allow the characteristics of the strain (i.e., line) to be determined and scientific investigations to be concluded. This chapter addresses the establishment and maintenance of lines of GM rodents and their phenotypic characterisation. Very often lines are unique to an institute and of extreme importance to the success of its research and development programmes. The authenticity and uniformity of the lines should therefore be monitored and as far as possible be conserved in breeding and preservation programmes. These subjects are also discussed.

KEY FACTORS

- The laboratory animal facility should be compliant with national/international guidelines and regulations.
- Animal care takers and biotechnicians should be trained to at least FELASA B level.
- Animals should be maintained at a sufficiently high health status (FELASA recommendations 2002 including Murine Norovirus) and on clearly defined genetic backgrounds.
- Only internationally recognised, standardised nomenclature should be used.
- Maintain close, ongoing liaison between researchers and personnel responsible for husbandry and breeding.
- Careful, accurate and comprehensive record keeping.

9.1 BREEDING

9.1.1 BREEDING SCHEMES

The basis of every genetically modified (GM) strain is a founder animal that carries and transmits the genetic modification to the next generation. Its genetic makeup can be homogeneous or a mix of different genetic backgrounds. For example, in the case of the production of a knock out mouse, the genetic makeup of the founder is determined by the origin of the ES cell line, i.e., by the donor of the blastocysts. A decision has to be made about which genetic background is preferred for maintaining and investigating the genetic modification. Today there seems to be an increasing preference for the C57BL/6 genetic background. When founder animals already have the preferred genetic background, a decision has to be made about the most appropriate breeding scheme for propagating the genetic modification.

Starting with a mixed genetic background, repeated backcross breeding is a necessary part of the breeding scheme, at least until the modification has been fixed onto a homogeneous genetic background. A well-defined, stable and highly characterised genetic background is essential before an animal model can be used reliably as a model for (human) disease (Linder 2006). Initially it is advisable to breed the genetic modification onto two different genetic backgrounds in order to evaluate its expression and the resulting phenotype. Then an informed decision can be made about the most suitable genetic background for future studies. These basic principles also apply to breeding schemes aimed at combining multiple genetic modifications into one line, for example the production of double or triple knock outs and the production of conditional transgenic models.

9.1.1.1 Congenic Breeding

When a genetic modification has been produced on a mixed or unsuitable background, the modification has to be transferred onto the desired genetic background. The breeding scheme used is based on a "cross-backcross-cross" principle. A male carrying the genetic modification is crossed with a wild-type (WT) female. Male F1 animals carrying the gene of interest are then backcrossed with WT female animals of the desired genetic background to produce the N2 generation. This is repeated at least nine times. At least once, a transgenic female has to be bred to a WT male to replace the "mixed" X-chromosome with that of the WT background. After 10 generations the genetic background will be about 99% identical to that of the WT strain. Then N10 generation mice are intercrossed to produce the first generation of the established congenic mouse strain.

Using marker assisted breeding schemes, the number of backcross generations may be reduced (Armstrong, Brodnicki, and Speed 2006). In this process, the selection of animals for producing the next generation is based on information relating to the modified gene(s) and the background genome. DNA markers, for example single nucleotide polymorphisms (SNPs) or simple sequence length polymorphisms (SSLPs), are used to screen the hybrid background of potential breeders of the next generation. Animals that carry the highest proportion of the desired background are selected as breeders for the next generation. This process is known as "speed congenics".

Whatever method is used, the entire process of producing a congenic line takes a considerable amount of time. In a competitive scientific environment this often conflicts with the desire to publish experimental results as soon as possible after the GM line has been established. One alternative is to start a first temporary production colony after only a limited number of backcross generations have been produced, while continuing additional backcrosses for to establish the definitive colony in parallel. Both control and experimental animals are derived from this temporary production colony. After an additional number of backcross generations, a new temporary production colony is set up from that generation and the previous one is terminated. These steps are repeated until the congenic line has been fully established.

A transgene may cause lethality and/or considerably impair the welfare of homozygous (Tg/Tg) offspring. When that is the case the transgenic line should be maintained by mating hemizygous (Tg/0) animals with WT (+/+) animals.

Strains are maintained as foundation stock from which production colonies can be expanded according to need. The number of foundation stock should be kept as small as practicable and they should be monitored closely, both for genetic integrity and health status. The number required is dependent on the reproductive performance of the strain, although three breeding pairs are usually sufficient. There are several appropriate breeding schemes. In the case of homozygous foundation stock, it is advisable to outcross to the WT background strain every few generations in order to minimise genetic drift and inbreeding depression.

9.1.1.2 Pair Mating

One male and one female occupy a single cage as a stable pair for several generations. This is the preferred animal configuration for breeding foundation stock because it allows optimal monitoring of reproductive performance. The couples should be replaced as soon as there are indications of a decline in reproductive performance.

The permanent housing of a male and female in the same cage allows copulation to take place during postpartum oestrous. This way the interval between pregnancies is reduced to a minimum. However, over time, continuous breeding may adversely affect animal welfare and breeding results.

9.1.1.3 Trio Mating

One male and two females are caged together. This configuration may be employed to expand a line. The dams can be left to litter in the same cage and that allows cross fostering to occur. However, it is more difficult to unequivocally determine maternity. On the other hand individual housing of females should be avoided as much as possible. After pregnancy has been confirmed the male can be removed or left in the cage to allow postpartum coitus to take place.

9.1.1.4 Harem Mating

One male is caged with several female animals (2-4). The females are removed as soon as pregnancy is detected or when they have been with the male for approximately 14 days. This is used for expansion of the production colony, so establishing maternity is not regarded as essential, because females can be allowed to litter in the same cage. The advantages of this configuration are that fewer male animals are needed and it allows expansion of the colony in a relatively short period of time. However, attention should be given to the welfare of the male in particular, because this breeding scheme may lead to him "burning-out".

9.1.2 Breeding Records

The breeding performance of GM animals may differ from that of WT counter-parts, and consequently it should be carefully monitored. Careful observation and record keeping are essential instruments in this process, especially during the phase of congenic breeding; however, these should not be limited to this phase of breeding. Recording breeding performance should go hand in hand with recording genotypic and phenotypic data. Even when a genetic modification is maintained in a homozygous state on a fixed background, genotyping is still essential although it can be done intermittently to check the genetic status. It is possible for breeding mistakes to go unnoticed for a long time when apparently phenotypically identical animals with different genetic backgrounds are mated (Benavides et al. 1998; Nitzki et al. 2007).

9.1.2.1 Records of Individual Animals

Records of production data are useful for evaluating the performance of individual animals and of the colony. In the female animal the following indices may be recorded:

- Average litter size over time
- Average number of weaned pups
- Sex ratio of litter after birth/after weaning
- Average whelping interval (= average time between litters)
- Average weaning interval
- Production index (= average number of weaned pups across productive period)

The following indices may be recorded for the male:

- Plug rate (= number of plugged females per number of matings)
- Pregnancy rate (= number of pregnancies per number of plugged females)

Records of animal characteristics should include data on:

- Strain designation (nomenclature according to international guidelines: http://www.informatics.jax.org/nomen/
- Unique breeder ID number;
- Date of birth;
- Ancestry (at least information on parental animals);
- Genotype;
- Phenotype.

In practise, recording of phenotype is usually done only at the level of strain unless an individual animal demonstrates a deviant phenotype.

9.1.2.2 Lineage Records

It is essential to keep accurate documentation of ancestry (i.e., lineage and genotype). Software for breeding registration programmes or colony management programmes (CMPs) are available commercially. However, often institutes use CMPs developed in-house according to their specific needs. Because this market is continuing to develop, it is advisable to search the World Wide Web using key words such as "colony management programme" and to consult with established institutes that have a CMP in place.

9.1.3 Genotyping

9.1.3.1 Techniques

Genotyping is an essential part of the establishment of a GM line. The screening method chosen should be tested and validated. It should be able to identify a single copy of the modified DNA against the genetic background. The following molecular techniques are available:

- Southern blotting
- Polymerase chain reaction (PCR)
- Q(uantitative)-PCR
- Fluorescent *in situ* hybridisation (FISH)
- Dot blotting

For further information see Chapter 8.

Southern blotting is still the method of choice for determining transgene copy number and integration site number. It is not prone to false positive and negative results once it has been validated. However, with the technical advances made and the availability of quantitative real time PCR in most institutes, the latter technology is rapidly replacing Southern blotting for this purpose (Mitrecic et al. 2005; Soliman et al. 2007; Wishart et al. 2007; Joshi et al. 2008) so that it has become the method of choice for routine genotyping. Only small quantities of DNA are required and reliable results can be obtained relatively quickly (Pinkert 2003).

It is important that the size and nature of the sample are compatible with the molecular technique employed; for example Southern blotting requires much larger quantities of DNA than PCR. At the time of sampling, animals should be given a unique identifier in order to link genotyping results with the appropriate animal. Several methods exist to identify individual animals (Weyand 1998; Sorensen et al. 2007). Some of these are invasive and include the removal of tissue, for example ear notching. These tissue samples can be collected and, where possible, should be used to isolate the DNA needed for genotyping to avoid the need for further invasive procedures. Less invasive sampling methods are collection of stool and buccal swaps. In general, the choice of appropriate detection technique should take account of scientific requirements combined with a sampling technique that causes least discomfort to the animal (Castelhano-Carlos et al. 2009).

9.1.3.2 DNA Markers

Minimal genotyping of animals essentially focuses on the gene of interest and its aim is to establish whether it is present or absent. It was pointed out in the section on congenic breeding that production of a congenic line can be enhanced by including screening for additional loci, preferably on different chromosomes. Of these, the most frequently used are SNPs and SSLPs or microsatellites (Benavides et al. 1998; Petkov et al. 2004; Armstrong, Brodnicki, and Speed 2006; Nitzki et al. 2007).

9.1.4 Phenotyping

Every time a new GM-animal is generated it should be regarded as a "new strain". The phenotypic consequences of a genetic modification cannot always be predicted and, consequently, characterisation of the phenotype of these animals should be carried out. The phenotyping strategy should be such as to optimise the chance of detecting unexpected effects of the mutation as well as the expected effects (Brown, Hancock, and Gates 2006). Lack of phenotyping increases the risk that unidentified traits may cause extraneous variability and may increase the risk of systematic errors at the same time (Marques et al. 2007). A reliable decision about the model of choice for a particular research question can only be made when sufficient knowledge about the characteristics of an animal model has been acquired. It also assists in deciding on optimal housing and maintenance conditions to ensure the animal's welfare.

The success of an ENU (N-ethyl-N-nitrosourea) mutagenesis project (see Chapter 8) is totally dependent on effective phenotyping. Mutations created by ENU are random point mutations created in the genes of premeiotic spermatogonia. After breeding, the successful mutations are identified by extensive phenotyping of offspring (Brown, Chambon, and de Angelis 2005).

With the myriad of tests available, it is important to decide the most appropriate tests to use. It may be decided to use only one test, chosen on the basis of existing knowledge of the modified gene. In other cases, for example when there is no or limited knowledge about the mutation(s), as is the case with ENU mutagenesis, the selection may be extended to a battery of tests covering different phenotypic characteristics, such as behaviour, and overall pathology. The number of animals to be tested may range from one animal per genetic modification to several animals and there is a need to address possible differences in zygosity with respect to the modified gene(s), both sexes of these possible genotypes, and different stages of development.

9.1.4.1 Phenotyping Strategy

The term "comprehensive characterisation" is often used to indicate that an extensive battery of tests has been carried out. However, this term should be reserved to indicate that different variables of the animals have been thoroughly examined by a suitably broad combination of tests. Comprehensive characterisation is here understood to include both variables regarding the test animal and the tests comprising the test battery, including extensive pathology. Basic characterisation often refers to the use of published test batteries for a limited number of the most frequently encountered behavioural characteristics. The most straightforward but least complete phenotyping approach, is one that focuses only on the expected phenotypic effect of the genetic modification.

The outcome of all tests is phenotypic information. This information may shed light on possible impairment of welfare of the strain under investigation, and in such cases measures should be introduced to minimise the impact of this impairment.

As a rule of thumb it is important to carry out a thorough literature/ database search for existing (phenotypic) information about the background strain. In addition, information about strains carrying similar mutations and/ or mutations in the same area of the genome can be of great value when gathering information about the phenotype.

At the time of writing, there is no comprehensive database at which results from all GM animals are listed. Web sites describing screening protocols and databases for phenotypic characteristics can be found in the literature (Bolon 2006; Hancock et al. 2007). However, the reader is urged to make his/her own search of the web as well.

The Mouse Phenome Database (MPD) (http://phenome.jax.org/) and the EuroPhenome resource (http://www.europhenome.eu) are central databases holding substantial information about phenotypes (Grubb, Churchill, and Bogue 2004; Mallon, Blake, and Hancock 2008; Grubb et al. 2009). The MPD contains information about both inbred and mutant strains. The EuroPhenome resource holds the results of large-scale phenotyping projects conducted in accordance with the EMPReSS protocols. Both databases are annotated according to the Mammalian Phenotype (MP) Ontology and include tools for browsing and searching for phenotypes (Gkoutos et al. 2005; Smith, Goldsmith, and Eppig 2005). The Mouse Resource Portal of the Wellcome Trust Sanger Institute (http://www.sanger.ac.uk/cgi-bin/modelorgs/mgc/index.cgi) presents phenotypic information generated from the institute's primary screen.

The diversity of phenome databases using different systems is being addressed by the InterPhenome project (http://www.interphenome.org). A working group, the Mouse Phenotype Database Integration Consortium (MPDIC), has been formed and represents large parts of the phenotyping community; it is striving to link these databases, to set standards for the data, and to establish mechanisms to facilitate data exchange between, and search facilities across the various phenome databases (Hancock et al. 2007).

9.1.4.1.1 *The Hierarchical Approach*

A range of clinical tests can be used to uncover the phenotypic characteristics of a specific, GM mouse and a variety of SOPs for these tests can be found in published protocols. Considering the huge number of possible tests, it is worthwhile putting the tests into some order of priority to indicate how relevant results can be obtained as quickly and efficiently as possible. This prioritising is dependent on:

- The selection of tests:
  - Comprehensive;
  - Basic characterisation standard battery of tests;
  - Specific characterisation aimed at the expected phenotype and based on existing knowledge about gene function.

- The animals to be characterised:
  - Gender;
  - Zygosity status;
  - Developmental stage;
  - Control vs. test animals;
  - Numbers to test.

9.1.4.1.2 *Selection of Tests*

A practical approach to gaining useful information about a mouse strain is first to apply a battery of relatively unsophisticated tests—a primary screen—such as those comprising the SHIRPA test (Rogers et al. 1997). These provide a superficial but broad assessment of the mouse's phenotype. A number of test batteries have been proposed that employ this hierarchical approach, focusing either on specific functional domains, for example, behaviour (Crawley and Paylor 1997; Crawley 2003), or employing a wider selection of screens (Rogers et al. 1997; Murray 2002). Results from these primary screens should serve as a pointer to the other tests to be applied. In this way phenotypes of interest that have been identified by the primary screen are followed up by more time-consuming and in-depth sophisticated secondary screens.

Tests results, especially those of behavioural testing, may be influenced by environmental factors that may range from cage environment to diet and health status (Barthold 2004). On the other hand, specific triggers may have to be applied to bring phenotypic variants to the surface (e.g., inducing diabetes by feeding a high-fat diet).

Another important environmental variable is the phenotyping test itself. Most tests cannot be applied without affecting the animal, so it is reasonable to expect that the phenotyping test itself can significantly influence the result; even the order in which the individual tests are performed can influence the outcome of consecutive tests. This has been investigated for behavioural tests and it has been reported that the results of some of these are sensitive to the test order while others are resistant (McIlwain et al. 2001). Mice that have become experienced through participation in some tests change their behaviour significantly in subsequent tests in comparison with naïve mice (Voikar, Vasar, and Rauvala 2004). For this reason it is recommended to start with tests that assess anxiety and exploratory activity and to perform the cognitive tests at a later stage.

In general, results need to be comparable and replicable. Because of this, environmental conditions should be standardised as much as possible and should document the conditions under which the results were obtained (Brown, Hancock, and Gates 2006). It has been shown that the environment of individual laboratories can also influence behavioural results even when attempts have been made to standardise conditions between them (Crabbe, Wahlsten, and Dudek 1999; Mandillo et al. 2008). On the other hand, it should be realised that systematic variation of genetic and environmental backgrounds, rather than excessive standardisation, is needed to ensure robustness of the results and to reveal biologically relevant interactions between the mutation and the genetic and environmental background of the animals (Wurbel 2002). Despite this, relevant control animals should always be included.

It is advisable to consult with specialised centres before embarking on a phenotyping project. This is true for all screening platforms in all fields of research. There is an open international network on behavioural phenotyping, the Mutant Mice Behaviour network (MMB), and the specific environmental challenges it offers have been established (http://www.medizin.uni-koeln.de/mmb-network/index.html; Arndt and Surjo 2001).

9.1.4.1.3 Animals to Test

Some phenotypes are subtle and can be compensated for by other mechanisms; these may be age dependent. Other phenotypes are limited to certain developmental stages and/ or linked with specific physiological conditions (e.g., pregnancy). Therefore it is advisable to characterise GM animals at different stages of their lives: young, adult, and geriatric (Morton and Hau 2003; Wells et al. 2006). Some phenotypes are sex dependent and consequently there may be a need to test both male and female animals. In addition, both homo and heterozygous animals should be tested within every group and they should be of similar age (age matched). Some traits may not reveal themselves until after a couple of generations, and if this is suspected, consideration should be given to conducting phenotyping for several generations.

As mentioned earlier, the initial genetic background may not be homogeneous and/ or not be the genetic background of choice. The preferred homogeneous genetic background is only reached after several generations of backcrossing and it is only then that characterisation will give the most reliable results. However, it is time-consuming to establish this point and may also be contrary to welfare considerations if there is impairment of welfare.

When phenotyping GM animals, the question of controls needs to be dealt with. The most obvious choice is to use WT siblings as controls (Charsa, Knoblauch, and Ladiges 2003). When the strain is maintained by homozygous breeding and no WT siblings are available, the corresponding inbred strain is recommended as control strain. However, it should be realised that after many generations of separation between the GM line and the WT inbred strain, genetic drift may have introduced genetic differences between the two lines in addition to the genetic modification.

The obvious aim of phenotyping an animal is to identify new traits either for welfare assessment or scientific use. In order to optimise the chance of detecting these, a sufficient number of animals must be tested. It is important to give careful consideration to the number of animals required, taking account of animal welfare, time and expense. It hardly makes sense to suggest clear cut recommendations because the size of the group may be influenced by many factors, including genetic background, penetration expression of the modification and welfare implications. However, it has been recommended that starting with approximately 10 WT, 10 hemi- or hetero-, and 10 homozygous animals (50% males and 50% females), it would be possible to carry out statistical evaluation and this should be regarded as the minimum number of animals to test. Depending on the results of the tests, more animals could be included (Mertens and Rulicke 2000; Bailey, Rustay, and Crawley 2006). The issue of sample size has also been addressed in relation to the "background noise" (defined as the difference in response to the same test by mice of the same strain but from different laboratories/suppliers). This intra-strain variation of response to a specific test affects the number of mice recommended for each test, because the background noise could mask phenotypic differences. Other parameters such as the variance of the trait in question (subtle/strong phenotype) also have an effect. Therefore, it is advisable to carry out a pilot-study in order to establish the values of the above parameters and to calculate sample sizes accordingly (Meyer et al. 2007).

9.1.4.2 Phenotyping Protocols

The SHIRPA protocols (SmithKline Beecham, Harwell, MRC, Imperial College School of Medicine at St. Mary's, Royal London Hospital Phenotype Assessment) were among the first applied for phenotyping GM animals (Rogers et al. 1997). The SHIRPA protocols have been in use for many years (sometimes in a modified form) and they have profoundly influenced more recent and elaborate phenotyping protocols.

9.1.4.2.1 The Mouse Phenome Project

The Mouse Phenome Project was initiated in 2000 by the Jackson Laboratories with the objective of developing a comprehensive database. The MPD (http://phenome.jax.org/pub-cgi/phenome/mpd-cgi) encompasses phenotypes from inbred mouse strains. It now also includes results of GM mice. The MPD aims to provide scientists with appropriate information about mice for (1) physiological testing, (2) drug discovery, (3) toxicology studies, (4) mutagenesis, (5) modelling human diseases, (6) QTL analyses and identification of new genes and (7) determining the influence of environment on genotype (Grubb et al. 2009).

9.1.4.2.2 Eumorphia

Eumorphia (European Union Mouse Research for Public Health and Industrial Applications) was an integrated research programme funded by the European Commission. It was involved in the development of new approaches to phenotyping, mutagenesis and informatics and has led to improved characterisation of mouse models for better understanding of human physiology and disease. Eighteen laboratories from eight European countries formed the consortium, which developed over 150 SOPs for phenotyping between 2002 and 2006 (http://www.eumorphia.org). The SOPs are standardised and validated on a cohort of inbred strains across a number of laboratories.

A primary screening protocol EMPReSS (European Mouse Phenotyping Resource for Standardised Screens) has been suggested for large-scale phenotyping, (http://www.empress.har.mrc.ac.uk; Brown, Chambon, and de Angelis 2005). EMPReSS covers the major body systems including: clinical chemistry, hormonal and metabolic systems, cardiovascular system, allergy and infection, renal function, sensory function, neurological and behavioural function, cancer, bone/cartilage and respiratory function, as well as generic approaches to pathology and gene expression. In addition the Eumorphia project has established the EuroPhenome database to hold data on phenotypes obtained from the EMPReSS SOPs (http://www.europhenome.org). They include SOPs for a broad spectrum of phenotyping tests and employ a hierarchical approach. Tests are validated and standardised to a high degree so that results are comparable over laboratories and time. However, the EMPReSS protocol does not include assessment of welfare. This important aspect needs to be addressed separately (Committee on Recognition and Alleviation of Distress in Laboratory Animals 2008).

9.1.4.2.3 Eumodic

The European Mouse Disease Clinic EUMODIC (http://www.eumodic.eu) is a project funded by the European Commission, which started in 2007. EUMODIC is undertaking a primary phenotypic assessment of up to 650 mouse mutant lines, as a first step towards a comprehensive functional examination of the mouse genome. In addition, a number of these mutant lines will be subject to a more in-depth, secondary phenotypic assessment. The EUMODIC consortium is made up of 18 laboratories across Europe that are experts in the field of mouse functional genomics and phenotyping and have a track record of successful collaborative research in Eumorphia. EUMODIC has elaborated a selection of the EMPReSS screens, EMPReSSslim, which is structured for comprehensive, primary, high throughput phenotyping of large numbers of mice. The EMPReSS tests are run in two "pipelines" in order to minimise the number of tests that each animal undergoes and to allow the testing to be completed over a six week period from 9 to

14 weeks of age. The sequence of tests in the two pipelines is designed to allow each test to be run at relevant ages whilst preventing tests from unacceptably influencing the outcome of those which follow.

9.1.4.2.4 The MuTrack System

The MuTrack system was developed for the Tennessee Mouse Genome Consortium in connection with the National Institute of Health's neuromutagenesis programme (Baker et al. 2004). The protocols for phenotyping can be found at http://www2.tnmouse.org/index.php. Tests are divided into 10 primary screens: aggression, aging, auditory, drug abuse, ethanol, epilepsy, eye, general behavioural, neurohistology and a social behaviour domain. In addition, they include three secondary screens focusing on drug abuse, nociception and learning/memory. The protocols for screening are individually designed by different scientists and contain detailed recommendations for screening purposes.

9.1.4.2.5 The PhenoSITE

The protocols available at the PhenoSITE were developed in connection with the Japanese mouse mutagenesis programme at RIKEN Genomic Sciences Centre (GSC; http://www.gsc.riken.go.jp). Their basic phenotype screen includes a modified SHIRPA and further tests: haematology and urine tests, X-ray, seizure induction, home cage activity, open field test, passive avoidance and late onset fundus imaging, blood pressure, hearing and tumour development. The mice are tested between the ages of 8 and 12 weeks.

9.1.4.2.6 Other Protocols

A protocol with a slightly different and broader objective than the programmes just described has been developed (Mertens and Rulicke 2000). It is based on two comprehensive standardised forms and includes elements relating to registration of animals, recommendations for husbandry/breeding, phenotyping, welfare assessment and animal certificate/passport. The first form, "Data Record Form", includes score sheets for litters and individual health and development, based on monitoring from birth until death or euthanasia. The tests are based on observation or other very simple tests. The second form, "Characterisation of Genetically Modified Animal Lines: Standard Form", includes general information about the genetic modification, phenotype and clinical burden, ethical evaluation; information about breeding, husbandry and transportation; and detailed information on genotype and phenotype, which goes into detail with new traits/observations that are specific for the particular strain.

9.1.4.2.7 The Concept of Mouse Clinic

The need of facilities and expertise for phenotyping is growing constantly and it is for this reason that the concept of mouse clinics has emerged. However, there are still relatively few mouse genetics institutes around the world that have broad experience in phenotyping and can be categorised as phenotyping centres or Mouse Clinics. One example is the German Mouse Clinic (GMC) at the German Research Centre for Environmental Health, Munich (http://www.helmholtz-muenchen.de/en/ieg/gmc/). It is an open platform for phenotyping and offers examination of mouse mutants to external scientists using a broad, standardised phenotypic screen of more than 300 parameters. The screens address specific areas including: behaviour, bone and cartilage development, neurology, clinical chemistry, eye development, immunology, allergy, steroid metabolism, energy metabolism, lung function, vision, pain perception, molecular phenotyping, cardiovascular analyses and pathology. Mouse mutants accepted at the GMC are first assessed using a primary screen. A standardised workflow has been developed for screening a cohort of 40 mice (10 males and 10 females of mutant and WT, respectively) in accordance with EMPReSS protocols and takes about 10 weeks to complete. An application form and instructions for submission of a strain for free phenotyping can be found at the above homepage. In addition, several commercial companies offer phenotyping services.

9.2 MANAGEMENT

9.2.1 HUSBANDRY, ANIMAL CARE AND WELFARE

The husbandry and care of GM animal is in essence no different from their WT equivalents. However, depending on the modification and its manifestation, special attention should be given to animal welfare assessment, water and food consumption, reproductive performance, development and behaviour. Deviations from the requirements of the WT counterpart may identify a need for:

- Specialised housing conditions, for example, IVC cages in the case of immune compromised animals
- Use of other types of bedding, nesting material, and/or environmental enrichment
- Special diets, for example, softer pellets or higher protein content
- Additives in water, for example, antibiotics
- Higher numbers of breeding animals for line maintenance
- Extending the period from birth to weaning and/or special food provision for weaned offspring
- More frequent assessments of welfare

9.2.2 HOUSING AND TRANSPORT

The housing and transport of GM animals must be carried out within the framework of European and national laws and regulations. At the European level, the Commission Recommendation of 18 June 2007 provides guidelines for the accommodation and care of animals used for experimental and other scientific purposes (http://eur-lex.europa.eu/LexUriServ/LexUriServ.do?uri = OJ:L:2007:197:0001:0089:EN:PDF).

The Transport Working Group of the UK Laboratory Animal Science Association (LASA) has published standards for laboratory animal transport (Swallow et al. 2005). These guidelines present good practice and the group will update them to incorporate new knowledge, technologies, and legislation as necessary.

9.2.3 HEALTH MONITORING

The production of GM lines involves considerable effort, time, numbers of animals and financial costs. After it has been produced, a GM line is considered invaluable, at least until an improved model becomes available. Loss of a unique line because of health problems is not acceptable and health monitoring programmes have been designed to track the health status of the colony over time. Internationally accepted recommendations for the health monitoring of colonies of rats and mice have been published (Nicklas et al. 2002; Nicklas 2008) and at the time of writing are being reviewed and updated by an expert FELASA working group (http://www.felasa.eu).

GM models may be immunocompromised. In those cases sentinel animals should be used to monitor for infection with micro-organisms (at least for those screened for by serology). Sentinel animals are animals of known health status. They can be used either as contact sentinels; that is, housed in the same cage as the GM animals, or as non-contact sentinels that may be exposed to dirty bedding (and water) of the GM animals concerned. This method is less stringent than the former and infection with certain micro-organisms may go undetected (Ike et al. 2007; Shek 2008). In individually ventilated cage (IVC) systems, exposure of sentinel animals to a sample of exhaust air from all cages of a single rack has not yet proven to be a successful strategy probably because not all micro-organisms are transmitted efficiently by air (Compton, Homberger, and MacArthur Clark 2004; Brielmeier et al. 2006).

The number of animals that need to be tested are determined by colony size and estimated prevalence of the infection in the colony. The prevalence of an infection is dramatically reduced when animals are housed in IVCs and handled under appropriate biocontainment conditions. In theory the prevalence could be as low as the animals in a single cage; in this case the absence of infection could only be verified by testing all cages and confidence is proportionally lower when only limited numbers of cages are screened. An alternative strategy is to house contact sentinels with retired breeders and to combine this with inter-current testing of animals.

Special attention should be given to the introduction of animals from other establishments into a colony. New arrivals may be accepted subject to undergoing a quarantine period during which they are housed together with sentinel animals. Alternatively, animals may be accepted for re-derivation through embryo transfer or aseptic hysterectomy (Baker 1988; Reetz et al. 1988; Suzuki et al. 1996a; Morrell 1999; Rall et al. 2000; Van Keuren and Saunders 2004; Fray et al. 2008). Embryo transfer is the more reliable way of avoiding vertical transmission of bacteria and viruses and it does not involve the fostering of live pups with risk of rejection by the foster mother.

9.2.4 Preservation and Recovery

There are four main reasons for cryopreserving embryos, sperm and/ or ovarian tissue: (1) to archive (unique) lines currently not in demand (2) to prevent loss of strains due to disease or accident (3) to minimise genetic drift and (4) to facilitate the exchange of genetic material between institutes.

9.2.4.1 General Aspects

Since the first successful cryopreservation of a mouse embryo (Whittingham, Leibo, and Mazur 1972), embryo freezing has been the procedure of choice for archiving GM mouse lines. However, technical developments have made it possible not only to successfully preserve embryos at different developmental stages, but also gametes (spermatozoa and oocytes) and ovarian tissue. In order to determine the most appropriate method, material and quantity to be frozen, it is necessary to consider several aspects of frozen strain recovery, including: fertility characteristics of the GM line, the availability of the GM line across research institutes, the expected frequency of revitalisation, the approach to risk management of infectious diseases and the technical equipment and methodology required for assisted reproduction.

The genetic material of unique lines should be stored in at least two, physically separated locations. If disaster strikes one location, the material may be recovered from the other. In the case of two locations, the total number of frozen aliquots should be divided by two to determine the number of frozen aliquots available for restoring the line. Depending on the techniques used, revitalisation of an archived strain and its subsequent expansion into a colony requires a great deal of time. This should be taken into consideration when the ultimate purpose of cryo-banking is to discontinue a breeding line of animals.

Continuous breeding is associated with genetic drift and/or the risk of epigenetic modifications of transgene expression. These events may accumulate from one generation to the next and lead to sub-line formation (i.e., the development of a line genotypically distinct from the original). When deemed necessary the original line can be revitalised from frozen stock.

There is increasing exchange of GM lines between institutes, whether locally or in different countries across the globe. At the same time, laws and regulations as well as health and welfare issues place increasing constraints on the transport of live animals. Shipping frozen material is a good alternative.

In order to successfully retrieve archived material from a cryo-bank, a proper registration system is essential. The information that should be stored on each specimen includes:

- Strain/line identifier (i.e., name according to internationally standardised nomenclature)
- Nature of sample: embryo, sperm, ovary
- Location and number of samples in cryo-bank
- Date of storage

- Genotype
- Cryopreservation method
- Recommended thawing method

It is imperative to use strain designations that unequivocally identify strains/lines and are in accordance with international rules and guidelines of standardised nomenclature (http://www.informatics.jax.org/nomen/.). Consistent application of nomenclature rules for genes, alleles and strains is fundamental to the administration of both archived and GM lines in breeding. Correct strain designation provides key information about the mutation (for example: spontaneous or induced, gain of function or loss of function, constitutive or conditional random or targeted) and also about the origin of the transgenic sequences, the involvement of reporter genes and on the original as well as on the current genetic background of the mutation.

9.2.4.2 Public and Private Repositories

Both privately and publicly funded repositories have been set up to archive strains/lines of interest to the scientific community. Such repositories and archives may select material based on a scientific review of the lines in order to choose mutants that represent a biomedical model that will be valuable to the scientific community. Most archives do not accept species other than mice for cryopreservation. The Federation of International Mouse Resources (FIMRe; Davisson 2006) is a collaborating group of mouse repository and resource centres worldwide whose collective goal is to archive and provide strains of mice as cryopreserved embryos and gametes, ES cell lines and live breeding stock to the research community.

9.2.4.3 Cryopreservation of GM Strains

The decision to archive a transgenic model generates questions such as: what should be cryopreserved and what method is most appropriate under the given circumstances (Mazur, Leibo, and Seidel 2008)?

Cryopreservation of mouse embryos has been the procedure of choice for a long time and is currently regarded as the gold standard. The recovery rate after freezing depends largely on the genetic background of the strain as well as on the techniques used for freezing and thawing (Rall et al. 2000).

Recovery of frozen sperm and production of offspring has proven to be very difficult in certain GM mouse lines. Recently, the Jackson Laboratory introduced an adapted *in vitro* fertilisation (IVF) protocol (http://cryo.jax.org/ivf.html), which is said to improve recovery results with frozen and thawed sperm (Ostermeier et al. 2008).

Oocytes have also been successfully preserved. However, this approach provides no advantage over embryos because only matured, ovulated oocytes, for example, have been successfully preserved. Attempts to preserve oocytes at earlier stages of development after their isolation directly from ovaries have not been successful. It has been shown experimentally that live offspring can be obtained also from frozen and thawed ovarian tissue (Dorsch et al. 2007).

Live offspring have also been obtained after transfer of nuclei of preserved somatic cells into enucleated oocytes (Wakayama et al. 2008; see also Chapter 8).

Material to be frozen should be collected under sanitary conditions. Even though re-derivation by transfer of embryos has been used successfully to eradicate infectious diseases, any cryopreservation process that allows survival of eukaryotic cells also preserves prokaryotic contaminants such as bacteria, mycoplasmas, or viruses (Baker 1988; Reetz et al. 1988; Suzuki et al. 1996a; Van Keuren and Saunders 2004). In view of this, the recommendations of the International Embryo Transfer Society (IETS), which has been accepted by the World Organisation for Animal Health (http://www.oie.int/eng/normes/mcode/code2007/en_chapitre_3.3.4.htm), should be followed.

Sometimes the availability of homozygous animals for cryopreservation of embryos is restricted or is possible only after a long period of breeding. An alternative is to evaluate cryopreservation of heterozygotes. Heterozygote embryos can be obtained by mating transgenic males with superovulated WT females purchased from a commercial source. This strategy speeds up the cryopreservation process considerably, although it takes longer to bring the transgene back to homozygosity after recovering the line.

9.2.4.3.1 Embryo Cryopreservation and Recovery

In principle, all pre-implantation embryonic stages can be used for archiving. However, the eight-cell stage embryo appears to be particularly robust and can be used successfully with all freezing techniques.

Only high-quality embryos should be selected for freezing. It can be difficult to establish the minimum number of embryos that need to be frozen down before breeding of the strain can safely be discontinued. However, as rule of thumb 400–500 embryos are considered to be sufficient. This allows 200–250 embryos to be stored in two locations for restoring the line; assuming there is an average 15% efficiency of recovery after thawing and transfer, only 30–37 pups will be born. In the case of a heterozygote, only 15–18 pups would have the right genotype of which half are male and half female. Production of 500 embryos requires a significant number of donor females. The efficiency is substantially improved by inducing superovulation by serial injection of pregnant mare serum gonadotropin (PMSG) or follicle stimulating hormone (FSH) and human chorionic gonadotropin (hCG). Although general protocols have been published, the optimal dosages and intervals of injections vary between different strains and species (Suzuki et al. 1996b; Vergara et al. 1997).

Like other living cells, embryonic blastomeres are composed mainly of water. Without cryoprotectant these solidify and crystallise at subzero temperatures. One basic principle to accomplish successful cryopreservation of any living cell is to minimise intracellular ice formation. This can be achieved either by reducing the amount of intracellular water or by lowering the temperature so rapidly that crystallisation cannot take place and intracellular water becomes a vitreous solid instead. The thawing process should be such that recrystallisation cannot occur. Combinations of freezing and thawing procedures have been compared (Macas et al. 1991).

There are three basic methods for freezing embryos: (1) the slow or equilibrium method (Whittingham, Leibo, and Mazur 1972), (2) the quick freezing method (Szell and Shelton 1986) and (3) the vitrification method (Rall and Fahy 1985).

9.2.4.3.1.1 Slow or Equilibrium Method With the slow-cooling or equilibrium method, it is necessary to achieve physico-chemical conditions that allow most internal water to leave the cell. The procedure seeks to achieve a super-cooled state of the intracellular water, in which internal water remains unfrozen in the presence of external ice. Under these conditions, internal water leaves the cell to establish osmotic equilibrium with the hypertonic extracellular environment. This procedure requires that the temperature of the intracellular water is reduced slowly to allow its efflux through the cell membrane before the temperature reaches a point at which intracellular crystallisation takes place. The process is enhanced by adding permeating cryoprotectant agents, such as: glycerol, ethylene glycol, propanediol (PROH) or dimethyl sulfoxide (DMSO) to the fluids in which the cells are bathed. These molecules are able to go through the cell membrane and to substitute for intracellular water.

The slow or equilibrium methodology is based on addition of approximately 1.5 M of penetrating cryoprotectants and a slow-cooling rate from room temperature to between −32°C and −40°C before plunging into liquid nitrogen. This procedure requires strict control of the rate of cooling, for example by using a programmable freezer, but ensures reproducible results in inexperienced hands. Furthermore, the concentration of cryoprotectant needed is low and is considered to offer little threat of toxicity. In addition, the geometry of the embryo container is less critical than in other methods and both cryotubes and plastic insemination straws can be used.

9.2.4.3.1.2 Quick Freezing Method In the quick or fast two-step method, the freezing procedure is speeded up by adding non-permeating macromolecules such as sucrose to the medium. Macromolecules are unable to cross the cell membrane and increase water efflux from the cytoplasm by raising the extracellular osmotic pressure, leading to shrinkage of the cell. The procedure can be completed without use of a programmable freezer.

9.2.4.3.1.3 Vitrification Method Vitrification requires the use of very high concentrations of cryoprotectants that are potentially highly toxic to the embryos. Consequently, freezing and thawing have to be performed very rapidly to minimise cell damage. This allows little margin for variations in the duration of cooling and to some extent also complicates handling. However, the technique reduces the time required to freeze embryos from hours to minutes and the need for automated coolers is avoided. During vitrification intracellular crystals do not have time to form and, instead, a solid vitreous mass is formed. The entire procedure can be performed with very simple equipment found in every laboratory and in experienced hands survival rates are similar to the thermostat-added methodology. Vitrification has been used successful with mice and rats.

Because vitrification requires a very high rate of cooling, the embryo container must make allowance for this. Care should be taken selecting the right material and size of straws. Consistently good results have been obtained with open pulled straws (Kong et al. 2000; Zhao et al. 2007) although, because these are not sealed, there is interchange with liquid nitrogen in the storage tank, which should be regarded as a biosecurity shortcoming (Bielanski et al. 2003).

9.2.4.3.1.4 Embryo Transfer Regardless of the system used to preserve a mutant line, embryo transfer is always an indispensable step for its reestablishment. *In vitro* culture always reduces embryo viability, so transfer has to be scheduled in such a way as to keep the period of *in vitro* culture as brief as possible. On the other hand, the period of *in vitro* culture may be used to select those embryos for transfer that have the most viable appearance. The Jackson Laboratory advises that embryos be transferred immediately after they have been thawed (http://jaxmice.jax.org). Prolonged culture may progressively compromise further development and should be avoided.

Embryo transfer procedures have been described in detail elsewhere (Johnson et al. 1996; Morrell 1999; Nagy et al. 2003) and information provided here serves to highlight only those aspects critical to success of the embryo transfer procedure.

In either mice or rats, synchronised pseudopregnant females should be used as recipients. Even if freshly fertilised oocytes are transferred, the ovarian cycle of 4–5 days is too short for them to develop sufficiently for implantation before natural regression of the cyclic corpora lutea. The solution to this problem is to induce pseudopregnancy by mating with a vasectomised male. This serves as a trigger for the production of prolactin, increased blood concentrations of which prolong secretion of progesterone by the corpora lutea. Circulating progesterone concentrations remain high for 11–12 days and are independent of fertilisation. The uterus of (pseudo) pregnant surrogate mothers remains receptive to an implanting embryo till the fourth day of (pseudo) pregnancy.

The time of onset of pseudopregnancy and stage of embryonic development must be synchronised. This means that the stage of pseudopregnancy of the surrogate mother should be equivalent to, or somewhat earlier than that of the embryos to be transferred.

During embryonic development, the embryo transits through two very different anatomical regions of the female reproductive tract—the uterine horn and the oviduct. An ovulated oocyte, surrounded by its cumulus cells is fertilised in the ampulla of the oviduct. Gradually the granulosa cells become detached from the zygote and the embryo begins to move towards the uterine horn. In mice, embryos spend 60–70 hours in the oviduct before they reach the compacted morula stage. Then they cross the uterotubal junction and enter the uterine horn where implantation takes place.

Embryos should be transferred into the anatomical area corresponding to their stage of development, although there is some degree of flexibility. Rodent embryos are able to enter a state called "diapause", a physiological condition that allows the embryo to down-regulate its metabolism till the correct environment for implantation is available. In mice, a developmental delay of up to 12 days for transferred embryos has been observed (Rulicke et al. 2006). As a rule of thumb, in a synchronic transfer, the embryo should be more advanced than the area to which it is inserted; for example, blastocysts can be transferred into the oviduct, but one cell stage embryos should never be transferred to the uterus.

Depending on the strain, mice and rats produce litters that may vary in size from 1 or 2 to 15 or more pups. The physiological limit to the number of foetuses was established in the early 1960s, when McLaren demonstrated that super foetation started to compromise litter size only when extremely high numbers of foetuses were carried at the same time. No significant decrease in viability was observed even with litter sizes of 18 or 19 in a single horn (McLaren and Biggers 1958). The ability of uterine horns to expand and provide shelter to a large number of implantations diminishes concern about transferring too may embryos into a recipient and wasting them. However, the insertion of excessively large numbers of embryos is not advisable because mice have only 10 mammary glands and some small offspring may not be able to survive. In practise no more than 30 embryos are transferred to one uterus horn.

The strain of the foster mother can differ from that of the embryos that are being transferred. Selection of the foster strain should be based on good maternal care and robustness, for example, B6CBAF1 mice. However, it has been demonstrated that both foster maternal strain and pup strain are key determinants of maternal behaviour (van der Veen et al. 2008). Consequently, the observed behaviour of offspring may differ from predictions based on their genetic background.

Unlike the situation in domestic species such as pigs, migration of embryos from one uterus horn into the other horn happens only extremely rarely in rodents (Rulicke et al. 2006). The anatomical explanation for this phenomenon is that in rodents the uterine lumen is divided into two separate cavities, the two horns. This union between the uterine lumens is in the upper part of the cervical canal, which makes trans-uterine migration nearly impossible.

Embryos are transferred to either one or both uterine horns; both result in successful pregnancies. For animal welfare reasons, one-sided transfer is preferred, but there is little scientific rationale for preferring one over the other. There is a report that the uterine horns differ regarding the ability to support embryonic development and that the right side is more successful in Swiss-Webster and a hybrid strain of mice (Wiebold and Becker 1987). Another study has shown that there may be a position effect within a given uterine horn; embryos were lighter at both the ovarian and cervical extremities (von Domarus, Louton, and Lange-Wuhlisch 1986). Without additional scientific evidence, both unilateral and bilateral transfer procedures are acceptable.

Embryo transfer is a surgical procedure that can take 10–15 minutes in expert hands. Inhalation anaesthesia is preferred. The most controversial aspect of surgical transfer is the use of analgesia, which has been said to interfere with embryo implantation. However, it has recently been found that opiates did not interfere with embryonic development following uterine transfer into day 2.5 recipients (Krueger and Fujiwara 2008).

9.2.4.3.2 Spermatozoa Cryopreservation and Recovery

In contrast to the situation in livestock, where cryopreservation of spermatozoa is not incompatible with high *in vivo* fertilisation rates, cryopreservation of mice spermatozoa has presented a technical challenge. The morphology of these cells is quite different from the pear-like shape of spermatozoa of many other mammals. This appears to limit membrane elasticity and ability to survive freezing. In addition, cryoprotective agents are extremely toxic to mice spermatozoa (Willoughby et al. 1996). The first successful cryopreservation of murine spermatozoa was achieved in 1990 (Tada et al. 1990). Since then several authors have reported that these protocols have poor reproducibility and there is considerable strain-to-strain variation. The most widely used protocol is based on the use of raffinose and skim milk (Takeshima, Nakagata, and Ogawa 1991). Protocols have been published recently which are designed specifically to improve fertility of frozen and thawed C57BL/6 sperm using cryoprotective agents such as cycloheximide and monothioglycerol (Ostermeier et al. 2008; Takeo et al. 2008). These cryoprotective agents stabilise the sperm membrane during freezing and thawing, although their full impact on the methodology has yet to be evaluated.

9.2.4.3.2.1 In Vitro Fertilisation (IVF) For IVF, frozen spermatozoa are less efficient than fresh semen. However, an improved protocol has been published that consistently yields fertilisation rates of 70% ± 5% with C57BL/6 (Ostermeier et al. 2008). Many different protocols, media and methods have been described in the literature but all of them follow the same basic principles. Spermatozoa need a period of incubation before they are added to the cultured oocytes; during this time they become capacitated, characterised by several physiological changes including an influx of calcium into the cell, a certain destabilisation of the membranes and development of a hyperkinetic status that allows them to penetrate the zona pellucida and to release enzymes from the acrosome when in contact with the oocyte so as to permit fertilisation.

Oocytes must be harvested before they experience ageing associated with so-called hardening of the zona pellucida, which inhibits fertilisation. The highest fertilisation rates are achieved with oocytes collected 13–14 hours after hCG administration. Older females produce fewer oocytes and these appear to age more quickly than those harvested from younger donors. It has not been established whether embryos produced by juvenile females have the same developmental competence as those isolated from mature females.

9.2.4.3.2.2 Intracytoplasmic Sperm Injection (ICSI) Intracytoplasmic sperm injection (ICSI) can result in fertilisation even with completely immobile spermatozoa following thawing (Wakayama and Yanagimachi 1998; Szczygiel et al. 2002; Hirabayashi et al. 2005; Li et al. 2009). ICSI may then offer the last opportunity for saving a line, but it requires considerable skill, expertise and special equipment. Intracytoplasmic injection is a method of assisted reproduction that in mice has allowed recovery of lines from spermatozoa frozen even without cryoprotectant, from lyophilised sperm heads and in extreme cases from dead sperm cells collected from the cauda epididymis of animals frozen at –20°C for more than 20 years. However, it is preferable to use cells frozen with cryoprotectants or maintained in liquid nitrogen, because DNA is subject to degradation and fragmentation when kept in suboptimal conditions and that compromises development of the resulting zygote (Perez-Crespo et al. 2008). As mentioned before, the introduction of ICSI made sperm preservation the easiest and most accessible way of preserving the genetic material of transgenic lines. ICSI in mice faces particular technical problems resulting from the elastic nature of the zona pellucida and egg membrane, the relatively small quantity of cytoplasm, and the distinctive shape of the head of rodent sperms. When use of the piezo drill was described, the methodology became more widely practised (Kimura and Yanagimachi 1995; Ergenc et al. 2008). Alternatively, a laser beam can be used for the same purpose. Both require considerable operator skill.

9.2.4.3.3 Oocyte and Ovarian Tissue Cryopreservation

There are no real technical advantages in preserving mature oocytes rather than embryos because their collection from donors is just as demanding. However, it is the only available way of preserving the genotype of a line when there is only one founder or a unique female available. After collection, additional treatment is necessary to separate the ovulated oocytes from sticky cumulus cells. Vitrification is the freezing method of choice, because oocytes are larger than blastomere cells so depletion of their cytoplasmic water is more difficult.

Successful generation of live offspring from xenografted ovaries in mice (Snow et al. 2002) and rats (Dorsch et al. 2007) suggests that cryopreservation of ovaries could be considered a reliable method provided that orthotopical transplantation into syngenic recipients is feasible. However, further improvement of the method is necessary to make it available as routine.

9.2.4.4 Genome Resource Banking Management

Although comprehensive and accurate record keeping is essential in any animal facility, this is especially true for a genetic resource bank. The technician who freezes the sample may not be the same as the one who thaws it some considerable time later. For this reason records have to provide extremely accurate information about: (1) the material frozen, (2) the method used for freezing, including the recommended thawing procedure, (3) sample location and identification codes.

During storage, care should be taken to avoid: (1) sample temperature oscillations when searching for a specific straw or cryotube; (2) unnecessary evaporation of liquid nitrogen; and (3) unnecessary long-term exposure of personnel to cold. Cryocontainers, whether they are straws or tubes, should be clearly marked with at least a unique identification code. The markings and/or labels used should be of sufficient quality to withstand low temperatures. Extended records should be kept to serve as an inventory either on written cards and/or as a computer database. The IETS provides recommendations in its handbook that should be used as a reference (http://www.iets.org/).

9.3 CONCLUSIONS

Technological advances and demands for more complex animal models pose challenges for the field of laboratory animal science. The demand for high-quality, GM animal models of defined genetic constitution, and micro-biological status and assurance of their well-being requires specialist knowledge about, and expertise in their breeding and maintenance. Accurate genotyping, phenotyping and determination of the micro-biological status are the basis of effective characterisation of a transgenic model. Facility structures and housing equipment must meet the requirements for housing transgenic rodents, some aspects of which have been translated into European and national law. Generation of transgenic lines is expensive and demands considerable investment in personnel as well as in infrastructure, research and development. For these reasons, GM animals represent a considerable financial investment quite apart from their scientific value, and the risk of loss should be minimised as far as possible; this involves establishing procedures for assured preservation and rederivation programmes for genetic material.

9.4 QUESTIONS UNRESOLVED

There is an urgent need for a comprehensive, highly accessible database listing GM strains, their genotypes and phenotypes, and other relevant characteristics.

Recommendations for health monitoring lead to exclusion of micro-organisms from colonies and this has generated phenotypic changes of model systems. Research is needed to investigate the effects that individual and combined species of micro-organisms exert on the phenotype of at least the major WT inbred strains.

A means must be found of efficiently combining tissue sampling for genotyping and animal identification that has the least impact on animals' well-being.

Clear advice is needed about how comprehensive phenotypic screens should be of conditional knock out mice.

REFERENCES

Armstrong, N. J., T. C. Brodnicki, and T. P. Speed. 2006. Mind the gap: Analysis of marker-assisted breeding strategies for inbred mouse strains. *Mammalian Genome* 17:273–87.

Arndt, S. S. and D. Surjo. 2001. Methods for the behavioural phenotyping of mouse mutants. How to keep the overview. *Behavioural Brain Research* 125:39–42.

Bailey, K. R., N. R. Rustay, and J. N. Crawley. 2006. Behavioral phenotyping of transgenic and knockout mice: Practical concerns and potential pitfalls. *ILAR Journal* 47:124–31.

Baker, D. G. 1988. Natural pathogens of laboratory rats, mice, and rabbits and their effects on research. *Clinical Microbiology Reviews* 11:231–66.

Baker, E. J., L. Galloway, B. Jackson, D. Schmoyer, and J. Snoddy. 2004. MuTrack: A genome analysis system for large-scale mutagenesis in the mouse. *BMC Bioinformatics* 5:11.

Baker, H. J. 1988. Rederivation of inbred strains of mice by means of embryo transfer. *Laboratory Animal Science* 38:661–62.

Barthold, S. W. 2004. Genetically altered mice: Phenotypes, no phenotypes, and faux phenotypes. *Genetica* 122:75–88.

Benavides, F., D. Cazalla, C. Pereira, A. Fontanals, M. Salaverri, A. Goldman, V. Buggiano, G. Dran, and E. Corley. 1998. Evidence of genetic heterogeneity in a BALB/c mouse colony as determined by DNA fingerprinting. *Laboratory Animals* 32:80–85.

Bielanski, A., H. Bergeron, P. C. Lau, and J. Devenish. 2003. Microbial contamination of embryos and semen during long term banking in liquid nitrogen. *Cryobiology* 46:146–52.

Bolon, B. 2006. Internet resources for phenotyping engineered rodents. *ILAR Journal* 47:163–71.

Brielmeier, M., E. Mahabir, J. R. Needham, C. Lengger, P. Wilhelm, and J. Schmidt. 2006. Microbiological monitoring of laboratory mice and biocontainment in individually ventilated cages: A field study. *Laboratory Animals* 40:247–60.

Brown, S. D., P. Chambon, and M. H. de Angelis. 2005. EMPReSS: Standardized phenotype screens for functional annotation of the mouse genome. *Nature Genetics* 37:1155.

Brown, S. D. M., J. M. Hancock, and H. Gates. 2006. Understanding mammalian genetic systems: The challenge of phenotyping in the mouse. *PLoS Genetics* 2:e118.

Castelhano-Carlos, M. J., N. Sousa, F. Ohl, and V. Baumans. 2010. Identification methods in newborn C57BL/6 mice: A developmental and behavioural evaluation. *Laboratory Animals* 44:88–103.

Charsa, R., S. Knoblauch, and W. Ladiges. 2003. Phenotypic characterization of genetically engineered mice. In *Handbook of laboratory animal science*, eds. J. Hau and G. L. Van Hoosier 205–30. Vol. I. New York: CRC Press.

Committee on Recognition and Alleviation of Distress in Laboratory Animals. 2008. *Recognition and alleviation of distress in laboratory animals,* 115. Washington, DC: National Research Council.

Compton, S. R., F. R. Homberger, and J. MacArthur Clark. 2004. Microbiological monitoring in individually ventilated cage systems. *Laboratory Animal Europe* 4:28–35.

Crabbe, J. C., D. Wahlsten, and B. C. Dudek. 1999. Genetics of mouse behavior: Interactions with laboratory environment. *Science* 284:1670–2.

Crawley, J. N. 2003. Behavioral phenotyping of rodents. *Comparative Medicine* 53:140–46.

Crawley, J. N. and R. Paylor. 1997. A proposed test battery and constellations of specific behavioral paradigms to investigate the behavioral phenotypes of transgenic and knockout mice. *Hormones and Behavior* 31:197–211.

Davisson, M. 2006. FIMRe: Federation of international mouse resources: Global networking of resource centers. *Mammalian Genome* 17:363–64.

Dorsch, M. M., D. Wedekind, K. Kamino, and H. J. Hedrich. 2007. Cryopreservation and orthotopic transplantation of rat ovaries as a means of gamete banking. *Laboratory Animals* 41:247–54.

Ergenc, A. F., M. W. Li, M. Toner, J. D. Biggers, K. C. Lloyd, and N. Olgac. 2008. Rotationally oscillating drill (ros-drill) for mouse ICSI without using mercury. *Molecular Reproduction and Development* 75:1744–51.

Fray, M. D., A. R. Pickard, M. Harrison, and M. T. Cheeseman. 2008. Upgrading mouse health and welfare: Direct benefits of a large-scale rederivation programme. *Laboratory Animals* 42:127–39.

Gkoutos, G. V., E. C. Green, A. M. Mallon, J. M. Hancock, and D. Davidson. 2005. Using ontologies to describe mouse phenotypes. *Genome Biology* 6:R8.

Grubb, S. C., G. A. Churchill, and M. A. Bogue. 2004. A collaborative database of inbred mouse strain characteristics. *Bioinformatics* 20:2857–59.

Grubb, S. C., T. P. Maddatu, C. J. Bult, and M. A. Bogue. 2009. Mouse phenome database. *Nucleic Acids Research* 37:D720–D730.

Hancock, J. M., N. C. Adams, V. Aidinis, A. Blake, M. Bogue, S. D. Brown, E. J. Chesler, et al. 2007. Mouse phenotype database integration consortium: Integration [corrected] of mouse phenome data resources. *Mammalian Genome* 18:157–63.

Hirabayashi, M., M. Kato, J. Ito, and S. Hochi. 2005. Viable rat offspring derived from oocytes intracytoplasmically injected with freeze-dried sperm heads. *Zygote* 13:79–85.

Ike, F., F. Bourgade, K. Ohsawa, H. Sato, S. Morikawa, M. Saijo, I. Kurane, et al. 2007. Lymphocytic choriomeningitis infection undetected by dirty-bedding sentinel monitoring and revealed after embryo transfer of an inbred strain derived from wild mice. *Comparative Medicine* 57:272–81.

Johnson, L. W., R. J. Moffatt, F. F. Bartol, and C. A. Pinkert. 1996. Optimization of embryo transfer protocols for mice. *Theriogenology* 46:1267–76.

Joshi, M., H. K. Pittman, C. Haisch, and K. Verbanac. 2008. Real-time PCR to determine transgene copy number and to quantitate the biolocalization of adoptively transferred cells from EGFP-transgenic mice. *Biotechniques* 45:247–58.

Kimura, Y., and R. Yanagimachi. 1995. Intracytoplasmic sperm injection in the mouse. *Biology of Reproduction* 52:709–20.

Kong, I. K., S. I. Lee, S. G. Cho, S. K. Cho, and C. S. Park. 2000. Comparison of open pulled straw (OPS) vs glass micropipette (GMP) vitrification in mouse blastocysts. *Theriogenology* 53:1817–26.

Krueger, K. L., and Y. Fujiwara. 2008. The use of buprenorphine as an analgesic after rodent embryo transfer. *Laboratory Animals (NY)* 37:87–90.

Li, M. W., B. J. Willis, S. M. Griffey, J. L. Spearow, and K. C. Lloyd. 2009. Assessment of three generations of mice derived by ICSI using freeze-dried sperm. *Zygote* 17:239–51.

Linder, C. C. 2006. Genetic variables that influence phenotype. *ILAR Journal* 47:132–40.

Macas, E., M. Xie, P. J. Keller, B. Imthurn, and T. Rulicke. 1991. Developmental capacities of two-cell mouse embryos frozen by three methods. *Journal of In Vitro Fertilization and Embryo Transfer* 8:208–12.

Mallon, A. M., A. Blake, and J. M. Hancock. 2008. EuroPhenome and EMPReSS: Online mouse phenotyping resource. *Nucleic Acids Research* 36:D715D718.

Mandillo, S., V. Tucci, S. M. Holter, H. Meziane, M. A. Banchaabouchi, M. Kallnik, H. V. Lad, et al. 2008. Reliability, robustness, and reproducibility in mouse behavioral phenotyping: A cross-laboratory study. *Physiological Genomics* 34:243–55.

Marques, J. M., H. Augustsson, S. O. Ögren, and K. Dahlborn. 2007. *Refinement of mouse husbandry for improved animal welfare and research quality*. Sweden: Swedish University of Agricultural Sciences.

Mazur, P., S. P. Leibo, and G. E. Seidel, Jr. 2008. Cryopreservation of the germplasm of animals used in biological and medical research: Importance, impact, status, and future directions. *Biology of Reproduction* 78:2–12.

McIlwain, K. L., M. Y. Merriweather, L. A. Yuva-Paylor, and R. Paylor. 2001. The use of behavioral test batteries: Effects of training history. *Physiology & Behavior* 73:705–17.

McLaren, A., and J. D. Biggers. 1958. Successful development and birth of mice cultivated in vitro as early as early embryos. *Nature* 182:877–78.

Mertens, C., and T. Rulicke. 2000. A comprehensive form for the standardized characterization of transgenic rodents: Genotype, phenotype, welfare assessment, recommendations for refinement. *Altex* 17:15–21.

Meyer, C. W., R. Elvert, A. Scherag, N. Ehrhardt, V. Gailus-Durner, H. Fuchs, H. Schafer, M. Hrabe de Angelis, G. Heldmaier, and M. Klingenspor. 2007. Power matters in closing the phenotyping gap. *Die Naturwissenschaften* 94:401–6.

Mitrecic, D., M. Huzak, M. Curlin, and S. Gajovic. 2005. An improved method for determination of gene copy numbers in transgenic mice by serial dilution curves obtained by real-time quantitative PCR assay. *Journal of Biochemical and Biophysical Methods* 64:83–98.

Morrell, J. M. 1999. Techniques of embryo transfer and facility decontamination used to improve the health and welfare of transgenic mice. *Laboratory Animals* 33:201–6.

Morton, D. B., and J. Hau. 2003. Phenotypic characterization of genetically engineered mice. In *Handbook of laboratory animal science*, eds. J. Hau and G. L. Van Hoosier, 457–81. Vol. I. New York: CRC Press.

Murray, K. A. 2002. Issues to consider when phenotyping mutant mouse models. *Laboratory Animals (NY)* 31:25–29.

Nagy, A., M. Gertsenstein, K. Vintersten, and R. Behringer. 2003. *Manipulating the mouse embryo. A laboratory manual*, 3rd ed. Cold Spring Harbor, NY: Cold Spring Harbor Laboratory Press.

Nicklas, W. 2008. International harmonization of health monitoring. *ILAR Journal* 49:338–46.

Nicklas, W., P. Baneux, R. Boot, T. Decelle, A. A. Deeny, M. Fumanelli, and B. Illgen-Wilcke. 2002. Recommendations for the health monitoring of rodent and rabbit colonies in breeding and experimental units. *Laboratory Animals* 36:20–42.

Nitzki, F., A. Kruger, K. Reifenberg, L. Wojnowski, and H. Hahn. 2007. Identification of a genetic contamination in a commercial mouse strain using two panels of polymorphic markers. *Laboratory Animals* 41:218–28.

Ostermeier, G. C., M. V. Wiles, J. S. Farley, and R. A. Taft. 2008. Conserving, distributing and managing genetically modified mouse lines by sperm cryopreservation. *PLoS ONE* 3:e2792.

Perez-Crespo, M., P. Moreira, B. Pintado, and A. Gutierrez-Adan. 2008. Factors from damaged sperm affect its DNA integrity and its ability to promote embryo implantation in mice. *Journal of Andrology* 29:47–54.

Petkov, P. M., M. A. Cassell, E. E. Sargent, C. J. Donnelly, P. Robinson, V. Crew, S. Asquith, R. V. Haar, and M. V. Wiles. 2004. Development of a SNP genotyping panel for genetic monitoring of the laboratory mouse. *Genomics* 83:902–11.

Pinkert, C. A. 2003. Transgenic animal technology: Alternatives in genotyping and phenotyping. *Comparative Medicine* 53:126–39.

Rall, W. F., and G. M. Fahy. 1985. Ice-free cryopreservation of mouse embryos at −196 degrees C by vitrification. *Nature* 313:573–75.

Rall, W. F., P. M. Schmidt, X. Lin, S. S. Brown, A. C. Ward, and C. T. Hansen. 2000. Factors affecting the efficiency of embryo cryopreservation and rederivation of rat and mouse models. *ILAR Journal* 41:221–27.

Reetz, I. C., M. Wullenweber-Schmidt, V. Kraft, and H. J. Hedrich. 1988. Rederivation of inbred strains of mice by means of embryo transfer. *Laboratory Animal Science* 38:696–701.

Rogers, D. C., E. M. Fisher, S. D. Brown, J. Peters, A. J. Hunter, and J. E. Martin. 1997. Behavioral and functional analysis of mouse phenotype: SHIRPA, a proposed protocol for comprehensive phenotype assessment. *Mammalian Genome* 8:711–13.

Rulicke, T., A. Haenggli, K. Rappold, U. Moehrlen, and T. Stallmach. 2006. No transuterine migration of fertilised ova after unilateral embryo transfer in mice. *Reproduction, Fertility, and Development* 18:885–91.

Shek, W. R. 2008. Role of housing modalities on management and surveillance strategies for adventitious agents of rodents. *ILAR Journal* 49:316–25.

Smith, C. L., C. A. Goldsmith, and J. T. Eppig. 2005. The mammalian phenotype ontology as a tool for annotating, analyzing and comparing phenotypic information. *Genome Biology* 6:R7.

Snow, M., S. L. Cox, G. Jenkin, A. Trounson, and J. Shaw. 2002. Generation of live young from xenografted mouse ovaries. *Science* 297:2227.

Soliman, G. A., R. Ishida-Takahashi, Y. Gong, J. C. Jones, R. L. Leshan, T. L. Saunders, D. C. Fingar, and M. G. Myers, Jr. 2007. A simple qPCR-based method to detect correct insertion of homologous targeting vectors in murine ES cells. *Transgenic Research* 16:665–70.

Sorensen, D. B., C. Stub, H. E. Jensen, M. Ritskes-Hoitinga, P. Hjorth, J. L. Ottesen, and A. K. Hansen. 2007. The impact of tail tip amputation and ink tattoo on C57BL/6JBomTac mice. *Laboratory Animals* 41:19–29.

Suzuki, H., K. Yorozu, T. Watanabe, M. Nakura, and J. Adachi. 1996a. Rederivation of mice by means of in vitro fertilization and embryo transfer. *Experimental Animals* 45:33–38.

Suzuki, O., T. Asano, Y. Yamamoto, K. Takano, and M. Koura. 1996b. Development in vitro of preimplantation embryos from 55 mouse strains. *Reproduction, Fertility, and Development* 8:975–80.

Swallow, J., D. Anderson, A. C. Buckwell, T. Harris, P. Hawkins, J. Kirkwood, M. Lomas, et al. 2005. Guidance on the transport of laboratory animals. *Laboratory Animals* 39:1–39.

Szczygiel, M. A., H. Kusakabe, R. Yanagimachi, and D. G. Whittingham. 2002. Intracytoplasmic sperm injection is more efficient than in vitro fertilization for generating mouse embryos from cryopreserved spermatozoa. *Biology of Reproduction* 67:1278–84.

Szell, A., and J. N. Shelton. 1986. Role of equilibration before rapid freezing of mouse embryos. *Journal of Reproduction and Fertility* 78:699–703.

Tada, N., M. Sato, J. Yamanoi, T. Mizorogi, K. Kasai, and S. Ogawa. 1990. Cryopreservation of mouse spermatozoa in the presence of raffinose and glycerol. *Journal of Reproduction and Fertility* 89:511–16.

Takeo, T., T. Hoshii, Y. Kondo, H. Toyodome, H. Arima, K. Yamamura, T. Irie, and N. Nakagata. 2008. Methyl-beta-cyclodextrin improves fertilizing ability of C57BL/6 mouse sperm after freezing and thawing by facilitating cholesterol efflux from the cells. *Biology of Reproduction* 78:546–51.

Takeshima, T., N. Nakagata, and S. Ogawa. 1991. Cryopreservation of mouse spermatozoa. *Jikken Dobutsu. Experimental Animals* 40:493–97.

van der Veen, R., D. N. Abrous, E. R. de Kloet, P. V. Piazza, and M. Koehl. 2008. Impact of intra- and interstrain cross-fostering on mouse maternal care. *Genes, Brain, and Behavior* 7:184–92.

Van Keuren, M. L., and T. L. Saunders. 2004. Rederivation of transgenic and gene-targeted mice by embryo transfer. *Transgenic Research* 13:363–71.

Vergara, G. J., M. H. Irwin, R. J. Moffatt, and C. A. Pinkert. 1997. In vitro fertilization in mice: Strain differences in response to superovulation protocols and effect of cumulus cell removal. *Theriogenology* 47:1245–52.

Voikar, V., E. Vasar, and H. Rauvala. 2004. Behavioral alterations induced by repeated testing in C57BL/6J and 129S2/Sv mice: Implications for phenotyping screens. *Genes, Brain, and Behavior* 3:27–38.

von Domarus, H., T. Louton, and F. Lange-Wuhlisch. 1986. The position effect in mice on day 14. *Teratology* 34:73–80.

Wakayama, S., H. Ohta, T. Hikichi, E. Mizutani, T. Iwaki, O. Kanagawa, and T. Wakayama. 2008. Production of healthy cloned mice from bodies frozen at −20 degrees C for 16 years. *Proceedings of the National Academy of Sciences of the United States of America* 105:17318–322.

Wakayama, T., and R. Yanagimachi. 1998. Development of normal mice from oocytes injected with freeze-dried spermatozoa. *Nature Biotechnology* 16:639–41.

Wells, D. J., L. C. Playle, W. E. Enser, P. A. Flecknell, M. A. Gardiner, J. Holland, B. R. Howard, et al. 2006. Assessing the welfare of genetically altered mice: Laboratory environments and rodents' behavioural needs. *Laboratory Animals* 40:111–14.

Weyand, M. E. 1998. Methods for the individual identification of laboratory mice and rats. *Laboratory Animals* 27:42–49.

Whittingham, D. G., S. P. Leibo, and P. Mazur. 1972. Survival of mouse embryos frozen to −196 degrees and −269°C. *Science* 178:411–14.

Wiebold, J. L., and W. C. Becker. 1987. Inequality in function of the right and left ovaries and uterine horns of the mouse. *Journal of Reproduction and Fertility* 79:125–34.

Willoughby, C. E., P. Mazur, A. T. Peter, and J. K. Critser. 1996. Osmotic tolerance limits and properties of murine spermatozoa. *Biology of Reproduction* 55:715–27.

Wishart, T. M., S. H. Macdonald, P. E. Chen, M. J. Shipston, M. P. Coleman, T. H. Gillingwater, and R. R. Ribchester. 2007. Design of a novel quantitative PCR (QPCR)-based protocol for genotyping mice carrying the neuroprotective wallerian degeneration slow (wlds) gene. *Molecular Neurodegeneration* 2:21.

Wurbel, H. 2002. Behavioral phenotyping enhanced: Beyond (environmental) standardization. *Genes, Brain, and Behavior* 1:3–8.

Zhao, X. M., G. B. Quan, G. B. Zhou, Y. P. Hou, and S. E. Zhu. 2007. Conventional freezing, straw, and open-pulled straw vitrification of mouse two pronuclear (2-PN) stage embryos. *Animal Biotechnology* 18:203–12.

10 Impact of Handling, Radiotelemetry, and Food Restriction

Timo Nevalainen, Finland; Marlies Leenaars, The Netherlands; Vladiana Crljen, Croatia; Lars Friis Mikkelsen, Denmark; Ismene Dontas, Greece; Bart Savenije, The Netherlands; Carlijn Hooijmans, The Netherlands; and Merel Ritskes-Hoitinga, The Netherlands

CONTENTS

Objectives	228
Key Factors	229
10.1 Handling and Restraint	230
10.1.1 Factors Promoting Adjustment of Animals to Handling and Restraint	230
10.1.1.1 Frequent Handling	230
10.1.1.2 Familiarity of the Environment	230
10.1.1.3 Communication	230
10.1.1.4 Cage Complexity	230
10.1.2 Handling Methods	231
10.1.2.1 The Person	231
10.1.2.2 The Animal's Prior Experience	231
10.1.2.3 The Comparisons	231
10.1.2.4 Habituation to Handling	233
10.1.3 Comparison of Restraint Methods	233
10.1.3.1 Species Differences	233
10.1.3.2 Habituation to Restraint	234
10.1.4 Comparison of Handling and Restraint to Other Common Procedures	235
10.2 Implantable Telemetry	236
10.2.1 Refinement Possibilities	236
10.2.1.1 Transmitter Size	236
10.2.1.2 Peri- and Post-operative Care	237
10.2.2 Reduction Possibilities	238
10.2.3 Scientific Integrity	238
10.2.3.1 Observer Effect	238
10.2.3.2 Accuracy and Precision of Telemetry	239
10.3 Food Restriction in Laboratory Rodents and Rabbits	240
10.3.1 Benefits of Food Restriction in Rodents	240
10.3.1.1 Reduced Mortality	240
10.3.1.2 Health	241
10.3.1.3 Uniformity	243

		10.3.1.4	Control	244
		10.3.1.5	Sensitivity	245
	10.3.2	Practical Aspects of Food Restriction		246
	10.3.3	Food Restriction in Rabbits		247
10.4	Conclusions			248
10.5	Questions Unresolved			249
References				250

OBJECTIVES

Virtually no procedure on laboratory animals can be conducted without concomitant handling or restraint. Yet, handling and restraint are potential stressors that have consequences for both the welfare of the animals, and the results of the study (Gattermann and Weinandy 1996; Clark, Rager, and Calpin 1997). For the purposes of refinement, it is necessary to identify the least stressful techniques and the ones to which animals have been proven to habituate.

Until now, the main emphasis when refining procedures has been directed at painful procedures; and methods to alleviate pain, suffering and distress, and quite correctly so. However, at the same time it is important to strive to improve the welfare of animals overall. In this respect, gains attainable at the level of an individual animal are much smaller, but when viewed in the context of a large number of animals, the total outcome may well be a considerable refinement. Basic animal handling and restraint are typical examples of these kinds of procedures; and the question is whether there are proven ways to refine them.

Radiotelemetry is a widely used method for registering electric signals without the need to connect wires or catheters; no handling or restraint is needed. Another major advantage of telemetry is that signals from animals can be obtained unobtrusively and there is no need to restrict movement of the animal. As such it has been considered as both a refinement and reduction alternative in experiments involving laboratory animals. However, in small animals the use of radiotelemetry involves surgical implantation of a transmitter, and thereafter the animals have to live with the implanted sensing and transmitting devices, both of which potentially compromise animal welfare. This chapter is based on published original studies and assesses the impact of the technology on refinement and reduction and examines its contribution to scientific progress.

The restriction of food intake, rather than *ad libitum* feeding, is reported to result in more "robust" animals (Hart, Neumann, and Robertson 1995), meaning that they are able to cope better with experimental stressors (Refinement), and for this reason is considered in this chapter. Ethical and technical aspects of food restriction have been a matter of discussion for over half a century. There are many reports that restricting food intake (75% of *ad libitum* intake) increases life span and leads to health improvements, in particular reduction in obesity, cardiovascular and renal degenerative disease and tumour incidence, not only in rodents but also many other species, including invertebrates (Keenan, Smith, and Soper 1994; Masoro 2005). These consequences are interesting for animal experimentation, because food restriction makes it possible to conduct long-term studies with fewer animals (Reduction), and fewer confounding effects on metabolism and pathologies can be expected. *Ad libitum* feeding is defined as ensuring that diet is available at all times, whereas under food restriction the amount of food (energy) is restricted although its nutritional adequacy is maintained; that is, all essential nutrients are supplied at adequate levels (Hart, Neumann, and Robertson 1995).

The fact that restricted food consumption can increase longevity and health in animals is analogous to the human situation. Despite these well known and largely reproducible effects, *ad libitum* feeding remains the norm for laboratory rodents. For other laboratory species *ad libitum* feeding is considered bad practise, and therefore is not used (Hart, Neumann, and Robertson 1995), unless the purpose of the study is to investigate obesity and its related problems. For bioassays, *ad libitum* feeding is considered the most poorly controlled variable, leading to lack of reproducibility

(Keenan, Laroque, and Dixit 1998) and needless suffering. For long-term toxicological studies at least 25 animals per group per sex need to survive at least two years; some contract laboratories have adopted a standard scheme of food restriction for rodents because many rodents did not survive until the age of 2 years under conditions of *ad libitum* feeding. Unfortunately, a consequence of this food restriction scheme has been the adoption of individual housing in order to ensure a standard individual food intake. For animal welfare reasons, social species ought to be housed socially whenever possible and, consequently, there is a need for a refined way of housing rodents under restricted feeding conditions.

KEY FACTORS

Unlike humans, animals cannot be interviewed, but their reactions to stressors can be measured and then interpreted. When animals are stressed, their attempts to cope can be assessed by monitoring the activity of the sympathetic nervous system, hypothalamic-pituitary axis (HPA) activity, or their behaviour (see Chapter 14). The best methods for assessment avoid disturbing the animals or invasive sampling and recording. The first part of this chapter summarises available scientific studies on the effects of handling and restraining the most commonly used species (i.e., mice, rats and rabbits), and interprets these in relation to animal welfare. More specifically, this chapter deals with research reports relevant to animal care routines that identify ways of enhancing animal welfare, comparisons of the methods available and habituation to handling procedures.

Webster's dictionary defines telemetry as the science of the use of an electric instrument adapted for recording at a distance the readings of other instruments (1972). Telemetry studies with animals are common, and approaches to refinement of procedures and animal housing have been reviewed extensively by the BVAAWF/FRAME/RSPCA/UFAW Joint Working Group on Refinement (Morton et al. 2003; Hawkins et al. 2004).

Telemetry can be viewed both as a problem and a solution in terms of alternatives to the use of animals. Two of the three Rs alternatives introduced by Russell and Burch 50 years ago (Russell and Burch 1959), refinement and reduction, are applicable to telemetry studies. Russell and Burch argued that application of the three Rs should not be detrimental to the scientific outcome of animal studies (Russell and Burch 1959). We propose that a more ambitious approach should be taken: why not seek alternatives that lead to improvements in scientific quality? Not all two Rs methods satisfy the conditions set by Russell and Burch, and some may be associated for example with wastage of animals. Ethical considerations concerning animal studies should go further and include scientific reasoning. Telemetry is an example of a method that offers not only refinement and reduction, but also concomitantly enhances scientific integrity. For the purposes of clarity these will be dealt with separately.

The effects of food restriction can be quite spectacular. In rats, food restriction has been reported to increase two-year survival rate up to three times as compared with *ad libitum* feeding. In addition, almost no neoplasms and non-neoplastic lesions were found on autopsy following restricted feeding, whereas they were a significant problem with *ad libitum* feeding (48.3% and 11.7% cause of death, respectively; Hubert et al. 2000). These results were found following quite severe food restriction (55%), which could lead to specific deficiencies and related problems, but moderate food restriction (75–85% of *ad libitum* food intake) also results in beneficial effects and does not cause deficiencies.

From a scientific point of view, it is important to understand the mechanisms by which food restriction leads to so many favourable health effects. These positive effects lead to improved welfare because of the reduced incidence of disease (Refinement). The challenge is to determine those feeding schemes (amount of food available at what time of day) and housing systems that ought to be used in order to assure enhanced animal welfare under restricted feeding conditions (Refinement). Because food restriction prolongs life, and standardising the food intake of each animal is likely to reduce inter-animal variation in experimental results, it is expected that by applying food restriction instead of *ad libitum* feeding to rodents, fewer animals will be needed for statistical assessment of effect (Reduction). There is a wealth of information available on the comparison of *ad libitum* versus restricted feeding in rodents. For rabbits, food restriction should be the normal feeding practice in order to prevent obesity (Hart, Neumann, and Robertson 1995). Issues surrounding food restriction in rabbits are slightly different from those in rodents, and therefore this topic will be dealt with separately.

10.1 HANDLING AND RESTRAINT

10.1.1 Factors Promoting Adjustment of Animals to Handling and Restraint

Animals' ability to adapt to handling or restraint is strongly influenced by their earlier experiences. These may be used to accustom animals to those procedures and thereby improve animal welfare. These early factors can be incorporated into the daily routines of a well-managed facility; others should be incorporated as an integral part of the procedure. There are surprisingly few studies on this topic even though there may be potential for considerable refinement.

10.1.1.1 Frequent Handling

Daily handling of rat pups at the age of 14 and 18 days appears to increase secretion of both noradrenaline and adrenaline, compared to minimally handled controls; this effect does not occur during the preceding week and it fades away by adulthood (McCarty et al. 1978). Similar handling of rabbit pups has a profound effect on their response to handling later in life; in this case the sensitive window is during the first week of life. In contrast to the situation in rats, in rabbits the effect is long lasting; that is, rabbits handled when young approach people more readily in adulthood, thereby demonstrating reduced fear (Bilko and Altbacker 2000).

10.1.1.2 Familiarity of the Environment

Familiarity with the surroundings in which restraint takes place appears to influence the HPA response. Rats housed in three different ways displayed the same HPA responses to acute restraint and to repeated restraint in their home room. However, if rats, which had been restrained daily for seven days in a familiar room, were exposed to the same procedure on the 8th day in a novel context, habituation was attenuated (Grissom et al. 2007). The effect may be modulated by communication between animals.

10.1.1.3 Communication

Communication between animals experiencing stress involves olfactory and auditory signals, and is clearly involved in responses to stressful procedures. A stress response in one animal has been shown to evoke physiological or psychological consequences in undisturbed animals housed in the same space. Exposure of adult mice and rats to handling, restraint, injection or killing of conspecifics results in increased blood corticosterone concentration (Smolensky et al. 1978; Pitman, Ottenweller, and Natelson 1988), hyperthermia (Kikusui et al. 2001), elevated heart rate (Sharp et al. 2003b), and heightened immunological responsiveness (Fernandes 2000) as compared to animals left undisturbed. When procedures are carried out sequentially on animals group housed in a cage, there seems to be an element of anticipation that results in biological variation as a function of the order in which they are picked up (Knott, Hutson, and Curzon 1977; Brodin et al. 1994).

10.1.1.4 Cage Complexity

Changes in the complexity of the internal environment of cages cause a variety of physiological responses (Lemaire and Mormede 1995; Sharp, Azar, and Lawson 2005). So far there have been few investigations into the role of housing refinements as a possible way of refining procedures. The results have been controversial at best. Housing large groups of rats in an enriched environment with three times the floor area had no effect on the anticipatory reaction to normal handling (Augustsson et al. 2002). In a similar study with mice, Moons, Van Wiele, and Odberg (2004), concluded that the provision of shelters did not make catching or handling more difficult. Moreover, groups of 10 rats housed in large polycarbonate cages containing a variety of environmental enrichment objects that were changed three times each week, developed higher resting plasma concentrations of corticosterone and larger adrenals compared to controls. In the case of rats housed in the complex cages, a single handling elicited lower ACTH, corticosterone and adrenaline responses, while repeated handling led to more rapid habituation than in control rats (Moncek et al. 2004). In female mice, both greater cage complexity and handling increased heart rate and core body temperature, but the blood corticosterone concentration, measured 90 minutes after restraint, suggested that animals from the enriched groups were less stressed (Meijer et al. 2006).

10.1.2 Handling Methods

Handling—holding the animal for less than a minute—can be either manual or involve the use of restraint device. Handling has both acute and long-lasting effects, but with the exception of habituation, the latter, have not been studied at all.

10.1.2.1 The Person

The person doing the actual handling has a major influence on the success of the procedure and its outcome with respect to both the welfare of the animal and to the precision of the experiment (Augustsson et al. 2002), and ideally, handling should be rewarding to the animal (Davis and Perusse 1988). When comparing responses to handling it is important that handlers are experienced if valid comparisons are to be made. Nonetheless, sources of variability arising from the handler remain largely unknown (Gartner et al. 1980; Dobrakovova et al. 1993; Thompson, Brannon, and Heck 2003; van Driel and Talling 2005).

10.1.2.2 The Animal's Prior Experience

One way of decreasing stress in rats is to familiarise them with the surroundings (Burman and Mendl 2004); gentle handling is also important. Michel and Cabanac (1999) measured the body weight set point of rats subjected to gentle handling, a procedure known to induce an emotional rise in body temperature. The body weight set point was estimated from the rat's food hoarding behaviour and calculated as the intersection of the regression line for hoarding with the body weight axis. When handled, the body weight set point declined, which is believed to result from an elevation of hypothalamic corticotrophin-releasing hormone (CRH; Michel and Cabanac 1999). Concomitantly, handling elevated body temperature and the rats ate less and defecated more frequently than controls.

Not surprisingly, handling rats for 5 seconds caused the blood concentration of corticosterone and prolactin to rise dramatically while growth hormone concentration fell. All the three hormones reached maximum change about 15 minutes after handling, and did not return to pre-handling values within the following 60 minutes. This suggests that hormone responses to brief handling are rather complex (Seggie and Brown 1975).

10.1.2.3 The Comparisons

At present radiotelemetry is the technique of choice for detecting small differences in cardiovascular parameters (Anderson et al. 1999; Kramer et al. 2001; Harkin et al. 2002). However, the high sensitivity and precision of telemetry can also present a problem. Krohn, Hansen, and Dragsted (2003) suggested that changes in cardiovascular parameters should be considered to have practical welfare implications only if they are greater than 6–7% in rats. Also, there seems not to be a linear relationship between recorded cardiovascular responses and the impact of a procedure on well-being, for example, inbred BN rats are much less responsive, both in terms of blood pressure and heart rate, than F344 rats. Perhaps a different index (e.g., strain specific night–day differences of the parameter) might provide a more reliable method of assessing welfare implications (Kemppinen et al. 2009).

Handling induces large perturbances in cardiovascular function, including an increase in heart rate, blood pressure, and body temperature that can last for up to 60 minutes (Irvine, White, and Chan 1997; Harkin et al. 2002; Sharp et al. 2003b). None of these studies compared responses to different methods of handling; instead, they only examined one method.

A recent study used telemetry to compare cardiovascular responses in Wistar and Sprague-Dawley (SD) stock rats in response to three different methods of restraint: scruffing, holding on the arm with the tail trapped and the body encircled, and a plastic cone (Baturaite et al. 2005). In that study, daily handling for three consecutive days revealed differences in heart rate and pressure responses between the methods; the cone was associated with the greatest change in cardiovascular parameters and so appeared to be the worst method. Although the differences were not large, they were statistically significant. However, this study does suggest that there is potential for refining handling techniques and that it is appropriate to select handling techniques with care. Although manual handling seems to be preferable to restraint based on technical devices, it must be emphasised that mastery of the handling technique used is of major importance.

The temperament and behaviour of different *stocks or strains* of animals differ and appreciation of this is important in determining the success that can be expected. In mice particularly, the vast number of strains means there is a wide variety of behavioural and physiological characteristics that influence the ease of handling (Wahlsten, Metten, and Crabbe 2003). Similar differences occur in rats; for example F344 inbred rats are social, but difficult to handle; whereas BN rats are aggressive and easy to handle (Rex et al. 1996; van der Staay and Blokland 1996). One consequence of these behavioural differences is that it is easy to elicit an immobilisation reflex by holding BN rats by the nape of the neck, an outcome impossible with F344 rats (Figure 10.1).

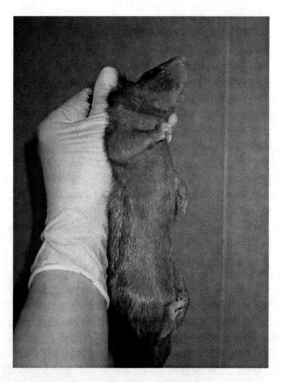

FIGURE 10.1 BN rat displaying reflex while held by the nape of the neck.

10.1.2.4 Habituation to Handling

All practical training emphasises the need for animals to be habituated; for example in standard tests for anxiety, ensuring that the animal is accustomed to the experimenters improves the consistency of the results (Bilko and Altbacker 2000; van Driel and Talling 2005). Repeated handling of mice on a daily basis reduced the rise in rectal temperature, which was interpreted as habituation of emotional origin (Cabanac and Briese 1992).

A few studies have attempted to compare the impact of different handling methods by measuring changes in the serum concentration of hormones such as ACTH, corticosterone, prolactin, adrenaline and noradrenaline (Dobrakovova and Jurcovicova 1984; Dobrakovova et al. 1993). Prolactin secretion in particular, has long been known to increase in response to stress, and can increase by threefold within just 5 seconds of handling (Seggie and Brown 1975). Nonetheless, collection of serum necessitates blood sampling, which in itself is a stressor and can interfere with interpretation of the results.

Repeated touching of a sensitive area of skin such as the nape of the neck leads quickly to habituation, as assessed by the amount of ultrasonic vocalisation. The long latencies, their repetition following a single touch, combined with rapid habituation to touch all suggest that 22 kHz vocalisation is a distress signal emitted in response to potential danger to the animal, not to physical discomfort or pain (Brudzynski and Ociepa 1992). Another study provided evidence of communication specific to the intensity of stress; in this, untreated rats were held in the same room as stressed rats, and subsequently were found to have elevated blood corticosterone concentrations similar to those of the stressed rats. Only rats housed with mildly stressed rats habituated completely within three days (Pitman, Ottenweller, and Natelson 1988).

The most common method of handling, known as scruffing, involves holding a large portion of the skin of the dorsum in one hand, or alternatively taking hold only of skin dorsal to the neck. The latter elicits a reflex simulating the mother rat carrying a pup (Webster et al. 1981; Mileikovsky, Verevkina, and Nozdrachev 1994; Wilson and Kaspar 1994). Over a period of three days rats seem to habituate only to repeated handling by this nape of the neck scruff method; the duration of the response is about 30% shorter on the 2nd and 3rd day. No such habituation was seen with methods involving encircling and restraining the tail on the arm or with the plastic cone. This finding points to a way of considerably enhancing the welfare of the animals (Baturaite et al. 2005), and is in line with the known high sensitivity of the nape of the neck in rats (Brudzynski and Ociepa 1992).

10.1.3 COMPARISON OF RESTRAINT METHODS

Often, mice and rats need to be restrained for long periods (>10 minutes) and this usually involves using tube-like devices, such as polycarbonate piping or the so called Bollman Cages (Bollman 1948). Such restraint is required for various purposes, for example, sampling, prolonged drug administrations and recordings and it has even been used as a method of producing stress in the animal. It is these stress studies that provide us with a relatively large body of information about the consequences of restraint on animal welfare; direct comparison of restraint methods has been much less extensively reported. One complicating factor remains: that is, the duration of restraint, which ranges from 10 minutes to 4 hours.

10.1.3.1 Species Differences

Restraint of mice causes an increase in the concentration of circulating corticosteroids, suppresses immune responsiveness (Blecha, Kelley, and Satterlee 1982), lowers lymphocyte and elevates neutrophil counts (Brehe and Way 2008), and decreases intestinal IgA secretion (Jarillo-Luna et al. 2007). Following a short (5 minute) period of restraint, heart rate declined whilst the animal was in the restraining tube, and rose again when it was returned to its home cage (Meijer et al. 2007).

As in the case with handling, rat strains and stocks seem to differ in their response to restraint. Several studies report that the increase in corticosterone concentration during restraint was larger in spontaneously hypertensive (SHR) rats than Wistar Kyoto rats (McCarty et al. 1978; McMurtry and Wexler 1981; McDougall et al. 2000) and also larger in F344 than in Wistar rats (Li et al. 2004). Kvetnansky and others (1992) showed that both handling and immobilisation of rats causes an increase in the plasma concentration of noradrenaline and adrenaline, with a peak at one minute of immobilisation (Kvetnansky et al. 1992).

Irvine, White, and Chan (1997) evaluated blood pressure responses to restraint in SHR and Wistar/Kyoto (WKY) control rats using telemetry. They found that heart rate and blood pressure can increase considerably as a result of restraining the animals by the method commonly used during blood pressure recording, a factor that raises questions about the reliability of the tail-cuff method.

10.1.3.2 Habituation to Restraint

Daily physical restraint of rapidly growing mice over a period of a week retarded their growth. However, there seemed to be a compensatory mechanism that resulted in subsequent full recovery of energy balance (Laugero and Moberg 2000). Repetition of the restraint for periods of one or three weeks did not result in habituation, as assessed from the adrenocortical response (Tuli, Smith, and Morton 1995).

The majority of studies of habituation have been carried out on rats and have been based on activity of the HPA-axis. It has been reported that HPA responses habituate with repeated exposure to restraint (Stamp and Herbert 1999; Cole et al. 2000; Gadek-Michalska and Bugajski 2003; Carter, Pinnock, and Herbert 2004; Bhatnagar, Lee, and Vining 2005). Nonetheless, other studies have failed to find evidence of habituation (Kant et al. 1985; Ling and Jamali 2003; Mikkelsen 2007). Table 10.1 summarises the findings from studies on habituation to restraint.

In rats, repeated restraint results in acute release of corticotropin-releasing factor (CRF), temporary hypophagia and chronic weight loss. Additional weight loss occurred following a second period of restraint, but additional periods of restraint daily for 10 days did not lead to further weight loss (Harris et al. 2002). The response of the HPA axis differed markedly between two inbred strains of rat: F344 and Lewis, with the former strain being more reactive. When assessed on the basis of increased blood catecholamine concentration, evidence for habituation was found on both the second and third occasion (DeTurck and Vogel 1980).

Repeated restraint has led to higher peak corticosterone concentrations in F344 rats than in Lewis or Wistar rats. By the time of the fourth episode of restraint, there was evidence of adaptation of the response in the F344 rats, but not in the other strains (Li et al. 2004). Clement and others (1998), using male Lewis rats, found that 5–10 minute periods of immobilisation, repeated daily, were enough to achieve habituation of serotonin metabolism although serum corticosterone concentration remained unchanged.

Restraint for a single period of two hours has been shown to reduce subsequent locomotion and increase defecation in male rats after they had been placed in an open field, though when the restraint was repeated daily these effects were no longer seen on the 5th day. This habituation was associated with enhanced sensitivity to a serotonin agonist. Female rats responded less to a single period of restraint but failed to adapt to the repeated stress (Kennett et al. 1986). Changes in plasma prolactin and pituitary cyclic AMP concentrations in rats subjected to 15 minutes of repeated restraint daily were attenuated on the 11th day, but not to a stressor that they had not previously experienced, suggesting that habituation was a consequence of behavioural experience (Kant et al. 1985).

Daily periods of restraint lasting one hour for 9 days resulted in marked tachycardia and transient hypothermia, both of which peaked at about 10 minutes and declined thereafter and both of which normalised during repeated restraint (Stamp and Herbert 2001). In another telemetry study, daily restraint for periods of 1 hour caused tachycardia and pressor responses in WKY and SHR rats, changes being larger in SHR rats. The duration of tachycardia in the WKY rats remained constant over 10 days, but a graded reduction occurred in the SHR rats (McDougall et al. 2000). Finally, a

TABLE 10.1
Summary of Studies on Habituation to Repeated Restraint in Mice and Rats

	Study	Stock/Strain: Sex	Type of Restraint	Restraint Duration	# of Daily Repetitions	Parameters	Habituation Yes/No
Mice	(Laugero and Moberg 2000)	C57BL/6 M	tube	4 hours	1, 3, & 7	Energy balance, corticosterone	No
	(Tuli, Smith, and Morton 1995)	BALB/c M	tube	1 hour	7 & 21	Adrenocortical response	No
Rats	(Albonetti and Farabollini 1993)	Wistar M	tight box	20–180 minutes	7	Behaviour	No
	(Casada and Dafny 1990)	Sprague-Dawley M	small box	1 hour	1 & 4	Sensory evoke potentials	No
	(Bhatnagar, Lee, and Vining 2005)	Sprague-Dawley F	tube	30 minutes	8	HPA-axis Heart rate	Yes
	(Carter, Pinnock, and Herbert 2004)	Lister Hooded M	tube	30 & 60 minutes	5	HPA-axis Heart rate	Yes
	(Cole et al. 2000)	Sprague-Dawley M	tube	1 hour	6	HPA-axis	Yes
	(Gadek-Michalska and Bugajski 2003)	Wistar M	tube	10 minutes	3 & 7	HPA-axis	Yes
	(Stamp and Herbert 1999)	Lister Hooded M	tube	60 minutes	1, 7, & 14	HPA-axis	Yes
	(Kant et al. 1985)	Sprague-Dawley M	tube	15 minutes	10	HPA-axis Prolactin	No
	(Ling and Jamali 2003)	Sprague-Dawley M	?	1 hour	4	HPA-axis	No
	(Mikkelsen 2007)	Sprague-Dawley F	Bollman cage and tube	3 hours	10	Blood pressure, heart rate	No
	(Kant et al. 1985)	Sprague-Dawley M	tube	15 minutes	10	HPA-axis Prolactin	Yes
	(Stamp and Herbert 2001)	Lister Hooded M	tube	60 minutes	9	Blood pressure Heart rate	Yes
	(McDougall, Lawrence, and Widdop 2005)	WKY SHR	tube	1 hour	10	Blood pressure Heart rate	Yes

more recent study found no evidence of habituation after 3 hours of daily restraint for 10 consecutive days (Mikkelsen 2007).

10.1.4 Comparison of Handling and Restraint to Other Common Procedures

Although all procedures on animals elicit a stress response, the key question is how can they be ranked in terms of animal welfare? There have been surprisingly few attempts to do this. Harkin and others (2002) concluded that in rats, procedures could be ranked in increasing order of stress as follows: white noise/confinement in a mouse cage, novel odour, encircling handling, subcutaneous injection, exposure to bedding soiled by another animal and housing with a new cage-mate. Gattermann and Weinandy (1996) used telemetry to measure heart rate and core temperature in golden hamsters and ranked the effect on these of several procedures. They reported that the stress responses during the light phase increased in the following order: handling, vaginal smear, intruder/resident confrontation, cage change and grouping. In both rankings, handling was located at the lower end of the scale of stressfulness.

10.2 IMPLANTABLE TELEMETRY

10.2.1 REFINEMENT POSSIBILITIES

10.2.1.1 Transmitter Size

When deer mice, weighing about 24 g, received intra-abdominal implants weighing 1, 2, or 3 g (4.1, 8.3, or 12.5% of body weight), they lost tissue mass, proportional to the mass implanted, although food intake decreased only with the 3 g implant. Removal of the weights restored body mass, but only transiently (Adams, Korytko, and Blank 2001). In laboratory mice, intra-abdominal implantation of dummy transmitters (10.7–12.5% of body weight) resulted in considerable weight loss during the first few days, but by two weeks they had regained the weight of the controls (Baumans et al. 2001).

Both the weight and volume of the transmitter may compromise behaviour, especially in mice. Baumans and others (2001) used an automated monitoring system to assess the behavioural consequences of inserting a transmitter into the abdominal cavity of BALB/c and 129/Sv mice. Climbing and locomotion were compromised immediately after implantation, while grooming and immobility increased, but by two weeks these behaviours had normalised, suggesting complete recovery.

Somps (1994) examined the effects of four implants, representing 15% of body weight, on behaviour of rats and found no differences in growth, food or water intake, body temperature, or activity rhythms. Moran and others (1998) examined the effect of implant size in young rats, and reported that subcutaneous implants weighing 35 g (20.6% of body weight) or more caused skin lesions, seromas, and retarded growth. In a subsequent study they compared implants of 12.5 g and 23.5 g (7.4% and 13.8% of body weight), and observed no ill effects, except for an increase in adrenal weight following the heavier implant, compared to naive or sham operated controls.

Leon and others (2004) studied the effects on body temperature, locomotor activity, growth, food and water intake and diurnal rhythm of implanting into the abdominal cavity, transmitters whose weight had a fixed ratio to body weight (16.6% for mice; 1.2% for rats); post-operative analgesia was not used. It required two weeks for the mice to return to their preoperative weight, but they never caught up with the controls, even though food intake returned to normal by 3 days. The rats responded differently; body weight did not decrease below starting levels and food intake normalised within 2 days. Induction of anaesthesia with isoflurane, whether or not followed by laparatomy, caused the body weight of mice to fall for 5–7 days, but for no more than 1 day in rats, illustrating the additional burden of implantation. In both species, diurnal temperature and activity rhythms returned to baseline levels within the first week. In summary, comparison of body weight/temperature ratios may be confounded by species effects, hence it is not conclusive. Furthermore, the results would have been more applicable to the practical situation if analgesics had been administered.

Modern transmitters are considerably smaller than those evaluated in the studies quoted above; transmitters for use in mice are about 8–19% of body weight, while those for rats are 4–6%. Transmitters developed for use in mice can be implanted into rats smaller than 175 g, though at the cost of shorter battery life (Braga and Prabhakar 2009). The impact of modern telemetry transmitters (2–3% of body weight) on well-being has been assessed in both F344 and BN rats. After a two-week recovery period, faeces were collected twice weekly over a 16-week period; in both strains, rats carrying transmitters displayed higher faecal corticosteroid concentration than controls although immunoglobulin A excretion remained unchanged (Kemppinen et al. Unpublished manuscript).

10.2.1.2 Peri- and Post-operative Care

Intraperitoneal implantation involves laparotomy with attendant risks of peritonitis and, if re-positioning of animals is necessary during surgery, asepsis may be compromised. A subcutaneous location may be preferable, and when combined with sterile skillfully conducted surgical techniques, and proper use of analgesic agents, it has been shown to result in fewer complications from infection or, seromas, and quicker recovery (Larsen and Bastlund 2006).

Johnston and others (2007) reported several refinement strategies applicable to telemetry studies in mice. They compared recovery and survival of nine different mouse strains following combined abdominal implantation and cranial surgery; the outcome correlated with initial body weight; 80–95% of mice over 26 g survived this double intervention, compared with only 44% animals weighing 24 g. Reversal of anaesthesia by administration of atipamezole was associated with increased mortality, but the provision of thermal support, which is often advocated, had little or no effect on recovery. The authors concluded that procedures designed to facilitate recovery that are based on data from other species may not always be appropriate in mice.

The need for pain alleviation following surgery has been evaluated both in mice and rats. Hayes and others (2000) assessed several analgesic regimens after intra-abdominal implantation of transmitters in mice. Mice received ibuprofen in the drinking water, parenteral administration of buprenorphine or remained untreated as controls; the time taken for recovery of food and water intake, locomotor activity and core temperature was determined. Animals receiving oral ibuprofen were more active than controls from day 2 onwards. However, mice given buprenorphine were even more active, maintained body temperature better, but had reduced food and water intake during both that period and also thereafter. Assessment of analgesic efficacy using physiologic and behavioural measures is useful and should be carried out at least once with the selected analgesic regime following every new surgical procedure (see Chapter 13 for more information).

The effects of analgesic agents and assessment of their efficacy on recovery of rats following transmitter implantation have been examined by several groups. Sharp and others (2003a) compared post-implantation recovery following three analgesic regimes (buprenorphine, ketoprofen, and butorphanol), supplemented by subcutaneously administered fluids (a control group received fluids only). In all groups, mean arterial blood pressure (MAP) had normalised by day two after surgery, and food consumption within 12 days. Heart rate, nocturnal activity, water consumption, and body weight gain recovered within one week. The authors were unable to explain the apparent lack of effectiveness of analgesics, but their findings underline the importance of carefully monitoring the effectiveness of analgesic regimes in each group of animals after surgical intervention.

Roughan and Flecknell (2001, 2004) have identified pain related behaviours that can be used to determine the efficacy of pain relief measures in rats following surgery such as laparotomy. They described quantifiable activities; back arching, horizontal stretching followed by abdominal writhing and twitching while inactive, and showed that the frequency of these behaviours can be reduced by treatment with analgesics. Ketoprofen and carprofen were equipotent in Wistar rats irrespective of dose, and provided analgesia for 4–5 hours at a dose of 5 mg/kg (Roughan and Flecknell 2001). In another study, they compared orally administered buprenorphine and subcutaneously administered carprofen or buprenorphine in F344 rats, and concluded that the most acute pain in this model lasted for between 270 and 390 minutes, and that all of these analgesic regimens alleviated this throughout its duration (Roughan and Flecknell 2004). In all, this way of assessing efficacy of pain alleviation provides a basis for recommending that it should be used routinely following abdominal surgery in rats.

In order to establish the time it takes for recovery after abdominal insertion of telemetry transmitters, Greene Clapp, and Alper (2007) assessed MAP, HR, body temperature, body weight, food consumption and activity in male SD rats for two weeks following surgery. The surgery was done under isoflurane anaesthesia using a single injection of carprofen the following morning. The MAP tended to be higher on day 3 and activity was less on day 3 than on days 8 or 15, whereas HR and body temperature displayed no change. Food consumption regained pre-surgery levels by day 9. Body weight on day 4 was the same as that pre-surgery and thereafter the rats gained weight. The authors concluded that these rats recovered from the surgery in approximately one week (Greene, Clapp, and Alper 2007).

Butz and Davisson (2001) used transmitters placed subcutaneously on the flank of female mice of two strains: C57/BL6 (C57) and BPH/5, to monitor blood pressure within the thoracic aorta. Although there was no morbidity or mortality, neither strain fully recovered from anaesthesia and surgery (as indicated by return of normal circadian rhythms and baseline MAP and HR levels), until 5–7 days after the surgery (Butz and Davisson 2001).

Under normal conditions, rats display regular rhythmic changes in body temperature, which on an average is one degree higher during the 12-hour dark cycle than in the light phase. Stress disrupts the rhythm and markedly decreases the night–day difference, with the disruption being most marked during the 1st days of stress (Kant et al. 1991). The circadian rhythm of temperature (CRT) has been used to monitor the post-surgical recovery of rats. One week after insertion of an intra-abdominal transmitter connected to EMG electrodes on a neck muscle and EEG electrodes on the skull, the amplitude of CRT was low and average temperature was elevated. The baseline CRT was not re-established before the third week, suggesting that rats may require at least three weeks for full recovery (Bastlund et al. 2004).

10.2.2 Reduction Possibilities

The principal opportunity for reduction arising from the use of telemetry is that the same animals can be used in longitudinal studies, in which measurements are conducted repeatedly; and indeed the effect is considerable in comparison with the number of animals needed with traditional methods. Another, less obvious benefit arises from the ability to make measurements from undisturbed, freely moving animals that consequently are less variable. Ultimately, both these aspects result in more information from fewer animals.

10.2.3 Scientific Integrity

The use of telemetry avoids the need for handling and restraint of animals during recording; it has been shown that these latter result in profound effects on many parameters that can be recorded with telemetry. Moreover, not only the animal being manipulated, but also other animals in the cage are affected when they become aware that a procedure is being done on their cage mates, as discussed below.

10.2.3.1 Observer Effect

The sensitivity of telemetry has been illustrated by monitoring HR and body temperature in mice for 60 minutes after a cage mate had been weighed or restrained for 60 minutes, then returned to the cage. Both of these parameters increased in the observer mice during both restraint and recovery, but the response during restraint was considerably greater than after weighing. Once the animals were placed back into the cage after either procedure, HR and body temperature were significantly higher in observer animals than in those manipulated. As judged by the same measures, repeating the procedures on days 1, 3, 7 and 14 led to habituation in the observer mice following restraint (Gilmore, Billing, and Einstein 2008).

Observer effects of decapitation, cage change and restraint combined either with subcutaneous injection or with tail-vein injection have been studied in SD rats. Male rats, witnessing decapitation (whether associated with necropsy or not) developed small increases in HR and MAP only in singly housed animals; in solitary or pair-housed rats these parameters increased during routine cage changing. Both indices increased transiently when rats witnessed animals being restrained and given a tail-vein injection (Sharp et al. 2002). Female rats, housed in pairs or larger groups, displayed lower HR and MAP responses to acute husbandry and experimental procedures than those housed singly (Sharp et al. 2003b). Thus, it appears that male SD rats are not stressed when observing common experimental procedures (Sharp et al. 2002). Female SD rats may be marginally stressed when witnessing decapitation, but similar reactions occur when observing common husbandry and experimental procedures. In both sexes, these responses were less in group-housed rats (Sharp et al. 2002, 2003b).

10.2.3.2 Accuracy and Precision of Telemetry

An observer effect combined with changes induced by manipulation may impact on both the accuracy and precision of experimental findings. Whitesall and others (2004) compared telemetrically recorded systolic blood pressure (SBP) with values obtained simultaneously using a tail-cuff in conscious mice. In their hands, tail-cuff measurements did not always accurately reflect induced increases in blood pressure or circadian changes. They could not explain the observed disagreement between operator and instrument-specific tail-cuff measurements.

The action of antihypertensive drugs in telemetry studies is affected by animal manipulation. In rats, evidence concerning the accuracy of telemetry is conflicting. Both spontaneously hypertensive and WKy rats displayed elevated heart rate and blood pressure values when restrained in the manner used for tail-cuff readings (Irvine, White, and Chan 1997) indicating that telemetry has significant advantages in comparison with methods that require handling and restraint. In contrast with the findings by Irvine, White, and Chan (1997), another study concluded that telemetrically obtained measurements of SBP, MAP, diastolic blood pressure (DBP), and HR in SD rats were all higher than values obtained using a tail-cuff (Abernathy, Flemming, and Sonntag 1995). Both of these studies used animals trained specially for this procedure. The coefficient of variation for measurements using the tail-cuff was between 1.75 and 3 times larger than those obtained using telemetry (Abernathy, Flemming, and Sonntag 1995). In summary, cardiovascular telemetry has more power to detect small statistical differences, or to reveal an effect with fewer animals (i.e., it has the potential to reduce the number of animals required).

Two reference substances were used in a validation study; one of these markedly stimulates locomotor activity at low to moderate doses and induces stereotypy at higher doses, and the other triggers a well-defined behavioural syndrome and a dose-dependent hypothermia. The results confirmed that telemetry is a valid method for measuring activity, core temperature, and cardiovascular parameters. Moreover, the rats were less stressed than those restrained for more invasive measurements and it was claimed that the technique could be combined with other observations to produce a more complete ethogram of the animal's responses to pharmacological challenges (Deveney et al. 1998). Likewise, EEG signals derived by telemetry from a freely-moving rat can be obtained during the expression of behavioural tasks in a limited space (Cotugno et al. 1996). Indeed, the ability to simultaneously record behaviour and gather telemetric data from freely moving animals is a major advantage compared to other methods (Diamant et al. 1993).

Telemetric recording of pressure need not be confined to the cardiovascular system, and can be applied to measuring pressure also in other body compartments (e.g., in the aqueous humour of the eyeball), thereby circumventing the blood-pressure elevating effect of traditional measuring techniques based on tonometry (McLaren, Brubaker, and FitzSimon 1996). Using rabbits as their study animals, Dinslage, McLaren, and Brubaker (1998) compared the effect of tonometry, animal handling and water drinking on intraocular pressure and found that changes induced by these activities were of the similar magnitude to the effects of compounds being studied, which makes it difficult to attribute the effects to either one.

The use of telemetry to record core body temperature and activity has been shown to more accurately assess the temporal effects of various pharmaceutical agents and to be more efficient and less labour intensive than use of a rectal temperature probe (Clement, Mills, and Brockway 1989).

Telemetry has been used to monitor gastrointestinal motility in freely moving, unstressed animals. The electromyographic (EMG) signals were picked up by two electrodes sutured to the jejunum of rats and connected to an implantable EMG transmitter; both fasting and post-prandial jejunal activity were registered. The resting, phasic contraction pattern was disrupted 15 minutes after the start of feeding and lasted for 2 hours (Meile and Zittel 2002).

10.3 FOOD RESTRICTION IN LABORATORY RODENTS AND RABBITS

10.3.1 Benefits of Food Restriction in Rodents

10.3.1.1 Reduced Mortality

Food restriction usually leads to a significantly reduced mortality rate. Figure 10.2 depicts a typical survival graph of rats in a two-year study (Hubert et al. 2000). Under *ad libitum* feeding conditions, mortality rapidly becomes a serious problem after the age of 1 year, and is generally higher in males than in females. Table 10.2 summarises the findings of a survey of 2-year survival figures reported in rat and mice studies.

The data in the graph and table illustrate three things clearly in relation to moderate levels of food restriction: (1) the two-year mortality rate can be reduced greatly by restricting food intake, and higher levels of food restriction are associated with a lower mortality rate; (2) the level of food

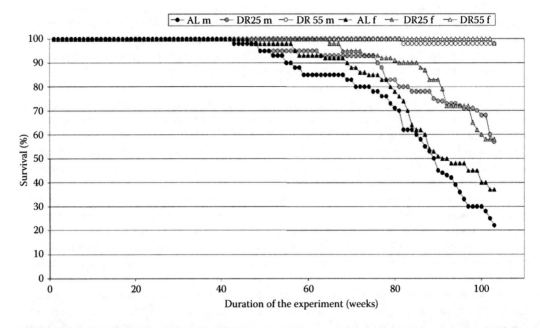

FIGURE 10.2 Survival rates of Sprague-Dawley rats during a two-year study involving three levels of dietary restriction. AL = *Ad libitum*, DR25 = 75% of *ad libitum* food is given, DR55 = 45% of *ad libitum* food is given in male (m) and female (f) rats. (From Hubert, M. F., Laroque, P., Gillet, J. P., and Keenan, K. P., *The Journal of Toxicological Sciences,* 58, 195–207, 2000. With permission from Oxford Press.)

TABLE 10.2
Two-Year Survival Rates for Several Rat and Mice Strains and Stocks Fed *Ad Libitum* and at Different Levels of Food Restriction

Species/Strain	Food Restriction (% of *Ad Lib* Given)	Two-Year Survival Rate (Per Gender)	Source
B6C3F1 mice	Ad libitum	83% (m)	(Leakey et al. 2003)
	80%	94% (m)	
C3B10F1 mice	Ad libitum	85% (f)	(Weindruch 1996)
	85 kcal/week	92% (f)	
	50 kcal/week	98% (f)	
	40 kcal/week	98% (f)	
C57BL/6	Ad libitum	64% (m)	(Forster, Morris, and Sohal 2003)
	60%	89% (m)	
B6D2F1	Ad libitum	59% (m)	
	60%	91% (m)	
DBA/2	Ad libitum	42% (m)	
	60%	63% (m)	
Sprague-Dawley rats	Ad libitum	22% (m), 37% (f)	(Hubert et al. 2000)
	75%	57% (m), 58% (f)	
	45%	98% (m), 98% (f)	
Sprague-Dawley rats	Ad libitum	18% (m), 18% (f)	(Molon-Noblot et al. 2003)
	76%	44% (m), 40% (f)	
	70%	68% (m), 56% (f)	
	48%	78% (m), 82% (f)	
Fischer 344 rats	Ad libitum	45% (m)	(Yu et al. 1982)
	60%	84% (m)	

restriction is related inversely to mortality rate; and (3) the effect of food restriction can vary according to species, strain and gender. Caution is needed in the case of severe food restriction schedules, because nutrient deficiencies can occur and would adversely affect the health and welfare of the animals and the outcome of the study. Moderate food restriction (i.e., 75–85% of ad libitum food intake) does not lead to nutrient deficiencies. For more severe food restriction regimes it is essential to calculate the intake of all nutrients, and where necessary, to provide supplements to assure intake of the minimum required amounts of essential nutrients (National Research Council 1995; http://www.nap.edu).

In rats, there is evidence that food restriction reduces mortality rate and thus prolongs life expectancy, although in mice this effect does not appear to be as clear. In order to determine whether life is indeed prolonged, it is possible to evaluate mortality-doubling-time when fitting a Gompertz curve to the entire data set, or just to the oldest 10% of animals (Masoro 2005).

10.3.1.2 Health

Food restriction is an important issue in toxicity and carcinogenicity studies (Hart, Neumann, and Robertson 1995). Body weight, tumour incidence, time of first occurrence and rate of progress are important read-out parameters in such testing and reference values for these parameters in control animals have been reported (Cheung et al. 2003; Fukuda and Iida 2003; Son and Gopinath 2004; Brix et al. 2005). In particular, the increased incidence of tumours in control animals fed *ad libitum* confounds interpretation of agent-induced tumours (see Figure 10.3). If the variability of endpoints such

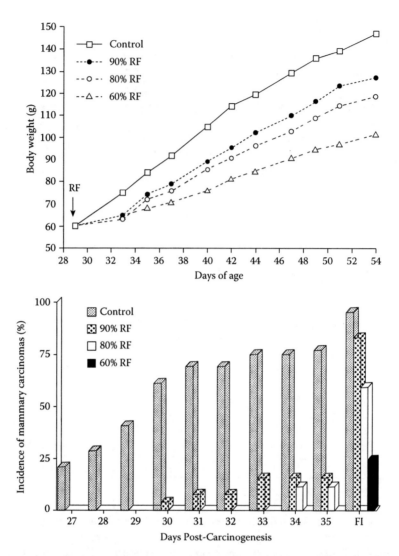

FIGURE 10.3 Effect of restricted feeding (RF) on body weight (top graph) cumulative and final incidences (FI) of mammary carcinomas (bottom graph) following i.p. administration of 50 mg/kg 1-methyl-1-nitrosourea (MNU) to 21-day-old female Sprague-Dawley rats. Differences in latency of tumour development were evaluated using a life table and differences in overall incidence were evaluated by chi-square analyses. RF was associated with a dose-dependent prolongation of the latency before tumours developed ($P < .01$) and a dose-dependent reduction in final tumour incidence ($P < .01$). (Data from Zhu, Z., Haegele, A. D., and Thompson, H. J., *Carcinogenesis*, 18(5), 1007–12, 1997. With permission from Oxford Press.)

as survival, background tumour incidence and body weight increases, there is a reduction in the reliability of safety assessment studies (Allaben et al. 1996). Caloric restriction can significantly reduce the variability in body weight, the incidence of spontaneous or induced tumour and delay the time of their first occurrence, although the progress of tumour development is not much affected (Zhu, Haegele, and Thompson 1997; Hubert et al. 2000; Leakey, Seng, and Allaben 2003). In a meta-analysis across studies, animals fed energy restricted diets were found to develop 55% fewer mammary tumours than the *ad libitum* fed control groups (Dirx et al. 2003).

The use of food restriction regimes instead of *ad libitum* feeding for studies into carcinogenesis can raise a dilemma—not only is the development of background tumours suppressed, but also the growth of induced tumours. Although it can be argued that the expression of induced tumours in the absence of spontaneous tumour development is more representative of the effect of the substance tested, the longer duration of experiments to evaluate the carcinogenic potential of a compound may raise welfare and resource considerations in the highly competitive field of drug development (T. Coenen 2008, personal communication) and may delay scientific progress in other areas. However, there is also a need to establish whether individual variation in food intake of animals maintained on *ad libitum* feeding might itself compromise the validity of scientific findings more than supplying a standardised restricted food intake to each individual.

Other health benefits associated with food restriction, as compared to *ad libitum* feeding, include a clearly reduced incidence of renal and cardiovascular degenerative diseases (Keenan, Smith, and Soper 1994) and reduction of age-related muscle pathologies (Molon-Noblot et al. 2005). Caloric restriction also attenuates ameloid-β deposition in plaques in the brains of transgenic mice, which spontaneously develop pathology related to Alzheimer's disease (Patel et al. 2005). There is increasing evidence suggesting that food restriction enhances mechanisms associated with glucose-insulin metabolism, for example, it has been reported to exert protective effects on diabetes pathologies (Bluher, Kahn, and Kahn 2003; Keenan et al. 2005).

10.3.1.3 Uniformity

Many treatment effects are judged only by the average value of the results (the mean). The variation around the mean, usually expressed as Standard Deviation or Coefficient of Variation (standard deviation expressed as a percentage of the mean), is often used only for statistical calculations, and frequently ignored as a valuable source of information in itself. From a laboratory animal science point of view, reduced variation in results leads to fewer animals being needed in order to establish whether there is a statistically significant effect (Reduction).

As a result of the general focus on mean values, background variation in response to the treatment is often not even reported in publications, and even when it is, the importance of this variability is usually ignored. Food restriction results in lower body weights, and a lower coefficient of variation as compared to *ad libitum* feeding. Table 10.3 shows the effects of *ad libitum* feeding and moderate (25%) and severe (55%) food restriction during a two-year experiment.

Based on these data, power calculations were carried out to determine how many animals would be needed to significantly detect a treatment difference in body weight of 10% between 29 week old

TABLE 10.3
Body Weights (Grams) of Control Sprague-Dawley Rats in a Two-Year Study

	Week 29		Week 53		Week 104	
	Mean ± SD	CoV (%)	Mean ± SD	CoV (%)	Mean ± SD	CoV (%)
AL group						
Males	677 ± 78	11.5	772 ± 109	14.1	782 ± 82	10.5
Females	348 ± 45	12.9	432 ± 76	17.6	556 ± 125	22.5
DR 25%						
Males	498 ± 25	5.0	575 ± 28	4.9	592 ± 46	7.8
Females	364 ± 16	6.1	292 ± 22	7.5	315 ± 32	10.2
DR 55%						
Males	288 ± 19	6.6	322 ± 20	6.2	325 ± 22	6.8
Females	169 ± 13	7.7	189 ± 13	6.9	196 ± 17	8.7

Source: Hubert, M. F., Laroque, P., Gillet, J. P., and Keenan, K. P., *The Journal of Toxicological Sciences*, 58, 195–207, 2000.
DR = dietary restriction.
Coefficients of Variation (CoV) have been calculated from those data and are reported here.

AL, DR25 and DR55 fed animals. The number of animals required was 11, 2 and 4 male and 13, 3 and 5 female rats, respectively, $(1-\beta = 0.80, \alpha = 0.05)$. Compared to AL feeding, both food restriction regimes (particularly DR 25%) reduced the number.

The increased uniformity resulting from food restriction as compared to *ad libitum* feeding applies to many parameters, but not to all. Hubert et al. (2000) assayed a wide variety of immunological and chemical serum analytes; some of these were more uniform following food restriction, whereas others were not. The Coefficient of Variation of a few values increased following severe food restriction, which could be due in part to nutrient (sub)deficiencies. Usually, blood composition is closely regulated, but this can break down if the animal is confronted with nutrient deficiencies. Because Hubert et al. (2000) provided only 45% of the ration consumed during *ad libitum*, nutrient deficiencies are quite likely.

The increased uniformity following restricted feeding schedules during lactation also carries over to the next generation. Carney and others (2004) reported the effects on body weight of rat pups from CD dams fed *ad libitum* or at 90%, 70% or 50% of the *ad libitum* food intake. The results are shown in Figure 10.4. The Coefficient of Variation was reduced by moderate food restriction (90 and 70% of *ad libitum* intake) but severe restriction (50% of *ad libitum* intake) doubled it (data not shown). Also, the severe dietary restriction reduced the average body weights of pups to only 70% of that of pups whose dams received 70% of *ad libitum* intake. This suggests serious deficiencies in either the yield or composition of milk, and clearly indicates that 50% of *ad libitum* food intake was nutritionally inadequate.

10.3.1.4 Control

Ad libitum feeding has the advantage that it requires very little human labour. To restrict the food intake of laboratory rodents, they are often housed and fed individually, or at least are separated during feeding periods. The advantage of this labour-intensive method is that all animals are checked individually daily and that any animal not consuming its daily food allowance is detected immediately and can be examined clinically for health problems.

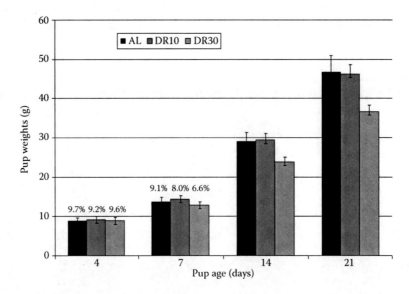

FIGURE 10.4 Weights of CD rat pups from dams fed during lactation *ad libitum* (AL), or at 90% (DR10) or 70% (DR30) of the *ad libitum* food intake. DR30 pups weighed slightly less from day 7. Coefficients of variation are calculated from these data and are reported above the weight bars. (From Carney, E. W., Zablotny, C. L., Marty, M. S., Crissman, J. W., Anderson, P., Woolhiser, M., and Holsapple, M., *Toxicological Sciences*, 82, 237–49, 2004. With permission from Oxford Press.)

Impact of Handling, Radiotelemetry, and Food Restriction

When a feeding station is used (Ritskes-Hoitinga 2006), food consumption of each individual can be registered automatically at each visit to the feeding station. This not only provides data about daily food consumption, but also feeding patterns and food consumed per visit. If combined with automated data capture, this strategy provides an additional, sensitive monitoring system for animal behaviour and health. The development and implementation of such feeding systems in rodent facilities provides more accurate scientific information and also enables animals to be housed socially. Moreover, the amount of diet and timing of meals can be modified according to the animals' behaviour and experimental needs.

10.3.1.5 Sensitivity

The effects of moderate food restriction described above contribute to developing more uniform experimental animals and the removal of confounding factors. To ascertain whether this leads to better animal models and experimental outcomes, it is necessary to compare animals fed *ad libitum* and those on restricted feeding regimens in both control and treatment groups. The report must provide means and standard deviations of output parameters to enable proper comparison. A review of the literature on this topic reveals that most scientific publications lack essential details of the experimental design, or results or, both so that it is impossible to establish whether food restriction would indeed have enabled the number of animals needed to be reduced. Because of this, Ritskes-Hoitinga (2006) found that of several dozens of articles on *ad libitum* and food restricted feeding, only a study by Leakey and others (2003) provided sufficient information to allow further analysis. The shortcomings were related either to the lack of one feeding group resulting in asymmetry of *ad libitum* and restricted feeding regimes in control and treatment groups, and/or failure to report standard deviations of means. In the study by Leakey and others (2003) several liver enzymes were measured following administration of different quantities of chloral hydrate in *ad libitum* and diet restricted B6C3F1 mice. Power calculations were performed on these data and the results are shown in Figure 10.5.

In addition to altering the liver-to-body weight ratio and concentrations of several hepatic enzymes, chloral hydrate caused a dose-dependent increase in the incidence of hepatocellular adenomas or carcinomas over a two-year period in the restricted fed group only. No dose-dependent relationship could be demonstrated in the *ad libitum* fed group because there was already a high incidence of neoplasia at the lower doses. Carcinogenic effects had not been found in an earlier study that used *ad libitum* feeding only, which suggests that the restricted fed mouse is a more sensitive model.

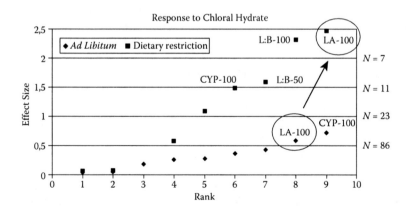

FIGURE 10.5 Response of male B6C3F1 mice fed *ad libitum* or at about 80% of *this intake*, in lauric acid ω-hydroxylase activity (LA) following a challenge with 100 mg/kg chloral hydrate. Mice fed the restricted ration showed a more distinct and uniform effect than those fed *ad libitum*, reducing the number of mice required to demonstrate a significant effect by 92%. CYP4A mice showed a similar, though less pronounced effect.

TABLE 10.4
Group Size Calculations for Determining a 10% Change in the Mean Level of the Reported Parameters by Food Restriction, as Compared to *Ad Libitum* Feeding

White Blood Cell Count ($10^3/\mu l$)	Mean (N = 100)	Standard Deviation	Coefficient of Variation	N for $\Delta = 10\%$
Males	12.0	2.5	21.2%	45
Females	8.1	2.5	31.0%	99
Total cholesterol (mg/dl)				
Males	71.3	17.7	25.1%	72
Females	81.3	14.7	18.2%	36
Total serum protein (g/dl)				
Males	6.8	0.3	3.9%	2
Females	7.4	0.4	5.9%	4

Source: Hubert, M. F., Laroque, P., Gillet, J. P., and Keenan, K. P., *The Journal of Toxicological Sciences,* 58, 195–207, 2000.

Note: For white blood cell counts, the original group size (N = 100) is considered appropriate for females. For total cholesterol a potential reduction of 28% in the case of males and 64% in the case of females, and for total serum protein a potential reduction of 96% in the numbers of female animals needed and 98% for males, may be possible when food restriction is applied instead of *ad libitum* feeding.

However, such effects need to be established on a case-to-case basis because they may depend on the total metabolism of the animal. Table 10.4. shows calculations to determine group size, based on data by Hubert et al. (2000), indicating various levels of potential for reduction by means of food restriction. For some parameters there is also a gender difference.

10.3.2 Practical Aspects of Food Restriction

Beynen (1992) has discussed the advantages and disadvantages of using food restriction versus *ad libitum* feeding. Many of the benefits have been described above, such as increased life span, reduced incidence of background tumour and other pathologies, reduced variation, and improved model sensitivity. Controlling food intake makes it easier to monitor the amount eaten by the animals, thus adding an additional health check to the routine daily inspections. It was seen earlier that food restriction delays the onset of tumour initiation and while some argue that this makes the model less suitable for carcinogenicity studies, Leakey and others (2003) were able to demonstrate a relationship between the dose and tumour development only on the restricted feeding schedule. Because food restriction also increases longevity, animals can be exposed to a test substance for a longer period, which may more reliably test its oncogenic potential (Hart, Neumann, and Robertson 1995). Similarly, in experiments using animals on a food restriction regime, body weights are better controlled and as a result of this, a decrease in body weight caused by the test substance is more likely to be detected than under *ad libitum* feeding conditions. The variation in food intake in *ad libitum* fed animals makes it difficult to determine how much tumour development is influenced by the test substance and how much by variation in food intake. Restricting food intake leads to more uniform consumption, which makes identification of the effect of the test substance more reliable.

The picture would not be complete without reviewing the disadvantages of food restriction as well. A reduction in tumour incidence and an increased life span prolongs experiments designed to demonstrate the effects of a drug. Altered sensitivity to a drug also means that the results are no longer directly comparable to previous studies. No standard protocols or feeding schedules have as yet been accepted for general use, and this represents a source of variation between studies, making it difficult to compare results. At present, restricted feeding often involves housing animals individually, which is an undesirable stressor for social species. Additionally, individual restricted feeding requires more work on the part of the animal caretaker. Finally, restricting the availability of food causes the animal to go without food for a period each day. Care has to be taken to match food provision to the animal's chronobiological needs in order to minimise the consequences for physiology and behaviour (Kalsbeek and Strubbe 1998; Krohn, Ritskes-Hoitinga, and Svendsen 1999; Ritskes-Hoitinga and Strubbe 2004). Feeding rats (a nocturnal species) during daytime altered and reduced bile flow compared with feeding during the dark hours (Ritskes-Hoitinga and Strubbe 2004). When rabbits were fed a restricted ration a few hours before the dark period rather than at the start of the light period, less stereotypic behaviour occurred (Krohn, Ritskes-Hoitinga, and Svendsen 1999). While the NRC guidelines for nutritional requirements on laboratory animals (National Research Council 1995) may be used for most rat and mice strains used in experimentation, the requirements of most transgenic animal strains have not been properly determined (Jegstrup et al. 2003).

Currently, new, more welfare-friendly methods are being developed to allow restricted feeding of laboratory rodents (http://www.mbrose.dk/products.htm). One of these involves use of a feeding station in conjunction with animals that have been equipped with an ID tag; individuals are registered when they enter the station and the food consumed during each visit is measured and logged. Feeding stations have been used for many years with farm animals, but more research is needed to adapt the approach to the species specific needs and behaviour of laboratory animals.

Another approach to restricted feeding, which simultaneously refines their environment, is to make the animals work for their food (Ritskes-Hoitinga and Strubbe 2004). The diet board is one device by which this can be accomplished, and its use is compatible with group housing in rats. Food pellets are pressed tightly into holes drilled in a wooden board and can be accessed by animals gnawing their way through the board (Kemppinen et al. 2008; Kasanen et al. 2009a, 2009b). Use of the diet board for young rats led to reduced body weight gain as compared to *ad libitum* feeding (Kasanen et al. 2009a), but was also associated with higher serum corticosterone concentration and lower faecal secretion of Immunoglobulin A, suggesting that working for food this way was stressful (Kasanen et al. 2009b); variability between animals was no less than on *ad libitum* feeding. Because the food in the diet board is available 24 hours per day, individual animals decide for themselves the amount of food they eat. In other studies food has been scattered among the litter material so that animals had to search for it (Ritskes-Hoitinga and Strubbe 2004).

In the meantime, more interest is being shown in restricted feeding. Several professional bodies have issued statements to this effect. The Society for Toxicology and Pathology has announced that whenever scientifically appropriate, dietary optimisation should be employed (Nold et al. 2001) and the European Agency for the Evaluation of Medicinal Products recommends that dietary restriction should be considered when a low survival rate is anticipated. The FDA refers to dietary control amongst "points to consider" as an approach to greatly reducing variability both within and between studies (Allaben et al. 1996). For some contract toxicological studies in the United States, moderate dietary restriction is standard practise for long-term (2-year) studies to ensure that sufficient animals survive the experimental period.

10.3.3 FOOD RESTRICTION IN RABBITS

Restricted feeding has become a token of good veterinary practise in order to prevent obesity in adult laboratory rabbits (Hart, Neumann, and Robertson 1995). During growth, rabbits are usually fed *ad libitum*, although restriction of food intake has been recommended as a way of reducing mortality from enteritis among young rabbits around the time of weaning (Nikkels, Mullink, and van Vliet 1976). As with rodents, it is important to establish (a) the amount of food that needs to be fed in relation to the age and intended use of the animal, and (b) at what time of the day this needs to be supplied.

In one study young, growing male New Zealand White rabbits (6–22 weeks of age) intended for meat production, were fed either *ad libitum*, or 70% of *ad libitum* consumption (Fodor et al. 2003). At 22 weeks of age the body weight of the rabbits fed restrictedly was about 86% that of the *ad libitum* fed animals. Feed restriction reduced the libido of males and delayed sexual maturity, so this level of restricted feeding was not recommended for rearing breeding stock (Fodor et al. 2003). Also in young females (6–22 weeks of age) sexual maturity was delayed under restricted feeding conditions (70% of *ad libitum*) as compared to *ad libitum* feeding (Fodor et al. 2001). Moreover, this restricted feeding schedule reduced body weight and body fat content and distorted development of body parts (allometric growth), suggesting that 70% of *ad libitum* feeding during the early growth phase of does was deficient (Fodor et al. 2001).

The rabbit is normally active nocturnally although, if food is provided for only 4 hours during the 12-hour light period, activity increases around the time of food availability (Jilge 1991). It seems that exposure to this feeding schedule during the light period makes the rabbit behave as a diurnal animal (Jilge 1991; van den Buuse and Malpas 1997). It is speculated that this is caused by a feeding-entrainable oscillator, which co-exists with the light-dark entrainable circadian oscillator system (Jilge, Hornicke, and Stahle 1987).

When three-month-old laboratory rabbits were fed a ration of 60% of *ad libitum* intake just before onset of the dark phase so as to stimulate the situation in nature, where most food is eaten during the dark period (Hornicke et al. 1984), the incidence of stereotypic behaviours declined in comparison to rabbits fed either a restricted amount of food in the morning or fed *ad libitum* (Krohn, Ritskes-Hoitinga, and Svendsen 1999). The presence of stereotypic behaviour is thought to be an indicator of reduced well-being, so the finding that it is reduced if restricted feeding takes place at a natural time point of the diurnal cycle suggests that welfare may be improved in such circumstances compared to *ad libitum* regimes. Both the time at which food is given and the scheme of food provision appear to be important. Feeding adult rabbits a restricted amount of food (60% of *ad libitum*) at the beginning of the light phase (7:30 am) resulted in significantly increased food conversion and water intake compared to rabbits that had food freely available between 7:30 am and 2:30 pm (at which time food remaining was removed), even though the total food intake in grams per day was equal (Ritskes-Hoitinga and Schledermann 1999). The rabbits that had food freely available from 7:30 am to 2:30 pm had calmer behaviour and ate the food throughout the period that the food was available. The total amount of food provided as the restricted amount at 7:30 am was immediately eaten, probably leaving the rabbits to search for more food the rest of the day. These examples clearly illustrate that in rabbits the restricted feeding schedule can influence experimental results. In general, it is important to consider species-specific characteristics and behaviour when the aim is to improve animal welfare.

10.4 CONCLUSIONS

- Although handling and restraint are potential sources of stress, they are unavoidable procedures. Accustoming animals to handling or restraint prior to the test procedure is advantageous to the handler also, because he/she becomes more familiar with what to expect during the actual experiment.
- Frequent and gentle handling of rabbit pups during a sensitive window—the first week after birth—is advisable; unfortunately there have not been enough studies to draw similar conclusions for mice and rats.
- Familiarity of animals with the housing environment and personnel is crucial to minimising the stress of handling and restraint, at least in rats. Animals should not associate either handling or restraint with painful procedures, it may be better to reward the animal after the procedure.
- The fewest possible number of animals should be present to witness any procedure because stress levels are communicated between animals and may cause bias; the sampling or recording order of the animals should be randomised.

- It is a prerequisite for successful handling that personnel are highly competent; the method of handling used should take account of the stock or strain of animals concerned, the procedures to be performed and personal preferences. Inexperienced or nervous handlers should seek the assistance of competent persons or may sometimes use device-based methods.
- The best way of habituating rats to handling is to make use of the immobilisation reflex (in strains where this is present) and to handle them by the nape of the neck.
- In general, methods of handling and restraint must take account of differences attributable to stock/strain and sex.
- Always allow the animals "the benefit of the doubt" when habituating them to handling and restraint; if an animal does not exhibit detectable habituation, you will become acquainted with it and its reactions.
- Avoid handling when major changes are taking place in the animal's cage; for example, arrival of new animals or the period immediately after cage change.
- Current transmitter weights are relatively large for mice, but are unlikely to cause untoward effects if suitable analgesia regimes are used and two week recovery is allowed. It is advisable to restrict telemetry studies to larger mice strains.
- There are established and effective analgesia regimens for post-operative pain alleviation suitable for both mice and rats (see Chapter 13).
- Specific and reliable behaviours associated with pain and distress in rats have been described.
- The effectiveness of the analgesic regime proposed should be verified at least once for each type of operation and for each stock and strain.
- It is advisable to allow recovery of at least 14 days after the implantation surgery before animals are subjected to further procedures.
- Group housing of rats makes them less reactive to an observer effect.
- Telemetry produces more accurate and reliable results, at least in cardiovascular studies, than traditional methods.
- *Ad libitum* feeding of experimental animals has a negative impact on their long-term health, welfare and survival time and restricted feeding is preferable providing welfare is assured.
- Novel methods of restricted feeding are emerging, which are compatible with normal diurnal rhythms, group housing and monitoring food consumption (patterns); once implemented these have a powerful effect on reduction.

10.5 QUESTIONS UNRESOLVED

The question of whether improved housing environments reduce the stress associated with procedures remains open. Very few studies have examined this and the methods and techniques applied have been inconsistent.

Modern microelectronic technology allows fabrication of on-chip co-integrated transmitters and sensors, which are ideal for implantation and do not need battery replacement because they can be charged externally by applying a fluctuating electromagnetic field. Their size is considerably smaller (e.g., 2.5 × 2.5 mm) than those used currently for implantable telemetry in small rodents. Applications, available now or in the near future, include measurement of temperature, blood pressure or intraocular pressure. Because transmitters have diagnostic or therapeutic potential in human patients they will be produced in high volumes, which may render them low-cost devices (Mokwa and Schnakenberg 2001).

It is not yet clear how general is the effect of restricted feeding on many parameters, and few publications include sufficient information to enable this to be assessed. Food restriction can make animals less vulnerable to experimental stressors (i.e., they are more robust) and better able to adjust to physiological and biochemical perturbations. Also, it often results in animals that are better experimental models because they are more uniform and show a higher response (effect size) to an experimental treatment. These characteristics mean that more information can be obtained using fewer animals. It remains to be established whether food restriction is appropriate for carcinogenicity studies, for example.

REFERENCES

Abernathy, F. W., C. D. Flemming, and W. B. Sonntag. 1995. Measurement of cardiovascular-response in male Sprague-Dawley rats using radiotelemetric implants and tailcuff sphygmomanometry—A comparative-study. *Toxicology Methods* 5:89–98.

Adams, C. S., A. I. Korytko, and J. L. Blank. 2001. A novel mechanism of body mass regulation. *The Journal of Experimental Biology* 204:1729–34.

Allaben, W. T., A. Turturro, J. E. Leakey, J. E. Seng, and R. W. Hart. 1996. FDA points-to-consider documents: The need for dietary control for the reduction of experimental variability within animal assays and the use of dietary restriction to achieve dietary control. *Toxicological Pathology* 24:776–81.

Anderson, N. H., A. M. Devlin, D. Graham, J. J. Morton, C. A. Hamilton, J. L. Reid, N. J. Schork, and A. F. Dominiczak. 1999. Telemetry for cardiovascular monitoring in a pharmacological study: New approaches to data analysis. *Hypertension* 33:248–55.

Augustsson, H., L. Lindberg, A. U. Hoglund, and K. Dahlborn. 2002. Human-animal interactions and animal welfare in conventionally and pen-housed rats. *Laboratory Animals* 36:271–81.

Bastlund, J. F., P. Jennum, P. Mohapel, V. Vogel, and W. P. Watson. 2004. Measurement of cortical and hippocampal epileptiform activity in freely moving rats by means of implantable radiotelemetry. *Journal of Neuroscience Methods* 138:65–72.

Baturaite, Z., H. M. Voipio, O. Ruksenas, M. Luodonpaa, H. Leskinen, N. Apanaviciene, and T. Nevalainen. 2005. Comparison of and habituation to four common methods of handling and lifting of rats with cardiovascular telemetry. *Scandinavian Journal of Laboratory Animal Science* 32:137–48.

Baumans, V., J. A. Bouwknecht, H. Boere, K. Kramer, H. A. van Lith, H. A. van de Weerd, and H. van Herck. 2001. Intra-abdominal transmitter implantation in mice: Effects on behaviour and body weight. *Animal Welfare* 10:291–302.

Beynen, A. C. 1992. *Ad libitum* versus restricted feeding: Pros and cons. *Scandinavian Journal of Laboratory Animal Science* 19:19–22.

Bhatnagar, S., T. M. Lee, and C. Vining. 2005. Prenatal stress differentially affects habituation of corticosterone responses to repeated stress in adult male and female rats. *Hormones and Behavior* 47:430–38.

Bilko, A., and V. Altbacker. 2000. Regular handling early in the nursing period eliminates fear responses toward human beings in wild and domestic rabbits. *Developmental Psychobiology* 36:78–87.

Blecha, F., K. W. Kelley, and D. G. Satterlee. 1982. Adrenal involvement in the expression of delayed-type hypersensitivity to SRBC and contact sensitivity to DNFB in stressed mice. *Proceedings of the Society for Experimental Biology and Medicine (New York, NY)* 169:247–52.

Bluher, M., B. B. Kahn, and C. R. Kahn. 2003. Extended longevity in mice lacking the insulin receptor in adipose tissue. *Science* 299:572–74.

Bollman, J. L. 1948. A cage which limits the activity of rats. *The Journal of Laboratory and Clinical Medicine* 33:1348.

Braga, V. A., and N. R. Prabhakar. 2009. Refinement of telemetry for measuring blood pressure in conscious rats. *Journal of American Association for Laboratory Animal Science* 48:268–71.

Brehe, J., and A. L. Way. 2008. An endocrinology laboratory exercise demonstrating the effect of confinement stress on the immune system of mice. *Advances in Physiology Education* 32:157–60.

Brix, A. E., A. Nyska, J. K. Haseman, D. M. Sells, M. P. Jokinen, and N. J. Walker. 2005. Incidences of selected lesions in control female Harlan Sprague-Dawley rats from two-year studies performed by the national toxicology program. *Toxicologic Pathology* 33:477–83.

Brodin, E., A. Rosen, E. Schott, and K. Brodin. 1994. Effects of sequential removal of rats from a group cage, and of individual housing of rats, on substance P, cholecystokinin and somatostatin levels in the periaqueductal grey and limbic regions. *Neuropeptides* 26:253–60.

Brudzynski, S. M., and D. Ociepa. 1992. Ultrasonic vocalization of laboratory rats in response to handling and touch. *Physiology & Behavior* 52:655–60.

Burman, O. H. P., and M. Mendl. 2004. Disruptive effects of standard husbandry practice on laboratory rat social discrimination. *Animal Welfare* 13:125–33.

Butz, G. M., and R. L. Davisson. 2001. Long-term telemetric measurement of cardiovascular parameters in awake mice: A physiological genomics tool. *Physiological Genomics* 5:89–97.
Cabanac, A., and E. Briese. 1992. Handling elevates the colonic temperature of mice. *Physiology & Behavior* 51:95–98.
Carney, E. W., C. L. Zablotny, M. S. Marty, J. W. Crissman, P. Anderson, M. Woolhiser, and M. Holsapple. 2004. The effects of feed restriction during *in utero* and postnatal development in rats. *Toxicologic Pathology* 82:237–49.
Carter, R. N., S. B. Pinnock, and J. Herbert. 2004. Does the amygdala modulate adaptation to repeated stress? *Neuroscience* 126:9–19.
Cheung, S. Y., H. L. Choi, A. E. James, Z. Y. Chen, Y. Huang, and F. L. Chan. 2003. Spontaneous mammary tumors in aging noble rats. *International Journal of Oncology* 22:449–57.
Clark, J. D., D. R. Rager, and J. P. Calpin. 1997. Animal well-being. II. Stress and distress. *Laboratory Animal Science* 47:571–79.
Clement, H. W., M. Kirsch, C. Hasse, C. Opper, D. Gemsa, and W. Wesemann. 1998. Effect of repeated immobilization on serotonin metabolism in different rat brain areas and on serum corticosterone. *Journal of Neural Transmission* (Vienna, Austria: 1996) 105:1155–70.
Clement, J. G., P. Mills, and B. Brockway. 1989. Use of telemetry to record body temperature and activity in mice. *Journal of Pharmacological Methods* 21:129–40.
Cole, M. A., B. A. Kalman, T. W. Pace, F. Topczewski, M. J. Lowrey, and R. L. Spencer. 2000. Selective blockade of the mineralocorticoid receptor impairs hypothalamic-pituitary-adrenal axis expression of habituation. *Journal of Neuroendocrinology* 12:1034–42.
Cotugno, M., P. Mandile, D. D'Angiolillo, P. Montagnese, and A. Giuditta. 1996. Implantation of an EEG telemetric transmitter in the rat. *Italian Journal of Neurological Sciences* 17:131–34.
Davis, H., and R. Perusse. 1988. Human-based social-interaction can reward a rat's behavior. *Animal Learning & Behavior* 16:89–92.
DeTurck, K. H., and W. H. Vogel. 1980. Factors influencing plasma catecholamine levels in rats during immobilization. *Pharmacology, Biochemistry, and Behavior* 13:129–31.
Deveney, A. M., A. Kjellstrom, T. Forsberg, and D. M. Jackson. 1998. A pharmacological validation of radiotelemetry in conscious, freely moving rats. *Journal of Pharmacological and Toxicological Methods* 40:71–79.
Diamant, M., L. Van Wolfswinkel, B. Altorffer, and D. De Wied. 1993. Biotelemetry: Adjustment of a telemetry system for simultaneous measurements of acute heart rate changes and behavioral events in unrestrained rats. *Physiology & Behavior* 53:1121–26.
Dinslage, S., J. McLaren, and R. Brubaker. 1998. Intraocular pressure in rabbits by telemetry II: Effects of animal handling and drugs. *Investigative Ophthalmology & Visual Science* 39:2485–89.
Dirx, M. J., M. P. Zeegers, P. C. Dagnelie, T. van den Bogaard, and P. A. van den Brandt. 2003. Energy restriction and the risk of spontaneous mammary tumors in mice: A meta-analysis. *International Journal of Cancer* 106:766–70.
Dobrakovova, M., and J. Jurcovicova. 1984. Corticosterone and prolactin responses to repeated handling and transfer of male rats. *Experimental and Clinical Endocrinology* 83:21–27.
Dobrakovova, M., R. Kvetnansky, Z. Oprsalova, and D. Jezova. 1993. Specificity of the effect of repeated handling on sympathetic-adrenomedullary and pituitary-adrenocortical activity in rats. *Psychoneuroendocrinology* 18:163–74.
Fernandes, G. A. 2000. Immunological stress in rats induces bodily alterations in saline-treated conspecifics. *Physiology & Behavior* 69:221–30.
Fodor, K., L. Zoldag, S. G. Fekete, A. Bersenyi, A. Gaspardy, E. Andrasofszky, M. Kulcsar, F. Eszes, and M. Shani. 2003. Influence of feeding intensity on the growth, body composition and sexual maturity of male New Zealand white rabbits. *Acta Veterinaria Hungarica* 51:305–19.
Fodor, K., S. G. Fekete, L. Zoldag, A. Bersenyi, A. Gaspardy, E. Andrasofszky, M. Kulcsar, and F. Eszes. 2001. Influence of feeding intensity on corporeal development, body composition and sexual maturity in female rabbits. *Acta Veterinaria Hungarica* 49:399–411.
Fukuda, S., and H. Iida. 2003. Life span and spontaneous tumors incidence of the Wistar mishima (WM/MsNrs) rat. *Experimental Animals* 52:173–78.
Gadek-Michalska, A., and J. Bugajski. 2003. Repeated handling, restraint, or chronic crowding impair the hypothalamic-pituitary-adrenocortical response to acute restraint stress. *Journal of Physiology and Pharmacology* 54:449–59.
Gartner, K., D. Buttner, K. Dohler, R. Friedel, J. Lindena, and I. Trautschold. 1980. Stress response of rats to handling and experimental procedures. *Laboratory Animals* 14:267–74.

Gattermann, R., and R. Weinandy. 1996. Time of day and stress response to different stressors in experimental animals. Part I: Golden hamster (*Mesocricetus auratus* waterhouse, 1839). *Journal of Experimental Animal Science* 38:66–76.

Gilmore, A. J., R. L. Billing, and R. Einstein. 2008. The effects on heart rate and temperature of mice and vas deferens responses to noradrenaline when their cage mates are subjected to daily restraint stress. *Laboratory Animals* 42:140–48.

Greene, A. N., S. L. Clapp, and R. H. Alper. 2007. Timecourse of recovery after surgical intraperitoneal implantation of radiotelemetry transmitters in rats. *Journal of Pharmacological and Toxicological Methods* 56:218–22.

Grissom, N., V. Iyer, C. Vining, and S. Bhatnagar. 2007. The physical context of previous stress exposure modifies hypothalamic-pituitary-adrenal responses to a subsequent homotypic stress. *Hormones and Behavior* 51:95–103.

Harkin, A., T. J. Connor, J. M. O'Donnell, and J. P. Kelly. 2002. Physiological and behavioral responses to stress: What does a rat find stressful? *Laboratory Animal* 31:42–50.

Harris, R. B., T. D. Mitchell, J. Simpson, S. M. Redmann, Jr., B. D. Youngblood, and D. H. Ryan. 2002. Weight loss in rats exposed to repeated acute restraint stress is independent of energy or leptin status. *American Journal of Physiology. Regulatory, Integrative and Comparative Physiology* 282:R77–R88.

Hart, R. W., D. A. Neumann, and R. T. Robertson. 1995. *Dietary restriction—Implications for the design and interpretation of toxicity and carcinogenicity studies*, eds. R. W. Hart, D. A. Neumann, and R. T. Robertson. Washington, DC: ILSI Press.

Hawkins, P., D. B. Morton, D. Bevan, K. Heath, J. Kirkwood, P. Pearce, L. Scott, G. Whelan, A. Webb, and Joint Working Group on Refinement. 2004. Husbandry refinements for rats, mice, dogs and non-human primates used in telemetry procedures. Seventh report of the BVAAWF/FRAME/RSPCA/UFAW joint working group on refinement, part B. *Laboratory Animals* 38:1–10.

Hayes, K. E., J. A. Raucci, Jr., N. M. Gades, and L. A. Toth. 2000. An evaluation of analgesic regimens for abdominal surgery in mice. *Contemporary Topics in Laboratory Animal Science* 39:18–23.

Hornicke, H., G. Ruoff, B. Vogt, W. Clauss, and H. J. Ehrlein. 1984. Phase relationship of the circadian rhythms of feed intake, caecal motility and production of soft and hard faeces in domestic rabbits. *Laboratory Animals* 18:169–72.

Hubert, M. F., P. Laroque, J. P. Gillet, and K. P. Keenan. 2000. The effects of diet, ad libitum feeding, and moderate and severe dietary restriction on body weight, survival, clinical pathology parameters, and cause of death in control Sprague-Dawley rats. *The Journal of Toxicological Sciences* 58:195–207.

Irvine, R. J., J. White, and R. Chan. 1997. The influence of restraint on blood pressure in the rat. *Journal of Pharmacological and Toxicological Methods* 38:157–62.

Jarillo-Luna, A., V. Rivera-Aguilar, H. R. Garfias, E. Lara-Padilla, A. Kormanovsky, and R. Campos-Rodriguez. 2007. Effect of repeated restraint stress on the levels of intestinal IgA in mice. *Psychoneuroendocrinology* 32:681–92.

Jegstrup, I., R. Thon, A. K. Hansen, and M. R. Hoitinga. 2003. Characterization of transgenic mice—A comparison of protocols for welfare evaluation and phenotype characterization of mice with a suggestion on a future certificate of instruction. *Laboratory Animals* 37:1–9.

Jilge, B. 1991. The rabbit: A diurnal or a nocturnal animal? *Journal of Experimental Animal Science* 34:170–83.

Jilge, B., H. Hornicke, and H. Stahle. 1987. Circadian rhythms of rabbits during restrictive feeding. *The American Journal of Physiology* 253:R46–R54.

Johnston, N. A., C. Bosgraaf, L. Cox, J. Reichensperger, S. Verhulst, C. Patten, Jr., and L. A. Toth. 2007. Strategies for refinement of abdominal device implantation in mice: Strain, carboxymethylcellulose, thermal support, and atipamezole. *Journal of the American Association for Laboratory Animal Science* 46:46–53.

Kalsbeek, A., and J. H. Strubbe. 1998. Circadian control of insulin secretion is independent of the temporal distribution of feeding. *Physiology & Behavior* 63:553–58.

Kant, G. J., R. A. Bauman, R. H. Pastel, C. A. Myatt, E. Closser-Gomez, and C. P. D'Angelo. 1991. Effects of controllable vs. uncontrollable stress on circadian temperature rhythms. *Physiology & Behavior* 49:625–30.

Kant, G. J., T. Eggleston, L. Landman-Roberts, C. C. Kenion, G. C. Driver, and J. L. Meyerhoff. 1985. Habituation to repeated stress is stressor specific. *Pharmacology, Biochemistry, and Behavior* 22:631–34.

Kasanen, I. H., K. J. Inhila, J. I. Nevalainen, S. B. Vaisanen, A. M. Mertanen, S. M. Mering, and T. O. Nevalainen. 2009a. A novel dietary restriction method for group-housed rats: Weight gain and clinical chemistry characterization. *Laboratory Animals* 43:138–48.

Kasanen, I. H., K. J. Inhila, O. M. Vainio, V. V. Kiviniemi, J. Hau, M. Scheinin, S. M. Mering, and T. O. Nevalainen. 2009b. The diet board: Welfare impacts of a novel method of dietary restriction in laboratory rats. *Laboratory Animals* 43:215–23.

Keenan, K. P., C. M. Hoe, L. Mixson, C. L. Mccoy, J. B. Coleman, B. A. Mattson, G. A. Ballam, L. A. Gumprecht, and K. A. Soper. 2005. Diabesity: A polygenic model of dietary-induced obesity from ad libitum overfeeding of Sprague-Dawley rats and its modulation by moderate and marked dietary restriction. *Toxicologic Pathology* 33:650–74.

Keenan, K. P., P. F. Smith, and K. A. Soper. 1994. Effect of dietary (caloric) restriction on aging, survival, pathology, and toxicology. In *Dietary restriction*, 609–628. Washington, DC: ILSI Press.

Keenan, K. P., P. Laroque, and R. Dixit. 1998. Need for dietary control by caloric restriction in rodent toxicology and carcinogenicity studies. *Journal of Toxicology and Environmental Health. Part B, Critical Reviews* 1:135–48.

Kemppinen, N., A. Meller, K. Mauranen, T. Kohila, and T. Nevalainen. 2008. Work for food—A solution to restricting food intake in group housed rats? *Scandinavian Journal of Laboratory Animal Science* 35:81–90.

Kemppinen, N., J. Hau, A. Meller, K. Mauranen, T. Kohila, and T. Nevalainen. Impact of intra-abdominal telemetry transponder on faecal corticoid and immunoglobulin A (IgA) excretion in rats. Unpublished Manuscript.

Kemppinen, N. M., A. S. Meller, K. O. Mauranen, T. T. Kohila, and T. O. Nevalainen. 2009. The effect of dividing walls, a tunnel, and restricted feeding on cardiovascular responses to cage change and gavage in rats (*Rattus norvegicus*). *Journal of the American Association for Laboratory Animal Science* 48:157–65.

Kennett, G. A., F. Chaouloff, M. Marcou, and G. Curzon. 1986. Female rats are more vulnerable than males in an animal model of depression: The possible role of serotonin. *Brain Research* 382:416–21.

Kikusui, T., S. Takigami, Y. Takeuchi, and Y. Mori. 2001. Alarm pheromone enhances stress-induced hyperthermia in rats. *Physiology & Behavior* 72:45–50.

Knott, P. J., P. H. Hutson, and G. Curzon. 1977. Fatty acid and tryptophan changes on disturbing groups of rats and caging them singly. *Pharmacology, Biochemistry, and Behavior* 7:245–52.

Kramer, K., L. Kinter, B. P. Brockway, H. P. Voss, R. Remie, and B. L. Van Zutphen. 2001. The use of radiotelemetry in small laboratory animals: Recent advances. *Contemporary Topics in Laboratory Animal Science* 40:8–16.

Krohn, T. C., A. K. Hansen, and N. Dragsted. 2003. Telemetry as a method for measuring the impact of housing conditions on rats' welfare. *Animal Welfare* 12:53–62.

Krohn, T. C., J. Ritskes-Hoitinga, and P. Svendsen. 1999. The effects of feeding and housing on the behaviour of the laboratory rabbit. *Laboratory Animals* 33:101–7.

Kvetnansky, R., D. S. Goldstein, V. K. Weise, C. Holmes, K. Szemeredi, G. Bagdy, and I. J. Kopin. 1992. Effects of handling or immobilization on plasma levels of 3,4-dihydroxyphenylalanine, catecholamines, and metabolites in rats. *Journal of Neurochemistry* 58:2296–2302.

Larsen, K., and J. Bastlund. 2006. Subcutaneous versus intraperitoneal placement of radiotelemetry transmitters for long-term recording of electroencephalography. *Scandinavian Journal of Laboratory Animal Science* 33:43–44.

Laugero, K. D., and G. P. Moberg. 2000. Energetic response to repeated restraint stress in rapidly growing mice. *American Journal of Physiology. Endocrinology and Metabolism* 279:E33–E43.

Leakey, J. E., J. E. Seng, J. R. Latendresse, N. Hussain, L. J. Allen, and W. T. Allaben. 2003. Dietary controlled carcinogenicity study of chloral hydrate in male B6C3F1 mice. *Toxicology and Applied Pharmacology* 193:266–80.

Leakey, J. E., J. E. Seng, and W. T. Allaben. 2003. Body weight considerations in the B6C3F1 mouse and the use of dietary control to standardize background tumor incidence in chronic bioassays. *Toxicology and Applied Pharmacology* 193:237–65.

Lemaire, V., and P. Mormede. 1995. Telemetered recording of blood pressure and heart rate in different strains of rats during chronic social stress. *Physiology & Behavior* 58:1181–88.

Leon, L. R., L. D. Walker, D. A. DuBose, and L. A. Stephenson. 2004. Biotelemetry transmitter implantation in rodents: Impact on growth and circadian rhythms. *American Journal of Physiology. Regulatory, Integrative and Comparative Physiology* 286:R967–R974.

Li, X. F., J. Edward, J. C. Mitchell, B. Shao, J. E. Bowes, C. W. Coen, S. L. Lightman, and K. T. O'Byrne. 2004. Differential effects of repeated restraint stress on pulsatile lutenizing hormone secretion in female Fischer, Lewis and Wistar rats. *Journal of Neuroendocrinology* 16:620–27.

Ling, S., and F. Jamali. 2003. Effect of cannulation surgery and restraint stress on the plasma corticosterone concentration in the rat: Application of an improved corticosterone HPLC assay. *Journal of Pharmacy & Pharmaceutical Sciences* 6:246–51.

Masoro, E. J. 2005. Overview of caloric restriction and ageing. *Mechanisms of Ageing and Development* 126:913–22.

McCarty, R., R. Kvetnansky, C. R. Lake, N. B. Thoa, and I. J. Kopin. 1978. Sympatho-adrenal activity of SHR and WKY rats during recovery from forced immobilization. *Physiology & Behavior* 21:951–55.

McDougall, S. J., J. R. Paull, R. E. Widdop, and A. J. Lawrence. 2000. Restraint stress: Differential cardiovascular responses in Wistar-Kyoto and spontaneously hypertensive rats. *Hypertension* 35:126–29.

McLaren, J. W., R. F. Brubaker, and J. S. FitzSimon. 1996. Continuous measurement of intraocular pressure in rabbits by telemetry. *Investigative Ophthalmology & Visual Science* 37:966–75.

McMurtry, J. P., and B. C. Wexler. 1981. Hypersensitivity of spontaneously hypertensive rats (SHR) to heat, ether, and immobilization. *Endocrinology* 108:1730–36.

Meijer, M. K., B. M. Spruijt, L. F. van Zutphen, and V. Baumans. 2006. Effect of restraint and injection methods on heart rate and body temperature in mice. *Laboratory Animals* 40:382–91.

Meijer, M. K., R. Sommer, B. M. Spruijt, L. F. van Zutphen, and V. Baumans. 2007. Influence of environmental enrichment and handling on the acute stress response in individually housed mice. *Laboratory Animals* 41:161–73.

Meile, T., and T. T. Zittel. 2002. Telemetric small intestinal motility recording in awake rats: A novel approach. *European Surgical Research* 34:271–74.

Michel, C., and M. Cabanac. 1999. Opposite effects of gentle handling on body temperature and body weight in rats. *Physiology & Behavior* 67:617–22.

Mikkelsen, L. F. 2007. The effect of training on long term restraint of rats evaluated by telemetry. MSc., Faculty of Life Sciences, University of Copenhagen.

Mileikovsky, B., S. V. Verevkina, and A. D. Nozdrachev. 1994. Effects of stimulation of the frontoparietal cortex and parafascicular nucleus on locomotion in rats. *Physiology & Behavior* 55:267–71.

Mokwa, W., and U. Schnakenberg. 2001. Micro-transponder systems for medical applications. *IEEE Transactions on Instrumentation and Measurement* 50:1551–55.

Molon-Noblot, S., M. F. Hubert, C. M. Hoe, K. Keenan, and P. Laroque. 2005. The effects of *ad libitum* feeding and marked dietary restriction on spontaneous skeletal muscle pathology in Sprague-Dawley rats. *Toxicologic Pathology* 33:600–08.

Moncek, F., R. Duncko, B. B. Johansson, and D. Jezova. 2004. Effect of environmental enrichment on stress related systems in rats. *Journal of Neuroendocrinology* 16:423–31.

Moons, C. P., P. Van Wiele, and F. O. Odberg. 2004. To enrich or not to enrich: Providing shelter does not complicate handling of laboratory mice. *Contemporary Topics in Laboratory Animal Science* 43:18–21.

Moran, M. M., R. R. Roy, C. E. Wade, B. J. Corbin, and R. E. Grindeland. 1998. Size constraints of telemeters in rats. *Journal of Applied Physiology* (Bethesda, MD: 1985) 85:1564–71.

Morton, D. B., P. Hawkins, R. Bevan, K. Heath, J. Kirkwood, P. Pearce, L. Scott, et al. 2003. Refinements in telemetry procedures. Seventh report of the BVAAWF/FRAME/RSPCA/UFAW joint working group on refinement, part A. *Laboratory Animals* 37:261–99.

National Research Council. 1995. *Nutrient requirements of laboratory animals*, eds. Subcommittee on laboratory animal nutrition, Committee on animal nutrition, Board on agriculture, and National Research Council. 4th revised ed. Washington, DC: National Academy Press.

Nikkels, R. J., J. W. Mullink, and J. C. van Vliet. 1976. An outbreak of rabbit enteritis: Pathological and microbiological findings and possible therapeutic regime. *Laboratory Animals* 10:195–98.

Nold, J. B., K. P. Keenan, A. Nyska, and M. E. Cartwright. 2001. Society of toxicologic pathology position paper: Diet as a variable in rodent toxicology and carcinogenicity studies. *Toxicologic Pathology* 29:585–86.

Patel, N. V., M. N. Gordon, K. E. Connor, R. A. Good, R. W. Engelman, J. Mason, D. G. Morgan, T. E. Morgan, and C. E. Finch. 2005. Caloric restriction attenuates a beta-deposition in Alzheimer transgenic models. *Neurobiology of Aging* 26:995–1000.

Pitman, D. L., J. E. Ottenweller, and B. H. Natelson. 1988. Plasma corticosterone levels during repeated presentation of two intensities of restraint stress: Chronic stress and habituation. *Physiology & Behavior* 43:47–55.

Rex, A., U. Sondern, J. P. Voigt, S. Franck, and H. Fink. 1996. Strain differences in fear-motivated behavior of rats. *Pharmacology, Biochemistry, and Behavior* 54:107–11.

Ritskes-Hoitinga, J. 2006. *Does a rat have boeddha nature?* Radboud University Nijmegen.

Ritskes-Hoitinga, J., and C. Schledermann. 1999. A pilot study into the effects of various dietary restriction schedules in rabbits. *Scandinavian Journal of Laboratory Animal Science* 26:66–74.

Ritskes-Hoitinga, J., and J. Strubbe. 2004. Nutrition and animal welfare. In *The welfare of laboratory animals*, ed. E. Kaliste, 51–80. Dordrecht: Kluwer Academic publishers.

Roughan, J. V., and P. A. Flecknell. 2001. Behavioural effects of laparotomy and analgesic effects of ketoprofen and carprofen in rats. *Pain* 90:65–74.

Roughan, J. V., and P. A. Flecknell. 2004. Behaviour-based assessment of the duration of laparotomy-induced abdominal pain and the analgesic effects of carprofen and buprenorphine in rats. *Behavioural Pharmacology* 15:461–72.

Russell, W. M. S., and R. L. Burch. 1959. *The principles of humane experimental technique*. Potters Bar, England: Special edition, Universities Federation for Animal Welfare.

Seggie, J. A., and G. M. Brown. 1975. Stress response patterns of plasma corticosterone, prolactin, and growth hormone in the rat, following handling or exposure to novel environment. *Canadian Journal of Physiology and Pharmacology* 53:629–37.

Sharp, J., T. Azar, and D. Lawson. 2005. Effects of a cage enrichment program on heart rate, blood pressure, and activity of male Sprague-Dawley and spontaneously hypertensive rats monitored by radiotelemetry. *Contemporary Topics in Laboratory Animal Science* 44:32–40.

Sharp, J., T. Zammit, T. Azar, and D. Lawson. 2002. Does witnessing experimental procedures produce stress in male rats? *Contemporary Topics in Laboratory Animal Science* 41:8–12.

Sharp, J., T. Zammit, T. Azar, and D. Lawson. 2003a. Recovery of male rats from major abdominal surgery after treatment with various analgesics. *Contemporary Topics in Laboratory Animal Science* 42:22–27.

Sharp, J., T. Zammit, T. Azar, and D. Lawson. 2003b. Stress-like responses to common procedures in individually and group-housed female rats. *Contemporary Topics in Laboratory Animal Science* 42:9–18.

Smolensky, M. H., F. Halberg, J. Harter, B. Hsi, and W. Nelson. 1978. Higher corticosterone values at a fixed single timepoint in serum from mice "trained" by prior handling. *Chronobiologia* 5:1–13.

Somps, C. J. 1994. *Biotelemetry implant volume and weight in rats: A pilot study report.* 108812.

Son, W. C., and C. Gopinath. 2004. Early occurrence of spontaneous tumors in CD-1 mice and Sprague-Dawley rats. *Toxicologic Pathology* 32:371–74.

Stamp, J. A., and J. Herbert. 1999. Multiple immediate-early gene expression during physiological and endocrine adaptation to repeated stress. *Neuroscience* 94:1313–22.

Stamp, J., and J. Herbert. 2001. Corticosterone modulates autonomic responses and adaptation of central immediate-early gene expression to repeated restraint stress. *Neuroscience* 107:465–79.

Thompson, C. I., A. J. Brannon, and A. L. Heck. 2003. Emotional fever after habituation to the temperature-recording procedure. *Physiology & Behavior* 80:103–8.

Tuli, J. S., J. A. Smith, and D. B. Morton. 1995. Effects of acute and chronic restraint on the adrenal gland weight and serum corticosterone concentration of mice and their faecal output of oocysts after infection with *Eimeria apionodes*. *Research in Veterinary Science* 59:82–86.

van den Buuse, M., and S. C. Malpas. 1997. 24-hour recordings of blood pressure, heart rate and behavioural activity in rabbits by radio-telemetry: Effects of feeding and hypertension. *Physiology & Behavior* 62:83–89.

van der Staay, F. J., and A. Blokland. 1996. Behavioral differences between outbred Wistar, inbred Fischer 344, Brown Norway, and hybrid Fischer 344 × Brown Norway rats. *Physiology & Behavior* 60:97–109.

van Driel, K. S., and J. C. Talling. 2005. Familiarity increases consistency in animal tests. *Behavioural Brain Research* 159:243–45.

Wahlsten, D., P. Metten, and J. C. Crabbe. 2003. A rating scale for wildness and ease of handling laboratory mice: Results for 21 inbred strains tested in two laboratories. *Genes, Brain, and Behavior* 2:71–79.

Webster, N. 1972. *Webster's new twentieth century dictionary*. Cleveland, OH: Dorset & Baber.

Webster, D. G., T. H. Lanthorn, D. A. Dewsbury, and M. E. Meyer. 1981. Tonic immobility and the dorsal immobility response in twelve species of muroid rodents. *Behavioral and Neural Biology* 31:32–41.

Whitesall, S. E., J. B. Hoff, A. P. Vollmer, and L. G. D'Alecy. 2004. Comparison of simultaneous measurement of mouse systolic arterial blood pressure by radiotelemetry and tail-cuff methods. *American Journal of Physiology. Heart and Circulatory Physiology* 286:H2408–H2415.

Wilson, C., and A. Kaspar. 1994. Changes in immobility responses in rat pups with maternal stimuli. *The Journal of General Psychology* 121:111–20.

Zhu, Z., A. D. Haegele, and H. J. Thompson. 1997. Effect of caloric restriction on pre-malignant and malignant stages of mammary carcinogenesis. *Carcinogenesis* 18:1007–12.

11 Basic Procedures: Dosing, Sampling and Immunisation

*Ismene Dontas, Greece; Jann Hau, Denmark;
Katerina Marinou, Greece; and Timo Nevalainen, Finland*

CONTENTS

Objectives 258
Key Factors 258
11.1 Administration Routes 259
    11.1.1 Oral (per os-PO) Administration 259
    11.1.2 Intragastric (IG) Gavage 261
    11.1.3 Subcutaneous (SC) Administration 263
    11.1.4 Intramuscular (IM) Administration 263
    11.1.5 Intraperitoneal (IP) Administration 264
    11.1.6 Intravenous (IV) Administration 265
    11.1.7 Inhalation 265
    11.1.8 Dermal Application 265
    11.1.9 Intradermal (ID) Administration 266
11.2 Improving Sampling Procedures 266
    11.2.1 Blood Sampling 266
        11.2.1.1 Refinement Aspects in Mice 266
        11.2.1.2 Accuracy and Precision in Mice 269
        11.2.1.3 Approaches to Refinement in Rats 270
        11.2.1.4 Accuracy and Precision in Rats 272
    11.2.2 Sampling Urine and Faeces 273
    11.2.3 Other Sampling Procedures 274
11.3 Immunisation for Production of Polyclonal Antibodies (Pabs) 274
    11.3.1 The Antigen 274
    11.3.2 The Adjuvant 275
    11.3.3 The Animal 276
        11.3.3.1 Species 276
        11.3.3.2 Species/Strain-Stock 277
        11.3.3.3 Sex 277
        11.3.3.4 Age 277
    11.3.4 The Immunisation Protocol 277
        11.3.4.1 The Injection Route 277
        11.3.4.2 The Volume of Injection 278
        11.3.4.3 The Dose of Antigen 278
        11.3.4.4 Booster Immunisation 278
    11.3.5 Existing Guidelines on the Production of Polyclonal Antibodies 279

11.4	Conclusions	279
11.5	Questions Unresolved	280
References		280

OBJECTIVES

The use of laboratory animals in research entails a number of quite frequently used procedures, such as handling, restraint, administration of substances and withdrawal of blood or other bodily fluids. These procedures are carried out on a vast number of laboratory animals; even minor refinements can result in considerable improvement of animal welfare overall. In an ideal situation the animal would not notice the procedure at all—indeed such an outcome would also best serve the scientific objectives.

Implementation of the Three Rs is not only desirable for welfare of the animals, but is also a legal requirement in Europe (Council of Europe 2006; European Commission 2007a). Replacement is the primary aim, but when this is not possible, it is expected that refinement and reduction strategies will be applied. There appears to be considerable opportunity to refine procedures that cause mild to moderate pain, suffering or distress; also, when pain cannot be alleviated for scientific reasons, perhaps the greatest opportunity to introduce refinement arises during the conduct of procedures, whereas reduction strategies are principally envisaged at the level of experimental design. Some of these will be discussed in this chapter.

Second to handling and restraint, the commonest procedure carried out on laboratory animals is the administration of substances via a multiplicity of routes and for a variety of scientific purposes, ranging from fundamental biomedical research to safety evaluation. This chapter describes proven and tested refinements applicable to administration of substances to rodents, and identifies those technical procedures that are potentially stressful.

Perhaps, the next most frequently carried out procedure is withdrawal of a sample of blood or other fluid, or collection of urine or faeces. The impact of these techniques and the quality of samples obtained are inter-linked; the less stressful the procedure is for the animal and the greater the skill of the technician, the more likely it is that the sample will be truly representative of the animal's physiological condition and amenable to further processing.

The third section of this chapter brings together the discussion about administration and withdrawal of samples by reviewing the way they contribute to a practical example—the production of polyclonal antibodies for which, at present, there are no *in vitro* satisfactory methods. Although the number of animals used for this purpose in the member states of the European Union is relatively small, possibly about 1% of the total use for scientific purposes (European Commission 2007b; Home Office 2009), the example is pertinent because it requires substantial numbers of injections and repeated blood sampling. Improving these techniques, coupled with minimising the volumes of blood withdrawn, careful selection of volumes of agents injected and attention to the nature of the substances injected can have a major impact on animal welfare without compromising the scientific outcome.

KEY FACTORS

When compounds are administered to and samples taken from laboratory animals, it is important that conduct of the procedures does not itself interfere with the normal physiology or biochemistry of the animals. Unfortunately, reality is often different; in most cases it is necessary to handle, restrain, warm and, sometimes, anaesthetise the animal. Methods such as blood sampling are far more susceptible to interference associated with the procedure than others, for example, collection of faeces and urine. In this context, the critical question is: How much do administration and sampling procedures interfere with the results and their interpretation? The answer to this question has a bearing on the scientific validity of the process. Because of these considerations, dosing and sampling procedures should be conducted with the least possible compromise to animal welfare. Knowledge, attitude, technical skill and good facilities all contribute to achieving this objective.

11.1 ADMINISTRATION ROUTES

The most common ways of administering substances to rodents are by the oral (PO), intragastric (IG), subcutaneous (SC), intramuscular (IM), intraperitoneal (IP), intravenous (IV), respiratory, dermal and intradermal routes (Table 11.1). Many recommendations have been published on this topic. Substances of varying characteristics (solubility, viscosity, temperature, pH, particle size of suspensions, total volume, etc.) may be administered by various routes and for many purposes (Diehl et al. 2001; Morton et al. 2001) and several video presentations are freely available on the Internet showing the basic techniques (Smith 2008). The best route for administration of a substance to an animal depends on several considerations including the physical properties of the compound and its pharmacokinetic properties in the species concerned; these should be studied beforehand to ensure that it will be given in the most effective way (Wolfensohn and Lloyd 2003). Routes for administration of many pharmaceutical agents are stipulated by the manufacturers and departure from these requires careful justification.

The physiological effects of administering agents to mice by SC, IP and IM injection have been evaluated using radiotelemetric recordings of heart rate, body temperature and subsequent determination of plasma corticosterone levels, with a view to determining how much stress is caused. Following injection by each route, all these parameters increased for several minutes; heart rate, in particular, appeared to be a reliable indicator of acute animal stress (Meijer et al. 2006). These results indicate that there is scope for refinement of these and other commonly used administration procedures. Naturally, the competence of the person carrying out any procedure, from handling and administering substances to blood sampling, can significantly influence animal stress (van Herck et al. 1998; Rao, Peace, and Hoskins 2001; Meijer et al. 2006). Additionally, rewarding the animals after the procedure, for example giving a tasty treat or inserting an enrichment item into the cage, accustoms the animal to a pleasurable experience following the discomfort and constitutes a refinement.

11.1.1 ORAL (PER OS-PO) ADMINISTRATION

It has been suggested that the most refined method of administering substances orally (*per os:* PO) is to incorporate the substance into the animals' diet, as a percentage or, more correctly, calculated according to total daily food intake (Sun et al. 2003; Guhad 2005). This method may not always be scientifically appropriate and, although it causes no stress, the compound may change the taste or smell of the diet. It is advisable to introduce gradually increasing percentages, and to monitor food intake pre- and post-incorporation, in order to ensure the desired intake of the test compound. This may be the most refined way of PO administration to animals; however, one disadvantage to this method is that it is not possible to establish how much each animal has taken and, in addition, caretakers may be exposed to the substance, especially if it is distributed in powdered diet. Therefore, prior consideration of the hazard this substance might present to humans is necessary (M. Heimann, pers. comm.).

Soluble compounds can be added to the animal's drinking water, but as with incorporation in the food, it is difficult to ensure each animal receives the correct dose. Determination of the weight or volume of water consumed provides a measure of the amount of compound consumed, but it is not possible to pre-determine the quantity that the animal receives. In order to check influences on palatability arising from a change of taste or smell, water consumption should be monitored pre- and

TABLE 11.1
The Most Frequently Used Administration Routes, Recommended Maximal Volumes and Summary of their Respective Characteristics

Administration Routes	Benefits	Disadvantages	Maximal Volumes (Mice/Rats)	Precautions	Potential Pitfalls
Oral	Minimal stress	Least accurate	10/10 ml/Kg	Palatability; previous animal habituation	Buccal injury, waste of substance
Intragastric	Precise dosage	Stressful; mortality risk	10/10 ml/Kg	Catheter size	Intratracheal misplacement, tracheal, or esophageal rupture
Subcutaneous	Precise dosage	Minimal stress	10/5 ml/Kg		Intradermal administration
Intramuscular	Precise dosage	Painful	0.1/0.2 ml/site	Plunger withdrawal; neutral pH, isomolar and not cytotoxic	Intravascular administration, nerve injury
Intraperitoneal	Precise dosage	Risk of vital organ puncture	80/20 ml/Kg	Plunger withdrawal	Puncture of splanchnic organs; peritonitis
Intravenous	Precise dosage	Special training required; need for restraint and special preparation of animal	5/5 ml/Kg	Infuse slowly	Risk of extravasation, misplaced substance may cause irritation of neighbouring tissues
Respiratory	Rapid absorption	Special cage/chamber or stressful restraint; mortality risk	—	Irritancy and pH of substance; previous animal habituation; monitor animal for stress	Risk of distress and death
Dermal	Ease of application	Application area secure from animal (dressings, collars); stressful	0.2/0.5 ml	Previous animal habituation; non-abrasive shaving; prevention from licking	Skin irritation; oral ingestion
Intradermal	Precise dosage	Painful	0.05/0.1 ml	Slow injection; local anaesthetic cream	Subcutaneous administration

post-inclusion; dosing by this route should be discontinued if there is evidence that animals avoid drinking, which may lead to dehydration and deterioration of body condition (Wolfensohn and Lloyd 2003). On the other hand, there is also a danger that admixture of a palatable substance such as saccharine may increase drinking and encourage the animals to consume excessive quantities of the drug-water mixture.

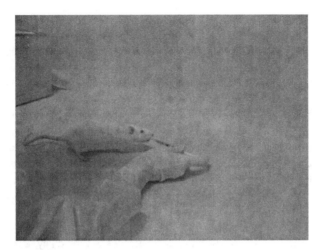

FIGURE 11.1 Oral administration: freely moving trained rat licking corn oil from an insulin syringe.

A third method of PO administration, not requiring intragastric gavage (IG), is the administration of liquids or pastes directly into the mouth of trained, freely moving or restrained animals (Figure 11.1). The stress induced by this procedure depends on the animals' habituation and the experimenter's competence. It is important to avoid allowing an inexperienced handler to attempt to dose an unaccustomed animal in this way. The taste and smell of the compound are also very important factors for the animal's acceptance, as it may develop aversive or defensive behaviour (M. Heimann, pers. comm.).

11.1.2 Intragastric (IG) Gavage

Intragastric gavage is widely used for the administration of compounds in pharmacological, pharmacokinetic and toxicological studies in laboratory animals. There is a risk of mortality if the substance is misplaced in the trachea or if accidental rupture of the oesophagus or trachea occurs, so the technique should be practised beforehand on models or anaesthetised animals. Fortunately, misdosing is a rare event (Morton et al. 2001; Murphy et al. 2001; Rao, Peace, and Hoskins 2001). IG elicits immediate changes in physiological parameters indicative of stress, such as: elevation of blood corticosterone, glucose, growth hormone and prolactin concentrations, and an increase in heart rate and blood pressure (Okva et al. 2006). Some of these changes (e.g., heart rate and blood pressure) are short-lived and values return to normal within an hour or two. Several ways of refining IG have been investigated, involving attention to the material and structure of the catheter (Wheatley 2002; Okva et al. 2006), viscosity and volume of solution administered (Brown, Dinger, and Levine 2000; Alban et al. 2001; Okva et al. 2006) habituation of animals to the procedure (Okva et al. 2006), and last, but certainly not least, the manual skill and experience of the person performing the procedure (Rao, Peace, and Hoskins 2001).

Okva et al. (2006) used telemetry to study habituation of rats to IG. During the first 3 days they found no evidence of habituation as measured by heart rate and blood pressure, but on the fourth day there was a significant decrease in both parameters during the 60 minutes following the procedure, providing evidence for habituation. The person carrying out the procedure benefits from training by gaining experience and learning how the animals are likely to respond (Rao, Peace, and Hoskins 2001; Okva et al. 2006; Kemppinen et al. 2009). The stress induced by IG appears to be strain-dependent also, for example Brown Norway rats housed in cages enriched with furniture and dividing walls were shown to be less stressed by IG than Fischer 344 rats in similar cages, as revealed by reduced cardiovascular responses (Kemppinen et al. 2009).

The stress response to IG is more intense with larger volumes and when lipid vehicles are used (Brown, Dinger, and Levine 2000). A stress test in which rats were placed on a platform surrounded by water for periods of 1 hour transiently increased intestinal epithelial permeability; chronic psychological stress results in longer-lasting changes that persist for at least 1 week (Yang et al. 2006). In addition to increased intestinal permeability and secretion, chronic stress has also been shown to result in reduced food intake and loss of body weight (Santos et al. 2000). Furthermore, administration of excessive volumes to rats may result in opening of the pyloric sphincter and passage of the material directly into the duodenum (Bonnichsen, Dragsted, and Hansen 2005). Taking the above into account, it is likely that IG is a scientifically unsuitable method of administration to evaluate the effects of compounds, which may alter intestinal permeability. This may be especially true if the animals were stressed, for example, if they had not been habituated to the procedure.

In rats, the maximum volume administered should be 5 ml/kg ideally and never more than 10 ml/kg (Brown, Dinger, and Levine 2000; Diehl et al. 2001) through a 15 or 16G catheter with a bulb of 2.4 mm diameter (Waynforth and Flecknell 1992). Obviously, excessive volumes may be stressful to the animals, whereas it is difficult to measure with precision volumes that are too small. Moreover, in rats there appears to be an optimal volume of 4–6 ml/kg, which results in the least increase in both heart rate and blood pressures during 60 minutes following the procedure (Figure 11.2; Okva et al. 2006). In mice, volumes administered should not exceed 10 ml/kg (Diehl et al. 2001). Other factors that should be considered include the viscosity of the substance and/or the vehicle being administered. Although aqueous-based vehicles are usually preferred, the physicochemical characteristics of the test material will to some extent determine whether the material is suitable for administration by IG (Gad et al. 2006).

Strategies to refine IG include consideration of the effects of excessive gastric distension, which may be minimised by administering only small volumes or, if this is not possible, by previous fasting of the animals. It has been reported that only 6 hours of fasting is sufficient for the stomach of a rat to become empty, and longer periods of food deprivation may constitute an unnecessary stress (Vermeulen et al. 1997). In contrast, water should always be available (Morton et al. 2001). Withholding food may also affect the rate of intestinal absorption (Diehl et al. 2001).

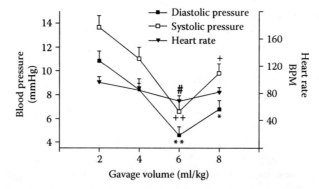

FIGURE 11.2 Changes in systolic and diastolic blood pressure (y1) and heart rate (y2) caused by the administration of different volumes by intragastric gavage (IG) in rats. (From Okva, K., Tamoseviciute, E., Ciziute, A., Pokk, P., Ruksenas, O., and Nevalainen, T., *Scandinavian Journal of Laboratory Animal Science*, 33, 243–52, 2006. With permission.) This figure shows the change from the baseline (60 minutes before the procedure) for 40 minutes following IG. Data presented are means ± SEM from groups of seven rats. * $P < .01$ versus 2 ml/kg; ** $P < .01$ versus 2 ml/kg and 4 ml/kg (diastolic pressure); + $P < .01$ versus 2 ml/kg; ++ $P < .01$ versus 2 ml/kg and 4 ml/kg (systolic pressure); # $P < .05$ versus 2 ml/kg (heart rate).

Basic Procedures: Dosing, Sampling and Immunisation

Not surprisingly, flexible catheters are less stressful to animals than metal ones (Wheatley 2002; Okva et al. 2006), although they are more susceptible to being distorted during the procedure, broken or damaged by biting (Wheatley 2002). If a flexible catheter is damaged by biting, the material being gavaged may be misplaced, although the likelihood of this can be minimised by using a mouth gag. Metal catheters are resistant to damage in this way, and in mice they are said to lead to better delivery of irritant substances to the stomach, thereby reducing adverse incidents arising from reflux (M. Heimann, pers. comm.).

In conclusion, IG techniques can be refined by ensuring that personnel are careful, well-trained and experienced, by using a flexible catheter and adhering to suitable administration volumes; habituation of the animal is another important aspect though it may take longer than 3 days (Okva et al. 2006). In the case of substances potentially hazardous to human health, additional safety precautions must be taken following administration by any method via the gastrointestinal tract, because small quantities of unchanged substance or its metabolites may be present in the faeces and constitute a threat to exposed personnel.

11.1.3 SUBCUTANEOUS (SC) ADMINISTRATION

Injection by the subcutaneous (SC) route is considered the parenteral route least likely to provoke pain, provided that the optimal volume of agent and a suitable needle size are used. In the study by Meijer et al. (2006), the injection procedure itself (without an inoculum) provoked an increase in heart rate similar to the IM and IP injection methods, although it returned to baseline values more quickly.

The way of administering a SC injection, most likely to avoid pain or inadvertent rupture of a vessel or muscle, is considered to be rapid and firm puncture of the skin. Furthermore, less discomfort is caused if the injection is given into an area with loose skin. The commonest injection site is the skin of the nape of the neck and the back (Morton et al. 2001).

In order to restrain an animal with the least possible stress, it has been recommended that, prior to injection, the rodent's head is covered with a towel so that it is in the dark; this appears to have a calming effect (Rasmussen and Ritskes-Hoitinga 1999). With a stress-free animal, the injection is carried out quickly and uneventfully and with high precision of dosing.

In rats, volumes per site should preferably not exceed 5 ml/kg and the maximum is 10 ml/kg (Diehl et al. 2001); the needle size is determined largely by the viscosity of the compound to be injected (injectate): 25G for aqueous solutions and 23G or 21G for viscous solutions or suspensions (Waynforth and Flecknell 1992). In mice, volumes per site should ideally be less than 10 ml/kg and no more than 40 ml/kg (Diehl et al. 2001). If it is necessary to administer larger volumes than the foregoing, it is advisable to divide the volume and administer it at two different sites either by puncturing the skin elsewhere or changing the path of the needle below the same entry point (Waynforth and Flecknell 1992).

11.1.4 INTRAMUSCULAR (IM) ADMINISTRATION

Intramuscular (IM) administration is very common in rodents, including for delivery of anaesthetic or antibiotic agents, although the use of IM injections in mice has been challenged because of the small muscle size (Morton et al. 2001; Zutphen van, Baumans, and Beynen 2001). Most information cited in the literature about this route concerns selection of muscle, volume and needle size.

It is necessary to have a good knowledge of the recommended intramuscular injection sites and muscle size, and to restrain the animal firmly to minimise trauma of the muscle and neighbouring vessels and nerves (Morton et al. 2001). The biceps femoris, the semitendinosus, quadriceps and gluteus maximus muscles are the sites of first choice (Stark and Ostrow 1990; Waynforth and Flecknell 1992). The injectate must not be cytotoxic, should have a neutral pH (i.e., about 7.3–7.4) and be isomolar. It is important to ensure that the needle tip is lodged in the muscle and not in a blood vessel, by applying slight tration to the syringe plunger to check for the absence of blood before injection. In rats, it is possible to administer volumes up to 0.1 ml preferably, and no more than 0.2 ml at any one site (Diehl et al. 2001) because the injected substance places the muscle fibres under tension and causes pain. For the same reasons, it is advisable to deliver the injections slowly. The ideal needle size is 25G (Morton et al. 2001; Wolfensohn and Lloyd 2003). Larger volumes should be injected as two separate aliquots into both thighs, respecting the limits indicated above. In mice, volumes up to 0.05 ml per site are preferred and no more than 0.1 ml (Diehl et al. 2001) using a 27G needle size (Wolfensohn and Lloyd 1994; Morton et al. 2001).

However, in mice, IM injections are considered to be more painful than most other routes of administration (Morton et al. 2001; Zutphen van, Baumans, and Beynen 2001). Indeed, IM administration of volumes as small as 0.02 ml of normal saline in mice prolonged the time taken for their heart rate to return to baseline values beyond that for SC and IP injections of 0.25 ml saline (Meijer et al. 2006), demonstrating that very small volumes (0.02 ml) of IM injections can produce more discomfort than tenfold greater volumes (0.25 ml) administered by the other two routes (SC and IP).

In conclusion, IM administration in mice is not recommended and should be used exclusively in cases when no other route is applicable. One refinement option for IM administration in rats consists in the covering of the rat's head with a towel so that it is in the dark during the injection, which has a calming effect (Rasmussen and Ritskes-Hoitinga 1999).

11.1.5 INTRAPERITONEAL (IP) ADMINISTRATION

Intraperitoneal (IP) injections should not be given more than once a day. If more frequent IP dosing is necessary, the implantation of an osmotic mini-pump avoids the stress associated with multiple injections and, additionally, circumvents the fluctuations in drug concentration, which may occur otherwise (Clarke 1993). With proper training and precautions, daily IP injections do not appear to be unduly stressful, although the consequences of a poorly conducted injection can be fatal. It has been estimated that 10–20% of IP injections fail in their intention, because of misplacement of the injected substance (Claassen 1994). Some authors recommend brief inhalation anaesthesia in order to avoid accidental puncture of the intestines or causing unnecessary stress to the animal (Morton et al. 2001). However, others argue that IP injections should be given to conscious animals either held in one hand with the head variously tilted up or down, or with the animal covered in a restraining towel (Rasmussen and Ritskes-Hoitinga 1999; Wolfensohn and Lloyd 2003).

Injection sites are the lower left or right quadrants of the abdomen. The preferred position should be the lower right side because, anatomically, the small intestine lies on that side, and because it has a thicker wall it is more resistant to needle punctures than the large intestine. A recent article verified that the caecum occupied the left side of the abdomen in 72% of Long-Evans and Wistar rats (Coria-Avila et al. 2007).

In rats, ideally, volumes of injectate given by IP injection should not exceed 10 ml/kg and the maximum is 20 ml/kg (Diehl et al. 2001); needle size may vary from 23 to 25G (Morton et al. 2001). In mice, volumes of 20 ml/kg are preferred, with a maximum of 80 ml/kg (Diehl et al. 2001) and a needle size varying from 25 to 27G (Morton et al. 2001). The issue of needle size is controversial; there are arguments that larger gauge needles are safer, because they are less likely to puncture the viscera. Syringe stoppers, which secure an appropriate injection depth (4–5 mm), have also been said to improve placement of the injectate; on the other hand, if needles are too short the material may be deposited in the abdominal muscles. Regarding the issue of pain, Meijer et al. (2006) reported that the delay in the return of heart rate to normal after IP puncture without injectate, demonstrated that puncture of the peritoneum is more painful than SC and IM punctures.

11.1.6 Intravenous (IV) Administration

Intravenous (IV) administration to rodents can be accomplished in either non-sedated or sedated or anaesthetised animals. Each approach includes a certain amount of stress. Non-sedated animals need to be immobilised in restraining devices, whereas sedated animals have been subjected previously to the stress of injection or inhalation. Rodent injection sites not requiring surgical exposure include the dorsal and lateral tail veins, the dorsal metatarsal vein, the dorsal penile vein and the sublingual veins.

In rats, IV injection volumes should be less than 5 ml/kg for a bolus injection and up to 25 ml/kg for slow injection (Diehl et al. 2001); recommended needle size varies from 25 to 27G (Morton et al. 2001). In mice, volumes should be less than 5 ml/kg for a bolus injection and 20 ml/kg for a slow injection (Diehl et al. 2001); needle sizes between 26 and 28G are recommended (Morton et al. 2001).

Repeated administration of pharmaceutical agents (or removal of blood) is greatly facilitated by providing continuous venous access. More or less successful devices of various sizes have been developed for permanent implantation into rats and mice (Koeslag, Humphreys, and Russell 1984; Chester, Weitzman, and Malt 1985); these are generally equipped with specially designed tethers and swivels, through which the venous catheter passes; this arrangement allows the animal free movement inside its cage. The ideal device should allow quick and easy IV access without the need to interfere with the animal's freedom or to sedate or anaesthetise it, thereby minimising stress but, at the same time, enabling continuous venous access for several weeks (de Wit et al. 2001). In addition, advances in catheter technology have led to the development of various heparin coatings that have anti-infective and anticoagulant properties and benefit these animal procedures by prolonging useful life and minimising the need of flushing or replacing. Improved intravascular catheter tips are less traumatic to the vessels' intimal lining and so decrease the animals' thrombogenic response. More recently, bio-implantable self-powered osmotic pumps and remotely activated miniature infusion pumps and even have been developed that constitute a significant refinement when frequent IV administrations would otherwise be required. The benefits they offer frequently outweigh the harm associated with their implantation when this is done by experienced and skilled anaesthetists and surgical technicians.

11.1.7 Inhalation

Administration of substances via the respiratory route is common in toxicology studies, especially when there is a need to assess the effects of a compound on the respiratory system or when rapid absorption is required. Consideration must be given to the pH and irritancy of the compound to the respiratory tract. Animals may be placed in appropriately constructed cages or in inhalation chambers and are exposed to vaporised or aerosolised substances, which they inhale; a potential pitfall with this method is that exposure in this way may be followed by the animals licking settled compound from their fur and its ingestion. Alternatively, rodents may be restrained individually in special conical tubes of appropriate size for nose-only inhalation exposure (Morton et al. 2001). This latter method causes additional distress to the animals, because they have to be restrained for some time without their conspecifics.

11.1.8 Dermal Application

Although used less frequently than other routes, this method, used in rodents and rabbits, has potential for refinement. Unless nude rats or mice are used, the animals must be carefully shaved, avoiding unintended abrasion of the skin and preferably done 24 hours prior to the application. The application area should be covered with a dressing to prevent the animal from licking; habituation is required before the application starts. The dose of compound applied should be no more than 2 g/kg, in volumes of less than 500 µl (Morton et al. 2001). Care must be taken to ensure that dressings are not excessively tight and that they are applied in the most painless way. Jackets have been used to protect applications in rodents, but animals may be partly immobilised by these and become stressed, even if they have been allowed a period for habituation. Neck collars seem to be a satisfactory alternative to jackets for rodents; they should be worn for a maximum of 20–22 hours after which time they should be removed for at least 2 hours to permit free movement, grooming and so on.

11.1.9 Intradermal (ID) Administration

Intradermal administration is used in immunology studies and in research into inflammation, skin sensitisation and cutaneous blood-flow diagnostic procedures (Leenaars et al. 1998; Morton et al. 2001). Intradermal injections are usually performed on the back of the animals and a fine-gauge needle should be held almost parallel to the skin surface and forwarded only a few millimetres. If there is a sudden loss of resistance, this means that the needle has passed subcutaneously and it is then necessary to withdraw the needle and retry. Successful administration should normally lead to a formation of a bleb, because there is very little interstitial space for the volume of material injected to occupy (Morton et al. 2001). Volumes of 0.05–0.1 ml may be administered depending on the thickness of the skin; multiple injections into neighbouring sites may be performed (Diehl et al. 2001). Intradermal administration may be painful, so the use of a local anaesthetic cream is recommended, particularly where larger volumes are to be injected.

11.2 IMPROVING SAMPLING PROCEDURES

Several handbooks provide detailed descriptions of various sampling techniques (e.g., sampling for blood, urine, faeces and cerebrospinal fluid); it is not necessary to repeat them here (Petty 1982; Semler, Gad, and Chengelis 1992; Iwarsson, Lindberg, and Waller 1994; Weiss et al. 2000; Koch 2006). This account focuses on validated methods of incorporating refinement and reduction strategies as well as issues relating to assuring sample quality.

11.2.1 Blood Sampling

Reports of working groups on Internet sites describe methods for blood withdrawal from laboratory rodents (BVA/FRAME/RSPCA/UFAW 1993; Diehl et al. 2001; NC3Rs 2009); they describe best practice and promote common sense refinement methods (e.g., how much can be safely withdrawn and how often samples can be taken); the characteristics of different sampling sites are summarised in Table 11.2. A variety of blood sampling sites have been proposed, some suitable only for puncture as a means of obtaining the sample, whereas others are suitable also for insertion of catheters for repeated sampling. In rodents, these sites include the carotid artery, aorta, ventral tail artery, cardiac puncture, sublingual vein, retro-orbital sinus, submandibular vein, jugular vein, vena cava, femoral vein, saphenous vein, dorsal metatarsal vein and dorsal and lateral tail veins (Petty 1982; Winsett, Townsend, and Thompson 1985; Horton, Olson, and Hobson 1986; Waynforth and Flecknell 1992; BVA/FRAME/RSPCA/UFAW 1993; Iwarsson, Lindberg, and Waller 1994; Weiss et al. 2000; Golde, Gollobin, and Rodriguez 2005; Jekl et al. 2005; Koch 2006; Smith 2008; Joslin 2009; Fukuta 2004/2010). The following presentation focuses on Refinement and Reduction alternatives that have been demonstrated to be effective, and reviews factors that impact on the suitability of the sample obtained for further accurate and precise analysis.

11.2.1.1 Refinement Aspects in Mice

In mice there appears to be an observer effect when blood samples are withdrawn from cage mates. Kugler, Lange, and Kalveram (1988) evaluated the influence of sampling order on corticosterone

TABLE 11.2
Simplified Characteristics of Common Blood Sampling Methods in Mice and Rats

Species	Method	Conditioning Verified	Welfare Impact	Anaesthesia Required	Habituation	Benefits	Disadvantages
Mouse	Saphenous vein	No	Low	No	No	Loose restraint: suitable for repeated glucose samples; easy to perform	
	Tail veins puncture/incision	No	Low	No	Yes?	Minimum restraint; daily sampling; incision suitable for repeated glucose samples	
	Submandibular vein	No	Medium	No	No	Easy to perform	Tight restraint
	Retro-orbital sinus	No	Considerable	Yes	No	Familiar surrounding effect	Accuracy problems; artefacts; tissue damage
	Jugular vein	No	Medium	Yes	No	High precision?	Accuracy problems
Rat	Saphenous vein	Yes	Low	No	Yes		
	Tail veins puncture/incision	Yes	Low	No	Yes?	Home cage effect; better precision than heart puncture	Minimum restraint; Low precision?; haemolysis
	Orbital sinus	No	Considerable	Yes	No	Home cage effect; better precision than heart puncture	Tissue damage possible; competence dependant; low precision?
	Jugular vein	No	Medium	Yes	No	High precision; fast recovery	
	Sublingual/gingival vein	No	Low	Yes	No		

Note: For more details, see text and articles referenced in the text.

release in female C57/BL6 mice housed in groups of five. The mice were successively caught and decapitated every 2 minutes. Corticosterone concentration in blood sampled in this way increased in the fourth and fifth mice in a cage (i.e., within 4–8 minutes). These results suggest that quiescent corticosterone levels can only be determined in samples obtained within 4 minutes following capture of the first mouse (Kugler, Lange, and Kalveram 1988). However, a similar study using BALB/c mice found no evidence for pheromone release or vocalisation during tail bleeding, because corticosterone concentration in the blood of other animals in the cage remained unchanged (Tuli, Smith, and Morton 1995). It thus remains unclear whether mice are affected by being in the vicinity of cage mates removed for experimental procedures such as blood sampling and anaesthesia followed by exsanguination.

Some blood sampling methods require the use of anaesthesia (BVA/FRAME/RSPCA/UFAW 1993), and it is important to appreciate the effects of different anaesthetic agents on the specific parameter being studied. This may be one reason for the large variability of corticosterone concentration in blood reported for mice. Shipp and Woodward (1998) withdrew blood samples from adult C57BL/6J mice of both sexes between 9 and 10 a.m., using either methoxyflurane or CO_2 anaesthesia under two different conditions; anaesthesia was induced either in the housing room or following transfer to an adjacent room (requiring about 30 seconds). The CO_2 anaesthesia involved placing the entire cage in a chamber flushed with CO_2. Blood corticosterone concentrations of mice anaesthetised with CO_2 in the holding room were significantly lower than those in the other three groups and appeared to resemble the values of quiescent laboratory mice. This suggests that in mice the combination of a suitable anaesthetic agent coupled with a familiar routine and environment constitutes a procedural refinement because it reduces acute distress preceding anaesthesia (Shipp and Woodward 1998).

Considering how frequently blood samples are taken from mice, surprisingly few studies have compared the impact on welfare of different blood sampling methods. Dürschlag et al. (1996) described modification of the traditional tail incision method, allowing rapid acquisition (1–1.5 minutes) of serial blood samples (40–150 µl) from conscious laboratory mice without pre-warming. In this modification, one hand holds the tail and at the same time restrains the animal by gently pressing on its back. The tail is bent upwards, held between thumb and forefinger while the ring finger stabilises the tail dorsal to the site of incision. Blood is obtained by lacerating ventral blood vessels at the level of base of the tail with a short oblique cut. Plasma corticosterone concentration in subsequent samples (1–9 days) remained remarkably constant, indicating tail incision may be of particular relevance to studies that involve relating physiological measures to behavioural responses in laboratory mice, and hence may be considered as a refinement method (Dürschlag et al. 1996).

Abatan, Welch, and Nemzek (2008) compared blood samples withdrawn by puncture of the common saphenous vein and a modified tail-clip technique in mice; corticosterone responses were similar, but the tail-clip method produced fewer behavioural reactions. This may be because the saphenous approach requires more restraint and the outcome may be influenced by the manual skills of the handler. When a series of four blood collections was repeated within 1 week, blood corticosterone concentration showed no evidence of habituation following either method, although attempts by mice to escape from the handler were more marked with the saphena method. Moreover, there were no stress-associated changes in the leukogram after serial blood collection (Abatan, Welch, and Nemzek 2008).

Grouzmann et al. (2003) studied responses of plasma catecholamine to stress elicited by blood sampling in mice. While reviewing the literature on the topic, basal noradrenalin and adrenaline concentrations varied greatly (i.e., from 4 to 140 nM depending on the method used for blood sampling). This was not the case with samples of blood obtained from arterial catheters, which they considered to be the most accurate method (i.e., adrenaline and noradrenalin levels: 1.1 ± 0.3 nM and 4.1 ± 0.5 nM, respectively). Such large variability prevents meaningful study comparison and may well conceal a catecholamine response to stress. Blood obtained by retro-orbital bleeding from mice anaesthetised with halothane also failed to reveal a rise in noradrenalin and adrenaline concentrations, although samples obtained following decapitation yielded values between 25 and

30 nM, indicating that the mice had been stressed. The authors concluded that sampling methods involving arterial catheterisation and retro-orbital bleeding appear to be the most reliable procedures for studying activation of the sympathetic nervous system in mice under both unstressed and stressed conditions (Grouzmann et al. 2003).

There have been few studies on the effects of blood sampling on the welfare of mice and, hence, only a few scientifically-based recommendations can be made. The impact of procedures such as whole body or local warming of the animal and the effect of the temperature and duration used is not known.

11.2.1.2 Accuracy and Precision in Mice

The influence of sampling technique on the accuracy and precision of subsequent determinations of blood composition in mice has been reviewed by Quimby (1989). Doeing, Borowicz, and Crockett (2003) studied the effect on haematological parameters of blood withdrawal from C57BL/6 mice anaesthetised with methoxyflurane, following puncture at four different sites: cardiac, tail, metatarsal and saphenous vein. The white blood cell counts (WBC) in samples obtained by cardiac puncture were significantly lower than for the other three sites, but within these there were no differences in leukocyte count (Doeing, Borowicz, and Crockett 2003). Nemzek et al. (2001) compared the influence of sampling from three sites on the haemogram of BALB/c mice under neurolept-anaesthesia; the sites were the tail, eye and heart. The WBC count was highest in blood withdrawn from the tail, lower from eye and significantly lower from the heart; overall tail and cardiac blood reflected changes in all cell types. Haematocrit, RBC and platelet counts were significantly higher in the tail compared to heart blood (Nemzek et al. 2001). Another study, comparing tail incision and saphenous puncture methods in conscious mice, reported higher WBC, neutrophil and lymphocyte counts from samples originating from the tail (Abatan, Welch, and Nemzek 2008). Although comparisons are complicated because different anaesthetics were used, it is clear that the site from which blood is withdrawn influences the haematological profile, so it is important to use the same sampling site within an experiment.

Another essential question when considering the most appropriate method of collecting blood samples is how best to avoid adventitious contamination of the sample (result artefacts). An example of such an artefact is a distinctive cytogram called BASO/lobularity channel, which has been shown to be associated with retro-orbital sinus puncture, but not with jugular vein or heart puncture in mice anaesthetised with CO_2, and which is believed to be a result of contamination by extraneous material derived from the orbital sinus (Olin, Diters, and Freden 1997).

An early study of Friedel et al. (1975) compared the serum concentration of 10 enzymes in blood samples taken from the jugular vein and ventral aorta. In mice the activities of lactate dehydrogenase, malate dehydrogenase, aspartate aminotransferase and myokinase were found to be higher in blood samples obtained from the ventral aorta and they concluded that the most accurate method for blood sampling in the mouse is by catheterisation of the jugular vein (Friedel et al. 1975). Schnell and others (2002) compared clinical pathology parameters resulting from three different blood collection methods in anaesthetised C57BL/6 mice. The methods used were cardiac puncture, caudal vena cava and retro-orbital bleeding. They found that blood withdrawn from the caudal vena cava showed least variation in both sexes and appeared slightly more consistent than the cardiac method for the parameters evaluated. The largest difference between groups was in transaminase concentration. While ALT values were similar in samples collected by cardiac and vena cava puncture, AST concentrations in the latter varied less than in samples from the vena cava (Schnell et al. 2002). Overall, these studies demonstrate the substantial impact of sampling site on clinical chemical characteristics of blood in mice.

Blood glucose and insulin concentration are often measured in animal model studies. Rogers and others (1999) investigated the impact of collection site on plasma glucose and insulin; the sites compared were the retro-orbital sinus and the tail vein in young conscious female C57BL/6 mice. In their study retro-orbital collection yielded blood of lower glucose concentration and higher insulin concentration than blood collected from the tail vein (Rogers et al. 1999). Tabata, Kitamura, and Nagamatsu (1998) examined changes in plasma glucose concentration after repeated bleeding in mice and found that each bleeding elicited a transient increase, but that it was possible to sample plasma glucose at 1-hour intervals without interference from the sampling procedure.

Christensen and others (2008) have recently evaluated the suitability of four different methods of repeated blood sampling for conducting the oral glucose test in mice, from the point of view of achieving more consistent results and minimising the number of animals needed. The four methods used were amputation of the tail tip, lateral tail incision, puncture of the tail tip and retro-orbital puncture. Four blood samples were drawn at 30-minute intervals and samples were assessed for haemolysis and whether the procedure induced a change in blood glucose over the 90-minute test period. Samples obtained by retro-orbital puncture were free of haemolysis, while puncture or amputation of the tail tip induced haemolysis in a significant proportion of samples. All methods, except for puncture of the tail tip, influenced blood glucose concentration. The authors argued that although lateral tail incision had some impact on blood glucose concentration, it was the method of choice for the oral glucose test, because it produced a clot-free, non-haemolysed sample; whereas, although retro-orbital sampling yielded a high quality sample, it induced such a dramatic change in blood glucose that it should not be used for the purpose in mice (Christensen et al. 2008). These studies relate only to blood glucose and insulin determinations, but it may well be that other assays are affected in different ways and similar background studies are needed to establish the best methods.

11.2.1.3 Approaches to Refinement in Rats

Some blood sampling methods require anaesthesia, while others can be performed without (BVA/FRAME/RSPCA/UFAW 1993). Considerable attention has been paid to determining the most suitable anaesthetic for rats; isoflurane has been suggested as a suitable alternative to CO_2 as an anaesthetic agent for blood withdrawal in rats. Fitzner Toft and others (2006) used radiotelemetry to compare physiological responses to jugular, retro-orbital and tail vein puncture, and sampling by jugular puncture whilst conscious and under anaesthesia induced by CO_2 and isoflurane. Retro-orbital puncture caused both heart rate and blood pressure to rise by the greatest amount and recovery time was longest. Collection of blood induced rapid fluctuations in both parameters and caused an increase in body temperature lasting 2–3 hours, although in the case of the tail vein it remained elevated for more than 30 hours. Isoflurane anaesthesia reduced the increase in blood pressure and heart rate and resulted in fewer fluctuations. The authors concluded that, of the methods examined, blood sampling by jugular puncture appeared to allow fastest recovery and isoflurane is a better anaesthetic agent than CO_2 for blood sampling (Fitzner Toft et al. 2006).

Altholtz and others (2006) compared the effects of CO_2 and isoflurane anaesthesia on corticosterone blood concentration during serial jugular blood sampling in rats. Both anaesthetics elicited a sharp increase in corticosterone concentration within 30 minutes of administration, with further elevation by 1 hour and a decline thereafter. Although isoflurane resulted in higher baseline corticosterone concentrations, the increase at subsequent time points was significantly higher in animals exposed to CO_2. The authors concluded that isoflurane reduces the effects of stress, while CO_2 does the opposite, and suggested that isoflurane is the better of the two anaesthetics for serial blood collection (Altholtz et al. 2006). O'Brien et al. (1993) used measures of urinary concentrations of corticosterone and testosterone to compare stress experienced by rats under four different anaesthetic regimens. They found that rats anaesthetised with ketamine-zylazine, hypnorm-midazolam and methoxyflurane, but not halothane, exhibited increased urinary corticosterone excretion suggesting that the anaesthesia was associated with stress (O'Brien et al. 1993).

Fomby and others (2004) combined CO_2/O_2 anaesthesia in F344 rats with two different animal handling techniques to establish a procedure suitable for collecting serum for corticosterone assay. When blood was obtained from the retro-orbital plexus under CO_2/O_2 anaesthesia, the concentration of corticosterone in serum of male and female rats, anaesthetised in their home cages, varied between about 25% and 50% compared with those anaesthetised in a separate chamber. This establishes that a minor change in the method of anaesthesia can result in more reliable baseline corticosterone levels; this constitutes both refinement and reduction (Fomby et al. 2004).

There is an ongoing, unresolved debate about the optimal way of obtaining a blood sample from a minimally stressed rodent, for determining stress hormone concentrations. Some investigators prefer the use of indwelling catheters whereas others claim that catheter implantation results in chronic stress and interferes with stress hormone measurements. Vachon and Moreau (2001) evaluated stress responses of rats by comparing the concentrations of glucose and corticosterone in blood after repeated blood sampling by venepuncture under short-term anaesthesia and sampling conscious rats with catheters implanted in their jugular vein. In both cases, blood glucose concentrations increased when blood sampling started, peaked at significantly increased concentrations at 30 minutes and decreased thereafter until the end of the assessment period. However, at the 1- and 2-hour time points the concentration of corticosterone was significantly lower in catheterised rats compared to those anaesthetised. Although both sampling methods were similar with regard to peak concentrations and time-to-peak of blood glucose and corticosterone, stress was arguably less in cannulated rats than those that underwent repeated anaesthesia (Vachon and Moreau 2001).

Later, Vahl and others (2005) compared the concentration of ACTH and corticosterone in blood samples obtained from rats via previously implanted vena cava catheters, tail vein nicks or excising the tip of the tail. Apart from the increase resulting from the initial stress of restraint, the plasma ACTH and corticosterone responses were similar for all sampling methods. The elevation in basal ACTH, which occurred in animals subjected to tail vein nick, suggests that sampling needs to be completed rapidly (<3 minutes) to avoid initiating the pituitary stress response. These results support the use of either chronic vascular catheters or withdrawing blood from a tail vein (Vahl et al. 2005).

Fluttert, Dalm, and Oitzl (2000) showed that up to 300 µl of blood can be sampled within 90 seconds from a small incision at the end of rats' tails; samples can be taken sequentially from this site. This method does not require anaesthesia, surgery or restraint and the low, basal concentration of corticosterone in the resultant blood samples, even sequential samples collected over a 3-hour period, indicate that the procedure can be considered stress-free. Indeed, the authors also reported a home cage effect—corticosterone concentration remained much lower when rats were housed throughout the 3-hour period in their own cage as compared to a novel cage (Fluttert, Dalm, and Oitzl 2000).

One question central to refinement is: do rats habituate to blood sampling? Haemisch, Guerra, and Furkert (1999) assessed the impact of repeated, frequent blood sampling by measuring blood concentrations of corticosterone and β-endorphin in Wistar rats. Blood samples were collected from the tail vein of well-accustomed conscious rats on four occasions over 2-hour periods, both at the summit and trough time of the diurnal corticosterone cycle. At both phases of the cycle, corticosterone concentration rose in response to the first sampling and returned towards baseline in successive samples; β-endorphin also increased initially, but remained elevated thereafter. The inhibition of corticosterone secretion associated with sustained elevation of β-endorphin concentration suggests that repeated blood sampling is perceived as a moderately stressful procedure (Haemisch, Guerra, and Furkert 1999).

Retro-orbital puncture is a widely used method of collecting blood samples; it enables rapid collection of relatively large volumes of blood but whilst it has no detrimental effect on the general health of the animal it has the potential to damage the eye. As an improvement over this method, Angelov and others (1984) suggested sampling from the sublingual vein; they claimed it was simple to perform in unanaesthetised rats, and yielded sufficient quantities of blood and left the eye unimpaired. Their results showed that withdrawal of blood from the sublingual vein did not affect body weight or food consumption, and there were no significant differences in haematological or serum chemistry values compared to retro-orbital samples. Much later Zeller and others (1998) refined the method by recommending general anaesthesia in response to suggestions of the 1993 Working Group (BVA/FRAME/RSPCA/UFAW 1993).

Van Herck and others (2001) studied behavioural responses to three different blood sampling techniques: retro-orbital puncture during ether anaesthesia, tail vein puncture under O_2/N_2O-halothane anaesthesia and saphenous venipuncture in conscious rats. After each sample had been taken, the behaviour of rats was monitored using automated behaviour assessment methodology (LABORAS 2009). The results indicated that retro-orbital puncture under ether anaesthesia caused more distress than the other two techniques. Additionally, haematological and plasma chemistry values varied significantly with different collection techniques (Van Herck et al. 2001). Mahl and others (2000) compared the use of retro-orbital and sublingual vein approaches in male rats anaesthetised with isoflurane. Simulating a pharmacokinetic study, blood was sampled six times during 1-day from non-fasted animals and, again after 3 weeks of recovery, from fasted animals on a single occasion. There was considerable inter-individual variability, but the mean prolactin concentration was higher on each occasion and corticosterone was elevated after a single sample in fasted animals. The findings indicate that blood sampling from the retro-orbital sinus was more stressful, although, because both groups of animals were anaesthetised, the responses relate to both anaesthesia and the sampling method (Mahl et al. 2000).

Blood sampling is a manual skill in which experience and expertise play a crucial role; this has been demonstrated by van Herck and others (1998) who examined clinical signs following retro-orbital blood sampling in rats. Four animal technicians collected blood from over 300 rats, using minor variations in technique. The person carrying out the procedure had a major influence on the type, frequency and outcome of clinical observations. Two experienced animal technicians carried out the technique without apparent adverse consequences, whereas a less experienced technician caused severe abnormalities. Using a Pasteur pipette or haematocrit capillary, or puncturing the lateral versus the medial canthus of the orbit were equally successful, whereas application of antibiotic eye ointment into the conjunctival sac significantly decreased the incidence of clinical findings. A skilled animal technician could sample four times from the same site at 14-day intervals without significantly increasing adverse findings (van Herck et al. 1998).

Automated blood sampling equipment allows the collection of blood samples without the animal being aware of this although it is necessary to establish a catheter connection to the animal. Abelson and others (2005) evaluated whether the frequency of blood sampling influences the plasma corticosterone concentration in rats. Blood samples (200 µl) were withdrawn at high (24 samples) or low (three samples) frequency during a 6-hour period immediately following insertion of the catheter. Blood corticosterone concentration remained at the high post-insertion level when blood was collected frequently, but there was a steady, significant decline during low-frequency sampling, suggesting that sampling too frequently interferes with determination of this stress hormone (Abelson et al. 2005).

11.2.1.4 Accuracy and Precision in Rats

Sources of variability in blood sample composition have been documented in a classical handbook (Meeks 1989). They can be subdivided into those related to the physiological status of the animal and its environment, those related to sampling procedure and those attributable to assay methods. Interestingly, this handbook states that a major source of variation is related to different sites from which the sample may be withdrawn; Meeks quotes a rather old study carried out with F344 rats sampled under methoxyflurane anaesthesia, in which clinical chemical determinations were made of 19 common blood constituents, including serum enzymes, cholesterol and triglycerides. Least variation was detected in blood collected into vacuum tubes directly from the heart; the largest variation was found in samples taken from the tail and retro-orbital sinus. Moreover, significant differences attributable to the sampling site were found for almost every parameter (Neptun, Smith, and Irons 1985). Suber and Kodell (1985) collected blood samples from female Sprague-Dawley rats by retro-orbital puncture, tail vein incision and cardiac puncture and assayed seven haematological and seven clinical chemical parameters. Six of the haematological parameters were significantly more variable in blood withdrawn directly from the heart than from the other sites; moreover there was a significant difference in variance of all clinical chemistry parameters between blood obtained by retro-orbital puncture and directly from the heart. However, over 60% of serum samples obtained by cardiac punctures were haemolysed, compared with about 25% of those from tail vein incision. They concluded that lack of haemolysis and the lower variance make the retro-orbital puncture technique the method of choice for collection of blood samples from rats (Suber and Kodell 1985). Both these studies used anaesthetised animals, and the latter was associated with a high incidence of haemolysis, so it is difficult to draw clear conclusions. Although there is a need for clear advice on site selection, both in the interests of scientific integrity and refinement, there have been few recent studies in this area.

Smith, Neptun, and Irons (1986) compared haematological parameters in samples of blood withdrawn from the right ventricle, abdominal aorta, abdominal vena cava and retro-orbital plexus of F344 rats anaesthetised with methoxyflurane, and from the tail of conscious rats. White blood cell counts from tail samples were approximately double than those from other sites. Compared with other sites, samples from the retro-orbital plexus and tail exhibited significantly more variability of most haematological values; blood samples obtained from the right ventricle were the least variable. Dameron et al. (1992) have shown that in Sprague-Dawley rats anaesthetised with ether, blood samples taken from the retro-orbital sinus exhibit prolonged prothrombin and partial thromboplastin times when compared to samples from the posterior vena cava. There were no haematological differences between samples but coagulation times and serum magnesium and phosphate showed biologically significant differences.

In the study by Mahl and others (2000) described in Section 11.2.1.3, blood samples were taken from the retro-orbital plexus and the sublingual vein of non-fasted rats on six occasions during 1-day and once more after 3 weeks, following fasting. Haematological evaluation showed that repeated blood sampling caused a gradual decrease in erythrocyte count, haemoglobin concentration, and haematocrit. Repeated withdrawal of approximately 10% of the total blood volume led to a decrease in these parameters by up to 10%; this limit should not be exceeded for reasons of animal welfare and in the interests of obtaining reliable data. The decrease in lymphocyte and increase in neutrophil counts after repeated sampling were slightly more pronounced in blood taken from the retro-orbital plexus than from the sublingual vein. There were significantly higher creatine kinase and aspartate aminotransferase activities in samples from the retro-orbital plexus. These findings also suggest that blood sampling from the retro-orbital plexus causes more tissue damage than from the sublingual vein (Mahl et al. 2000).

11.2.2 Sampling Urine and Faeces

Several handbooks carry good and rather detailed descriptions of techniques for sampling or collecting urine and faeces from rodents (Petty 1982; Semler, Gad, and Chengelis 1992; Iwarsson, Lindberg, and Waller 1994; Weiss et al. 2000; Koch 2006; Fukuta 2004/2010). Most rely on individual housing in metabolic cages, voluntary or induced urination, or fairly complicated and welfare-unfriendly cannulation.

Few refinements to the use of metabolic cages have been described, even though the technique requires individual housing, a departure from European guidelines, which requires specific scientific justification. One such refinement is an "enriched" metabolism cage, which has been shown to deliver the same quality and quantity of urine and faecal samples (Sörensen et al. 2008).

Qualitative urinalysis (e.g., urinary pH, protein, glucose, bilirubin, blood, ketone, urobilinogen, and creatinine), can be carried out using just a few drops of rodent urine that can be collected without the use of metabolic cages, particularly advantageous when several animals are to be sampled. The single animal method involves holding a mouse, which often causes it to expel a drop of urine, which can be collected on plastic film; where a group of animals is involved, each is transferred to an individual compartment, the floor of which is covered with plastic film. When urine is voided, each mouse is transferred back to its home cage and the sample is aspirated; volumes of 10–250 µl can be obtained. The method can be modified for rats (Kurien and Scofield 1999).

The collection of uncontaminated rodent urine samples is challenging. Fundamental aspects of urine collection to be considered are: (1) ease of collection; (2) quality of sample; (3) prevention of contamination; (4) impact on the animal of the procedures used; (5) levels of pain caused to the animal; and (6) refinement of methods to reduce stress, pain, or distress. Qualitative and quantitative sampling strategies for a variety of species including rodents and rabbits, have been reviewed recently from the point of view of animal welfare (Kurien, Everds, and Scofield 2004).

Faeces can be collected with anal cups (Petty 1982); earlier designs of these collection cups involved fixation with which tape frequently led to constriction and serious necroses of the rat tail, and hampered its growth. A modification of the cup resolved these problems; it has been tested in 118 rats for up to 4 weeks without untoward effects (Schaarschmidt et al. 1991).

11.2.3 Other Sampling Procedures

Other sample procedures include collection of cerebrospinal fluid, milk, saliva, peritoneal fluids and lymph. These are less frequently performed than those described above, and detailed descriptions of them can be found in handbooks (Petty 1982; Waynforth and Flecknell 1992; Iwarsson, Lindberg, and Waller 1994; Weiss et al. 2000; Koch 2006; Fukuta 2004/2010).

11.3 IMMUNISATION FOR PRODUCTION OF POLYCLONAL ANTIBODIES (PABS)

The production of polyclonal antibodies involves the administration of substances, usually parenterally, in order to provoke an immune response—the secretion of immunoglobulins—which can be detected and harvested from the blood. To follow progress of the response and to harvest the antibody it is usually necessary to withdraw blood samples. Rabbits are most frequently used for production of small quantities although both mice and rats are utilised also; however, in birds immunoglobulins are also secreted into the yolk of the egg and it is possible to extract them without the need for blood sampling. Reference will be made to the use of birds for polyclonal antibody production, but there are technical reasons (the antibody is an immunoglobulin Y, which differs in certain ways from the more commonly used immunoglobulin G) as well as lack of familiarity, why small mammals are still frequently used.

Scientists performing an immunisation procedure generally wish to achieve a high antibody titre, within a short time, and the antibody itself should be specific and of high affinity. Factors influencing these characteristics include: the nature of the antigen, the use of an adjuvant, the choice of animal and the immunisation protocol (route and volume of injection, boosting schedule, etc.).

11.3.1 The Antigen

Antigens can be single molecules, such as proteins, peptides and polysaccharides, or particulate complex multi-antigens including bacteria, viruses, parasites, protozoa, mammalian cells and artificial particles. Single antigens include proteins, peptides and polysaccharides. The intensity of the immune response is directly related to the size of the antigen and the degree of foreignness. Most particulate antigens are highly immunogenic and for single antigens there are generalities that "the bigger the better" and "the more foreign the molecule the better".

Small antigens (< 3–4 kDa) are unable to elicit an antibody response in themselves (Poulsen, Hau, and Kollerup 1987; Poulsen et al. 1990) and a protein-carrier molecule is required to activate T-cells. The antigen must possess at least one epitope (foreign antigenic determinant) that can be recognised by the cell surface receptor on B-cells, and it must have at least one surface structure that can be recognised simultaneously by a class II protein and a T-cell receptor.

Basic Procedures: Dosing, Sampling and Immunisation

The purity of the antigen is important. Impurities are often immunogenic in their own right and may prove to be immunodominant, so diminishing the immune response (Leenaars 1997). It can be time consuming and laborious to purify the antigen, but usually the effort is well worthwhile, because it is much easier to enhance the specificity of an antiserum only weakly contaminated with unwanted antibodies than to remove the latter by extensive absorption procedures.

Antigens may be biologically active molecules and have the potential to harm the animal immunised, or antigen preparations may be rendered toxic by contamination with endotoxins, residues of chemical used in preparation, or extremes of pH (Leenaars et al. 1999). Ideally, antigens should be prepared and stored in accordance with carefully described and validated procedures, and records kept of all key stages.

11.3.2 THE ADJUVANT

The ideal adjuvant stimulates development of high and sustainable antibody titres, is effective in a variety of species, is applicable to a broad range of antigens, can be easily and reproducibly combined with antigens to make a mixture that is easy to inject, is effective after a small number of injections, has low toxicity for the immunised animal and is not hazardous to the investigator. Unfortunately, the adjuvant that meets all these criteria remains to be identified.

There are over 100 known adjuvants, but only a few of these are used routinely (Table 11.3). The adjuvant–antigen mixtures most frequently used are water-in-oil (W/O) emulsions, based on oily products such as Freund's adjuvant (FA; Freund, Casals, and Hosmer 1937). Freund's Incomplete Adjuvant (FIA) consists of 85–90% mineral oil and 10–15% detergent. Freund's Complete Adjuvant (FCA) is FIA to which is added heat-killed, dried *Mycobacterium* species. Freund's complete adjuvant is widely used because it provokes long-lasting immune responses (Talmage and Dixon 1953), and booster doses (Hohmann 1998) are not always needed because of the depot effect of FCA.

When a water-based solution of antigenic material is injected into an animal, it is promptly and rapidly disseminated. Precipitation or adsorption of antigens using compounds such as alum (incorporated safely into many vaccines used in humans) prolongs retention of the antigen only slightly and antibody titres are usually unimpressive. However, if the same material is injected as a water-in-oil emulsion the oil vesicles remain at the site of injection for many months and antigen leaves the injection site with a half-life of approximately 14 days. The persistent cell-mediated reactions at the site of the original depot, may lead to tissue inflammation and granulomas, which may become necrotic (Leenaars et al. 1999), granulomas have also been described in lungs, kidneys, liver, heart, lymph nodes and skeletal muscle (Hau 1988; Leenaars 1997).

In chickens, FCA reduces the frequency of egg laying, and consequently antibody productivity, as compared to FIA (Bollen and Hau 1999b). The FA-induced arthritis constitutes the only well-validated animal model of chronic pain (Colpaert 1987; Erb and Hau 1994) and the side effects of adjuvants, in particular FCA, have led to serious concern for animal welfare. Effective alternative adjuvants are now available that also stimulate cell-mediated immunity (Herbert 1967; Lindblad and Hau 2000). More detailed information on adjuvants can be obtained from several review articles and books (Lindblad and Sparck 1987; Jennings 1995; Leenaars et al. 1998).

Freund's complete adjuvant constitutes a human health hazard because personnel who accidentally inject themselves while immunising animals experience very unpleasant complications (Freund 1947; Chapel and August 1976); consequently, the use of FCA should be discouraged. If used, there should be a limit to the volume injected and restrictions placed on the location of injection and the number of boost injections. The FCA is normally administered with the first immunisation and should never be injected twice into the same animal—repeated injections may cause severe tissue reactions and even anaphylactic shock if the animal has become sensitised to any of the bacterial immunogens. An additional reason for avoiding the use of these types of adjuvants is

TABLE 11.3
Overview of Categories of Adjuvants that may be Used for Routine Polyclonal Antibody Production

Category	Examples	Mode of action
Mineral salts	$Al(OH)_3$, $AlPO_4$	Vehicle, depot effect
Oil emulsions (W/O, O/W, W/O/W)	FIA, Specol	Vehicle, depot effect
	Montanide	Activation of macrophages
Microbial (like) products	LPS, MDP, MPL	Stimulation of B or T cells
	TDM	Activation of macrophages
		Enhanced antigen uptake
Saponins	Quil-A	Facilitate cell-cell interaction
		Aggregation of antigen
Synthetic products	DDA	Activation of macrophages and complement
	Iscoms	Facilitate cell-cell interaction
		Stimulation of T cells
		Aggregation of antigen
	Liposomes	Vehicle
		Enhanced antigen uptake
	NBP	Activation of macrophages and complement
Cytokines	IL-2, IL-1, IFN-γ	Growth and differentiation of B and T cells
Adjuvant formulations	FCA, TiterMax™ RIBI, Gerbu, Softigen	Combinations of the above

W/O = water-in-oil; O/W = oil-in-water; W/O/W = water-in-oil-in-water
FIA = Freund's incomplete adjuvant
LPS = lipopolysaccharide
MDP = muramyl dipeptide
MPL = monophosphoryl lipid A
TDM = trehalose dimycolate
DDA = dimethyldioctadecylammonium bromide
ISCOMs = immuno stimulating complexes
NBP = non-ionic block polymer
FCA = Freund's complete adjuvant.

that antibodies directed against bacterial components of complete adjuvants can interfere with use of the antiserum.

Preparation of antigen/adjuvant mixtures should be done aseptically, and when an oil emulsion is used, the stability and quality of the emulsion is critical (Lindblad and Sparck 1987; Bollen and Hau 1997). The time between preparation of the mixture and its administration should be short and if it is necessary to store the mixture this should be done in a freezer and the quality of the emulsion checked after thawing and before injection.

11.3.3 The Animal

11.3.3.1 Species

Several factors need to be considered when determining the most appropriate species for a specific purpose, including the volume of antibody or antiserum needed (hence the volume of blood required), whether there is a phylogenetic relationship between the recipient and the source of the antigen, the character of the antibody synthesised by the recipient species and the intended use of the antibody. When large quantities of antibody are needed, farm animals such as sheep or goats are good choices. When smaller amounts are required, mice and rabbits are preferred, particularly because these animal species are easy to bleed compared to guinea pigs and hamsters for example.

11.3.3.2 Species/Strain-Stock

The antibody response to most antigens seems to be under polygenic control involving histocompatibility antigens (Long 1957; Lifshitz, Schwartz, and Mozes 1980); often there is both a species and strain difference (Harboe and Ingild 1973; Mayo et al. 2005). If it is important to minimise between-animal variation, mice, rats, or chickens may be preferred because all are readily available as inbred strains. The immune response of outbred stocks such as rabbits varies markedly between individuals (Young and Atassi 1982).

Because of the continuous transovarian passage of antibodies from blood to egg yolk in birds (Bollen and Hau 1997), it is convenient to harvest and purify antibodies from the egg yolk of the domestic fowl and several fairly simple methods have been described (Jensenius et al. 1981; Svendsen et al. 1995; Schade et al. 1996; Svendsen Bollen et al. 1996). Chicken egg antibodies (IgYs) can be extracted from the egg in quantities of approximately 100 mg/egg, and a chicken produces much more antibody than a similarly sized mammal. Chicken antibodies, raised against a protein from one mammalian species, will often react against the analogous protein in other mammalian species (Hau et al. 1980, 1981).

The use of chickens for production of antibodies is attractive with respect to implementation of the Three Rs (BVA/FRAME/RSPCA/UFAW 1993). Mammals are replaced by a species with a lower degree of neurophysiological sensitivity and the number of animals needed is considerably reduced. Oral immunisation techniques are being developed (Hau and Hendriksen 2005) and, if successful, will eliminate the need for restraint and distress associated with administration of antigen and withdrawal of blood to harvest antibodies. Refinements in the methodology may make it possible to produce antibodies without causing pain, suffering or distress so that it would not constitute a procedure within the new Directive (as drafted at the time of writing).

11.3.3.3 Sex

Although there are no scientific reasons for not using male animals, females are preferred because they are generally more docile and therefore can be group-housed more easily, an important consideration because social stress may impair the initial immune response in immunised animals (Abraham et al. 1994). A sex difference in the magnitude of antibody response within inbred strains has been observed (Abraham et al. 1994) and antibody secretion is impaired during pregnancy in the mouse (Poulsen and Hau 1990).

11.3.3.4 Age

In general, young adult animals respond better, having been exposed to fewer environmental immunostimulatory agents. After young adulthood the immune response declines with advancing age. Chickens should be of egg-laying age (18 weeks) by the time antibody is harvested but there is no great difference in titres obtained in young egg-layers compared with older hens about to cease egg laying (Bollen and Hau 1999a). The following minimum ages are recommended for Pab production: mice and rats at 6 weeks; rabbits and guinea-pigs at 3 months; chickens at 3–5 months, goats at 6–7 months and sheep at 7–9 months.

11.3.4 THE IMMUNISATION PROTOCOL

11.3.4.1 The Injection Route

The location at which the antigen is deposited in part determines the lymphoid organs activated and the type of antibody response that follows. Other considerations when selecting the route of injection are the species, the adjuvant, the quantity of antigen/adjuvant mixture and the welfare of the animals.

There is general agreement that the SC and IM routes are preferred. Alternatives include oral administration, intranasal inoculation, footpad (subcutaneous) injection and intrasplenic injection. Oral administration is preferred when using specific adjuvants, such as *Lactobacillus*, and when the requirement is a peripheral as well as a mucosal immune response. Intranasal administration of antigen in aqueous solution is used to induce tolerance; it is usually necessary to anaesthetise the animals and the antigen preparation is given by pipette or micro-cannula. Footpad injection (particularly when an oil adjuvant is used) and intrasplenic injection raise serious animal welfare concerns, and these routes are not necessary for routine Pab production. Their use should be justified on a case-by-case basis. If footpad injection is performed, only one hind foot should be used and the animals should be housed on soft bedding.

11.3.4.2 The Volume of Injection

As a general rule, injection volumes should be as small as possible, both to limit side effects and also because the immune response is generally stronger against an antigen administered in high concentration than a similar quantity at a low concentration. Another important consideration is the adjuvant used. As discussed earlier, oil-based adjuvants, particularly those incorporating microbial products, and viscous gel adjuvants form a depot at the injection site and induce local inflammation, which has the potential to induce sterile abscess formation. For this reason emulsion antigens should be administrated in smaller volumes than aqueous ones. In the case of SC injection, small volumes can be injected at several sites to reduce the intensity of side effects. However, the number of injection sites should be limited (Erb and Hau 1994).

11.3.4.3 The Dose of Antigen

For each antigen there is a dose range called the "window of immunogenicity". Too much or too little antigen may induce suppression, sensitisation, tolerance or other unwanted immunomodulation effects (Hanly, Artwohl, and Bennett 1995). The optimal dose for immunisation depends on several factors, including its inherent properties, whether it is purified or a component in a mixture of antigens, the adjuvant used, the route and frequency of injection and so on. Typically, to assure a high-titre antibody response, quantities vary between 25 and 50 µg up to 1 mg for a protein antigen delivered with an adjuvant. Although smaller quantities may be used for smaller animals, the amount of antigen required does not decrease proportionally to body weight. A better approach is to consider the number of lymphoid follicles to which the antigen will be distributed. Recommended doses of antigen combined with FA are: 10–200 µg for mouse and 250–5000 µg for goat and sheep. Very low doses of antigen (less than 1–5 µg) are used to induce hypersensitivity (allergy); (Poulsen, Hau, and Kollerup 1987; Nielsen, Poulsen, and Hau 1989; Poulsen et al. 1990) and should be avoided in immunisation schedules, particularly because booster injections may result in an anaphylactic shock in the animals.

11.3.4.4 Booster Immunisation

Generally, titres are not satisfactory if animals are immunised only once, even when an adjuvant is used. The time interval after which booster immunisation is given has great consequence, but it is difficult to give fixed recommendations. As a rule of the thumb, a booster can be considered after the antibody titre has reached a plateau or begins to decline. Usually, this is about 3 weeks after primary immunisation using an aqueous antigen, and at least 4 weeks when a depot-forming adjuvant is used; small blood samples are often taken to determine the titre and so assist in establishing the most appropriate time to harvest the antibody. The number of booster immunisations should be limited to a maximum of three. If antibody responses are still insufficient, the experiment should normally be terminated. However, antigens of low molecular weight often require more boosters. Also, frequent booster immunisations might be necessary when animals are used as antibody donors

Basic Procedures: Dosing, Sampling and Immunisation

over a long period of time. If a water-in-oil emulsion had been administered in the initial immunisation, a depot of antigen will have been established in the animal and booster immunisations may be carried out without the use of an adjuvant. Because high doses of antigen activate a large number of low-affinity B-cell clones, whereas low doses activate only high-affinity B-cell clones, it is preferable to use low doses at booster immunisations. However, too frequent immunisations with very low amounts of antigen may lead to tolerance. Intermittent bleeding of a hyper-immunised animal or plasmaphoresis appears to help maintain a high serum antibody titre, but this needs to be balanced against the impact on the animal's welfare. Often exsanguination of the animal under anaesthesia will provide the equivalent of three or four "harvesting" blood withdrawals and this may be sufficient for the required scientific purpose.

Booster injections should never be given at the same site as previous injections, and certainly not into granulomas or swellings. Furthermore, booster immunisations might advantageously be given by a different route than the primary immunisation in order to ensure that as many memory cells as possible are stimulated.

11.3.5 Existing Guidelines on the Production of Polyclonal Antibodies

Guidelines addressing the production of Pabs have been published by several national control authorities, organisations and institutions such as the Canadian Council on Animal Care (CCAC 1991), the NIH (Grumstrupp-Scott and Greenhouse 1988), and the Danish (Hau, 1990) and Dutch (VPI, 1993) Inspectorates. An overview of these has been prepared by Leenaars et al. (1999). Others can be found on the Internet, including a video showing how to conduct terminal bleeding in a rabbit (Smith 2008).

There are differences between the guidelines, but most agree about key aspects. For example, there is general consensus that footpad injection is not necessary; the use of FCA is discouraged and use of alternative adjuvants is recommended. Furthermore, most guidelines set limits to the volumes of antigen/adjuvant that can be injected and the volume of blood samples withdrawn.

11.4 CONCLUSIONS

Most methods for administering materials or withdrawing samples from rodents are amenable to refinement. Also, they are often associated with subtle changes in physiology, endocrinology or metabolism that often go unnoticed, but which can have a profound effect on the animal and the scientific issue being addressed. The skill and competence of staff who carry out these techniques must be assured by training and supervision.

In general, the least invasive methods are simpler to carry out consistently and have the least impact on animal well-being. For example, the administration of substances orally by incorporation into food or drinking water (where this is possible because they have no taste and accurate metering is not necessary) is not at all stressful, whereas administration by gavage can have effects on intestinal permeability, can lead to gastric overload with premature pyloric opening, and is stressful. Routes that are painful, such as intramuscular injection in mice, intradermal injections and injection into the footpad are rarely necessary and alternative routes should be sought. Similarly, formulation of the compounds being administered should ensure that they possess the minimum cytotoxic or irritant properties compatible with the purpose of the study; for example, irritant adjuvants should be replaced with less aggressive ones wherever possible and the injection site and volume selected to cause least damage to tissues.

Similarly, the withdrawal of blood or collection of other samples is complicated by biological as well as scientific considerations. Collection of samples from animals that are stressed requires considerably more technical skill on the part of the technician, and the samples are unrepresentative and likely to be much more variable than those taken from unstressed animals. Induction of anaesthesia may minimise the disturbance to the animal and facilitate the collection of samples, but is itself stressful and may introduce an additional interfering factor affecting the quality of the sample and the interpretation of scientific findings.

11.5 QUESTIONS UNRESOLVED

- There is no consensus regarding the optimal method of administering an IP injection, including best method of restraint, position of the head or injection site; evidence that in some rats stocks the large intestine lies quite often on the right side of the abdomen complicates the establishment of a side-specific recommendation.
- The welfare and scientific consequences of warming rodents before withdrawing blood samples need to be determined.
- There has been little systematic study of the impact on animals of selection of blood withdrawal sites, injection routes, handling and other relatively simple procedures that nevertheless are carried out extremely frequently. This whole topic deserves much more attention with a view to minimising the impact on animals and assuring the validity of the scientific procedures themselves.

REFERENCES

Abatan, O. I., K. B. Welch, and J. A. Nemzek. 2008. Evaluation of saphenous venipuncture and modified tail-clip blood collection in mice. *Journal of the American Association for Laboratory Animal Science* 47:8–15.

Abelson, K. S., B. Adem, F. Royo, H. E. Carlsson, and J. Hau. 2005. High plasma corticosterone levels persist during frequent automatic blood sampling in rats. *In Vivo* 19:815–19.

Abraham, L., D. O'Brien, O. M. Poulsen, and J. Hau. 1994. The effect of social environment on the production of specific immunoglobulins against an immunogen (human IgG) in mice. In *Welfare and science*, ed. J. Bunyan, 165–70. London: Royal Society of Medicine Press.

Alban, L., P. J. Dahl, A. K. Hansen, K. C. Hejgaard, A. L. Jensen, M. Kragh, P. Thomsen, and P. Steensgaard. 2001. The welfare impact of increased gavaging doses in rats. *Animal Welfare* 10:303–14.

Altholtz, L. Y., K. A. Fowler, L. L. Badura, and M. S. Kovacs. 2006. Comparison of the stress response in rats to repeated isoflurane or CO2:O2 anesthesia used for restraint during serial blood collection via the jugular vein. *Journal of the American Association for Laboratory Animal Science* 45:17–22.

Angelov, O., R. A. Schroer, S. Heft, V. C. James, and J. Noble. 1984. A comparison of two methods of bleeding rats: The venous plexus of the eye versus the vena sublingualis. *Journal of Applied Toxicology* 4:258–60.

Bollen, L. S., and J. Hau. 1997. Immunoglobulin G in the developing oocytes of the domestic hen and immunospecific antibody response in serum and corresponding egg yolk. *In Vivo* 11:395–98.

Bollen, L. S., and J. Hau. 1999a. Comparison of immunospecific antibody response in young and old chickens immunized with human IgG. *Laboratory Animals* 33:71–76.

Bollen, L. S., and J. Hau. 1999b. Freund's complete adjuvant has a negative impact on egg laying frequency in immunised chickens. *In Vivo* 13:107–8.

Bonnichsen, M., N. Dragsted, and A. K. Hansen. 2005. The welfare impact of gavaging laboratory rats. *Animal Welfare* 14:223–27.

Brown, A. P., N. Dinger, and B. S. Levine. 2000. Stress produced by gavage administration in the rat. *Contemporary Topics in Laboratory Animal Science* 39:17–21.

BVA/FRAME/RSPCA/UFAW. 1993. Removal of blood from laboratory mammals and birds. First report of the BVA/FRAME/RSPCA/UFAW joint working group on refinement. *Laboratory Animals* 27:1–22.

CCAC. 1991. *Guidelines on acceptable immunological procedures.* Ottawa, ON, Canada: CCAC.

Chapel, H. M., and P. J. August. 1976. Report of nine cases of accidental injury to man with Freund's complete adjuvant. *Clinical and Experimental Immunology* 24:538–41.

Chester, J. F., S. A. Weitzman, and R. A. Malt. 1985. Implantable device for drug delivery and blood sampling in the rat. *Journal of Applied Physiology* (Bethesda, MD: 1985) 59:1665–66.

Christensen, S. D., L. F. Mikkelsen, J. J. Fels, T. B. Bodvarsdottir, and A. K. Hansen. 2008. Quality of plasma sampled by different methods for multiple blood sampling in mice. *Laboratory Animals* 43:65–71.

Claassen, V. 1994. Neglected factors in pharmacology and neuroscience research: Biopharmaceutics, animal characteristics, maintenance, testing conditions. *Techniques in the behavioral and neural sciences*, Vol. 12. Amsterdam, New York: Elsevier.

Clarke, D. O. 1993. Pharmacokinetic studies in developmental toxicology—Practical considerations and approaches. *Toxicology Methods* 3:223–51.

Colpaert, F. C. 1987. Evidence that adjuvant arthritis in the rat is associated with chronic pain. *Pain* 28:201–22.

Coria-Avila, G. A., A. M. Gavrila, S. Menard, N. Ismail, and J. G. Pfaus. 2007. Cecum location in rats and the implications for intraperitoneal injections. *Lab Animal* 36:25–30.

Council of Europe. 2006. *Appendix of the European convention for the protection of vertebrate animals used for experimental and other scientific purposes (ETS123). Guidelines for accommodation and care of animals (article 5 of the convention) approved by the multilateral consultation*. Strasbourg: Council of Europe.

Dameron, G. W., K. W. Weingand, J. M. Duderstadt, L. W. Odioso, T. A. Dierckman, W. Schwecke, and K. Baran. 1992. Effect of bleeding site on clinical laboratory testing of rats: Orbital venous plexus versus posterior vena cava. *Laboratory Animal Science* 42:299–301.

de Wit, M., A. Raabe, G. Tuinmann, and D. K. Hossfeld. 2001. Implantable device for intravenous drug delivery in the rat. *Laboratory Animals* 35:321–24.

Diehl, K. H., R. Hull, D. Morton, R. Pfister, Y. Rabemampianina, D. Smith, J. M. Vidal, C. van de Vorstenbosch, and European Federation of Pharmaceutical Industries Association and European Centre for the Validation of Alternative Methods. 2001. A good practice guide to the administration of substances and removal of blood, including routes and volumes. *Journal of Applied Toxicology* 21:15–23.

Doeing, D. C., J. L. Borowicz, and E. T. Crockett. 2003. Gender dimorphism in differential peripheral blood leukocyte counts in mice using cardiac, tail, foot, and saphenous vein puncture methods. *BMC Clinical Pathology* 3:3.

Dürschlag, M., H. Wurbel, M. Stauffacher, and D. Von Holst. 1996. Repeated blood collection in the laboratory mouse by tail incision—Modification of an old technique. *Physiology & Behavior* 60:1565–68.

Erb, K., and J. Hau. 1994. Monoclonal and polyclonal antibodies. *Handbook of Laboratory Animal Science* 1:293–309.

European Commission. 2007a. *Commission recommendation of 18 June 2007 on guidelines for the accommodation and care of animals used for experimental and other scientific purposes (2007/526/EC)*. Brussels: European Commission.

European Commission. 2007b. *Report from the commission to the council and the European parliament; fifth report on the statistics on the number of animals used for experimental and other scientific purposes in the member states of the European union*. Brussels: European Council, COM(2007)675 final.

Fitzner Toft, M., M. H. Petersen, N. Dragsted, and A. K. Hansen. 2006. The impact of different blood sampling methods on laboratory rats under different types of anaesthesia. *Laboratory Animals* 40:261–74.

Fluttert, M., S. Dalm, and M. S. Oitzl. 2000. A refined method for sequential blood sampling by tail incision in rats. *Laboratory Animals* 34:372–78.

Fomby, L. M., T. M. Wheat, D. E. Hatter, R. L. Tuttle, and C. A. Black. 2004. Use of CO2/O2 anesthesia in the collection of samples for serum corticosterone analysis from Fischer 344 rats. *Contemporary Topics in Laboratory Animal Science* 43:8–12.

Freund, J. 1947. Some aspects of active immunisation. *Annual Review of Microbiology* 1:291.

Freund, J., J. Casals, and E. P. Hosmer. 1937. Sensitization and antibody formation after injection of tubercle bacilli and parafin oil. *Proceedings of the Society for Experimental Biology and Medicine. Society for Experimental Biology and Medicine* (New York, NY) 37:509.

Friedel, R., I. Trautschold, K. Gartner, M. Helle-Feldmann, and D. Gaudssuhn. 1975. Effects of blood sampling on enzyme activities in the serum of small laboratory animals (author's transl.). *Zeitschrift Fur Klinische Chemie Und Klinische Biochemie* 13:499–505.

Fukuta, K. 2004/2010. Collection of body fluids. In *The laboratory mouse, Handbook of experimental animals*, 543–54. New York: Academic Press.

Gad, S. C., C. D. Cassidy, N. Aubert, B. Spainhour, and H. Robbe. 2006. Nonclinical vehicle use in studies by multiple routes in multiple species. *International Journal of Toxicology* 25:499–521.

Golde, W. T., P. Gollobin, and L. L. Rodriguez. 2005. A rapid, simple, and humane method for submandibular bleeding of mice using a lancet. *Lab Animal* 34:39–43.

Grouzmann, E., C. Cavadas, D. Grand, M. Moratel, J. F. Aubert, H. R. Brunner, and L. Mazzolai. 2003. Blood sampling methodology is crucial for precise measurement of plasma catecholamines concentrations in mice. *Pflugers Archiv: European Journal of Physiology* 447:254–58.

Grumstrupp-Scott, J., and D. D. Greenhouse. 1988. NIH intramural recommendations for the research use of complete Freund's adjuvant. *ILAR News* 30:9.

Guhad, F. 2005. Introduction to the 3Rs (refinement, reduction and replacement). *Contemporary Topics in Laboratory Animal Science* 44:58–59.

Haemisch, A., G. Guerra, and J. Furkert. 1999. Adaptation of corticosterone—But not beta-endorphin-secretion to repeated blood sampling in rats. *Laboratory Animals* 33:185–91.

Hanly, W. C., J. E. Artwohl, and B. T. Bennett. 1995. Review of polyclonal antibody production procedures in mammals and poultry. *ILAR Journal* 37:93–118.

Harboe, N., and A. Ingild. 1973. Immunization, isolation of immunoglobulins, estimation of antibody titre. *Scandinavian Journal of Immunology. Supplement* 1:161–64.

Hau, J. 1988. The rabbit as antibody producer—Advantages and disadvantage. Paper presented at Symposium über Zucht, Haltung und Ernährung des Kaninches. Das Kanin als Modell in der Biomed. Forshung.

Hau, J. 1989. The Danish guidelines. In *Danish State Inspectorate Annual Report 1989*, 66–70. Copenhagen, Denmark: Animal Experiments Inspectorate.

Hau, J., and C. F. Hendriksen. 2005. Refinement of polyclonal antibody production by combining oral immunization of chickens with harvest of antibodies from the egg yolk. *ILAR Journal* 46:294–99.

Hau, J., J. G. Westergaard, P. Svendsen, A. Bach, and B. Teisner. 1980. Comparison of the pregnancy-associated murine protein-2 and human pregnancy-specific beta 1-glycoprotein. *Journal of Reproduction and Fertility* 60:115–19.

Hau, J., J. G. Westergaard, P. Svendsen, A. Bach, and B. Teisner. 1981. Comparison of pregnancy-associated murine protein-1 and human pregnancy zone protein. *Journal of Reproductive Immunology* 3:341–49.

Herbert, W. J. 1967. Methods for preparation of water-in-oil, and multiple, emulsions for use as antigen adjuvants; and notes on their use in immunisation procedures. In *Handbook of experimental immunology*, ed. D. M. Weir. Oxford and Edinburg (UK): Blackwell Scientific Publications.

Hohmann, A. W. 1998. Freund's adjuvant and immune responses. In *The use of immuno-adjuvants in animals in Australia and New Zealand*, ed. G. Osmond. Australia: ANZCCART.

Home Office. 2009. *Statistics of scientific procedures on living animals. Great Britain 2008*. London: The Stationery Office, HC 800.

Horton, M. L., C. T. Olson, and D. W. Hobson. 1986. Femoral venipuncture for collection of multiple blood samples in the nonanesthetized rat. *American Journal of Veterinary Research* 47:1781–82.

Iwarsson, K., L. Lindberg, and T. Waller. 1994. Common non-surgical techniques and procedures. In *Handbook of laboratory animal science*, eds. P. Svendsen and J. Hau, 229–72, Volume I. Boca Raton, FL: CRC Press.

Jekl, V., K. Hauptman, E. Jeklova, and Z. Knotek. 2005. Blood sampling from the cranial vena cava in the Norway rat (*Rattus norvegicus*). *Laboratory Animals* 39:236–39.

Jennings, V. M. 1995. Review of selected adjuvants used in antibody production. *ILAR Journal* 37:119–25.

Jensenius, J. C., I. Andersen, J. Hau, M. Crone, and C. Koch. 1981. Eggs: Conveniently packaged antibodies. Methods for purification of yolk IgG. *Journal of Immunological Methods* 46:63–68.

Joslin, J. O. 2009. Blood collection techniques in exotic small mammals. *Journal of Exotic Pet Medicine* 18:117–39.

Kemppinen, N. M., A. S. Meller, K. O. Mauranen, T. T. Kohila, and T. O. Nevalainen. 2009. The effect of dividing walls, a tunnel, and restricted feeding on cardiovascular responses to cage change and gavage in rats (rattus norvegicus). *Journal of the American Association for Laboratory Animal Science* 48:157–65.

Koch, M. A. 2006. Experimental modelling and research methodology. In *The laboratory rat*, eds. M. A. Suckow, S. H. Weisbroth, and G. L. Franklin, 587–625. Amsterdam: Elsevier.

Koeslag, D., A. S. Humphreys, and J. C. Russell. 1984. A technique for long-term venous cannulation in rats. *Journal of Applied Physiology: Respiratory, Environmental and Exercise Physiology* 57:1594–96.

Kugler, J., K. W. Lange, and K. T. Kalveram. 1988. Influence of bleeding order on plasma corticosterone concentration in the mouse. *Experimental and Clinical Endocrinology* 91:241–43.

Kurien, B. T., N. E. Everds, and R. H. Scofield. 2004. Experimental animal urine collection: A review. *Laboratory Animals* 38:333–61.

Kurien, B. T., and R. H. Scofield. 1999. Mouse urine collection using clear plastic wrap. *Laboratory Animals* 33:83–86.

LABORAS. 2009. Animal behavior observation registration and analysis system. [Cited December 31, 2009]. Available from http://www.metris.nl/laboras/laboras.htm

Leenaars, P. P. A. M. 1997. *Adjuvants in laboratory animals: Evaluation of immunostimulating properties and side effects of Freund's complete adjuvant and alternative adjuvants in immunization procedures*. Doctoral Thesis. Rotterdam: Erasmus University.

Leenaars, P. P. A. M., C. F. M. Hendriksen, W. A. de Leeuw, F. Carat, P. Delahaut, R. Fischer, M. Halder, et al. 1999. The production of polyclonal antibodies in laboratory animals—The report and recommendations of ECVAM workshop 35. *Alternatives to Laboratory Animals* 27:79–102.

Leenaars, P. P., M. A. Koedam, P. W. Wester, V. Baumans, E. Claassen, and C. F. Hendriksen. 1998. Assessment of side effects induced by injection of different adjuvant/antigen combinations in rabbits and mice. *Laboratory Animals* 32:387–406.

Lifshitz, R., M. Schwartz, and E. Mozes. 1980. Specificity of genes controlling immune responsiveness to (T,G)-A--L and (phe,G)-A--L. *Immunology* 41:339–46.

Lindblad, E. B., and J. Hau. 2000. Escaping from the use of Freund's complete adjuvant. Paper presented at Progress in the reduction, refinement and replacement of animal experimentation. Proceedings from the 3rd world congress on alternatives and animal use in the life sciences 1999. Bologna, Italy, August 31–September 2, 1999.

Lindblad, E. B., and J. V. Sparck. 1987. Basic concepts in the application of immunological adjuvants. *Scandinavian Journal of Laboratory Animal Science* 14:1–13.

Long, D. A. 1957. Influence of the thyroid gland upon immune responses of different species to bacterial infection. *Endocrinology* 10:287.

Mahl, A., P. Heining, P. Ulrich, J. Jakubowski, M. Bobadilla, W. Zeller, R. Bergmann, T. Singer, and L. Meister. 2000. Comparison of clinical pathology parameters with two different blood sampling techniques in rats: Retrobulbar plexus versus sublingual vein. *Laboratory Animals* 34:351–61.

Mayo, S. L., M. Tufvesson, H. E. Carlsson, F. Royo, S. Gizurarson, and J. Hau. 2005. RhinoVax is an efficient adjuvant in oral immunisation of young chickens and cholera toxin B is an effective oral primer in subcutaneous immunisation with Freund's incomplete adjuvant. *In Vivo* 19:375–82.

Meeks, R. G. 1989. The rat. In *The clinical chemistry of laboratory animals*, eds. W. F. Loeb and F. W. Quimby, 19–25. New York: Pergamon Press.

Meijer, M. K., B. M. Spruijt, L. F. van Zutphen, and V. Baumans. 2006. Effect of restraint and injection methods on heart rate and body temperature in mice. *Laboratory Animals* 40:382–91.

Morton, D. B., M. Jennings, A. Buckwell, R. Ewbank, C. Godfrey, B. Holgate, I. Inglis, et al. 2001. Refining procedures for the administration of substances. Report of the BVAAWF/FRAME/RSPCA/UFAW joint working group on refinement. British Veterinary Association Animal Welfare Foundation/Fund for the Replacement of Animals in Medical Experiments/Royal Society for the Prevention of Cruelty to Animals/Universities Federation for Animal Welfare. *Laboratory Animals* 35:1–41.

Murphy, S. J., P. Smith, A. B. Shaivitz, M. I. Rossberg, and P. D. Hurn. 2001. The effect of brief halothane anesthesia during daily gavage on complications and body weight in rats. *Contemporary Topics in Laboratory Animal Science* 40:9–12.

NC3Rs. 2009. Blood sampling microsite. In NC3Rs [database online]. [Cited September 15, 2009]. Available from http://www.nc3rs.org.uk/bloodsamplingmicrosite/page.asp?id=313

Nemzek, J. A., G. L. Bolgos, B. A. Williams, and D. G. Remick. 2001. Differences in normal values for murine white blood cell counts and other hematological parameters based on sampling site. *Inflammation Research*. 50:523–27.

Neptun, D. A., C. N. Smith, and R. D. Irons. 1985. Effect of sampling site and collection method on variations in baseline clinical pathology parameters in Fischer-344 rats. 1. Clinical chemistry. *Fundamental and Applied Toxicology* 5:1180–85.

Nielsen, B. R., O. M. Poulsen, and J. Hau. 1989. Reagin production in mice: Effect of subcutaneous and oral sensitization with untreated bovine milk and homogenized bovine milk. *In Vivo* 3:271–74.

O'Brien, D., P. Tibi Opi, G. Stodulski, P. Saibaba, and J. Hau. 1993. Comparison of stress induced in rats by four different anaesthetic regimens as recorded by urinary concentrations of corticosterone and testosterone. *Scandinavian Journal of Laboratory Animal Science* 20:113–16.

Okva, K., E. Tamoseviciute, A. Ciziute, P. Pokk, O. Ruksenas, and T. Nevalainen. 2006. Refinements for intragastric gavage in rats. *Scandinavian Journal of Laboratory Animal Science* 33:243–52.

Olin, M. R., R. W. Diters, and G. O. Freden. 1997. The effect of collection site on the cytogram of the BASO/lobularity channel of a hematology system in the mouse. *Veterinary Clinical Pathology* 26:63–65.

Petty, C., ed. 1982. *Research techniques in the rat*, 66–70. Springfield, IL: Thomas.

Poulsen, O. M., B. R. Nielsen, A. Basse, and J. Hau. 1990. Comparison of intestinal anaphylactic reactions in sensitized mice challenged with untreated bovine milk and homogenized bovine milk. *Allergy* 45:321–26.

Poulsen, O. M., and J. Hau. 1990. Suppression of humoral antibody response during pregnancy in mice. *In Vivo* 4:381–84.

Poulsen, O. M., J. Hau, and J. Kollerup. 1987. Effect of homogenization and pasteurization on the allergenicity of bovine milk analysed by a murine anaphylactic shock model. *Clinical Allergy* 17:449–58.

Quimby, F. W. 1989. The mouse. In *The clinical chemistry of laboratory animals*, 3–18. New York: Pergamon Press.

Rao, G. N., T. A. Peace, and D. E. Hoskins. 2001. Training could prevent deaths due to rodent gavage procedure. *Contemporary Topics in Laboratory Animal Science* 40:7–8.

Rasmussen, C., and M. Ritskes-Hoitinga. 1999. An alternative method for rat fixation when giving subcutaneous, intramuscular and intraperitoneal injections ("Camilla's method"). *Scandinavian Journal of Laboratory Animal Science* 26:156–59.

Rogers, I. T., D. J. Holder, H. E. McPherson, W. R. Acker, E. G. Brown, M. V. Washington, S. L. Motzel, and H. J. Klein. 1999. Influence of blood collection sites on plasma glucose and insulin concentration in conscious C57BL/6 mice. *Contemporary Topics in Laboratory Animal Science* 38:25–28.

Santos, J., M. Benjamin, P. C. Yang, T. Prior, and M. H. Perdue. 2000. Chronic stress impairs rat growth and jejunal epithelial barrier function: Role of mast cells. *American Journal of Physiology. Gastrointestinal and Liver Physiology* 278:G847–G854.

Schaarschmidt, K., G. Muller, A. Heinze, L. Ruprecht, J. Stormann, U. Stratmann, G. H. Willital, and E. Unsold. 1991. Improved model of a fecal collection device for the prevention of coprophagia in the rat. *Journal of Experimental Animal Science* 34:67–71.

Schade, R., C. Staak, C. Hendriksen, M. Erhard, H. Hugl, G. Koch, A. Larsson, et al. 1996. The production of avian (egg yolk) antibodies: IgY—The report and recommendations of ECVAM workshop 21. *Alternatives to Laboratory Animals* 24:925–34.

Schnell, M. A., C. Hardy, M. Hawley, K. J. Propert, and J. M. Wilson. 2002. Effect of blood collection technique in mice on clinical pathology parameters. *Human Gene Therapy* 13:155–61.

Semler, D. E., S. C. Gad, and C. P. Chengelis. 1992. The rat. In *Animal models in toxicology*, 21–164. New York: Marcel Dekker Inc.

Shipp, K., and B. D. Woodward. 1998. A simple exsanguination method that minimizes acute pre-anesthesia stress in the mouse: Evidence based on serum corticosterone concentrations. *Contemporary Topics in Laboratory Animal Science* 37:73–77.

Smith, A. B. 2008. Films and slide shows. In The Norwegian Reference Center for Laboratory Animal Science & Alternatives [database online]. [Cited September 20, 2009]. Available from http://film.oslovet.veths.no/.

Smith, C. N., D. A. Neptun, and R. D. Irons. 1986. Effect of sampling site and collection method on variations in baseline clinical pathology parameters in Fischer-344 rats. II. Clinical hematology. *Fundamental and Applied Toxicology* 7:658–63.

Sörensen, D. B., K. Mortensen, T. Bertelsen, and K. Vognberg. 2008. Enriching the metabolic cage: Effects on rat physiology and behavior. *Animal Welfare* 17:395–403.

Stark, D. M., and M. E. Ostrow. 1990. Injection techniques. In *Laboratory animal technician*, Vol. II. Cordova: AALAS.

Suber, R. L., and R. L. Kodell. 1985. The effect of three phlebotomy techniques on hematological and clinical chemical evaluation in Sprague-Dawley rats. *Veterinary Clinical Pathology* 14:23–30.

Sun, D., A. Krishnan, K. Zaman, R. Lawrence, A. Bhattacharya, and G. Fernandes. 2003. Dietary n-3 fatty acids decrease osteoclastogenesis and loss of bone mass in ovariectomized mice. *Journal of Bone and Mineral Research* 18:1206–16.

Svendsen Bollen, L., A. Crowley, G. Stodulski, and J. Hau. 1996. Antibody production in rabbits and chickens immunized with human IgG. A comparison of titre and avidity development in rabbit serum, chicken serum and egg yolk using three different adjuvants. *Journal of Immunological Methods* 191:113–20.

Svendsen, L., A. Crowley, L. H. Ostergaard, G. Stodulski, and J. Hau. 1995. Development and comparison of purification strategies for chicken antibodies from egg yolk. *Laboratory Animal Science* 45:89–93.

Tabata, H., T. Kitamura, and N. Nagamatsu. 1998. Comparison of effects of restraint, cage transportation, anaesthesia and repeated bleeding on plasma glucose levels between mice and rats. *Laboratory Animals* 32:143–48.

Talmage, D. W., and F. J. Dixon. 1953. The influence of adjuvants on the elimination of soluble protein antigens and the associated antibody responses. *The Journal of Infectious Diseases* 93:176–80.

Tuli, J. S., J. A. Smith, and D. B. Morton. 1995. Corticosterone, adrenal and spleen weight in mice after tail bleeding, and its effect on nearby animals. *Laboratory Animals* 29:90–95.

Vachon, P., and J. P. Moreau. 2001. Serum corticosterone and blood glucose in rats after two jugular vein blood sampling methods: Comparison of the stress response. *Contemporary Topics in Laboratory Animal Science* 40:22–24.

Vahl, T. P., Y. M. Ulrich-Lai, M. M. Ostrander, C. M. Dolgas, E. E. Elfers, R. J. Seeley, D. A. D'Alessio, and J. P. Herman. 2005. Comparative analysis of ACTH and corticosterone sampling methods in rats. *American Journal of Physiology. Endocrinology and Metabolism* 289:E823–E828.

van Herck, H., V. Baumans, C. J. Brandt, A. P. Hesp, J. H. Sturkenboom, H. A. van Lith, G. van Tintelen, and A. C. Beynen. 1998. Orbital sinus blood sampling in rats as performed by different animal technicians: The influence of technique and expertise. *Laboratory Animals* 32:377–86.

Van Herck, H., V. Baumans, C. J. Brandt, H. A. Boere, A. P. Hesp, H. A. van Lith, M. Schurink, and A. C. Beynen. 2001. Blood sampling from the retro-orbital plexus, the saphenous vein and the tail vein in rats: Comparative effects on selected behavioural and blood variables. *Laboratory Animals* 35:131–39.

Vermeulen, J. K., A. De Vries, F. Schlingmann, and R. Remie. 1997. Food deprivation: Common sense or nonsense? *Animal Technology* 48:45–54.

VPHI. 1993. Code of Practice for the Immunisation of Laboratory Animals, 5. Rijswijk, The Netherlands: Veterinary Public Health Inspectorate.

Waynforth, H. B., and P. A. Flecknell. 1992. *Experimental and surgical technique in the rat,* 2nd ed. New York: Academic Press.

Weiss, J., G. R. Taylor, F. Zimmermann, and K. Nebendahl. 2000. Collection of body fluids. In *The laboratory rat*, ed. G. J. Krinke, 485–510. New York: Academic Press.

Wheatley, J. L. 2002. A gavage dosing apparatus with flexible catheter provides a less stressful gavage technique in the rat. *Lab Animal* 31:53–56.

Winsett, O. E., C. M. Townsend, Jr., and J. C. Thompson. 1985. Rapid and repeated blood sampling in the conscious laboratory rat: A new technique. *The American Journal of Physiology* 249:G145–G146.

Wolfensohn, S., and M. Lloyd. 1994. *Handbook of laboratory animal management and welfare.* Oxford: Oxford University Press.

Wolfensohn, S., and M. Lloyd. 2003. *Handbook of laboratory animal management and welfare*, 3rd ed. Oxford, UK: Blackwell Publishing.

Yang, P. C., J. Jury, J. D. Soderholm, P. M. Sherman, D. M. McKay, and M. H. Perdue. 2006. Chronic psychological stress in rats induces intestinal sensitization to luminal antigens. *The American Journal of Pathology* 168:104–14; quiz 363.

Young, C. R., and M. Z. Atassi. 1982. Genetic control of the immune response to myoglobin. IX. Overcoming genetic control of antibody response to antigenic sites by increasing the dose of antigen used in immunization. *Journal of Immunogenetics* 9:343–51.

Zeller, W., H. Weber, B. Panoussis, T. Burge, and R. Bergmann. 1998. Refinement of blood sampling from the sublingual vein of rats. *Laboratory Animals* 32:369–76.

Zutphen van, L. F. M., V. Baumans, and A. C. Beynen, eds. 2001. *Principles of laboratory animal science.* Amsterdam: Elsevier.

12 Imaging Techniques

Aurora Brønstad, Norway and Ismene Dontas, Greece

CONTENTS

Objectives..288
Key Factors 288
12.1 Animal Imaging Technology..288
    12.1.1 Imaging Techniques Based on Nuclear Magnetic Resonance 288
        12.1.1.1 Magnetic Resonance Imaging (MRI)......................................288
        12.1.1.2 Functional MRI (fMRI)...292
        12.1.1.3 Magnetic Resonance Spectroscopy (MRS) 292
    12.1.2 Imaging Techniques Using Ionising Radiation....................................292
        12.1.2.1 X-Radiography...293
        12.1.2.2 Computed Tomography (CT)...293
        12.1.2.3 Dual-Energy X-Ray Absorptiometry (DEXA).......................293
        12.1.2.4 Peripheral Quantitative Computed Tomography (pQCT) ...295
        12.1.2.5 Micro-Computed Tomography 297
        12.1.2.6 Radionuclear Imaging: PET and SPECT...............................297
    12.1.3 Imaging Techniques with High-Frequency Ultrasound Waves 299
        12.1.3.1 Ultrasonography...299
        12.1.3.2 Doppler Ultrasonography..300
    12.1.4 Imaging Techniques Detecting Light..300
        12.1.4.1 Fluorescence Optical Imaging and Bioluminescence Imaging 300
    12.1.5 Imaging Techniques Based on Infrared or Near Infrared Radiation (NIR) 301
        12.1.5.1 Near Infra-Red Spectroscopy (NIRS) 301
        12.1.5.2 Remote Thermography and Near Infrared Imaging.............302
12.2 Reduction and Refinement..302
    12.2.1 Severity Classifications 303
        12.2.1.1 Healthy Animal, No Invasive Procedure 304
        12.2.1.2 Healthy Animal, Invasive Procedure......................................304
        12.2.1.3 Non-Healthy Animals..304
        12.2.1.4 Non-Healthy Animals, Invasive Procedure 304
    12.2.2 Procedures that Impair Animal Welfare and Scientific Quality–Setting Targets for Refinement 305
        12.2.2.1 Anaesthesia...305
        12.2.2.2 Invasive Procedures 306
        12.2.2.3 Fluid Balance 306
    12.2.3 Other Issues, Quality Testing and Planning of the Imaging Facility....307
        12.2.3.1 Use of Genetically Modified Animals for *In Vivo* Studies of Functional Genomics...307
        12.2.3.2 Reduction, Refinement and Legitimising Animal Experiments..........308
12.3 Conclusions..308
12.4 Questions Unresolved 308
Appendix–List of Abbreviations..309
References...309

OBJECTIVES

The development and use of imaging technology in clinical practise has contributed to earlier diagnosis and the likelihood of a better outcome for several diseases, resulting in better evaluation of treatment effects and thereby contributing to improved patient survival for diseases such as cancer, brain disorders and cardiovascular disease. Morphological and functional information can be collected with both high spatial and temporal resolution and three-dimensional reconstructions of internal organs and structures can be made. The technology can also be used to study gene expression and protein interactions. Use of contrast agents and specific test substances further enhance the value of the techniques.

The advantages of earlier and more precise diagnosis apply also to the use of animals in experimental studies, because much of the progress in developing diagnostic tools for humans has depended on preliminary studies using animals. Technology, contrast agents and tracers are continuously being developed and improved so there is an ongoing need to refine preclinical studies in animals, where clinical monitoring has been based traditionally on clinical signs combined with collection of physiological data, biopsies, post-mortem anatomic examination and laboratory tests. Imaging technology allows real-time evaluation in the intact animal and is a powerful way of refining animal experiments by defining earlier and more precise humane endpoints (HEP). Thus, refinement of animal studies is the first benefit of current imaging technology. The second is reduction of the number of animals used. The ability to carry out repeated observations of the same animal makes it possible for each animal to serve as its own control. It is therefore possible to plan earlier interventions, establish earlier endpoints and gather quantitative data, or monitor tumour size or vascularisation at shorter intervals over time, all in the same animal.

This chapter is divided into two main parts. The first part presents an overview of different imaging technologies. Part two deals with challenges arising from the use of imaging technology insofar as these relate to animal welfare and reduction and refinement in studies. Anaesthesia is necessary when implementing most imaging techniques, but can modify physiological characteristics. An overview of the known effects of anaesthesia is presented in Table 12.1. Table 12.2 summarises the requirements of various techniques when applied to animals. Finally there is a short discussion about reduction versus refinement in technology-driven research. A list of abbreviations is added as an appendix at the end of the text.

KEY FACTORS

Optimally, any animal imaging procedure should have the least possible effect on the animal and concurrently produce a high-quality image, from which it is possible to extract meaningful data and valid conclusions. The key considerations when selecting an imaging method are the effects on the animal (anaesthesia, radiation, duration, frequency) and the adequacy of the method to reveal the desired tissues or changes.

12.1 ANIMAL IMAGING TECHNOLOGY

All imaging techniques are based on detection of radiation. The signal can be either of electromagnetic or mechanical nature, giving rise to different contrast in tissues. This section will give an overview of commonly used imaging technologies and describe the basic principles and common areas of use, based on the physical nature of the different imaging techniques.

12.1.1 IMAGING TECHNIQUES BASED ON NUCLEAR MAGNETIC RESONANCE

12.1.1.1 Magnetic Resonance Imaging (MRI)

Magnetic resonance imaging (MRI) depends on variations in proton spin (Figure 12.1a). In a strong magnetic field, protons become aligned in parallel or anti-parallel directions relative to the

Imaging Techniques

TABLE 12.1
Cardiovascular Effects of Anaesthesia in Mice

	Blood Pressure	Heart Rate	Cardiac Output	Stroke Volume	Myocardial Blood Flow	Reference	Comment
Ketamine–xylazine	↓	↓	↓	—	↓	Kober et al. 2005	Compared to isoflurane
		↓	↓			Janssen et al. 2004	Compared to baseline
		↓	↓			Chaves et al. 2000	Compared to halothane
		↓↓				Roth et al. 2002	Timeline study
Ketamine–midazolam		↓(↑)	↓	↑		Kober et al. 2004	Timeline study
Ketamine–medetomidine–atropine	(↑)	↓				Roth et al. 2002	Strain differences
Pentobarbital	↓	↓	↓			Zuurbier et al. 2002	Baseline
Isoflurane 1%	↓	↓	↓			Janssen et al. 2004	Baseline
Isoflurane 1.25%		↑	↑			Janssen et al. 2004	Baseline
						Kober et al. 2004	Compared to ketamine/xylazine
		↓				Roth et al. 2002	Timeline study
Isoflurane 1.5–2%	(↓)	(↓)	↓	↓		Zuurbier et al. 2002	Strain differences
Isoflurane 2%	↓	↑	↑			Janssen et al. 2004	Baseline
		↑	↑		↑↑	Kober et al. 2004	Compared to ketamine/xylazine
						Kober et al. 2005	Compared to ketamine/xylazine
Fentanyl–fluanisone–midazolam	↓↓	↓				Zuurbier et al. 2002	Strain differences, not recommended
Urethane	(↓)		↓			Janssen et al. 2004	Baseline
Tribromoethanol		—				Roth et al. 2002	Toxicity mortality reported

↑ Increased
↓ Reduced
(↓) Slightly reduced
— No effect

TABLE 12.2
The Impact of Different Imaging Techniques on Animals

	Restraint Necessary	Radiation	Acquisition Time for Imaging	Injection Reporter Probe Mandatory	Injection of Reporter Probe Optional
MRI/MRS	Anaesthesia	No	Minutes–hours	No	Yes
PET	Anaesthesia	Yes	Minutes	Yes	—
SPECT		Yes		Yes	—
Ultrasound Doppler	Manual	No	Minutes	No	Yes
X-ray	Manual	Yes	Minutes	No	Yes
CT	Anaesthesia	Yes	Minutes	No	Yes
DEXA	Anaesthesia	Yes	Minutes	No	No
pQTC	Anaesthesia	Yes	Minutes	No	No
µCT	Anaesthesia	Yes	Hours	No	Yes
NIRS	Anaesthesia	No	Minutes	No	No
Remote thermography and NIR	No	No	Minutes	No	No
Fluorescent optical imaging	Anaesthesia	No	Minutes	No	Yes
Bioluminescence imaging	Anaesthesia	No	Minutes	Yes	—

field (Figure 12.1b). When a radio frequency (RF) pulse is applied, with the resonant frequency of the proton spin, the proton absorbs energy and become energised (Figure 12.1c). This is referred to as proton excitation and the protons are said to be "flipped" out of the organised (parallel/antiparallel) state in the magnetic field (Westbrook and Kaut 1993; Berry and Bulpitt 2009; Figure 12.1d). When the RF pulse ceases, the protons recover or "relax" to their original orientation in the magnetic field and the time this takes is measured. Recovery is detected as an electrical signal, using a receiver coil—the MRI signal. Recovery in the longitudinal plane (along the magnetic field or z-axis) is referred to as T_1 time, while decay in the transverse (x-y) plane is known as T_2 time (Pautler 2004; Westbrook and Kaut 1993). The values of T_1 and T_2 vary, depending on properties of the different tissues. For example, T_2 decay for fat is about 80 ms while T_2 for water is 200 ms (Westbrook and Kaut 1993). Imaging technology can also be used to generate new types of quantitative data, such as "T_2 time" in MRI, which reflects the water content of tissue.

By using a well-defined magnetic field gradient and a powerful computer to process the signals, it is possible to reconstruct two- and three-dimensional images of internal organs based on proton resonance signals received from the tissues. Differences in signal strength influence image contrast. MRI is excellent for differentiating soft tissues and pathological changes within them, because different tissues relax at different rates. Relaxation times are also influenced by pathological conditions. For example fat (rich in hydrogen and carbon) has low molecular mobility and therefore a rapid recovery and short T_1 time. Water (hydrogen and oxygen) has a long recovery and gives a weaker T_1 signal (Westbrook and Kaut 1993). Determination of soft tissue contrast by MRI is superior compared to other techniques (Berry and Bulpitt 2009; Westbrook and Kaut 1993).

Contrast agents, which alter the relaxation times and modify contrast between different tissues (Westbrook and Kaut 1993) may enhance the discrimination ability of MRI. Gadolinium (Gd^{3+}), manganese (Mn^{2+}) and iron (Fe^{3+}) are commonly used contrast agents for experimental purposes.

With MRI, it is the tissue itself that generates the signal and the volume of the tissue influences the strength of the signal. The animal is placed inside a cylindrical magnet and exposed to brief

Imaging Techniques

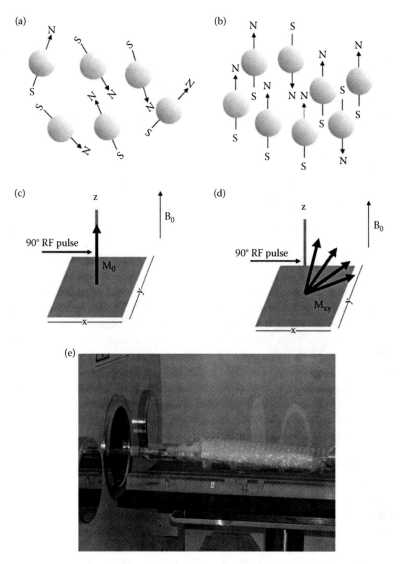

FIGURE 12.1 (a) Spinning protons behave like small magnets, with a north (N) and south (S) pole. Normally these are randomly oriented. (b) The protons organise in the magnetic field (B_0) in a parallel or anti-parallel order along the Z axis. (c) A radiofrequency (RF) pulse is used to flip protons out of the order in the magnetic field (Z axis). and (d) When the RF pulse is turned off, protons will reorientate in the magnetic field releasing a signal that can be detected by a coil close to the tissue. The recovery in Z directions is called T_1. The decay in the x-y plane is called T_2. (e) Anaesthetised rat prepared for MR imaging. The rat receives inhalation anaesthesia by a mask and is wrapped in isolation material to keep warm.

RF pulses (Figure 12.1e). Small masses, such as the body of a mouse, need stronger magnetic fields than a larger mass and magnets used in such cases usually have fields of approximately 4.7 or 7 Tesla (T), whereas the usual clinical scanner for human patients is 1.5 T. There is continuing progress in the development of magnets with stronger fields and of contrast agents optimised to different magnet field strengths. Better receiver coil systems also contribute significantly to improving image quality.

MRI is used for morphological studies of internal organs and tumours; for example, it allows differentiation between grey and white matter in the brain using T_2 weighted sequences. There is equipment suitable for use with mice and very suitable for rats (Pautler 2004). MRI is used to detect and follow pathological processes in the brain including metabolic disease, vascular disease, neurogenerative disease, demyelination and tumours. For example, the volume of tumours can be estimated or vascular perfusion can be evaluated from the uptake of contrast agents.

Magnetic resonance imaging is also suitable for evaluating cardiac function; for example, determining heart wall thickness and studying coronary arteries and heart valves. Dynamic characteristics such as the velocity, magnitude and direction of changes in myocardial ventricular volume can also be studied (Pautler 2004). By use of very fast sequences it is possible to make a movie (cine MRI) of the beating heart *in situ*, showing chamber contractions, heart filling and flow (Hiba et al. 2006; Hiller et al. 2008).

For renal studies, MRI has been used for morphological diagnosis of kidney conditions, such as polycystic kidney disease. As a supplement to renal angiography, MRI can be used also to examine functional defects, such as renal tube damage (Pautler 2004).

The size and growth of tumours can be determined—not only their volume but also properties related to malignancy such as angiogenesis (Pautler 2004) as well as dynamic interactions such as vascularisation and physiological processes in tumour masses and surrounding tissue (Gillies et al. 2000). These are parameters that may change as the tumour grows and interacts with host tissue. In fact, MRI is used for functional or dynamic studies in a wide range of circumstances and often involves the use of contrast agents. For example, wash out (clearance) or uptake of contrast labelled substances, like albumin, in organs and tissues can be estimated by taking sequences of MRI images over time (Gillies et al. 2000).

MRI has also been used for whole body screening to detect anatomical abnormalities when phenotypically characterising transgenic mice (Weissleder 2001).

Magnetic objects or implants must not be used in MRI studies because they disturb the magnetic field and give rise to artefacts. Also control may be lost of metal objects that can become magnetised in the strong magnetic field and have the potential to cause severe damage to animals, personnel or other material (Westbrook and Kaut 1993). Monitoring equipment, syringe pumps and other apparatus must be placed outside the magnetic field.

12.1.1.2 Functional MRI (fMRI)

Functional MRI (fMRI), is a type of specialised MRI scan, used to detect hemodynamic changes in the brain or spinal cord. It is assumed that these arise from neural activity in specific locations in the central nervous system that increases oxygen demand. Oxygenated and deoxygenated haemoglobin have different magnetic properties and give rise to different MR signals—the so-called blood-oxygen-level dependent (BOLD) contrast (Eichele et al. 2005). The BOLD signal is interpreted as an indirect measurement of neural activity (Logothetis et al. 2001). Functional MRI has been used to study brain function in pathological conditions (Hugdahl et al. 2004), during different activities and cognitive challenges (Rösler 2009).

12.1.1.3 Magnetic Resonance Spectroscopy (MRS)

Magnetic resonance spectroscopy (MRS) depends on the production of spectral signatures of different biochemical substances *in vivo* and provides a non-invasive way of identifying chemical substances (Dager et al. 2008). The spectra can be used to evaluate biochemical changes in pathological conditions (Herynek et al. 2004), in studies of metabolic disease processes such as non-invasive evaluation of metabolic activity for example in tumours (Thorsen et al. 2008), and studies of psychiatric disorders (Dager et al. 2008).

12.1.2 IMAGING TECHNIQUES USING IONISING RADIATION

This section describes imaging methods that use ionising radiation. These methods include (a) x-radiography, (b) computed tomography (CT), (c) dual energy x-ray absorptiometry (DEXA), (d) peripheral quantitative computed tomography (pQCT), and (e) micro-computed tomography (μCT).

12.1.2.1 X-Radiography

X-rays are produced by accelerating electrons using a high voltage 50–150 kV field, and are generated when the electrons hit the anode (Bremsstrahlung). X-rays are classified as ionising radiation and appropriate precautions have to be taken to protect personnel and animals from incidental exposure.

X-ray beams partly penetrate and are partly scattered or absorbed by tissues of different density and composition. A radiographic or CT image is a shadow of those different tissues (Figure 12.2). In general four different densities of tissue are differentiated using conventional radiography; gas, fat, soft tissues and calcified structures. Gasses do not absorb x-rays and therefore appear black on the radiographic image. The skeleton has the highest x-ray absorption and looks white in images. Fat absorbs fewer x-rays and therefore looks a little darker than the other soft tissues. Normally soft tissues cannot be distinguished, although surrounding fat tissue or gas (for example in the intestine) may provide some contrast (Armstrong, Wastie, and Rockall 2004). Silver-based photographic emulsions have been replaced largely by digital image capture techniques (Armstrong, Wastie, and Rockall 2004).

12.1.2.2 Computed Tomography (CT)

Computed tomography (CT) is also based on differential x-ray transmission through the body. The x-ray detection system is usually more sensitive than in conventional radiography and smaller differences in x-ray density can be distinguished. By using a powerful computer it is possible to also reconstruct slices and three-dimensional images of structures.

CT and conventional x-radiography are fast techniques. Several multi-modal imaging devices may also incorporate CT to deliver morphological images, for example in a PET/CT (Armstrong, Wastie, and Rockall 2004).

12.1.2.3 Dual-Energy X-Ray Absorptiometry (DEXA)

DEXA is a non-invasive method used to measure bone mass, bone mineral density (BMD), or bone mineral content (BMC), as well as fat and lean mass of the total body, axial or peripheral skeleton. Serial *in vivo* measurements can be made and provide information regarding increases or decreases of BMD in osteoporosis research, as well as information relating to fracture healing, total body composition, and periprosthetic BMD. Before absorptiometry was applied to laboratory animals, bone mineral density measurements could only be made *ex vivo* involving ash weight, analytical chemistry, bone histomorphometry, and biomechanical testing, which required use of a considerable number of animals. This non-invasive and painless imaging technique provides opportunities for reduction and refinement, allowing the use of fewer animals, which can be individually followed throughout a study.

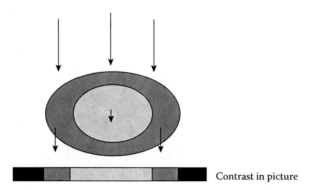

FIGURE 12.2 Principle of x-ray imaging. X-rays pass through tissue and image is generated due to different attenuation of x-ray radiation.

First generation instruments utilised an x-ray pencil-beam to scan the body, but recent developments have led to the introduction of a fan-beam. During the scan, the source passes over the subject and the detector, located under the animal, measures x-ray absorption at two different energy levels, allowing representation of bone, fat and lean tissue. Algorithms are used to calculate the quantity and type of tissue in each pixel scanned. The sum of all pixel values provides an estimate of the whole body composition in terms of BMC, fat and lean tissues, in grams as well as percentages (Figure 12.3).

A rat whole body scan takes about 10 minutes, depending on the animal's size and the position of its tail. The animal should be anaesthetised in order to ensure its immobility during the scan, and preferably positioned in ventral recumbency with the limbs abducted from the body at an angle of 90° in order to capture data applying to peripheral regions, in addition to total body data (Figures 12.4 and 12.5).

Correct positioning of the body is important for valid serial measurements, because DEXA converts three-dimensional structure into a two-dimensional image and consequently elliptical bones may provide a different areal BMD measurement, depending on the relative angle of the bone (Grier, Turner, and Alvis 1996). Accurate identification of the region of interest (ROI) is also important, because anatomical landmarks must be consistently used if serial measurements of the same area are to be analysed over time (Karahan et al. 2002). Small animal software provides separate ROIs for the head, trunk, arms and legs. For determination of bone loss in the rat animal model, skeletal site sub-regions incorporating trabecular bone are preferred, such as the proximal tibia or distal femur.

As an alternative to DEXA, which is an excellent non-invasive technique for measuring BMD and BMC in the rat, pQCT is sometimes preferred for more sensitive and accurate measurements of trabecular bone, as distinct from cortical bone (Rosen et al. 1995).

Patient:	ALEX 37,	Facility ID:			
Birth Date:	27/9/2007 0,7 years	Referring Physician:	ISMINI		
Height/Weight:	41,0 cm 0,3 kg	Measured:	2/7/2008	11:08:12 nμ	(11,40)
Sex/Ethnic:	Female White	Analyzed:	13/1/2009	12:07:47 μμ	(11,40)

BODY COMPOSITION[12]

Region	Tissue (%Fat)	Region (%Fat)	Tissue (g)	Fat (g)	Lean (g)	BMC (g)	Total Mass (kg)
Left Arm	66,0	65,8	369	243	125	1,1	0,370
Left Leg	64,3	63,3	158	101	56	2,3	0,160
Left Trunk	61,6	59,5	113	69	43	4,1	0,117
Left Total	64,5	63,7	703	454	250	9,4	0,713
Right Arm	66,0	65,8	353	233	120	1,1	0,354
Right Leg	64,3	63,1	161	103	57	2,9	0,164
Right Trunk	61,6	59,6	103	63	39	3,4	0,106
Right Total	64,5	63,8	675	436	239	8,2	0,684
Arms	66,0	65,8	722	477	245	2,2	0,724
Legs	64,3	63,2	318	204	114	5,2	0,323
Trunk	61,6	59,5	215	133	83	7,5	0,223
Total	64,5	63,7	1.379	890	489	17,7	1,396

FAT MASS RATIOS

Trunk/Total	Legs/ Total	(Arms+Legs)/ Trunk
0,15	0,23	5,17

FIGURE 12.3 Whole body composition of a 300 g female 9-month-old Wistar rat. Bone mineral content, fat and lean muscle mass in grams are calculated per body region and totally scanned by DEXA, small animal software.

Patient:	ALEX 37,	Facility ID:			
Birth Date:	27/9/2007 0,7 years	Referring Physician:	ISMINI		
Height/Weight:	41,0 cm 0,3 kg	Measured:	2/7/2008	11:08:12 nμ	(11,40)
Sex/Ethnic:	Female White	Analyzed:	13/1/2009	12:07:47 μμ	(11,40)

ANCILLARY RESULTS [Small Animal Body]

Region	BMD[12] (g/cm^2)	Young–Adult (%)	T-Score	Age-Matched (%)	Z-Score	BMC[12] (g)	Area (cm^2)
Head	0,238	–	–	–	–	2,8	12
Left Arm	0,107	–	–	–	–	1,1	10
Left Leg	0,097	–	–	–	–	2,3	24
Left Trunk	0,143	–	–	–	–	4,1	29
Left Total	0,134	–	–	–	–	9,4	70
Right Arm	0,107	–	–	–	–	1,1	10
Right Leg	0,104	–	–	–	–	2,9	28
Right Trunk	0,129	–	–	–	–	3,4	26
Right Total	0,120	–	–	–	–	8,2	69
Arms	0,107	–	–	–	–	2,2	20
Legs	0,101	–	–	–	–	5,2	52
Trunk	0,136	–	–	–	–	7,5	55
Ribs	0,102	–	–	–	–	2,0	20
Pelvis	0,135	–	–	–	–	2,5	19
Spine	0,180	–	–	–	–	2,9	16
Total	0,127	–	–	–	–	17,7	139

[12] Chemical/Ash calibration in use.
Filename: x0dd3k6ggh.dfz

For investigational use in laboratory animals or rather tests that do not involve human subjects.

FIGURE 12.4 Bone mineral density, bone mineral content and the area of each region measured as well as the total body measurements of a 300 gram female 9-month-old Wistar rat scanned by DEXA, small animal software.

12.1.2.4 Peripheral Quantitative Computed Tomography (pQCT)

Peripheral quantitative computed tomography (pQCT) is a similar system that is based on an x-ray scanner used according to the translation–rotation principle. It has a relatively small gantry with a diameter of approximately 8 cm and a resolution of approximately 70 μ. It provides a non-invasive way of separately analysing cortical and trabecular bone in slices of long bones from the peripheral skeleton (usually the tibia), even in rodents as small as mice (Breen et al. 1998). The animals must be anaesthetised and are placed in a purpose-built plexiglass cradle, in order to enable the limb to be positioned in a plexiglass tube before scanning (Figure 12.6). The initial scanogram is taken along the length of the tibia with slices 1 mm apart. The image produced on the screen is colour-coded with lighter areas depicting bone and darker ones depicting muscle, fat and water. This "scout" view makes it possible to set a reference line on the tibia's proximal end (knee joint; Figure 12.7) and then select the slices for final examination at 2.5 to 4 mm from the line (for evaluation of trabecular bone) and approximately 15 mm (for evaluation of cortical bone; Figure 12.8). The precision is high (0.6% for cortical bone and 0.3% for trabecular) and the amount of ionising radiation emitted for each slice is low (typically 30 μSievert/slice).

pQCT provides information non-invasively on micro-architecture, volumetric bone density in g/cm^3, and also gives a calculated estimate of bone strength, the strength strain index (SSI). Non-invasive calculated SSI *in vivo* and bone strength after three-point bending *ex vivo* have been shown

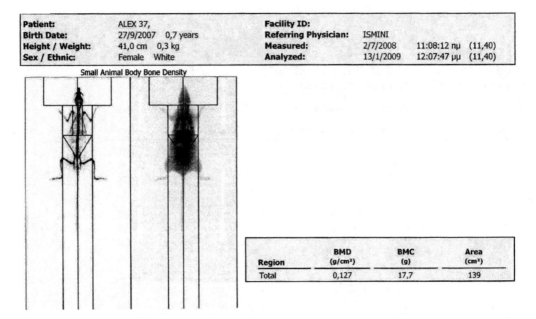

FIGURE 12.5 Printout of a DEXA whole body scanning of a rat correctly positioned, indicating total values of bone mineral density, bone mineral content and area.

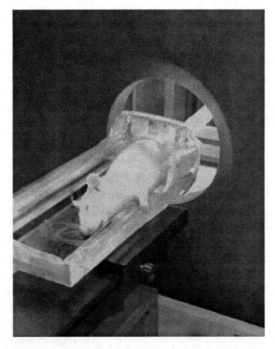

FIGURE 12.6 An anesthetised rat positioned for a pQCT scan of its left tibia, in the special plexiglass cradle, being inserted in the scanner's gantry.

to be well correlated (Jamsa et al. 1998). pQCT can be used to obtain precise morphometry measurements, such as periosteal and endosteal bone perimeters, cortical as well as trabecular bone density and area, total and cross-sectional bone area, trabecular thickness, number and separation, and so on (Lelovas et al. 2008).

Imaging Techniques

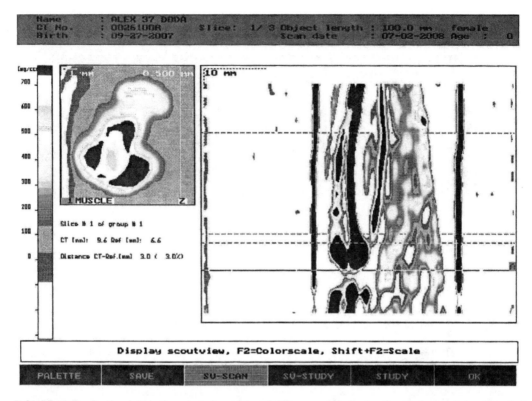

FIGURE 12.7 On the right is the scout view of the pQCT scanned limb, showing the knee joint and most of the tibial length. The reference line (continuous) for scanning sections of the tibia is positioned on the proximal end of the tibia. On the left is the first slice of the tibia taken at 3 mm distance from the reference line.

12.1.2.5 Micro-Computed Tomography

Micro-CT (μCT) is distinguished from other imaging x-ray techniques by its ability to acquire high-resolution images (~20 μ) based on the physical density of tissues. It is often used in skeletal research to provide a non-invasive *in vivo* three-dimensional image of bone tissue including details of its morphology and micro-architecture, such as trabecular thickness and separation. μCT can reveal the quality of osseointegration of a biomaterial implant within neighbouring biological tissue (bone). It also shows density gradients, which allow the distinction of fat from other tissues, fluids, and cavities without the need to use contrast agents. The resolution of *in vivo* μCT scans can be selected to fall into an isometric voxel range of approximately 10–200 μm, which means that it can identify and quantify very small volumes of fats cells in separate body deposits (visceral or subcutaneous), as well as measure the total volume of adipose tissue in an animal (Luu et al. 2009). It can also detect abnormal cells within neoplasms, both quantitatively and qualitatively. There are however limitations to the size of the specimen that can be scanned, depending on the μCT instrument used.

12.1.2.6 Radionuclear Imaging: PET and SPECT

Positron emission tomography (PET) and single photon emission computed tomography (SPECT) use radionucleotides to image functional processes. PET is based on detecting the coincidence of the two 511 kilo electro volt (keV) annihilation photons that have originated from β+ emitting sources (the patient). SPECT uses γ emitting isotopes in tracers such as ^{99m}Tc or ^{111}In, while PET uses higher energy positron emitting tracers (short living radioactive sources or nucleotides) such as ^{18}F or ^{11}C (Park and Gambhir 2005; Saha 2005). Radionucleotides decay because of the unstable combination of neutrons and protons, or excess energy, by emitting radiation such as α,

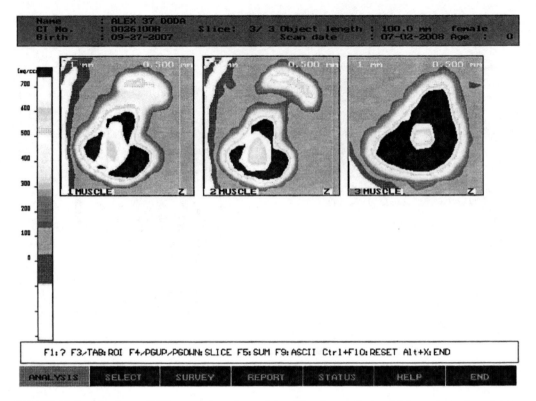

FIGURE 12.8 The three pQCT scanned sections of the tibia taken at 3, 4 and 15 mm from the reference line.

β^- or β^+ particles. Radionucleotides used for medical diagnostics are short lived. In general, the isotopes used in PET have higher energy and shorter half-lives than in SPECT (Park and Gambhir 2005); this improves the signal-to-noise ratio. Consequently, PET has higher spatial resolution and higher sensitivity than SPECT (Park and Gambhir 2005). The radionucleotides are coupled to a biologically active molecule, which is administered to the animal; they then serve as reporters for functional processes such as the consequences of gene therapy on metabolism. "Micro-PET" is a special adaptation for small animals and is capable of resolutions approaching 1 mm^3 (Gambhir 2002; Franc et al. 2008).

Positron emission tomography radio nucleotides can be incorporated directly into a biological molecule and provide a faithful representation of biological processes after *in vivo* administration. Uptake of biological substances can be measured by γ detectors and transformed to an image.

One biologically active molecule used in this way is fluorodeoxyglucose (FDG), a derivative of glucose, which has been used to demonstrate increased uptake by tumours with increased metabolism. Glucose uptake increases the concentration of the tracer that is associated with a γ-ray signal that can be detected with the γ detector (Figure 12.9).

The half-life of a radionucleotide is defined as the time required for the initial activity to decline to one-half and is unique for each radionucleotide. When radiopharmaceuticals are injected into an animal, they are eliminated from the body by biological processes such as faecal and urinary excretion. The biological half-life is defined as the time it takes to eliminate half of the administered activity from the body.

When needing to conduct nuclear medicine studies involving precise anatomical localisation, physicians always prefer to have a comparison between high resolution CT or MR images and

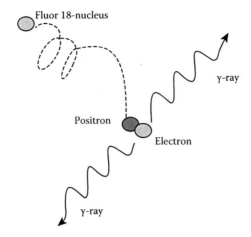

FIGURE 12.9 Positron and electron annihilate and emit two gamma rays in opposite direction.

low resolution PET or SPECT images. Combined PET–MRI solutions have been developed (Wehrl et al. 2009) and offer the advantage of superior soft tissue contrast of the MRI compared to CT. Alternatively, alignment techniques are used to establish one-to-one spatial correspondence between the two images.

Because of the increased sensitivity, specificity and accuracy in detecting various tumours, fusion images using the PET/CT modality have become the state-of-the-art technique in the imaging field. Animals can remain in the same position and accurate alignment and fusion of the CT (anatomical) and PET (functional) images assures greatly improved detection of lesions. PET/CT scanning is widely used in diagnostic oncology. Also, PET-MRI, allows simultaneous collection of data, which gives great opportunities for multi-parametric and multi-functional imaging including options like PET–fMRI or PET–MRS (Wehrl et al. 2009).

PET demands substantial capital outlay and large running costs, including highly competent personnel. To use short half-life radionucleotides, the PET scanner has to be located close to the cyclotron, which may present logistical challenges.

12.1.3 Imaging Techniques with High-Frequency Ultrasound Waves

These imaging techniques use high-frequency sound, ultrasound and acoustic, non-ionising radiation for imaging. Non-invasive contact with the body is necessary. These methods also include ultrasonic Doppler monitoring.

12.1.3.1 Ultrasonography

Ultrasonography is a method of studying the soft structures of the body without the use of ionising radiation. It uses high-frequency sound waves, which pass through tissues and are reflected back at interfaces to be detected by a receiver probe (Torres, Norcutt, and Dutton 2003; Armstrong, Wastie, and Rockall 2004). The time needed for the echo to return to the transducer is proportional to the distance travelled (Armstrong, Wastie, and Rockall 2004). Echoes returning from deeper structures are weaker than those arising from superficial levels because of increased sound attenuation. The reflected signals are converted to electrical impulses, amplified and displayed on a screen. There are three modes of echo display: A-mode (amplitude mode) is mainly used for ophthalmic or other examinations requiring precise length or depth measurements. B-mode (brightness mode) displays the returning echoes as dots along a single line; their brightness or greyness depends on the amplitude of the returned echoes and their position corresponds to the depth at which the echoes originated. M-mode (motion mode) is used in echocardiography for precise cardiac chamber and wall measurements and quantitative evaluation of valve or wall motion over time (Nyland et al. 2002).

Image contrast arises from differences in tissue density and the absorption and reflection of the ultrasound beam (Armstrong, Wastie, and Rockall 2004). A contact gel is used as a conductive agent (Torres, Norcutt, and Dutton 2003; Armstrong, Wastie, and Rockall 2004) and the skin has to be shaved to achieve good signal coupling. The procedure is painless but slow and the animal has to be immobilised or restrained during the procedure.

Fluid is a favourable medium for propagation of sound and therefore ultrasound is excellent for diagnosis of cysts or fluid-filled structures like the urinary bladder and the foetus in its amniotic sac, which produces large echoes from the wall. Air, bone and calcified tissue absorb ultrasound and give poor echoes back to the transducer (Armstrong, Wastie, and Rockall 2004). Parameters used are broadband ultrasound attenuation (BUA) and speed of sound (SOS). Contrast agents containing small air-bubbles have recently been developed for ultrasonography (Armstrong, Wastie, and Rockall 2004).

12.1.3.2 Doppler Ultrasonography

If the reflecting structures are moving, the frequency of the returned signal is modulated, corresponding to the speed of movement. This phenomenon is known as the Doppler effect and can be utilised for example to image blood flow, to identify structures by the presence or absence of flow, to ascertain the direction of flow, and to document flow disturbances associated with pathology such as stenosis or occlusion in the circulation (Armstrong, Wastie, and Rockall 2004). The rate of blood flow can be estimated by measuring the velocity using Doppler ultrasonography and multiplying it by the vessel's cross-sectional area determined by two-dimensional imaging (Nyland et al. 2002).

Pulsed wave Doppler ultrasonography offers depth discrimination, because the site of origin of the echo can be determined precisely from the time taken for return, which corresponds to the depth of the structure. In continuous wave Doppler ultrasonography, sampling is continuous and the time of return of the echoes is not monitored; consequently this technique does not permit depth discrimination, but it can be used to determine the direction of flow and to measure higher flow velocities than pulsed wave Doppler. The region of interest must be the only site of blood flow within the beam. This technique has also been applied to record high velocities distal to stenoses. Colour flow Doppler is a pulse wave modality in which velocity characteristics are encoded as a colour display, enabling visualisation of the source of velocities within the heart and large vessels (Brown and Gaillot 2008).

More recently, the development of intravascular ultrasound (IVUS) has allowed diagnostic imaging of the vessel wall from inside its lumen, using ultrasound technology. IVUS is a minimally invasive imaging methodology that requires intravascular catheterisation. The tip of a catheter is equipped with a miniature ultrasound probe and its other end is attached to computerised ultrasound equipment. IVUS is used especially for investigating arterial atherosclerotic plaques or vascular stenosis in atherosclerosis research (Kinlay 2001) and provides real-time, high-resolution images allowing precise tomographic assessment of plaque size, composition of artery segment and lumen area. The disadvantage of the method is that it requires anaesthesia and vascular catheterisation but this is offset by its ability to provide information on the vessels' endothelium *in vivo*, without requiring euthanasia of the animal for histological examination. Using IVUS, plaque distribution and volume within the artery wall can be determined, features that cannot be seen by angiography.

12.1.4 Imaging Techniques Detecting Light

12.1.4.1 Fluorescence Optical Imaging and Bioluminescence Imaging

In fluorescence optical imaging, laser light of a specific wavelength illuminates the specimen and induces the emission of fluorescence that is detected by an ultra sensitive CCD camera (Ntziachristos, Bremer, and Weissleder 2003; Figure 12.10). Implanted cells and tissues, which express fluorescence, can be tracked *in vivo* non-invasively to establish whether there is migration or metastasis (Park and Gambhir 2005). Genes coding for green fluorescent protein (GFP) derived from bioluminescent

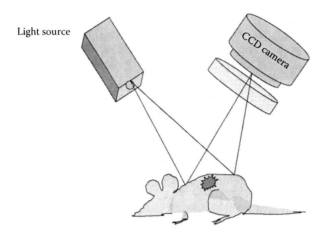

FIGURE 12.10 Fluorescence optical imaging. A laser light source enlightens fluorescent tissue that is recognised by a CCD camera.

jellyfish *Aequorea victoria* (Prasher et al. 1992) or red fluorescent protein (RFP; Hoffman 2002) can be introduced to cells in culture as well as germ line cells in genetically modified animals and used as reporter genes. Other markers such as cyan fluorescent protein (CFP), yellow fluorescent protein (YFP), and enhanced green fluorescent protein (EGFP) enable multi-protein labelling and can be used to report different activities within the same cell (Park and Gambhir 2005). Fluorescence imaging has been used to trace transfected cancer cells, stem cell based therapy and regeneration of neural tissue (Sutton et al. 2008).

In bioluminescence imaging, no input light is needed, because light is emitted by the enzyme luciferase when the substrate luciferine is available, by conversion of chemical energy to a photon (Park and Gambhir 2005; Sutton et al. 2008). The luciferase gene from the firefly can be introduced to cell lines and to germline cells resulting in genetically modified animals that express the luciferase gene (Contag et al. 1997; Park and Gambhir 2005). The luciferase gene can also be used as a reporter gene with expression of bioluminescence when the substrate luciferine is administered.

Bioluminescence has a higher signal-to-noise ratio than fluorescent optical imaging, because no background bioluminescence occurs (Park and Gambhir 2005; Fomchenko and Holland 2006)—standard rodent chow can contain autofluorescent components that may contribute to background noise. Animals should be fed special diets before scanning in an optical fluorescence imager (Sutton et al. 2008).

Fluorescence optical imaging and bioluminescent imaging is inexpensive compared to MRI or PET (Park and Gambhir 2005). It is not necessary to have special experience in order to be able to run the machine (Park and Gambhir 2005). For both fluorescence and bioluminescence, a reporter gene is required.

One major limitation of optical fluorescence and bioluminescence techniques is that tissue penetration is limited to only a few centimetres (Sutton et al. 2008). Shaving animals improves the quality of the image by preventing scattering of light by the fur. Fluorescence mediated tomography has been developed to better detect fluorescence in deeper tissues (Ntziachristos, Bremer, and Weissleder 2003; Zacharakis et al. 2005).

12.1.5 Imaging Techniques Based on Infrared or Near Infrared Radiation (NIR)

12.1.5.1 Near Infra-Red Spectroscopy (NIRS)

NIRS is a very useful method for monitoring regional tissue oxygenation. It is based on the property of haemoglobin, myoglobin and cytochrome oxydase to absorb near infrared radiation (NIR) differentially, depending on their concentration and their oxygenated state. Although it has been used previously as a research tool in animal models and also clinically to monitor cerebral oxygenation during complex cardiovascular operations and neurosurgery, the technique has recently been more enthusiastically used in clinical and experimental practise as a non-invasive means of assessing tissue oxygenation at the end-organ level.

The basic equipment required for an NIRS monitoring system includes: (1) light sources capable of generating multiple wavelengths or a single light source with the necessary filters to produce the required wavelengths in the NIR region; (2) fibreoptic bundles to transmit light from the source to the probe and from the probe to a detector; (3) a tissue probe containing an optode for the NIR light emission and an optode to collect light returning from the tissue; (4) photon detection hardware; (5) a processing computer with software capable of comparing delivered and recovered light; and (6) an information display system.

The device is capable of calculating the tissue content of total haemoglobin, oxyhaemoglobin and oxidised Cytochrome C. The advantages of this method are that the equipment is simple and easy to use, non-invasive and painless; results are obtained in real-time and independent of influences of pulsatile vascular flow. The method suffers from the disadvantage that different devices use different algorithms, there are no baseline values for NIRS parameters, and there is little information about how different mediators or drugs influence the values obtained.

Regional perfusion monitors are now receiving greater interest because it has been known for some time that changes in regional perfusion can occur significantly earlier than the global indices that have been measured traditionally. The technique of NIRS is exciting because it has the potential to derive information non-invasively about all of the major components of oxygen transport, ranging from bulk transport of oxygen to its cellular utilisation at the level of the mitochondria (Wolf, Ferrari, and Quaresima 2007).

12.1.5.2 Remote Thermography and Near Infrared Imaging

Skin temperature changes in a variety of pathological conditions and thermal radiation can be measured with an infrared sensitive camera (Jiang et al. 2005; Merla and Romani 2006). The process of imaging temperature differences is named thermography. The basis for development of this technology was the application of heat searching for military purposes.

Skin temperature has been known since antiquity to be influenced by pathological conditions such as fever and it is one of the cardinal signs of inflammation described by Hippocrates around 480 BC (Aldini, Fini, and Giardino 2008). Changes in muscular tone can affect circulation and surface skin temperature (Merla and Romani 2006) and muscular activity increases metabolic rate, which raises the temperature in surrounding tissues such as skin.

Near infrared cameras have been used for the diagnosis of breast cancer to detect circulatory and metabolic changes in tumour cells, together with mammography to support clinical findings (Merla and Romani 2006).

NIR thermography is a simple and cheap technique in comparison with methods such as MRI and PET, and does not demand such highly trained staff. No radiation or physical exposure of the tissue is involved so the method is totally non-invasive (Jiang et al. 2005).

However, it is important that image acquisition is done under conditions of controlled temperature, humidity and air circulation. The subject should be allowed to rest some time before imaging to avoid artefacts due to stress reactions. The camera also has to be acclimatised to the conditions in the room where it is used (Jiang et al. 2005).

12.2 REDUCTION AND REFINEMENT

The use of imaging in laboratory animal studies can substantially reduce the number of animals used in research. Moreover, imaging is claimed to offer animal welfare advantages because it is

Imaging Techniques

non-invasive and because it makes it possible to terminate studies earlier or to set early HEPs. Despite this, the use of animal imaging raises ethical and practical issues, which make it important to carefully consider refinement and reduction together. Fortunately, technical advances in equipment and procedures continue to make the technology easier to implement and more robust, thereby improving the efficacy and quality of research.

The ability to make repeated measurements in the same animal, which is a feature of most imaging techniques, has several advantages. Firstly, it reduces the number of animals needed because each animal can be imaged many times, and high-quality data collected without killing it. Secondly, each animal serves as its own control, which avoids variability associated with individual differences between animals (Figure 12.11).

Reuse of animals is regulated in most European countries and acceptable ethically only for minor procedures. Simply scanning a healthy animal is regarded as a minor procedure. When animal handling and anaesthesia are optimised, so as to cause only minimal stress or damage to the animal, it should be acceptable to use animals for several imaging uptakes. The alternative to this approach is to use more animals.

Despite this, frequently repeated imaging under anaesthesia is not without stress for the animal. If the experiment also involves invasive procedures or imaging non-healthy animals, the risk of anaesthetic death increases and such procedures should no longer be classified as minor.

12.2.1 Severity Classifications

Severity associated with imaging of animals can be classified according to the impact of anaesthesia on the animal undergoing the scanning. This impact arises from manipulation, injection of agents and anaesthetic risk. The health of the animal and other procedures performed will contribute to possible complications and the impact of anaesthesia (Table 12.3).

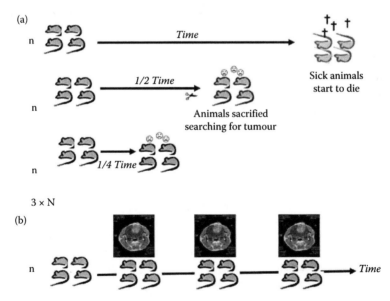

FIGURE 12.11 Example of how imaging can be used to reduce the number of animals compared to histological investigation. (a) Before MRI: Experiments must be repeated for all tumour cell lines, time frames, treatments or interventions, and combinations of these. This uses a large number of animals. (b) By use of MRI it is possible to do repeated measurements in the same animals, so that each animal can be studied over time; monitor treatments effects in same animal; euthanize before severe clinical symptoms (humane endpoints).

TABLE 12.3
Relationship Between the State of Health of the Animal and the Impact of Procedures on it

	Animal's Condition	
	Healthy	Poor Health Conditions
Non-invasive procedures	Impact of anaesthesia	Impact of anaesthesia ↑
Invasive procedures	Impact of anaesthesia ↑	Impact of anaesthesia ↑↑↑

■ Low Risk ■ Medium Risk ■ High Risk
Note: Invasive procedures increase the likelihood of complications.

12.2.1.1 Healthy Animal, No Invasive Procedure

Examples of such studies include quality control and development or optimisation of new imaging protocols and sequences when no procedure is carried out on the animal before imaging. It would also be relevant for morphological studies, for example of growth and development of genetically altered animals. The only stress the animal experiences results from handling and anaesthesia.

12.2.1.2 Healthy Animal, Invasive Procedure

These studies might involve simple administration of test substances percutaneously, or moderately invasive procedures on healthy animals. They might apply to the testing of substances and use of contrast agents. The animal has to receive an injection, and for repeated injections or infusions it may be necessary to implant catheters, resulting in relatively minor tissue trauma and the risk of complications.

12.2.1.3 Non-Healthy Animals

Imaging can assist with setting early HEPs in chronic progressive disease conditions. There is an increased risk of anaesthetic complications or death as a result of their illness. The anaesthetic risk is further increased if repeated anaesthesia is necessary, especially if only short recovery times are possible between imaging. For example, brain tumours can impair the central respiratory centre, which may compromise respiratory function and increase anaesthetic risk. The vitality of genetically modified animals might also be impaired, thereby increasing the anaesthetic risk.

12.2.1.4 Non-Healthy Animals, Invasive Procedure

The conduct of invasive procedures on severely diseased animals further increases the impact of the anaesthetic in an animal already compromised by poor health. The risk of infection (particularly of immunodeficient animals) and other complications is increased by surgical interventions such as implantation of catheters for administration of contrast agents. These procedures need careful justification and extra attention should be given to refinement, optimisation and quality control procedures.

For all these procedures a quality management system should be adopted: imaging is a complex procedure beset with many pitfalls. Critical processes must be identified and monitored, and if procedures fail to meet the requirements, prompt action must to be taken to improve performance. For example, unacceptable mortality as a result of poor anaesthetic technique raises ethical issues and also reduces productivity and efficiency.

12.2.2 Procedures that Impair Animal Welfare and Scientific Quality—Setting Targets for Refinement

Imaging technology constitutes a non-invasive diagnostic tool that makes it possible to collect data repeatedly from the same animal. In many experiments, access to a vein is needed to administer contrast agents or other test substances and in such cases, aseptic techniques should be used to minimise tissue trauma, reduce complications and increase the quality of research.

12.2.2.1 Anaesthesia

Although scanning is not painful for animals, it is necessary to immobilise them whilst the image is being generated and if they are going to be imaged frequently, animals need sufficient time to recover between scans. In the ideal situation, the anaesthetic chosen should be easily administered, provide reversible and controlled loss of consciousness, be safe for the animal and for the operator, and result in no mortality. Furthermore, the anaesthetic should not disturb physiological functions (or at least, influence the parameters being investigated) and cause minimal homeostatic changes. If surgical intervention is required, the anaesthetic should ensure optimal conditions for surgery; that is, relief of pain, narcosis and muscle relaxation.

The induction of anaesthesia is stressful, although optimisation and quality control of the protocol reduces its negative impact on the animals. Induction should be gentle and safe and a stable level of anaesthesia maintained during scanning, following which recovery should be smooth and quick. Special precautions should be taken to avoid hypothermia in anesthetised animals; for example, the use of insulating wraps or heating mats, as demonstrated in Figure 12.1e. A rapid and uneventful recovery is particularly important to allow animals to resume normal behaviour, including eating, grooming and social interactions, without any long-term impact on their physiological functions. A fuller account of best practices in anaesthesia is provided in Chapter 13.

Although usually regarded as a quick technique, imaging takes time. Depending on the nature of the study, it can take minutes to several hours (Table 12.2). This latter is a long time for some anaesthetic protocols and may lead to complications especially in small animals. Good planning and logistics are therefore important to minimise the time for which animals are anesthetised. Animals with advanced disease conditions such as cancer or circulatory failure are regarded as "at risk" according to the American Society of Anesthesiologists (ASA) Physical Status Classification System and require careful review of anaesthetic routines and monitoring to minimise anaesthetic risk.

Imaging techniques are commonly used to investigate cardiovascular function. Anaesthesia affects physiological functions and parameters and the cardiovascular system is extremely sensitive to anaesthesia (Yang et al. 1999; Chaves, Weinstein, and Bauer 2001; Roth et al. 2002; Zuurbier, Emons, and Ince 2002; Janssen et al. 2004; Kober et al. 2004, 2005; Table 12.1). These effects can be enhanced if the animal is stressed. Moreover, if the adrenergic system becomes activated, the distribution of body fluids may change and unnecessary stress during the induction of anaesthesia causes dramatic changes in the circulatory system, which may result in death. A summary of the influence of anaesthesia on cardiovascular functions is presented in Table 12.1. For functional studies of the heart and circulatory function it is most important to control these anaesthetic effects. Both pentobarbital and ketamine–xylazine have negative effects on circulation (Janssen et al. 2004), whereas ketamine–xylazine regimes reduce heart rate and myocardial blood flow (Kober et al. 2005) relative to isoflurane (Kober et al. 2004, 2005).

Movements caused by respiration or vascular pulsation impair the quality of the image. Respiration sensors and electrocardiogram (ECG) triggering are used to synchronise heartbeats and imaging. Mounting ECG probes takes time and prolongs anaesthesia sleeping time. Software developments such as "retrospective gating" for cardio cine MRI reduce movement artefacts of this sort during imaging independent of gating by the ECG probes. Retrospective gating reduces distortion, thereby shortening imaging and anaesthesia time (Bishop et al. 2006; Johnson 2008) and constitutes a major refinement of cardio-imaging.

Special considerations are necessary if artificial ventilation is required. In small animals this usually relies on intermittent positive pressure ventilation (IPPV), dependant on air being forced into the thorax rather than being drawn in under negative pressure as during spontaneous inhalation. Normally the negative intrathoracic pressure during spontaneous inhalation promotes venous filling of the heart. Artificial ventilation therefore compromises cardiac filling and this is usually compensated for by using an inspiration:expiration ratio of 1:3–1:2. It is important to be aware of these changes especially in relation to cardiovascular studies. IPPV may have to be considered for animals with increased anaesthesia risk or for long-term scanning studies.

Some imaging, such as Doppler/ultrasound investigations, can be performed on fully conscious animals, which avoids the cardiovascular influences of anaesthetic agents (Yang et al. 1999). However, restraint and immobilisation during imaging can stress non-anaesthetised animals, which causes activation of the autonomic nervous system and this affects several physiological parameters. It can be difficult to predict or control these effects, so it is preferable to control them by conducting imaging procedures under general anaesthesia. When volatile anaesthetics, such as isoflurane are used, the effect of anaesthesia is not more confounding than the use of non-anaesthetised animals (Yang et al. 1999).

Volatile anaesthesia provides a more stabile platform over time, allowing repeated and reproducible measurements and is therefore preferable to parenteral anaesthesia (Chaves, Weinstein, and Bauer 2001). Although vapour anaesthesia is regarded as the best controlled form of anaesthesia, there are studies indicating that its repeated use can cause morphological changes in the brain (Xie et al. 2008). To maximise recovery Johnson recommends not maintaining anaesthesia for more than 3 hours (2008).

Genetic differences can be responsible for variations in response to anaesthetic agents (Baran et al. 1975; Lovell 1986; Pick et al. 1991; Homanics, Quinlan, and Firestone 1999; Sonner et al. 1999; Zuurbier, Emons, and Ince 2002). Strain differences have been found between Swiss, CD-1 BALB/c and C57Black mice although these differences were less prominent under volatile anaesthesia (Zuurbier, Emons, and Ince 2002).

Deaths under anaesthesia are unacceptable for ethical reasons. In addition, the repetition of experiments incurs economic costs and is labour and time consuming and leads to complications with statistical evaluation of experimental data. It may be difficult to establish whether death was caused by anaesthetic complications, the experimental interventions, or a combination of both. If the data are just eliminated from the study and replaced with data obtained from other animals, this will cause an overestimation of treatment results. A strategy needs to be developed for treating these drop-outs.

12.2.2.2 Invasive Procedures

If the animals are placed inside a scanner it might be necessary to have placed an indwelling catheter in a vein previously. This should be done aseptically, using the least tissue traumatic approach, and ideally employing a superficial vein that can be penetrated easily transcutaneously. The catheter should be firmly fixed throughout the scanning period to prevent its dislodgement. This is important not only to ensure accurate delivery into the circulation, but also to avoid inadvertent leaking of irritant test substances or contrast agents into perivascular spaces, which might cause oedema, severe inflammation and pain, and in worst cases, tissue necrosis. To prevent clotting, it is advisable to flush the catheter using a sterile heparinised physiological solution such as Ringer's or 0.9% saline. Catheters for administration of substances must always be checked using a flush of non-irritant physiological solution (like sterile Ringer's or 0.9% NaCl; Zuurbier, Emons, and Ince 2002).

12.2.2.3 Fluid Balance

For long-term scanning it is advisable to administer fluids to maintain fluid balance (Zuurbier, Emons, and Ince 2002). This can be done intravenously if access is available. With small animals it is important to avoid overhydration and oedema by using excessive volumes for flushing catheters; it is recommended that a syringe pump is used to ensure accurate volumes of fluid are delivered.

Alternatively, sterile physiological fluid can be given subcutaneously, using aseptic routines. Fluids should be warmed to body temperature to avoid iatrogenic induction of hypothermia. An overview of the preparation necessary for different imaging techniques is presented in Table 12.2.

12.2.3 Other Issues, Quality Testing and Planning of the Imaging Facility

Fluorescence imaging has limited depth of tissue penetration; in addition, hair causes scattering of light and it is necessary to shave the animals to capture good images. The shaving procedure should be performed without traumatising the animal's skin and should not be so extensive as to induce hypothermia. Although shaving seems like a minor intervention, removal of fur should be considered in ethical review because of its physiological functions for the animal.

The performance of imaging equipment should be checked regularly. This can be done using standardised phantoms and should always be completed before animals are anesthetised. When it is necessary to test equipment using living animals, use should be made of surplus animals (e.g., from breeding genetically modified animals) whenever possible. These animals would be killed otherwise without any scientific benefit, and other animals would have had to be bred for this purpose only.

The imaging facility should be located close to the animal facility. Transport is a well-established stressor for animals, which need time to adapt to a new environment afterwards. Often it is not possible to locate the equipment adjacent to the core animal facility because of the need for proximity to other installations. For example, a PET scanner has to be close to the cyclotron in which short active nucleotides are made. The imaging equipment must be easily accessed for service and maintenance, for example, cryogene filling in MRI; it is advisable to preclude such technical activities from the animal house. Many imaging machines are sophisticated tools and require careful attention to features such as stability of the ground or need shielding from disturbing electromagnetic noise.

Often it is necessary to establish a micro-satellite animal accommodation unit close to the imaging equipment. In such cases, animals should be acclimatised as normal before they are anaesthetised and imaged. In the case of PET and SPECT it might be necessary to keep the animals in a shielded room for some days after experiments to allow isotopes to decay. Appropriate health and safety precautions and environmental considerations must be considered, though they are not further addressed here.

There is increased interest in the use of multi-modal imaging, or scanning of animals using more than one technique. Software packages allow the different images to be fused together so as to construct a more meaningful visual representation of the organ studied. This approach makes it possible to get the best out of each technique, although it extends the time for which the animal needs to be anaesthetised.

12.2.3.1 Use of Genetically Modified Animals for *In Vivo* Studies of Functional Genomics

The mouse is the principal animal model for exploring phenotypic changes caused by gene modifications and developing gene therapies. Non-invasive imaging techniques allow the acquisition of objective morphologic and functional data to support these activities.

The establishment and running of advanced imaging technology centres are very expensive and researchers are encouraged to cooperate to best utilise these resources and investments. International cooperation is often one criterion considered by research funding bodies. Many genetically modified animal models are unique and not available through commercial animal suppliers. However, increased cooperation between research institutions and the consequent movement of animals between facilities risk spreading infections. This issue is discussed further in Chapter 9. Non-commercial suppliers of laboratory animals often have poor quality control systems and produce animals of indeterminate health status; such animals must be regarded as potential vectors for contagious diseases and when being scanned, measures must be taken to protect subsequent healthy animals exposed to the equipment. One challenge is that new pathogens may be unrecognised, for example, *Helicobacter* discovered in 1994 (Whary and Fox 2006) and *norovirus* in 2003 (Henderson 2008).

12.2.3.2 Reduction, Refinement and Legitimising Animal Experiments

Imaging technology allows repeated imaging in the same animal and in this way may contribute to reducing the number of animals required for a study. In addition, non-invasive imaging is less traumatic for the animal than invasive techniques, and consequently experiments in which it is used may be more acceptable because the harm:benefit ratio is reduced (Bateson 1986). From this perspective imaging seems to be good for laboratory animals. However, as in many cases, technological development also creates new ethical dilemmas. Firstly, development of the technology itself often requires the use of animals and therefore contributes, albeit very modestly, to increases in the number of animals used in research. The paradox of whether to sacrifice a few individuals with the objective of saving many more in the future is a universal moral dilemma in deontological ethics. From a utilitarian perspective, the proposition can be accepted on the basis that it conforms to "the greatest good for the greatest number". There is no "universal moral law" that gives ethical answers for all cases (Emanuel Kant's Categorical Imperative; Rossouw 2008). In the context of animal imaging, the question arises about whether it is acceptable to sacrifice animals to develop a new technology that in the end will increase the need for more animals to answer a greater number of scientific questions.

The Three Rs, Replacement, Refinement and Reduction identified by Russell and Burch (1959) have been raised to the status of "universal moral laws". Russell and Burch (1959) argued that all three perspectives should be considered when planning and performing animal experiments. However, there is no universal moral law weighing the Three Rs against each other, and in applied ethics different ideals or "universal moral laws" often conflict with each other.

12.3 CONCLUSIONS

Imaging technology is used in the clinic to allow better evaluation of patients. The fact that imaging is done non-invasively is a great advantage. Development of new and better imaging tools is an important area of research. In animal research, imaging tools can be used to refine animal experiments by setting earlier, more accurate and humane endpoints. Also, imaging contributes to reduction, because the same animal can be investigated several times. Imaging opens many new opportunities in animal research, but also presents some pitfalls. The first part of this chapter gives an overview of different imaging techniques. The second part addresses the impact of these techniques on the animals. Anaesthesia is necessary for most imaging techniques and the choice of anaesthetic agent has an impact on animal welfare as well as on the phenomenon studied. Other issues such as hygiene considerations in imaging centres and increased transport of animals must be addressed. Finally, welfare implications and the impact of the procedures on each animal taking part in a particular experiment must be carefully considered.

12.4 QUESTIONS UNRESOLVED

One of the challenges remaining in imaging technology is to find ways to further decrease the time required for high-quality imaging. As mentioned above, the less time an animal is restrained or anaesthetised for imaging, the faster it recovers. Additionally, it would be advantageous to be able to minimise animal exposure time to ionising and even non-ionising sources during imaging. It is hoped that new generation imaging instruments will offer increased precision and throughput and permit the collection of more precise and more specific qualitative and quantitative data.

APPENDIX – LIST OF ABBREVIATIONS

CFP = Cyan Fluorescent Protein
CT = Computed Tomography
DEXA = Dual-energy X-ray absorptiometry
ECG = Electro Cardio Gram
EGFP = Enhanced Green Fluorescent Protein
fMRI = Functional Magnetic resonance imaging
GFP = Green Fluorescent Protein
HEP = Humane End Point
kV = kilo Volt (high voltage)
keV = kilo electron Volt (energy unit)
mCT = Micro-Computed Tomography
MRI = Magnetic resonance imaging
MRS= Magnetic resonance spectroscopy
NIR = Near Infrared Radiation
NIRS = Near Infra-Red Spectroscopy
NMR = Nuclear Magnetic Resonance
PET = Positron Emission Tomography
pQCT = peripheral Quantitative Computed Tomography
RFP = Red Fluorescent Protein
SOP = Standard Operation Procedures
SPECT = Single Photon Emission Computed Tomography
YFP = Yellow Fluorescent Protein

REFERENCES

Aldini, N. N., M. Fini, and R. Giardino. 2008. From Hippocrates to tissue engineering: Surgical strategies in wound treatment. *World Journal of Surgery* 32:2114–21.
American Society of Anesthesiologists Physical Status Classification System. Available from http://www.asahq.org/clinical/physicalstatus.htm
Armstrong, P., M. L. Wastie, and A. G. Rockall. 2004. *Diagnostic imaging,* 5th ed. Oxford: Blackwell Publishing.
Baran, A., L. Shuster, B. E. Eleftheriou, and D. W. Bailey. 1975. Opiate receptors in mice: Genetic differences. *Life Sciences* 17:633–40.
Bateson, P. 1986. When to experiment on animals. *New Scientist* 109:30–32.
Berry, E., and A. J. Bulpitt. 2009. *Fundamentals of MRI: An interactive learning approach.* Boca Raton, FL: CRC Press.
Bishop, J., A. Feintuch, N. A. Bock, B. Nieman, J. Dazai, L. Davidson, and R. M. Henkelman. 2006. Retrospective gating for mouse cardiac MRI. *Magnetic Resonance in Medicine* 55:472–77.
Breen, S. A., B. E. Loveday, A. J. Millest, and J. C. Waterton. 1998. Stimulation and inhibition of bone formation: Use of peripheral quantitative computed tomography in the mouse in vivo. *Laboratory Animals* 32:467–76.
Brown, D., and H. Gaillot. 2008. Heart. In *Atlas of small animal ultrasonography*, ed. D. Penninck, 151–216. Iowa: Blackwell Publishing.
Chaves, A. A., D. M. Weinstein, and J. A. Bauer. 2001. Non-invasive echocardiographic studies in mice: Influence of anesthetic regimen. *Life Sciences* 69:213–22.
Contag, C. H., S. D. Spilman, P. R. Contag, M. Oshiro, B. Eames, P. Dennery, D. K. Stevenson, and D. A. Benaron. 1997. Visualizing gene expression in living mammals using a bioluminescent reporter. *Photochemistry and Photobiology* 66:523–31.
Dager, S. R., N. M. Oskin, T. L. Richards, and S. Posse. 2008. Research applications of magnetic resonance spectroscopy to investigate psychiatric disorders. *Topics in Magnetic Resonance Imaging* 19:81–96.
Eichele, T., K. Specht, M. Moosmann, M. L. Jongsma, R. Q. Quiroga, H. Nordby, and K. Hugdahl. 2005. Assessing the spatiotemporal evolution of neuronal activation with single-trial event-related potentials and functional MRI. *Proceedings of the National Academy of Sciences of the United States of America* 102:17798–803.

Fomchenko, E. I., and E. C. Holland. 2006. Mouse models of brain tumors and their applications in preclinical trials. *Clinical Cancer Research* 12:5288–97.

Franc, B. L., P. D. Acton, C. Mari, and B. H. Hasegawa. 2008. Small-animal SPECT and SPECT/CT: Important tools for preclinical investigation. *Journal of Nuclear Medicine* 49:1651–63.

Gambhir, S. S. 2002. Molecular imaging of cancer with positron emission tomography. *Nature Reviews. Cancer* 2:683–93.

Gillies, R. J., Z. M. Bhujwalla, J. Evelhoch, M. Garwood, M. Neeman, S. P. Robinson, C. H. Sotak, and B. Van Der Sanden. 2000. Applications of magnetic resonance in model systems: Tumor biology and physiology. *Neoplasia* 2:139–51.

Grier, S. J., A. S. Turner, and M. R. Alvis. 1996. The use of dual-energy x-ray absorptiometry in animals. *Investigative Radiology* 31:50–62.

Henderson, K. S. 2008. Murine norovirus, a recently discovered and highly prevalent viral agent of mice. *Laboratory Animal* (NY) 37:314–20.

Herynek, V., M. Burian, D. Jirak, R. Liscak, K. Namestkova, M. Hajek, and E. Sykova. 2004. Metabolite and diffusion changes in the rat brain after leksell gamma knife irradiation. *Magnetic Resonance in Medicine* 52:397–402.

Hiba, B., N. Richard, M. Janier, and P. Croisille. 2006. Cardiac and respiratory double self-gated cine MRI in the mouse at 7 T. *Magnetic Resonance in Medicine* 55:506–13.

Hiller, K. H., C. Waller, A. Haase, and P. M. Jakob. 2008. Magnetic resonance of mouse models of cardiac disease. *Handbook of Experimental Pharmacology* 245–57.

Hoffman, R. 2002. Green fluorescent protein imaging of tumour growth, metastasis, and angiogenesis in mouse models. *The Lancet Oncology* 3:546–56.

Homanics, G. E., J. J. Quinlan, and L. L. Firestone. 1999. Pharmacologic and behavioral responses of inbred C57BL/6J and strain 129/SvJ mouse lines. *Pharmacology, Biochemistry, and Behavior* 63:21–26.

Hugdahl, K., B. R. Rund, A. Lund, A. Asbjornsen, J. Egeland, L. Ersland, N. I. Landrø, et al. 2004. Brain activation measured with fMRI during a mental arithmetic task in schizophrenia and major depression. *The American Journal of Psychiatry* 161:286–93.

Jämsä, T., P. Jalovaara, Z. Peng, H. K. Väänänen, and J. Tuukkanen. 1998. Comparison of three-point bending test and peripheral quantitative computed tomography analysis in the evaluation of the strength of mouse femur and tibia. *Bone* 23:155–61.

Janssen, B. J., T. De Celle, J. J. Debets, A. E. Brouns, M. F. Callahan, and T. L. Smith. 2004. Effects of anesthetics on systemic hemodynamics in mice. *American Journal of Physiology. Heart and Circulatory Physiology* 287:H1618–H1624.

Jiang, L. J., E. Y. Ng, A. C. Yeo, S. Wu, F. Pan, W. Y. Yau, J. H. Chen, and Y. Yang. 2005. A perspective on medical infrared imaging. *Journal of Medical Engineering & Technology* 29:257–67.

Johnson, K. 2008. Introduction to rodent cardiac imaging. *ILAR Journal* 49:27–34.

Karahan, S., S. A. Kincaid, S. D. Lauten, and J. C. Wright. 2002. In vivo whole body and appendicular bone mineral density in rats: A dual energy X-ray absorptiometry study. *Comparative Medicine* 52:143–51.

Kinlay, S. 2001. What has intravascular ultrasound taught us about plaque biology? *Current Atherosclerosis Reports* 3:260–66.

Kober, F., I. Iltis, P. J. Cozzone, and M. Bernard. 2005. Myocardial blood flow mapping in mice using high-resolution spin labeling magnetic resonance imaging: Influence of ketamine/xylazine and isoflurane anesthesia. *Magnetic Resonance in Medicine* 53:601–6.

Kober, F., I. Iltis, P. J. Cozzone, and M. Bernard. 2004. Cine-MRI assessment of cardiac function in mice anesthetized with ketamine/xylazine and isoflurane. *Magma* (New York, NY) 17:157–61.

Lelovas, P. P., T. T. Xanthos, S. E. Thoma, G. P. Lyritis, and I. A. Dontas. 2008. The laboratory rat as an animal model for osteoporosis research. *Comparative Medicine* 58:424–30.

Logothetis, N. K., J. Pauls, M. Augath, T. Trinath, and A. Oeltermann. 2001. Neurophysiological investigation of the basis of the fMRI signal. *Nature* 412:150–57.

Lovell, D. P. 1986. Variation in pentobarbitone sleeping time in mice. 1. Strain and sex differences. *Laboratory Animals* 20:85–90.

Luu, Y. K., S. Lublinsky, E. Ozcivici, E. Capilla, J. E. Pessin, C. T. Rubin, and S. Judex. 2009. In vivo quantification of subcutaneous and visceral adiposity by micro-computed tomography in a small animal model. *Medical Engineering & Physics* 31:34–41.

Merla, A., and G. L. Romani. 2006. Functional infrared imaging in medicine: A quantitative diagnostic approach. *Conference proceedings: Annual International Conference of the IEEE Engineering in Medicine and Biology Society* 1:224–27.

Ntziachristos, V., C. Bremer, and R. Weissleder. 2003. Fluorescence imaging with near-infrared light: New technological advances that enable in vivo molecular imaging. *European Radiology* 13:195–208.
Nyland, T. G., J. S. Matoon, E. J. Herrgesell, and E. R. Wisner. 2002. Physical principles, instrumentation, and safety of diagnostic ultrasound. In *Small animal diagnostic ultrasound*, eds. T. G. Nyland and J. S. Matoon, 1–8. Philadelphia: Saunders.
Park, J. M., and S. S. Gambhir. 2005. Multimodality radionuclide, fluorescence, and bioluminescence small-animal imaging. *Proceedings of IEEE* 2005:771–83.
Pautler, R. G. 2004. Mouse MRI: Concepts and applications in physiology. *Physiology* (Bethesda) 19:168–75.
Pick, C. G., J. Cheng, D. Paul, and G. W. Pasternak. 1991. Genetic influences in opioid analgesic sensitivity in mice. *Brain Research* 566:295–98.
Prasher, D. C., V. K. Eckenrode, W. W. Ward, F. G. Prendergast, and M. J. Cormier. 1992. Primary structure of the aequorea victoria green-fluorescent protein. *Gene* 111:229–33.
Rosen, H. N., S. Tollin, R. Balena, V. L. Middlebrooks, W. G. Beamer, L. R. Donohue, C. Rosen, A. Turner, M. Holick, and S. L. Greenspan. 1995. Differentiating between orchiectomized rats and controls using measurements of trabecular bone density: A comparison among DXA, histomorphometry, and peripheral quantitative computerized tomography. *Calcified Tissue International* 57:35–39.
Rösler, F. 2009. *Neuroimaging of human memory: Linking cognitive processes to neural systems*, 1st edn. New York: Oxford university press.
Rossouw, D. 2008. Practising applied ethics with philosophical integrity: The case of business ethics. *Business Ethics: A European Review* 17:161–70.
Roth, D. M., J. S. Swaney, N. D. Dalton, E. A. Gilpin, and J. Ross, Jr. 2002. Impact of anesthesia on cardiac function during echocardiography in mice. *American Journal of Physiology. Heart and Circulatory Physiology* 282:H2134–H2140.
Russell, W. M. S., and R. L. Burch. 1959. *The principles of humane experimental technique*. Potters Bar, England: Special edition, Universities Federation for Animal Welfare.
Saha, G. B. 2005. *Basics of PET imaging: Physics, chemistry, and regulations*. New York: Springer.
Sonner, J. M., D. Gong, J. Li, E. I. Eger II, and M. J. Laster. 1999. Mouse strain modestly influences minimum alveolar anesthetic concentration and convulsivity of inhaled compounds. *Anesthesia and Analgesia* 89:1030–34.
Sutton, E. J., T. D. Henning, B. J. Pichler, C. Bremer, and H. E. Daldrup-Link. 2008. Cell tracking with optical imaging. *European Radiology* 18:2021–32.
Thorsen, F., D. Jirak, J. Wang, E. Sykova, R. Bjerkvig, P. O. Enger, A. van der Kogel, and M. Hajek. 2008. Two distinct tumor phenotypes isolated from glioblastomas show different MRS characteristics. *NMR in Biomedicine* 21:830–38.
Torres, L. S., T. L.–W. Norcutt, and A. G. Dutton. 2003. *Basic medical techniques and patient care in imaging technology*, 6th ed. Philadelphia: Lippincott Williams & Wilkins.
Wehrl, H. F., M. S. Judenhofer, S. Wiehr, and B. J. Pichler. 2009. Pre-clinical PET/MR: Technological advances and new perspectives in biomedical research. *European Journal of Nuclear Medicine and Molecular Imaging* Suppl 1:S56–S68.
Weissleder, R. 2001. Scaling down imaging: Molecular mapping of cancer in mice. *Nature Reviews* 2:1–8.
Westbrook, C., and C. Kaut. 1993. *MRI in practice*. Oxford: Blackwell.
Whary, M. T., and J. G. Fox. 2006. Detection, eradication, and research implications of helicobacter infections in laboratory rodents. *Laboratory Animals* (NY) 35,7, 25–36.
Wolf, M., M. Ferrari, and V. Quaresima. 2007. Progress of near-infrared spectroscopy and topography for brain and muscle clinical applications. *Journal of Biomedical Optics* 12:062104.
Xie, Z., D. J. Culley, Y. Dong, G. Zhang, B. Zhang, R. D. Moir, M. P. Frosch, G. Crosby, and R. E. Tanzi. 2008. The common inhalation anesthetic isoflurane induces caspase activation and increases amyloid beta-protein level in vivo. *Annals of Neurology* 64:618–27.
Yang, X. P., Y. H. Liu, N. E. Rhaleb, N. Kurihara, H. E. Kim, and O. A. Carretero. 1999. Echocardiographic assessment of cardiac function in conscious and anesthetized mice. *The American Journal of Physiology* 277:H1967–H1974.
Zacharakis, G., J. Ripoll, R. Weissleder, and V. Ntziachristos. 2005. Fluorescent protein tomography scanner for small animal imaging. *IEEE Transactions on Medical Imaging* 24:878–85.
Zuurbier, C. J., V. M. Emons, and C. Ince. 2002. Hemodynamics of anesthetized ventilated mouse models: Aspects of anesthetics, fluid support, and strain. *American Journal of Physiology. Heart and Circulatory Physiology* 282:H2099–H2105.

13 Anaesthesia and Analgesia

Patricia Hedenqvist, Sweden and
Paul Flecknell, United Kingdom

CONTENTS

Objectives	313
Key Factors	314
13.1 Pre-Anaesthetic Consideration	314
13.2 Anaesthetic Agents	315
13.2.1 Inhalation Anaesthesia	315
13.2.1.1 Volatile Agents	317
13.2.1.2 Methods of Administration	317
13.2.1.3 Operator Safety	318
13.2.2 Injection Anaesthesia	318
13.2.2.1 Injectable Agents	318
13.2.2.2 Methods of Administration	321
13.3 Neuromuscular Blocking Agents (NMBAs)	322
13.4 Assisted Ventilation	323
13.5 Monitoring during Anaesthesia	323
13.5.1 Depth of Anaesthesia	323
13.5.2 Respiratory Function	324
13.5.3 Cardiovascular Function	324
13.5.4 Maintenance of Body Temperature	324
13.5.5 Maintenance of Fluid and Electrolyte Balance	325
13.6 Post-Anaesthetic Care	325
13.6.1 Warmth and Comfort	325
13.6.2 Fluid and Nutritional Support	326
13.7 Analgesia	326
13.7.1 Evaluation of Pain	326
13.7.1.1 Physiological Parameters	326
13.7.1.2 Behaviour	326
13.7.2 Drugs and Doses	327
13.7.2.1 Opioids	327
13.7.2.2 Non-Steroidal Anti-Inflammatory Agents (NSAIDs)	328
13.7.2.3 Local Anaesthetic Agents	328
13.7.3 Methods of Delivery	329
13.8 Conclusions	329
13.9 Questions Unresolved	329
References	329

OBJECTIVES

The use of anaesthesia and analgesia allows for surgery and other procedures to be undertaken in laboratory animals without the distress and pain they would otherwise cause. Anaesthesia is used in a variety of circumstances, for example when disease models are developed, for instrumentation of animals, and for collection of research data. The aims of this section are to outline what is generally accepted as current best practise, and to provide references to more detailed descriptions of specific techniques. Aside from the animal welfare issues mentioned above, providing anaesthesia that is appropriate for a specific research procedure is of considerable importance if meaningful data is to be obtained. Inappropriate selection of anaesthetic agents or failure to provide high standards of pre-, intra-, and post-operative care can all adversely affect the quality of data obtained from research animals. Similarly, the provision of effective post-operative analgesia is important both for reasons of animal welfare and to reduce the potentially major confounding factors caused by unalleviated pain.

A wide range of different anaesthetic techniques is available, and although this document seeks to promote "best practise", this is more often dependent upon making an appropriate and considered choice of anaesthetic, and application of good peri-operative care, rather than requiring the choice of specific anaesthetic agents. Inhalation, injection or local anaesthetic techniques may all be appropriate in particular circumstances and the choice depends on the aim, type and length of the procedure; the animal species; and the particular research objectives.

Analgesic drugs may be used as part of a balanced anaesthetic regimen, to treat post-operative pain or to relieve pain accompanying induced or spontaneous disease. Opioid drugs, non-steroidal anti-inflammatory drugs (NSAIDs), and local anaesthetics may be administered separately, or in combination, to achieve efficient pain relief.

KEY FACTORS

1. Whenever possible, painful procedures should be performed under general or local anaesthesia.
2. Stressful procedures should be performed under anaesthesia or sedation whenever possible.
3. Physiological conditions may be more stable, and reproducible between studies, when using inhalation anaesthesia compared with injection anaesthesia, and this may allow for a reduction of experimental group sizes, and a consequent reduction in animal use.
4. Anaesthetic and analgesic drugs should be chosen to fit the type and length of the procedure and the type and degree of pain expected.
5. Animals that may be at risk of experiencing pain should be examined and treated appropriately.
6. If there is doubt whether an animal is experiencing pain, the animal should be given the benefit of the doubt and treated accordingly.
7. It is not only analgesic drugs that can interfere with research objectives, but also pain itself.

13.1 PRE-ANAESTHETIC CONSIDERATION

Before research animals are anaesthetised they should be acclimatised to the housing conditions in the animal facility. Transport to the facility, and changes to caging, diet, environment and social group all influence the animals' physiology. Most of the consequent physiological changes in the cardiovascular, endocrine, immune and central nervous systems have returned to baseline within 7 days of transportation (Obernier and Baldwin 2006), although some effects may persist for longer periods.

During this period of acclimatisation it is beneficial to accustom the animals to handling and to monitor their body weight. Change in body weight is a good general indicator of health and may be used to monitor recovery after anaesthesia and surgery. Rodents are especially likely to lose body weight rapidly, because of their high-metabolic rate.

In contrast to many other animal species it is not necessary to withhold food from rodents and rabbits before induction of anaesthesia, because of their inability to vomit. Withholding food from small or young animals may even be hazardous, due to induction of hypoglycaemia. Withdrawal of food overnight may also cause loss of body mass, a decrease in blood fatty acid concentration and changes in water intake (Hedrich and Bullock 2004; Suckow, Weisbroth, and Franklin 2006).

Sedative or analgesic agents may be administered before induction of anaesthesia in order to reduce stress, to reduce the anaesthetic drug dose and as part of the management of postoperative pain. Reduction of anaesthetic drug dose also reduces dose-related side effects (see Table 13.1).

The use of pre-anaesthetic medication is often restricted to larger species, where there can be problems providing effective manual restraint. The reduction in stress is also important, however, and species such as rabbits that are easily stressed by handling are best sedated before they are removed from their cage or pen (Flecknell 2009). The majority of small rodents are anaesthetised either using an inhalational agent in an anaesthetic induction chamber, or by using injectable agents for anaesthetic induction. Handling the animal to administer a pre-anaesthetic sedative is therefore not a routine procedure. However, it is important to note that some analgesic agents require 30 minutes or more to achieve full effect, so treatment with these agents as part of pre-anaesthetic medication can ensure that more effective analgesia is provided.

Anticholinergic agents may be administered to prevent excessive salivation, bronchial secretion and bradycardia induced by opioids or vagal stimulation. Atropine is useful except in rabbits and some rat strains, which have high levels of atropine esterase (Olson et al. 1994). Glycopyrrolate is a longer-acting anticholinergic, which does not cross the blood–brain barrier or the placenta (Lemke 2007) and is not inactivated by atropine esterases.

Other pre-anaesthetic measures should include preparing procedures and equipment for intra-operative support; for example, heating pads must be switched on ahead of time so that they reach the required temperature. Emergency drugs must be readily accessible (e.g., antagonists to some of the anaesthetic agents that may be used, IV fluids, adrenaline).

Finally, it is important to keep an anaesthetic record for each animal. This will help identify any potential problems as they develop, and also aid in preparing reproducible protocols for subsequent groups of animals.

13.2 ANAESTHETIC AGENTS

13.2.1 Inhalation Anaesthesia

The benefits of using inhalation anaesthesia are manifold; ease of delivery for induction and maintenance, good control over depth of anaesthesia, low levels of drug metabolism (depending upon the agent used), limited effects on the physiology of the animal, and rapid recovery especially after short periods of anaesthesia. Another benefit is that when oxygen is used as the carrier gas, hypoxia during anaesthesia is prevented. The use of air alone as the carrier gas is not recommended, with the exception of very short procedures (<10 minutes). Almost all anaesthetics reduce lung ventilation and cardiac output, which results in poor tissue oxygenation, and therefore oxygen should be used to a minimum of 30% of the carrier gas (McDonell and Kerr 2007). Using pure oxygen as carrier gas renders a very high arterial blood oxygen partial pressure. More physiological levels are achieved by mixing oxygen with air, nitrogen or nitrous oxide.

One advantage of inhalational agents is that they can be administered in a reproducible and controlled manner. Although this is possible when using injectable agents given by the intravenous route (e.g., propofol + alfentanil) this is technically difficult to accomplish in very small animals such as rodents. The effects of administering injectable agents by the intraperitoneal (IP),

TABLE 13.1
Doses of Commonly Used Pre-Anaesthetic and Emergency Drugs in Rodents and Rabbits

	Drug	Mouse	Rat	Hamster	Guinea Pig	Rabbit
Anticholinergic	Atropine	0.04 mg/kg SC	0.04 mg/kg SC	0.04 mg/kg SC	0.05 mg/kg SC	1–2 mg/kg SC
	Glycopyrrolate	0.5 mg/kg SC				0.1 mg/kg IV, 0.5 mg/kg SC
Sedatives	Medetomidine	30–100 µg/kg, SC, IP	30–100 µg/kg, SC, IP	100 µg/kg, SC, IP		0.1–0.5 mg/kg SC
	Xylazine	5–10 mg/kg IP	1–5 mg/kg IP	5 mg/kg IP		2–5 mg/kg SC
	Midazolam	5 mg/kg IP			5 mg/kg IP, IM	0.5–2 mg/kg IV, SC
Analgesics	Buprenorphine	0.05–0.1 mg/kg SC	0.01–0.05 mg/kg SC		0.05 mg/kg SC	0.01–0.05 mg/kg SC, IV
	Carprofen	5 mg/kg SC	5 mg/kg SC			1.5 mg/kg po
Antagonists	Atipamezole	0.1–1 mg/kg IM, IP, SC, IV				0.1–1 mg/kg IM, SC, IV
	Naloxone	0.01–0.1 mg/kg IV, IM, IP				0.01–0.1 mg/kg IV, IM
Emergency drugs	Adrenaline	0.3ml/kg of 1:10,000 IV or intracardiac				
	Atropine	0.02 mg/kg IV or intracardiac				
	Lidocaine	2mg/kg IV or intracardiac				

Source: Flecknell, P. A., *Laboratory Animal Anaesthesia*, 3rd ed., Elsevier, London, 2009.

intramuscular (IM), or subcutaneous (SC) routes are much more variable, and hence physiological parameters may vary much less during inhalation anaesthesia compared to most injection anaesthesia regimes. This reduction in variation may substantially reduce the number of animals needed in order to achieve identical statistical sensitivity (Chaves, Weinstein, and Bauer 2001).

The potency of volatile anaesthetics is defined as minimum alveolar concentration (MAC). It is calculated as the concentration at which 50% of animals do not respond to a nociceptive (i.e., painful) stimulus and so is equivalent to the ED_{50} of the agent. To ensure most animals are unresponsive, between 1.2 and 1.4 MAC should be given (the ED_{95}). Higher concentrations are not usually needed, except when inducing anaesthesia, 2.0 MAC represents a deep level of anaesthesia and in some cases even an anaesthetic overdose (Steffey and Mama 2007). Pre-medication with tranquillisers, sedatives and opioid drugs, reduces MAC, as does increasing age.

13.2.1.1 Volatile Agents

Although several different agents have been used for laboratory animal anaesthesia, some are no longer available as anaesthetic agents in Europe (e.g., ether and methoxyflurane), and halothane is also no longer being supplied in many member states. Desflurane is used in medical anaesthetic practise, but is rarely used in a research animal setting. The two agents that can be recommended for routine use are isoflurane and sevoflurane.

13.2.1.1.1 Isoflurane

Isoflurane is a halogenated ether and the most commonly used volatile agent in the research setting today because of its high safety and efficiency. The MAC is 1.3–1.5% for isoflurane in the adult rat and mouse, and 2.1% in the rabbit (Steffey and Mama 2007). In 2-day-old Wistar rats, the MAC is 1.9% (Orliaguet et al. 2001). Induction of, and recovery from anaesthesia is very rapid and only 0.2% of the inhaled isoflurane undergoes biotransformation.

Like all volatile agents, isoflurane depresses respiration in a dose-dependent fashion. Cardiac output is preserved at clinically useful concentrations (Steffey and Mama 2007).

13.2.1.1.2 Sevoflurane

Sevoflurane has a lower blood:gas partition coefficient than isoflurane, which means that induction and recovery are even more rapid than with isoflurane. The MAC for sevoflurane is 2.7% in the mouse, 2.4–3% in the rat and 3.7% in the rabbit (Steffey and Mama 2007). The effects on respiration and circulation are similar to those of isoflurane. Approximately 2–3% of inhaled sevoflurane undergoes metabolism.

13.2.1.2 Methods of Administration

13.2.1.2.1 Vaporiser

Calibrated vaporisers are necessary to achieve safe delivery of modern volatile anaesthetics, which can otherwise reach dangerously high concentrations. Two main types of vaporisers can be used for rodent and rabbit anaesthesia: traditional vaporisers, in which the carrier gas passes through the agent to acquire the anaesthetic vapour (Hartsfield 2007), and a more recent type, which uses a syringe driver to inject the liquid anaesthetic into the carrier gas flow (http://www.univentor.com). The latter type allows for very low flows (50–999 ml/min) and is useful for animals with a body weight from 20 to 500 g and permits significant reduction in cost and environmental pollution. Anaesthetic vapour is then delivered to the animal either in an anaesthetic induction chamber or via a face mask. Further details of anaesthetic delivery systems are available in standard anaesthetic texts (Flecknell 2009).

13.2.1.2.2 Induction Chambers

For rodents, induction of inhalation anaesthesia is most conveniently accomplished with the use of a clear Plexiglas induction chamber. First the animal is placed in the chamber, and then the vaporiser is turned on. The gas flow to the chamber should be sufficient to fill up the chamber within a couple of minutes, and the gas should be introduced into the bottom of the chamber (because it is denser than air). A gas outlet is connected to the top of the chamber and connected to the exhaust system. The base of the chamber should be covered with synthetic sheepskin bedding, or paper tissue to soak up any urine during induction. Rodents that are housed together may be placed together in the chamber, as long as floor space is sufficient.

13.2.1.2.3 Mask Delivery

Most inhalational agents have strong odours, and animals may not inhale them from a face mask unless they have been sedated. Following appropriate sedation, anaesthesia can often be induced in a controlled manner using a face mask. Rabbits will breath-hold for extended periods when exposed to volatile agents even after sedation. Although vigorous struggling is prevented by use of sedatives, the procedure is probably distressing for the animal and best avoided (Flecknell and Liles 1996).

13.2.1.2.4 Open Jar

For very short procedures (<1 minute), it is possible to anaesthetise mice using isoflurane in an open jar. At normal room temperature (20°C) and pressure (1 atmosphere), 1 ml of liquid isoflurane will evaporate to approximately 182 ml of vapour. In a 500 ml jar, 0.11 ml of liquid isoflurane will therefore give rise to a vapour concentration of approximately 4%, which is safe for induction. The liquid must be allowed to evaporate and mix evenly in the jar before the mouse is introduced. After induction is complete and the mouse removed, recovery takes less than 60 seconds (personal observation). Use of a jar must take place in a ventilated hood or on a ventilated bench, and the concentration of anaesthetic produced can vary considerably. If a proper calculation is not performed, dangerously high concentration may be produced, so it is preferable to use a vaporiser to provide controlled delivery of the anaesthetic.

13.2.1.2.5 Endotracheal Intubation

Following induction of anaesthesia, animals lose their protective coughing and swallowing reflexes, and as a consequence may inhale material. To maintain and protect the airway, it may be advisable to pass an endotracheal tube. Well-established techniques have been described for most species (mouse: Hastings and Summers-Torres 1999, rat: Yasaki and Dyck 1991, rabbit: Alexander and Clark 1980), and intubation should always be considered, especially for longer procedures (e.g., those lasting for more than 15–30 minutes). Intubation also allows ventilation to be assisted, and the depressant effects of anaesthesia to be corrected. Some anaesthetic drugs depress ventilation more than others (e.g., opioids).

13.2.1.3 Operator Safety

Chronic exposure to low concentrations of volatile anaesthetics may present a risk to human health. Most health and safety agencies recommend that measures are taken to minimise exposure, and a number of systems are available to achieve this (Hartsfield 2007).

13.2.2 INJECTION ANAESTHESIA

13.2.2.1 Injectable Agents

13.2.2.1.1 Ketamine/Medetomidine (K/M)

Ketamine is a dissociative cyclohexamine that is an N-methyl-D-aspartic acid (NMDA) antagonist. It produces good analgesia in non-human primates, but has less effect in rabbits and rodents, in which it causes sedation and immobilisation, but insufficient analgesia for surgery (Green et al. 1981). Some of the side effects are salivation, muscle rigidity and spontaneous movements, which are a disadvantage if it is used as the sole anaesthetic agent. For this reason, ketamine is usually combined with a sedative, for example, medetomidine (an alpha-2-adrenergic agonist), which improves analgesia and muscle relaxation. The combination can be used for invasive procedures of short to medium duration in many species, including rabbits and rats.

The two compounds can be mixed in one syringe and administered IP in rodents and SC in rabbits (Flecknell 2009). The IM injection of K/M in rabbits shows no benefit over SC administration, is seemingly more painful, and therefore best avoided (Hedenqvist et al. 2001). Respiration is significantly reduced by K/M, and provision of supplemental oxygen is strongly recommended (Hellebrekers et al. 1997). In rabbits K/M preserves blood pressure better than the combination ketamine/xylazine (Henke et al. 2005).

Mice and guinea pigs do not consistently reach a plane of surgical anaesthesia when K/M is used (Green et al. 1981; Nevalainen et al. 1989). Female mice of some strains (e.g., Swiss Webster) may require a higher ketamine dose rate as part of the drug combination compared with male mice (Cruz, Loste, and Burzaco 1998) whereas some female rats (e.g., Sprague-Dawley) are more sensitive to K/M than males (Nevalainen et al. 1989).

The effect of medetomidine may be reversed by the administration of atipamezole (an alpha-2-adrenergic antagonist). Female Swiss Webster mice need a higher dose of atipamezole to reverse anaesthesia than male mice (Cruz, Loste, and Burzaco 1998). If surgery has been undertaken an analgesic drug should be administered before reversal. Medetomidine reduces insulin concentration and increases glucose concentration. There is a decrease of blood concentration of antidiuretic hormone and a direct effect on the kidney, so that medetomidine causes substantial fluid loss, which should be corrected by fluid administration (Hedenqvist and Hellebrekers 2003). Since K/M also produce cardiovascular and respiratory depression, reversal with atipamezole is strongly recommended (Flecknell 2009).

13.2.2.1.2 Ketamine/Acepromazine
This combination is useful for producing surgical anaesthesia in rabbits (Flecknell 2009) but may not produce surgical planes of anaesthesia in rodents.

13.2.2.1.3 Ketamine/Xylazine (K/X)
Ketamine may also be combined with the mixed alpha-1/alpha-2-adrenerigc agonist xylazine, to achieve surgical anaesthesia in rabbits and rodents. K/X has been reported to produce surgical anaesthesia in mice both reliably (Erhardt et al. 1984; Arras et al. 2001; Dorsch, Otto, and Hedrich 2004) and unreliably (Green et al. 1981; Buitrago et al. 2008). The differences are probably a consequence of variations in mouse strain, sex and dose, as well as other factors.

Some authors recommend adding the sedative acepromazine to the K/X combination to improve surgical anaesthesia in mice (Arras et al. 2001; Buitrago et al. 2008). Side effects caused by K/X are hypotension, reduction of cardiac output and a reversible oedema of the cornea (Hedrich and Bullock 2004); hypoxia is also produced and oxygen supplementation is recommended. Tissue damage has been reported after IM injection of K/X in rats, marmosets, and hamsters (Davy et al. 1987; Gaertner, Boschert, and Schoeb 1987; Smiler et al. 1990).

In guinea pigs, K/X unreliably produces surgical anaesthesia and its effectiveness may be improved by local anaesthetic infiltration or a small dose of isoflurane. Like ketamine/medetomdine, K/X produces polyuria and an increase in intraocular pressure. Atipamezole administration is recommended to reverse the effect of xylazine at the end of surgery. With or without acepromazine, K/X has been shown to cause nerve injury when administered intramuscularly to rabbits (Beyers, Richardson, and Prince 1991; Vachon 1999), whereas intravenous injection is effective and safe (Green et al. 1981). Analgesia is sometimes insufficient for surgery in pigmented rabbits given K/X and hypotension more pronounced than during K/M anaesthesia (Henke et al. 2005). In NZW rabbits on the other hand, surgical anaesthesia is often achieved (Green et al. 1981; Lipman, Marini, and Erdman 1990).

13.2.2.1.4 Fentanyl/Fluanisone/Midazolam

The combination of the opioid agonist (fentanyl) and the sedatives fluanisone and midazolam is perhaps the safest and most useful alternative to K/M for producing surgical anaesthesia in rodents and rabbits. To decrease the time to recovery, the effect of fentanyl can be reversed by administration of a mixed opioid agonist/antagonist such as butorphanol or buprenorphine, while still retaining analgesia. Fentanyl/fluanisone is sold under the trade name Hypnorm, and is at times difficult to acquire from retailers because it is produced only in low volumes.

When combining Hypnorm with midazolam, the two must first be mixed with water for injection, otherwise crystallisation may occur. The mixture can be administered by the IP or the SC routes in rodents. If the SC route is used, the dose needs to be approximately one-third lower than the IP dose, because immediate hepatic metabolism does not take place.

In rabbits Hypnorm is best administered by the SC route first, and after the rabbit is sedated (after 5–10 minutes), midazolam can be administered intravenously to effect (Flecknell 2009).

13.2.2.1.5 Propofol

Propofol is an alkylphenol that must be administered by the intravenous route to be effective, due to a high rate of metabolism in the liver. Maintenance of anaesthesia requires continuous intravenous infusion. After anaesthesia has been induced with a propofol bolus IV, the animal can be intubated or placed on a face mask, and anaesthesia maintained with a volatile agent. Slow injection is necessary at induction to avoid apnoea. Propofol undergoes rapid hepatic metabolism, which allows for very fast recovery once the infusion is stopped. High doses of propofol are needed to allow invasive procedures to be undertaken, so it is best used combined with an analgesic agent such as an opioid. If a potent opioid is used, intubation and mechanical ventilation are necessary, because of respiratory depression. Attempts have been made to administer a combination of propofol and an opioid to mice, by IP injection, but the anaesthesia produced was unreliable and associated with some mortality (Alves et al. 2007, 2009).

13.2.2.1.6 Barbiturates

Ultra-short acting barbiturates such as thiopental, may be used to induce anaesthesia by the intravenous route, to allow for maintenance of anaesthesia with volatile agents.

The medium long-acting barbiturate pentobarbital may be used to induce sleep in rodents after IP injection, but cannot safely be used to produce surgical anaesthesia. The dose that produces surgical planes of anaesthesia is dangerously close to the lethal dose and causes severe respiratory and cardiovascular depression (Skolleborg et al. 1990). Pentobarbital can be useful for non-survival surgical procedures, when administered by IV infusion and in combination with an opioid agonist (e.g., fentanyl) to achieve good analgesia. Mechanical ventilation is necessary because of severe respiratory depression. Pentobarbital is not suited for survival procedures, because it causes prolonged sedation from which there is very slow recovery. Barbiturates are not safe to use for any types of procedures in rabbits.

The long-acting barbiturate thiobutabarbital may be administered IP to produce prolonged anaesthesia in rats. It may be indicated in some diabetes research, because it causes less effect on blood glucose concentration than other anaesthetics (Hindlycke and Jansson 1992).

13.2.2.1.7 Local Anaesthetics

The use of local anaesthetics can be very beneficial to improve analgesia during and after surgery. For example, it may prevent some sensitisation of the nociceptive pathways, resulting in less postoperative pain (Skarda and Tranqulli 2007). The agents may be administered by infiltration of tissue, and for regional, epidural or spinal blocks. Shorter acting (e.g., lidocaine) and longer acting (e.g., bupivacaine) local anaesthetics are available, and the duration of action of the short-acting agents may be increased by the addition of adrenaline. However, adrenaline use is contraindicated in peripheral body parts (ears, nose, digits, penis), because of the risk of ischemia-induced necrosis.

Anaesthesia and Analgesia

The use of local anaesthesia to supplement general anaesthesia has several advantages. In old or debilitated animals, the general anaesthetic dose can be kept to a minimum and side effects thereby reduced. In small rodents a satisfactory plane of surgical anaesthesia is sometimes difficult to achieve with injectable anaesthetics, and the addition of a local anaesthetic can help. Local anaesthetic cream (EMLA) can be used to reduce pain caused by venipuncture in rabbits, cats, dogs and pigs (Flecknell 2009).

Maximum safe doses are similar in all species, and overdose may cause CNS and cardiovascular toxicity.

13.2.2.2 Methods of Administration

A variety of different routes can be used to administer anaesthetic agents by injection (Table 13.2).

13.2.2.2.1 Intravenous

Intravenous injection of anaesthetic agents is easily accomplished in larger species, but is more difficult in small rodents. An IV injection gives immediate effect (no absorption phase) and allows for

TABLE 13.2
Dose and Route of Adsministration of Some of the More Commonly Used Anaesthetic Agents

Animal Species	Anaesthetic Protocol	Dose and Method of Administration	Specific Remarks
Mouse	Fentanyl/Fluanisone + midazolam	10 ml/kg IP or 5–7ml/kg SC of a 1:1:2 mixture of Hypnorm, midazolam and water for injection	Drugs injected IP may partly undergo immediate hepatic metabolism, unlike drugs injected SC. Therefore a higher dose may be needed for IP than SC injection to achieve the same effect.
Rat	Fentanyl/Fluanisone + midazolam	2.7 ml/kg IP or 1.5–2 ml/kg SC of a 1:1:2 mixture of Hypnorm, midazolam and water for injection	See comment for mouse
	Ketamine + medetomidine	60–75 mg/kg + 0.25–0.5 mg/kg IP	
Guinea pig	Fentanyl/Fluanisone + midazolam	8 ml/kg IP of a 1:1:2 mixture of Hypnorm, midazolam and water for injection	
	Ketamine + medetomidine	40 mg/kg + 0.5 mg/kg IP	Avoid IM injection, can cause pain and muscle damage
Hamster	Fentanyl/Fluanisone + midazolam	4 ml/kg IP of a 1:1:2 mixture of Hypnorm, midazolam and water for injection	
	Ketamine + medetomidine	100 mg/kg + 0.25 mg/kg IP	
Rabbit	Fentanyl/Fluanisone + midazolam	0.3 ml/kg SC + 2 mg/kg SC or IV	Premedication SC with fentanyl/fluanisone is followed by IV injection of midazolam to effect
	Ketamine + medetomidine	15 mg/kg + 0.25 mg/kg SC	Rapid absorption after SC administration, no difference from IM injection in effect or duration
	Propofol	10 mg/kg injection bolus, 0.2–0.6 mg/kg/min infusion	Must be administered IV as an injection or infusion due to rapid hepatic metabolism

dosing to effect according to an individual response, unlike the situation with IM, SC or IP injection. Continuous intravenous infusion of short-acting drugs such as propofol, sufentanil or remifentanil enable accurate control over anaesthetic depth by adjustment of the infusion rate. Establishing intravenous access is also beneficial if emergency drugs need to be administered.

An IV injection/infusion also allows for rapid buffering of anaesthetic solutions that are acidic (ketamine) or alkaline (barbiturates), which, if injected by another route, may give rise to tissue damage and pain (Smiler et al. 1990; Branson 2001). Propofol can cause local pain upon intravenous injection, but the pain is minimised by premedication with an opioid or alpha-2-agonist (Branson 2001).

13.2.2.2.2 Intramuscular

This route of administration may be used in rabbits and larger species, but should be avoided in rodents, because of the risk of tissue damage. An IM injection of K/X may cause tissue damage and pain even in larger species (Davy et al. 1987; Gaertner, Boschert, and Schoeb 1987; Smiler et al. 1990; Beyers, Richardson, and Prince 1991).

13.2.2.2.3 Intraperitoneal

This injection route is commonly used in rodents because intravenous access may be difficult and they have a small muscle mass. The peritoneal cavity is richly vascularised and drug uptake is rapid after injection of small volumes. Part of the injected solution will be transported via the hepatic portal system to the liver before reaching the systemic blood circulation, which results in high first pass hepatic metabolism. A risk with IP injections is that part or all of the injected solution may be deposited in the gut or intra-abdominal fat and not be effective.

Some anaesthetic agents such as tribromoethanol or chloral hydrate cause inflammation and pain upon IP injection and therefore are best avoided (Vachon et al. 2000; Lieggi et al. 2005).

13.2.2.2.4 Subcutaneous

Subcutaneous administration of anaesthetics can be an alternative to IM or IP injection. The absorption rate of small volumes is often not very different between the routes and SC injection is seemingly less stressful or painful than IM injection in rabbits (Hedenqvist, Roughan, and Flecknell 2000). Anaesthetics that may be administered subcutaneously include ketamine/medetomidine in rabbits and rodents and fentanyl/fluanisone/midazolam in rats. The SC dose of the latter combination is approximately one-third of the IP dose (personal observations).

13.3 NEUROMUSCULAR BLOCKING AGENTS (NMBAS)

The NMBAs work by binding to the nicotinic acetylcholine receptor at the neuromuscular junction and block nerve conduction, thereby paralysing the animal (Martinez 2007). For ethical reasons, the use of NMBAs should be restricted to procedures for which they are absolutely necessary. They should not be used to increase muscle relaxation on a routine basis or to prevent the animal from breathing against the respirator. Greater muscle relaxation can be achieved by increasing the anaesthetic depth, and the respiratory drive can be depressed by hyperventilating the animal. Unlike the case in humans, animals are generally easy to intubate without the use of NMBAs.

If NMBAs need to be used, a number of steps must be taken to eliminate the risk of animals being insufficiently anaesthetised during paralysis. It is advisable to use doses of NMBAs that only partly paralyse the animal and thus still allow for movement in response to nociceptive stimulation. It is also advisable to allow the effects to subside before administering additional doses, in order to evaluate anaesthetic depth using withdrawal reflexes at regular intervals. Animals also need to be monitored for changes in blood pressure and heart rate, and inhalation and infusion devices should be equipped with alarms to indicate lack of function. Respiratory function must also be monitored (e.g., pulse oximetry and end tidal CO_2). Further, a familiar anaesthetic regimen that allows for a stable plane of anaesthesia should be used and stability must be established before the NMB is administered. Examples of agents available for use in laboratory animals are provided in Table 13.3.

13.4 ASSISTED VENTILATION

It might be necessary to artificially ventilate an animal to support respiration and maintain near normal physiological function, particularly during long-term anaesthesia. Assisted ventilation may also be needed during imaging, to synchronise respiratory movements with data collection (see Chapter 12).

If the animal is to recover consciousness, an endotracheal tube needs to be inserted, whereas for non-recovery procedures, a tracheostomy can be carried out. The animal can then be connected to a suitable ventilator. Tubes and intubation techniques must be refined and optimised to avoid trauma to the delicate structures in the pharynx and trachea and to minimise dead-space. The ventilator must be capable of delivering the volumes and frequencies appropriate for the species. During intermittent positive pressure ventilation (IPPV) air is forced into the thorax. This causes a positive pressure in the thorax in contrast to the negative pressure during spontaneous inhalation, which normally supports venous filling of the heart. Artificial ventilation therefore also affects the circulatory system. This effect can be reduced by allowing a long expiratory phase, and short period for inspiration, with an inspiration:expiration ratio of 1:3–1:4.

It is not necessary to use NMB agents in order to mechanically ventilate an animal, although a case can sometimes be made for their use during non-invasive imaging (Chapter 12). If an animal breathes spontaneously, the mechanical ventilator rate should be increased to produce moderate hypocapnia. The animal will usually stop making respiratory efforts at this point. The rate can then be reduced slowly, until normocapnia (measured using a capnograph) is achieved, if necessary.

13.5 MONITORING DURING ANAESTHESIA

Anaesthesia poses a risk to the animal's vital functions and consequently some form of monitoring must take place. Vital signs that need to be monitored include respiration, circulation, body temperature, acid–base status, kidney function and level of anaesthesia. In brief, low-risk procedures, monitoring may be simple, whereas long-duration, invasive and high-risk procedures require more sophisticated monitoring.

13.5.1 Depth of Anaesthesia

The depth of anaesthesia may be described as light, medium or deep. Light anaesthesia (immobility) involves loss of consciousness, as indicated by loss of the righting reflex. Muscular tone and response to noxious stimulation are gradually lost with increasing depth, together with a number of reflexes, forming a pattern that differs between animal species as well as the anaesthetic agents

TABLE 13.3
Neuromuscular Blocking Drugs for Use in Small Laboratory Animals

	Mouse	Rat	Guinea Pig	Rabbit
Alcuronium	—	—	—	0.1–0.2 mg/kg IV
Atracurium	—	—	—	—
Pancuronium	—	2 mg/kg IV	0.06 mg/kg IV	0.1 mg/kg IV
Tubocurarine	1 mg/kg IV	0.4 mg/kg IV	0.1–0.2 mg/kg IV	0.4 mg/kg IV
Vecuronium	—	0.3 mg/kg IV	—	—

Source: Flecknell, P. A., *Laboratory Animal Anaesthesia*, 3rd ed., Elsevier, London 2009.

used. Before surgery is undertaken, reactions to noxious stimulation such as toe-pinch (rat), ear-pinch (rabbit), or tail-pinch (mouse) must be absent. At a surgical plane of anaesthesia, noxious stimulation should cause only minimal changes in respiratory rate, heart rate or blood pressure (typically less than 10–15%).

Parameters and measurements that may help indicate the level of anaesthesia are the degree of muscle relaxation, the pattern and depth of respiration, and the heart rate and blood pressure. A change to marked abdominal movements with each breath signals a deeper level of anaesthesia in rodents. With lighter levels of anaesthesia, respiration rate, heart rate and blood pressure usually increase.

13.5.2 Respiratory Function

Respiratory function can be assessed by clinical observation of the respiratory rate and pattern. With severe respiratory depression, the skin and mucous membranes turn visibly cyanotic (blue or purple). Blood oxygenation, a function of respiration, can be monitored more accurately with a pulse-oximeter or by arterial blood gas analysis.

A pulse-oximeter measures the oxygenation of haemoglobin in arterial blood by placing a sensor across a capillary bed (e.g., across a digit, or the tongue, tail, or ear). In an awake healthy animal breathing room air, haemoglobin saturation is >95%. During anaesthesia, the aim is to keep oxygenation levels over 90%, which can usually be achieved by providing supplemental oxygen. Specialist instruments suitable for use even in very small animals are now available. Although a pulse-oximeter enables the degree of oxygenation to be assessed, no indication of carbon dioxide concentration or pH is obtained. Accurate measures of all three parameters can be obtained using a blood gas analyser, but this requires sampling of arterial blood. A good measure of blood and tissue carbon dioxide concentration is provided by measuring respiratory end tidal CO_2 concentration using a capnograph. Further details of monitoring devices can be found in standard veterinary anaesthetic text books (e.g., Flecknell 2009).

13.5.3 Cardiovascular Function

Measurements of heart rate and rhythm, blood pressure, ECG and capillary refill time are helpful in assessing cardiovascular function. Heart rate can be measured by palpating a peripheral pulse, or palpating the heart in small animals. Bradycardia may be caused by anaesthetic overdosage, opioids, alpha-2-agonists, vagal stimulation, hypothermia or hypoxia (Haskins 2007). Parasympatholytic agents (atropine, glycopyrrolate) may be used to correct bradycardia due to opioids or vagal stimulation. Tachycardia can be caused by hypovolemia and hyperthermia, or be a response to surgical stimulation in lightly anaesthetised animals. A simple estimate of peripheral perfusion can be obtained from the capillary refill time. The gums are usually the most accessible site, and the refill of capillaries following blanching by digital pressure can be observed in most larger species. In normal animals, following blanching by pressing with a finger, the mucous membranes regain their normal colour in less than a second. If refill is significantly delayed (> 1 second), it indicates poor peripheral tissue perfusion and possible circulatory failure.

Mean arterial blood pressure can be measured invasively or non-invasively, and should be maintained above 60–70 mm Hg, to ensure adequate tissue oxygenation and renal perfusion. Hypotension can be caused by hypovolemia, poor cardiac output, or vasodilation and hypertension by hyperthermia, renal failure, and light planes of anaesthesia.

The electrical activity of the heart can be measured non-invasively using an ECG. It should be remembered that ECG records the electrical activity of the heart and not mechanical performance.

13.5.4 Maintenance of Body Temperature

Anaesthesia induces heat loss by reducing metabolism and muscular activity and interference with hypothalamic thermostatic mechanisms. Many anaesthetics also cause peripheral vasodilatation, further increasing heat loss. Body temperature falls much faster in small animals than in large animals, because of their larger body surface to body weight ratio. Clipping fur, disinfecting the skin, contact with cold surfaces, opening of body cavities, and injection of cold fluids are actions that contribute to the development of hypothermia. Hypothermia may lead to over-dose of anaesthetic agent and prolonged recovery, because a drop in body temperature reduces anaesthetic need. Blood pressure and cardiac output fall during hypothermia, whereas peripheral vascular resistance increases (Branson 2001). Severe hypothermia may lead to cardiac failure caused by ventricular fibrillation or cardiac arrest.

To prevent hypothermia, warming must be initiated as soon as the anaesthetic drugs start to take effect. Rodents are best placed in a heating chamber (26–28°C) after administration of injectable anaesthetics, or on a heating pad after induction of inhalation anaesthesia. Heating must be continued during anaesthesia and recovery, until the animal is fully awake. At the same time care must be taken not to burn or overheat the animal. Best practise is to use a thermostatically regulated heating pad, which may be connected to a rectal thermometer. In any event, body temperature must be monitored throughout anaesthesia.

13.5.5 Maintenance of Fluid and Electrolyte Balance

Fluid and electrolyte imbalances can arise during and after anaesthesia and surgery. Body fluids may be lost as a result of haemorrhage and evaporation from the surgical site or expired air (Seeler 2007). Fasting before surgery and reduction of food and water intake after surgery additionally threaten homeostasis. The use of alpha-2-adrenergic agonists (e.g., medetomidine) contributes to loss of body fluids by causing diuresis and polyuria (Meyer and Fish 2008). Imbalances can occur in the tissues as well as the vascular space and need to be counteracted to maintain cardiovascular stability and proper tissue perfusion.

Fluids may be administered orally, subcutaneously, intraperitoneally, and in emergency, intravenously. Lactated Ringer's solution is an isotonic balanced electrolyte solution that may be used for maintenance or replacement purposes, and is better suited than isotonic saline (0.9%), which generally does not restore the animal's electrolyte requirements (Seeler 2007). For minor surgery, administration of 2.5 ml/kg/h of fluid SC or IV is recommended, whereas for severely traumatic procedures 10–15 ml/kg/h can be estimated. For blood loss, an extra volume of up to five times the estimated volume lost should be administered.

13.6 POST-ANAESTHETIC CARE

13.6.1 Warmth and Comfort

In survival studies, care must be taken to continue warming the animal until it has fully recovered from anaesthesia. For rodents, incubation chambers that can hold cages are preferable. Larger animals should be placed in rooms with increased temperature and covered with blankets. If the effect of any of the anaesthetic drugs can be reversed with an antagonist, recovery will be more rapid. Care must always be taken to provide analgesia if any post-procedure pain can be expected.

Rodents should not be placed directly on sawdust or wood shaving bedding during recovery, but in a cage with paper or tissue towels or synthetic sheep skin (e.g., Drybed). Once the animal becomes ambulatory it may be returned to its home cage, and if needed, food should be placed on the cage floor. To facilitate food intake, food pellets may be crushed and soaked in water and presented in a petri dish on the cage floor. Soft paper may also be placed in the home cage. Animals that are housed in familiar groups should be returned to the group after recovery. A familiar surrounding is helpful in reducing fear and stress, as are subdued lighting and low noise levels.

13.6.2 FLUID AND NUTRITIONAL SUPPORT

Animals that are dehydrated need supportive fluids given either by mouth or subcutaneously, intraperitoneally, or intravenously. Hypoglycemia should be corrected with glucose administered orally or IP (rodents) or dextrose solutions IV. Fluids should always be warmed to body temperature before being administered.

Rabbits have very sensitive intestinal tracts and may suffer from ileus (gut stasis) if eating does not resume soon after recovery from anaesthesia. If the rabbit does not eat voluntarily, it must be fed with a liquidised food replacement (e.g., critical care formula for rodents and rabbits) through a syringe. Canned baby food such as mashed carrots or mashed banana may also be fed in small amounts to stimulate appetite. Animals of other species need nutritional support if eating is not resumed soon after recovery. The need for (additional) analgesic treatment should always be considered in animals that lack appetite.

13.7 ANALGESIA

Alleviating pain reduces suffering, which is one of the most important aims when working with laboratory animals. Analgesic treatment in conjunction with surgery not only reduces suffering but improves recovery and reduces morbidity and mortality. Surgery is associated with neuroendocrine, metabolic, and immune alterations resulting from tissue damage, anaesthesia, and psychological stress (Shavit, Fridel, and Beilin 2006). Preoperative administration of analgesics provides more effective pain relief and may reduce the anaesthetic dose needed (Flecknell 2009). For the most effective pain relief, different drugs may be combined, for example, an opioid, a NSAID, and a local nerve block during an intervention, followed by repeated opioid and NSAID administration in the immediate postoperative period, and finally NSAID administration alone when pain is less severe. For doses of the most commonly used analgesic drugs see Table 13.4.

13.7.1 EVALUATION OF PAIN

13.7.1.1 Physiological Parameters

Physiological responses generally arise from changes in the sympatho-adrenal and the hypothalamic–pituitary–adrenal systems. Changes in heart and respiration rate, blood pressure and stress hormone concentrations in blood may indicate the presence of pain (Kent and Molony 2009). However, many of these changes may also occur in response to handling, eating, exercise and a number of stress factors. The link between pain scores and heart rate or respiratory rate was shown to be weak in dogs after orthopaedic surgery (Holton et al. 1998). Plasma cortisol/cortisone is elevated after surgery in calves and lambs, and is reversed by analgesic treatment (Kent, Molony, and Robertson 1993; Stafford et al. 2002), but appears to show a ceiling effect that suggests that the assessment of the relative severity of intense pain cannot be achieved. The association with chronic pain in sheep is low (Ley, Livingston, and Waterman 1991; Ley et al. 1994). In mice, faecal corticosterone concentrations were elevated after vasectomy and could be reduced by meloxicam treatment (Wright-Williams et al. 2007). Factors that may limit the value of corticosterone measurements for assessment of pain include individual variation, circadian changes, age and breed effects, and a variety of alternative stressors both pleasurable and stressful that activate the HPA-system (Kent and Molony 2009). These measures are useful only as research tools, because they do not allow rapid assessment of animals and administration of analgesic to control pain effectively.

13.7.1.2 Behaviour

Both spontaneous and evoked behaviour may be useful to assess pain in animals. Spontaneous behaviours include changes in posture, activity and vocalisation and evoked behaviours include reactions to handling and threshold testing to mechanical, chemical and thermal stimulation

(Kent and Molony 2009). Ongoing pain has been shown to elicit pain-related behaviour, which is species-specific and procedure-related. Rats, for example, show an increased frequency of back-arching and writhing after abdominal surgery (Roughan and Flecknell 2001) and similar behaviours can be observed in mice (Wright-Williams et al. 2007) and rabbits (Leach et al. 2009). Guarding of injured areas may also be present, for example guarding the hind foot after sciatic nerve lesion (Bennett and Xie 1988). These behaviours are reduced in frequency by treatment with analgesic drugs, which indicates that they may be related to pain.

Measurements of body weight and food and water intake have been proposed as indicators of post-operative pain and the efficacy of analgesic therapy (Liles et al. 1998). These latter measures are objective, but they are retrospective measures and so cannot be used to modify analgesic therapy for a particular animal. They can, however, be used as a simple measure of post-operative recovery, and as a means of adjusting future analgesic regimens for similar animals undergoing similar surgical procedures.

Unfortunately, there are few well-described and fully validated pain assessment techniques for laboratory animals. Those schemes that have been described (e.g., Roughan and Flecknell 2003) relate to particular types of surgery, so in many circumstances pain is difficult to assess. It is important to appreciate that the signs of pain in many animals are subtle, and quite difficult to detect, even by an experienced observer. It is therefore safest to assume that some pain will be present after any surgical procedure, and that analgesics will be required. Initial dosing with any of the analgesics described above will rarely cause undesirable side effects, and the positive effects on recovery from the procedure provide a strong justification for their routine use. What is also difficult, however, is to determine how long analgesic treatment should be continued. Prolonged treatment with opioids, especially when these are given at high doses, can have detrimental effects including reduction in food and water intake. At present, the following advice is offered:

1. After any major surgical procedure (e.g., laparotomy, thoracotomy, craniotomy), administer at least one dose of opioid analgesic (e.g., buprenorphine), combined with a single dose of NSAID. Less invasive procedures (e.g., vessel cannulation) probably require only a single dose of opioid or a single dose of NSAID.
2. Assess the animals as carefully as possible. Spend time assessing their normal behaviour pre-operatively, so that post-operative changes in behaviour can be identified. Monitor body weight and food and water consumption. If animals are failing to gain weight after 24 hours, administer a second dose of analgesic and if this produces an improvement, adopt this for all future procedures of this type.
3. Regularly review pain assessment and pain management schemes so that they can be updated as new information is published.

13.7.2 Drugs and Doses

13.7.2.1 Opioids

Opioids provide the most effective way of controlling pain caused by trauma or surgery. Pure mu opioid agonists (morphine, fentanyl, sufentanil, alfentanil) can be used intra-operatively as part of the anaesthetic regime and post-operatively by repeated injections (morphine) or IV infusion (fentanyl, sufentanil). These drugs may depress respiration, so mechanical ventilation is often necessary during anaesthesia and sometimes oxygen supplementation must be provided during recovery. Pure opioid agonists have no ceiling effect, which means that the higher the dose, the more effective the pain relief. Side effects (sedation, respiratory depression and reduced gastrointestinal peristalsis) are more severe with administration of pure agonists such as morphine than after partial opioid agonists such as buprenorphine or butorphanol, but these latter drugs do have a ceiling effect.

TABLE 13.4
Analgesics for Use in Small Laboratory Animals. Note that These are Only Suggestions Based on Clinical Experience and the Limited Published Data Which is Available. Dose Rates Should be Adjusted Depending upon the Clinical Response of the Animal

Analgesic	Mouse	Rat	Hamster	Guinea Pig	Rabbit
Buprenorphine	0.05–0.1mg/kg SC 8–12 hourly	0.05mg/kg SC 8–12 hourly	0.1mg/kg SC 8–12 hourly	0.05mg/kg SC 8–12 hourly	0.01–0.05mg/kg SC 6–12 hourly
Carprofen	5mg/kg SC uid	5mg/kg SC uid		4mg/kg SC uid	1.5mg/kg per os uid
Meloxicam	5mg/kg SC uid	1mg/kg SC uid		0.3mg/kg SC uid	0.6–1mg/kg SC uid
Ketoprofen	5mg/kg uid SC	5mg/kg uid SC			3mg/kg uid SC
Morphine	2.5mg/kg SC or IM 4 hourly	2.5mg/kg SC or IM 4 hourly		2–5mg/kg SC or IM 4 hourly	2–5mg/kg SC or IM 4 hourly

Source: Data adapted from Flecknell and Waterman-Pearson, 2000
Note: uid = once daily

Buprenorphine has a slow onset of action and reaches its peak about 60 minutes after SC injection (Dobromylskyj et al. 2000). The duration of action is relatively long, 6–8 hours, which is beneficial when treating postoperative pain. If administered before induction of anaesthesia, it reduces the need for isoflurane by approximately 20%. If a pure opioid agonist (e.g., fentanyl) is used as part of the anaesthetic regime, buprenorphine should not be administered beforehand, but may instead be used to reverse the effects of the pure agonist and provide postoperative pain relief. This will reduce the time to recovery.

Care must be taken if administering buprenorphine before K/M anaesthesia in rats, because this has been shown to increase mortality (Hedenqvist, Roughan, and Flecknell 2000) and it is advisable to administer this analgesic during recovery from this anaesthetic regime. For doses see table 13.4.

13.7.2.2 Non-Steroidal Anti-Inflammatory Agents (NSAIDs)

Non-steroidal, anti-inflammatory drugs are useful for treating mild to moderate post-operative pain. For more effective pain control, they can be combined with an opioid and/or a local anaesthetic drug. The following NSAIDs have been shown to have positive effects on behaviour and recovery after surgery in rodents: ibuprofen (Hayes et al. 2000), ketoprofen and carprofen (Cabre et al. 1998; Roughan and Flecknell 2001), meloxicam (Roughan and Flecknell 2004), and flunixine meglumine (Stewart and Martin 2003). Carprofen has found widespread use for post-operative pain control due to its long-lasting effect (24 hours) as well as low gastric and renal toxicity. Meloxicam has a similar duration of action and is palatable to rodents and non-human primates (Flecknell 2009). For doses see table 13.4.

13.7.2.3 Local Anaesthetic Agents

Local anaesthetics block sodium channels and thereby nerve conduction. Sensory nerves are more sensitive than motor nerves to their effect. Local anaesthetics may be administered topically on the skin, mucous membranes, eyes and ears, infiltrated into tissues by injection, deposited in the area of a nerve (local block), in body cavities or in the epidural or subarachnoidal space (epidural or spinal block). Lidocaine has a shorter duration of action, which may be increased by the addition of adrenaline. The combination with adrenaline must not be used in the peripheral body parts (e.g., tail, toes, ears), or irreversible ischemic damage may be caused. Bupivacaine has a longer duration of action than lidocaine, because it is more lipophilic in nature. The maximum doses are similar in all species (4 mg/kg for lidocaine and 2 mg/kg for bupivacaine). Toxicity results in CNS and cardiac effects. For doses see table 13.4.

13.7.3 METHODS OF DELIVERY

Even though administration of ibuprofen in the drinking water has been shown to be effective in mice after surgery (Hayes et al. 2000), this route of administration cannot be recommended. Animals often have highly variable water consumption after a surgical procedure and intake of a drug is very unpredictable. The more pain the animal experiences, the less it may drink. In addition, rodents may only drink during the dark phase of the photoperiod, so there may be a prolonged period during which no analgesic is ingested. Opioids and NSAIDs may be administered orally using a syringe, or mixed with palatable food (e.g., fruit jelly) and ingested voluntarily. Voluntary ingestion is likely to be more successful if the food or jelly has been presented before the surgical procedure. The dose must be adjusted accordingly to account for the first pass hepatic effect that occurs after oral administration. Generally, more reproducible and reliable results are obtained by administering analgesics by injection.

13.8 CONCLUSIONS

Standards of laboratory animal anaesthesia have increased dramatically over the last decade. Research workers are now more aware of the potential interactions between different anaesthetic agents and their animal models. They are also more aware of the problems that poor anaesthetic practise can cause, and the need to maintain high standards of peri-operative care. Anaesthetic regimens are now more often reported in more detail in the materials and methods sections of papers, and are also being published as short papers in specialist journals. These trends should be encouraged, since successful establishment of an animal model developed in another laboratory often requires careful attention to all aspects of the research protocol, including the anaesthetic methodology. As validated pain scoring systems become established for different species, and efficacy data for different analgesics is obtained, pain management will improve.

13.9 QUESTIONS UNRESOLVED

Despite the improvements that have been made in anaesthetic practise, older anaesthetic agents such as barbiturates are still widely used—sometimes appropriately, but in many instances they would be better replaced with more modern anaesthetic agents. It is essential that research workers consider the anaesthetic and peri-operative care procedures as an integral part of their research protocol.

Post-operative analgesic use is becoming more widespread, but the appropriate use of these agents is still limited by our poor ability to assess pain accurately in many laboratory species. Concerns related to the potential side effects and interactions of analgesics with particular research protocols need to be addressed by a careful and critical review of the relevant literature. In many instances, potential effects are likely to be of significance only when analgesics are administered at high-dose rates, for prolonged periods of time (Flecknell 2009).

REFERENCES

Alexander, D. J., and G. C. Clark. 1980. A simple method of oral endotracheal intubation in rabbits (*Oryctolagus cuniculus*). *Laboratory Animal Science* 30:871–73.

Alves, H. C., A. M. Valentim, I. A. Olsson, and L. M. Antunes. 2007. Intraperitoneal propofol and propofol fentanyl, sufentanil and remifentanil combinations for mouse anaesthesia. *Laboratory Animals* 41:329–36.

Alves, H. C., A. M. Valentim, I. A. Olsson, and L. M. Antunes. 2009. Intraperitoneal anaesthesia with propofol, medetomidine and fentanyl in mice. *Laboratory Animals* 43:27–33.

Arras, M., P. Autenried, A. Rettich, D. Spaeni, and T. Rulicke. 2001. Optimization of intraperitoneal injection anesthesia in mice: Drugs, dosages, adverse effects, and anesthesia depth. *Comparative Medicine* 51:443–56.

Bennett, G. J., and Y. K. Xie. 1988. A peripheral mononeuropathy in rat that produces disorders of pain sensation like those seen in man. *Pain* 33:87–107.
Beyers, T. M., J. A. Richardson, and M. D. Prince. 1991. Axonal degeneration and self-mutilation as a complication of the intramuscular use of ketamine and xylazine in rabbits. *Laboratory Animal Science* 41:519–20.
Branson, K. R. 2001. Injectable anaesthetics. In *Veterinary pharmacology and therapeutic*, ed. H. R. Adams, 213–67, 8th ed. Ames: Iowa State Press.
Buitrago, S., T. E. Martin, J. Tetens-Woodring, A. Belicha-Villanueva, and G. E. Wilding. 2008. Safety and efficacy of various combinations of injectable anesthetics in BALB/c mice. *Journal of the American Association for Laboratory Animal Science* 47:11–17.
Cabre, F., M. F. Fernandez, L. Calvo, X. Ferrer, M. L. Garcia, and D. Mauleon. 1998. Analgesic, antiinflammatory, and antipyretic effects of S(+)-ketoprofen in vivo. *Journal of Clinical Pharmacology* 38:3S–10S.
Chaves, A. A., D. M. Weinstein, and J. A. Bauer. 2001. Non-invasive echocardiographic studies in mice: Influence of anesthetic regimen. *Life Sciences* 69:213–22.
Cruz, J. I., J. M. Loste, and O. H. Burzaco. 1998. Observations on the use of medetomidine/ketamine and its reversal with atipamezole for chemical restraint in the mouse. *Laboratory Animals* 32:18–22.
Davy, C. W., P. N. Trennery, J. G. Edmunds, J. F. Altman, and D. A. Eichler. 1987. Local myotoxicity of ketamine hydrochloride in the marmoset. *Laboratory Animals* 21:60–67.
Dobromylskyj, P., P. A. Flecknell, B. D. Lascelles, P. J. Pascoe, P. Taylor, and A. Waterman-Pearson. 2000. Management of postoperative and other acute pain. In *Pain management in animals*, eds. P. A. Flecknell and A. Waterman-Pearson. London: W.B. Saunders.
Dorsch, M. M., K. Otto, and H. J. Hedrich. 2004. Does preoperative administration of metamizol (novalgin) affect postoperative body weight and duration of recovery from ketamine-xylazine anaesthesia in mice undergoing embryo transfer: A preliminary report. *Laboratory Animals* 38:44–49.
Erhardt, W., A. Hebestedt, G. Aschenbrenner, B. Pichotka, and G. Blumel. 1984. A comparative study with various anesthetics in mice (pentobarbitone, ketamine-xylazine, carfentanyl-etomidate). *Research in Experimental Medicine. Zeitschrift Fur Die Gesamte Experimentelle Medizin Einschliesslich Experimenteller Chirurgie* 184:159–69.
Flecknell, P. A. 2009. *Laboratory animal anaesthesia*, 3rd ed. London: Elsevier.
Flecknell, P. A., and J. H. Liles. 1996. Halothane anaesthesia in the rabbit: A comparison of the effects of medetomidine, acepromazine and midazolam on breath-holding during induction. *Journal of the Association of Veterinary Anaesthetists* 23:11–14.
Gaertner, D. J., K. R. Boschert, and T. R. Schoeb. 1987. Muscle necrosis in Syrian hamsters resulting from intramuscular injections of ketamine and xylazine. *Laboratory Animal Science* 37:80–83.
Green, C. J., J. Knight, S. Precious, and S. Simpkin. 1981. Ketamine alone and combined with diazepam or xylazine in laboratory animals: A 10 year experience. *Laboratory Animals* 15:163–70.
Hartsfield, S. M. 2007. Anesthetic machines and breathing systems. In *Lumb and Jones' veterinary anesthesia and analgesia*, eds. W. J. Tranquili, J. C. Thurmon, and K. A. Grimm, 453–94, 4th ed. Oxford: Blackwell Publishing.
Haskins, S. C. 2007. Monitoring anesthetized patients. In *Lumb and Jones' veterinary anesthesia and analgesia*, eds. W. J. Tranquili, J. C. Thurmon, and K. A. Grimm, 533–60, 4th ed. Oxford: Blackwell Publishing.
Hastings, R. H., and D. Summers-Torres. 1999. Direct laryngoscopy in mice. *Contemporary Topics in Laboratory Animal Science* 38:33–35.
Hayes, K. E., J. A. Raucci, Jr., N. M. Gades, and L. A. Toth. 2000. An evaluation of analgesic regimens for abdominal surgery in mice. *Contemporary Topics in Laboratory Animal Science* 39:18–23.
Hedenqvist, P., and L. J. Hellebrekers. 2003. Laboratory animal analgesia, anesthesia and euthanasia. In *Handbook of laboratory animal science*, eds. J. Hau and G. L. van Hoosier, 2nd ed. Boca Raton, FL: CRC Press.
Hedenqvist, P., J. V. Roughan, and P. A. Flecknell. 2000. Effects of repeated anaesthesia with ketamine/medetomidine and of pre-anaesthetic administration of buprenorphine in rats. *Laboratory Animals* 34:207–11.
Hedenqvist, P., J. V. Roughan, H. E. Orr, and L. M. Antunes. 2001. Assessment of ketamine/medetomidine anaesthesia in the New Zealand white rabbit. *Veterinary Anaesthesia and Analgesia* 28:18–25.
Hedrich, H. J., and G. R. Bullock. 2004. The laboratory mouse. *The handbook of experimental animal series*, ed. H. Hedrich, 555–70. New York: Elsevier Academic Press.
Hellebrekers, L. J., E. J. de Boer, M. A. van Zuylen, and H. Vosmeer. 1997. A comparison between medetomidine-ketamine and medetomidine-propofol anaesthesia in rabbits. *Laboratory Animals* 31:58–69.

Henke, J., S. Astner, T. Brill, B. Eissner, R. Busch, and W. Erhardt. 2005. Comparative study of three intramuscular anaesthetic combinations (medetomidine/ketamine, medetomidine/fentanyl/midazolam and xylazine/ketamine) in rabbits. *Veterinary Anaesthesia and Analgesia* 32:261–70.

Hindlycke, M., and L. Jansson. 1992. Glucose tolerance and pancreatic islet blood flow in rats after intraperitoneal administration of different anesthetic drugs. *Uppsala Journal of Medical Sciences* 97:27–35.

Holton, L. L., E. M. Scott, A. M. Nolan, J. Reid, E. Welsh, and D. Flaherty. 1998. Comparison of three methods used for assessment of pain in dogs. *Journal of the American Veterinary Medical Association* 212:61–66.

Kent, J. E., and V. Molony. 2009. *Guidelines for the recognition & assessment of pain in animals*. In Animal Welfare Research Group, Royal (Dick) School of Veterinary Studies, University of Edinburgh [database online]. Summerhall, Edinburgh, [Cited December 17, 2009]. Available from www.link.vet.ed.ac.uk/animalpain

Kent, J. E., V. Molony, and I. S. Robertson. 1993. Changes in plasma cortisol concentration in lambs of three ages after three methods of castration and tail docking. *Research in Veterinary Science* 55:246–51.

Leach, M. C., S. Allweiler, C. Richardson, J. V. Roughan, R. Narbe, and P. A. Flecknell. 2009. Behavioural effects of ovariohysterectomy and oral administration of meloxicam in laboratory housed rabbits. *Research in Veterinary Science* 87:336–47.

Lemke, K. A. 2007. Anticholinergics and sedatives. In *Lumb and Jones' veterinary anesthesia and analgesia*, eds. W. J. Tranquili, J. C. Thurmon, and K. A. Grimm, 203–40, 4th ed. Oxford: Blackwell Publishing.

Ley, S. J., A. Livingston, and A. E. Waterman. 1991. Effects of chronic lameness on the concentrations of cortisol, prolactin and vasopressin in the plasma of sheep. *The Veterinary Record* 129:45–47.

Ley, S. J., A. E. Waterman, A. Livingston, and T. J. Parkinson. 1994. Effect of chronic pain associated with lameness on plasma cortisol concentrations in sheep: A field study. *Research in Veterinary Science* 57:332–35.

Lieggi, C. C., J. D. Fortman, R. A. Kleps, V. Sethi, J. A. Anderson, C. E. Brown, and J. E. Artwohl. 2005. An evaluation of preparation methods and storage conditions of tribromoethanol. *Contemporary Topics in Laboratory Animal Science* 44:11–16.

Liles, J. H., P. A. Flecknell, J. Roughan, and I. Cruz-Madorran. 1998. Influence of oral buprenorphine, oral naltrexone or morphine on the effects of laparotomy in the rat. *Laboratory Animals* 32:149–61.

Lipman, N. S., R. P. Marini, and S. E. Erdman. 1990. A comparison of ketamine/xylazine and ketamine/xylazine/acepromazine anesthesia in the rabbit. *Laboratory Animal Science* 40:395–98.

Martinez, E. A. 2007. Muscle relaxants and neuromuscular blockade. In *Lumb and Jones' veterinary anesthesia and analgesia*, eds. W. J. Tranquili, J. C. Thurmon, and K. A. Grimm, 419–38, 4th ed. Oxford: Blackwell Publishing.

McDonell, W. N., and C. L. Kerr. 2007. Respiratory system. In *Lumb and Jones' veterinary anesthesia and analgesia*, eds. W. J. Tranquili, J. C. Thurmon, and K. A. Grimm, 117–52, 4th ed. Oxford: Blackwell Publishing.

Meyer, R. E., and R. E. Fish. 2008. Pharmacology of injectable anesthetics, sedatives, and tranquilizers. In *Anesthesia and analgesia in laboratory animals*, eds. R. Fish, M. J. Brown, P. J. Danneman, and A. Z. Karas, 27–82, 2nd ed. Amsterdam: Elsevier.

Nevalainen, T., L. Pyhala, H. M. Voipio, and R. Virtanen. 1989. Evaluation of anaesthetic potency of medetomidine-ketamine combination in rats, guinea-pigs and rabbits. *Acta Veterinaria Scandinavica. Supplementum* 85:139–43.

Obernier, J. A., and R. L. Baldwin. 2006. Establishing an appropriate period of acclimatization following transportation of laboratory animals. *ILAR Journal* 47:364–69.

Olson, M. E., D. Vizzutti, D. W. Morck, and A. K. Cox. 1994. The parasympatholytic effects of atropine sulfate and glycopyrrolate in rats and rabbits. *Canadian Journal of Veterinary Research = Revue Canadienne De Recherche Veterinaire* 58:254–58. Erratum in 1995 *Canadian Journal of Veterinary Research* 59:25.

Orliaguet, G., B. Vivien, O. Langeron, B. Bouhemad, P. Coriat, and B. Riou. 2001. Minimum alveolar concentration of volatile anesthetics in rats during postnatal maturation. *Anesthesiology* 95:734–39.

Roughan, J. V., and P. A. Flecknell. 2001. Behavioural effects of laparotomy and analgesic effects of ketoprofen and carprofen in rats. *Pain* 90:65–74.

Roughan, J. V., and P. A. Flecknell. 2003. Evaluation of a short duration behaviour-based post-operative pain scoring system in rats. *European Journal of Pain* (London, England) 7:397–406.

Roughan, J. V., and P. A. Flecknell. 2004. Behaviour-based assessment of the duration of laparotomy-induced abdominal pain and the analgesic effects of carprofen and buprenorphine in rats. *Behavioural Pharmacology* 15:461–72.

Seeler, D. 2007. Fluid and electrolyte therapy. In *Lumb and Jones' veterinary anesthesia and analgesia,* eds. W. J. Tranquili, J. C. Thurmon, and K. A. Grimm, 4th ed. Oxford: Blackwell Publishing.

Shavit, Y., K. Fridel, and B. Beilin. 2006. Postoperative pain management and proinflammatory cytokines: Animal and human studies. *Journal of Neuroimmune Pharmacology* 1:443–51.

Skarda, R. T., and W. J. Tranqulli. 2007. Local anesthetics. In *Lumb and Jones' veterinary anesthesia and analgesia,* eds. W. J. Tranquili, J. C. Thurmon, and K. A. Grimm, 4th ed. Oxford: Blackwell Publishing.

Skolleborg, K. C., J. E. Gronbech, K. Grong, F. E. Abyholm, and J. Lekven. 1990. Distribution of cardiac output during pentobarbital versus midazolam/fentanyl/fluanisone anaesthesia in the rat. *Laboratory Animals* 24:221–27.

Smiler, K. L., S. Stein, K. L. Hrapkiewicz, and J. R. Hiben. 1990. Tissue response to intramuscular and intraperitoneal injections of ketamine and xylazine in rats. *Laboratory Animal Science* 40:60–64.

Stafford, K. J., D. J. Mellor, S. E. Todd, R. A. Bruce, and R. N. Ward. 2002. Effects of local anaesthesia or local anaesthesia plus a non-steroidal anti-inflammatory drug on the acute cortisol response of calves to five different methods of castration. *Research in Veterinary Science* 73:61–70.

Steffey, E. P., and K. R. Mama. 2007. Inhalation anesthetics. In *Lumb and Jones' veterinary anesthesia and analgesia,* eds. W. J. Tranquili, J. C. Thurmon, and K. A. Grimm, 4th ed. Oxford: Blackwell Publishing.

Stewart, L. S., and W. J. Martin. 2003. Influence of postoperative analgesics on the development of neuropathic pain in rats. *Comparative Medicine* 53:29–36.

Suckow, M. A., S. H. Weisbroth, and C. L. Franklin, eds. 2006. *The laboratory rat.* New York: Elsevier.

Vachon, P. 1999. Self-mutilation in rabbits following intramuscular ketamine-xylazine-acepromazine injections. *The Canadian Veterinary Journal. La Revue Veterinaire Canadienne* 40:581–82.

Vachon, P., S. Faubert, D. Blais, A. Comtois, and J. G. Bienvenu. 2000. A pathophysiological study of abdominal organs following intraperitoneal injections of chloral hydrate in rats: Comparison between two anaesthesia protocols. *Laboratory Animals* 34:84–90.

Wright-Williams, S. L., J. P. Courade, C. A. Richardson, J. V. Roughan, and P. A. Flecknell. 2007. Effects of vasectomy surgery and meloxicam treatment on faecal corticosterone levels and behaviour in two strains of laboratory mouse. *Pain* 130:108–18.

Yasaki, S., and P. J. Dyck. 1991. A simple method for rat endotracheal intubation. *Laboratory Animal Science* 41:620–22.

14 Use of Humane Endpoints to Minimise Suffering

Coenraad Hendriksen, The Netherlands; David Morton, United Kingdom; and Klaus Cussler, Germany

CONTENTS

Objectives 333
Key Factors 334
14.1 Animal Suffering 334
    14.1.1 Pain 335
    14.1.2 Dystress 335
    14.1.3 Fear 335
    14.1.4 Lasting Harm 336
    14.1.5 Mental Distress 336
14.2 Recognising Adverse States 336
    14.2.1 Strategic Approach to Recognising Adverse Effects 337
14.3 Assessing Suffering 338
    14.3.1 Constructing a Score Sheet 340
    14.3.2 Using Score Sheets 340
14.4 Levels of Suffering that should not be Exceeded 342
14.5 The Development and Application of Humane Endpoints 343
    14.5.1 The Ws of Humane Endpoints 343
        14.5.1.1 What is a Humane Endpoint? 344
        14.5.1.2 Why Apply Humane Endpoints? 344
        14.5.1.3 When to Apply a Humane Endpoint 345
        14.5.1.4 Setting Humane Endpoints 346
    14.5.2 Responsibilities 347
        14.5.2.1 Attitude and Expertise 348
        14.5.2.2 Observation and Monitoring 349
14.6 Why are Humane Endpoints Not Always Used, Even When they can be Applied? 349
14.7 Why is Validation Needed? 349
14.8 Conclusions 350
14.9 Questions Unresolved 351
References 351

OBJECTIVES

Some suffering is usually inevitable in an experiment, no matter what measures are taken to minimise it. Suffering, which can be avoided or mitigated, is unethical. Any necessary or unavoidable suffering associated with achieving the scientific objective is the cost to the animals that is taken into account during ethical evaluation of the experiment (see Chapter 5). Moreover, as our knowledge of animal biology and experimental technique advance, improved guidelines and regulations lead to some previously necessary adverse effects becoming avoidable also.

Emphasis should be placed on preventing avoidable suffering throughout an animal's life and not just during the experiment. Those involved in animal experimentation should not simply aim for an absence of suffering, but try to enable animals to benefit in a positive sense from their environment. Everyone involved should engage in discussion about how best to satisfy scientific requirements whilst maintaining the animal and its welfare as a primary focus.

The word "animal" in this chapter should be interpreted as those (mainly) vertebrates, that are sentient (i.e., are able to experience negative and positive welfare states such as pain and pleasure). Finally, in all areas of animal research, any pain, suffering, distress and so on should always be reduced to the minimum compatible with achieving the scientific objective.

The objective of this chapter is to describe how pain, suffering, distress and lasting harm can arise during animal experimentation and how these states can be recognised avoided or alleviated. Consideration is also given to the ways in which these adverse states can be predicted and assessed (prospectively and retrospectively), so that humane experimentation can be undertaken by using early endpoints and so avoiding the need for animals to experience serious adverse states.

KEY FACTORS

All staff concerned with animal care and use must be competent at recognising signs of pain, distress, fear, lasting harm and mental distress. Not only this, it is important that they possess an attitude that promotes them to make suitable observations whilst monitoring procedures. All involved in setting humane endpoints must be aware of their responsibilities and communicate professionally with each other and other interested colleagues. The objective must be to establish endpoints that would be reached as early as possible in the course of the investigation, whilst ensuring that these are both relevant to the clinical progress of the animal and reliable and reproducible so that the outcomes of the investigation are not compromised. In addition, endpoints should be so defined that are independent of the laboratory or technicians responsible for the study so that they can be uniformly applied at different places and times.

14.1 ANIMAL SUFFERING

Animal suffering has both emotional and physical connotations. The word "animal" is used here to refer to vertebrates, although invertebrates may also express behaviour patterns that suggest that they experience it. Birds and mammals are capable of exhibiting behaviours that suggest anticipation of adverse events, such as a daily dosage regimen. Although it is not possible to measure direct feelings (or mental experiences) in animals, they can be deduced indirectly through behavioural and physiological observations. There is conflicting evidence about whether lower vertebrates such as fish, amphibians and reptiles experience pain and distress in ways qualitatively and quantitatively similar to mammals (EFSA 2009), although new evidence suggests that some species, often from later stages of foetal development, can experience adverse physiological and mental states (see Diesch et al. 2008). Higher primates, on the other hand, may experience feelings similar to humans.

Suffering can vary both in intensity and duration, and severity is a useful term to describe the product of these two dimensions. In a few countries some estimate of the severity incurred is recorded. A third dimension, considered by ethics committees, is the number of animals that experience suffering. Adverse states often occur in combination; for example, an animal that experiences chronic pain or is physically disabled may be anxious and fearful, even aggressive and consequently suffering mentally as well as physically. Some mammals in captivity show signs of boredom and frustration (mental distress), often revealed as abnormal or unnatural behaviours that are not normally seen in the wild and, depending on how long they have been present, may be reversed by enriching their environment. Strategies for refinement in experimentation should encompass measures for optimising the environment (see Chapter 4 and also Reese 1991; Poole 1997) and should both enhance animal welfare and reduce suffering (Morton 1992, 1998).

Both physiological and behavioural measures are used to recognise and assess the impact of a research protocol on animals and when and to what degree they are suffering. However, recognition is the key step, because failure to recognise when an animal is suffering means that no other action will be taken, and the animal may be left in that state (BVAAWF 1986; Duncan and Molony 1986; AVMA 1987; Smith and Boyd 1991; ILAR/NRC 1992; Townsend 1993; FELASA 1994; Morton and Townsend 1995).

Five distinct kinds of adverse state are recognised: pain, distress, fear, lasting harm and mental distress.

14.1.1 Pain

Pain is an unpleasant sensation arising from nociceptors that respond to thermal, mechanical or chemical stimuli (Melzack and Wall 1982). It has a protective function by warning of actual or potential tissue injury and from an evolutionary viewpoint has been highly conserved in the animal kingdom. Nerve impulses pass from nociceptors via afferent nerve pathways and relay centres to the cerebral cortex where signals are translated into the emotional feeling of pain, as well as its location. There also are descending pathways that modify transmission of afferent nerve impulses. Consequently, pain thresholds can be increased or decreased by higher neural influences. In some species, descending inhibitory fibres complete their development only shortly after birth, which has led some to suggest that neonates may experience more pain than adults (Fitzgerald 1994). However, Mellor et al. have disputed this with EEG evidence and some interesting comparisons with immature wallabies (Diesch et al. 2007; Mellor Personal Communication 2010). If pain in a mature nervous system persists, pain thresholds decrease, more neurones become activated and, as a consequence, more pain is felt over a greater area (hyperalgesia).

Although there is still debate about whether animals can suffer pain rather than simply feel it, animals and humans show similar physiological and behavioural responses to painful stimuli.

Although pain is the adverse effect most frequently associated with animal experimentation, it is usually an unwanted side effect rather than the scientific objective. However, if the scientific objective is to study pain itself, its intensity and duration is very carefully controlled and should be kept to a minimum compatible with obtaining the required scientific data (see Zimmermann 1983).

14.1.2 Dystress

This term was introduced by Hans Selye (but later mis-spelled as distress) to describe the situation when an animal can no longer adapt to external and internal stressors, such as excessively high or low environmental temperatures, transport over long distances, prolonged restraint, or inadequate or irregular food supply. Dystress is particularly likely to occur when an animal is unable to predict or control the stressor (Wiepkemea and Koolhaus 1993). Animals normally adapt appropriately to changes in their environment and this improves their fitness in a Darwinian sense; this state is called "eustress". This is why an animal should always be given adequate time to acclimatise/acclimate to changes in environmental conditions before it is subjected to any scientific procedure. The repetition of scientific procedures, which in themselves are not particularly stressful, may have a cumulative effect that can lead to dystress. Although some stress is unavoidable in research, it is usually possible to reduce it by strategies such as the use of positive reinforcement techniques (e.g., giving a reward after a procedure), allowing time for acclimation and habituation by training an animal.

14.1.3 Fear

Fear is probably the commonest adverse effect experienced by laboratory animals and involves activation of the sympathetic nervous system in preparation for defence strategies such as flight or aggression or, in some species, tonic immobility. Fear and anxiety are associated with increased heart rate and output, raised blood glucose concentration, elevated blood pressure, dilated pupils and urination and defecation; such changes may also impact on scientific outcomes. Fear occurs when an animal feels threatened and anticipates an adverse experience that it cannot avoid.

14.1.4 Lasting Harm

The term lasting harm is used here to describe outcomes of an experimental procedure that may not be painful or cause overt physical suffering but has the potential to cause fear, anxiety or chronic illness and may induce a state of mental distress (e.g., permanent paralysis, deafness, blindness).

14.1.5 Mental Distress

The term mental distress describes failure to cope and is an adverse mental state, for example arising from inability to forage or having insufficient space to carry out natural behaviours, but it can also occur separately. It is distinct from physical or physiological distress and is often revealed as abnormal behaviours such as stereotypies, which are usually repetitive and unvarying actions that seemingly have no obvious goal or function, may be novel for that animal or be a natural behaviour expressed in an exaggerated form (in terms of the effort or time spent carrying it out). Any behaviour not seen in that species in the wild should be investigated as a possible stereotypy. Broadly, these behaviours reflect mental states such as boredom, frustration and social isolation.

Behavioural patterns involving self-mutilation are particularly undesirable (chewing digits in monkeys, repeated gnawing leading to damage to the mouth and teeth), and reduce the ability of that animal to survive, although others involving less tissue damage may be more destructive mentally (weaving, inactivity, pacing). Such behaviours are seen in many mammals in zoos, in laboratories and can occur nearly in all species. Moreover, the scientific suitability of research animals showing stereotypic behaviours should be seriously questioned (Garner 2005).

14.2 RECOGNISING ADVERSE STATES

How can one tell when animals are in pain or are "suffering"? A good stockman may think it is obvious, but what exactly is being, or should be, observed? There is no single measure but the ability of humans to recognise adverse states in another animal depends partly on the species: It is easier to recognise adverse conditions in larger species, particularly if they are mammals. Many of us live closely with mammals such as dogs, cats and horses and are more familiar with signs of poor health and welfare in these than we are with mice and rats. Although we may feel a closer kinship with some species, we cannot simply assume that they suffer like humans and over-reliance on an anthropomorphic approach can be misleading. Despite this, a critical consideration of the animal's biology and biography (i.e., life's experience) may suggest adverse effects that should be looked for in an experiment (Morton, Burghardt, and Smith 1990). Similarly it may be helpful for the experimenter to imagine what he or she would experience in a comparable experimental situation (i.e., to take an empathetic attitude). For example, following a surgical procedure to repair a broken leg, one might feel pain and, therefore, limp, so a similar experimental procedure on an animal would be expected to be associated with limping as a result of pain post-operatively. Additional insight can be gained by discussion with doctors and nurses involved in the care of human patients with similar conditions. After all, if we use animals to model human clinical conditions, then we can use human clinical information to suggest clinical signs and palliative measures.

It is also important to take account of the reversed cycles of activity that laboratory rodents in particular possess in comparison with humans; they are active principally during the night and interference with their normal activity may be detected more easily then, instead of during the daytime. One useful behavioural test to assess inactive rats involves turning off the normal room light and observing the animals under a red light; normally they start to carry out nocturnal behaviour within a few minutes.

It is of fundamental importance that the scientist, animal technician and veterinarian can all recognise what is normal for a healthy animal of the species, strain, sex, age and background being used. The method of husbandry and the living conditions can also profoundly affect an animal's behaviour. Because behaviour often reflects how an animal is feeling, comparisons should always be made between its condition before the procedure is carried out and subsequently; appreciation of an animal's normal behaviour in the conditions in which it is being kept is arguably as important, if not more so, than biochemical and physiological measurements such as blood glucose, creatinine concentrations and so on.

It is important to avoid falling into the trap of "expecting the worst". Animals that are poorly socialised or unsuitably housed when young may develop abnormal behavioural patterns that become accepted as normal. Whilst the literature is a valuable source of information about what is normal, there is no substitute for direct observation of animals and guidance from skilled animal care staff. In order to develop an understanding of normal behaviours, researchers are encouraged to become involved in routine husbandry tasks such as feeding and watering, handling animals, cleaning the cages and so on.

As mentioned earlier many, if not most, owners of animals, particularly of companion animals, instinctively become accustomed to the animal's natural behaviour patterns and so come to know their animals as individuals. Although this is not practicable in large experiments involving tens or hundreds of laboratory rodents kept in groups, we still have a moral obligation to reduce suffering to a minimum in each animal. Although it is difficult to observe behaviours and to recognise suffering in small animals or to monitor individual animals (for example dietary intake or faecal and urinary output) when they are housed in groups, it is possible to observe their behaviour, posture and appearance and interactions with each other; for example they may withdraw from others in the social group. Transferring animals that appear to be in pain or distressed to a new cage may facilitate observation but will also be stressful.

14.2.1 STRATEGIC APPROACH TO RECOGNISING ADVERSE EFFECTS

Several years ago Morton and Griffiths considered the problem of recognising and assessing adverse effects in laboratory animals (Morton and Griffiths 1985). They investigated what keepers of animals were observing when they recognised an animal was "not right". As well as physiological and behavioural changes they reported: abnormal behavioural responses to a standard stimulus, clinical signs of overt disease (such as diarrhoea, vomiting, discharges from the nose and eyes) and abnormal postures. Observations such as these provide a basis for evaluating whether an animal is suffering although an overall strategy for recognition and assessment of pain in animals is required. The following system is recommended.

First, before approaching it, the animal should be carefully observed from a distance, particularly noting its *appearance, posture* and *behaviour.* Is its coat clean and shiny or dull and unkempt? Is it hunched in a corner, or lying stretched out? Is its behaviour normal for that species, strain or individual?

Having assessed the animal's appearance and behaviour, its reactions to a simple stimulus (such as a noise) should be tested to establish whether it is alert and responds in a predictable manner. Such a study of "provoked" behaviour might be continued whilst handling the animal for clinical examination to establish whether any areas of its body are particularly sensitive. Clinical signs of abnormality might also include staining around natural orifices (evidence of diarrhoea, blood in the urine, abnormal discharges, etc.). The animal can then be weighed to determine whether body weight has been lost because it is not eating or drinking normally. Careful interpretation might be necessary in circumstances in which body weight may not be lost but gained (e.g., tumour growth or ascites).

The six assessments that form the basis for evaluating adverse effects are:

- Appearance
- Posture
- Spontaneous behaviour
- Provoked behaviour
- Clinical signs
- Body weight

Additional assessments may be appropriate, such as: urine or blood assays and tests of organ function.

Interestingly, enrichment of the environment provides further opportunities for observing changes in behaviour; for example, if an animal is not using a running wheel, or not interacting as usual with a cage inclusion, this raises the question: Why not? The animal may be in pain, lethargic or suffering in some other way that causes its apparent lack of interest in the environment. Some of these alterations can be measured quantitatively (e.g., running wheel activity, food and water intake, body weight, body temperature, heart rate, blood pressure and respiration rate). Others are less easy to quantify, but are just as useful and relevant; for example, quality of respiration (e.g., laboured, noisy), lameness, ungroomed or dull coat, hair loss, hunched posture. Although these signs cannot be directly quantified, they can be scored so that important information is summarised. They can also be grouped according to their potential severity, for example, in the case of lameness, one can predict an increasing intensity of pain from a slight limp, to a marked limp, to where a foot hardly touches the ground when walking and, finally, when a foot is held off the ground and the animal is hopping.

14.3 ASSESSING SUFFERING

One approach to monitoring adverse states is to examine the experimental procedures involved and to score those in relation to their likely impact on the animal's well-being (Wallace 2000). This approach assumes that researchers are equally able to recognise these, and that all animals react in the same way. An alternative strategy is to assess what the animal is telling us, no matter what animal or who is carrying out the procedure by determining how far an animal has deviated from normality. This has the advantage that it relates directly to the impact of the experiment on each animal's well-being—the greater the deviation, the more the experimental procedures are presumed to have impacted on the animal and the more likely it is to be suffering—in other words, the severity. It is not necessary to establish why the animal is reacting in the way that has been observed (i.e., the cause of suffering whether it be pain, dystress, fear, boredom, frustration, mental distress, etc.). However the underlying cause has to be established for appropriate therapy to be implemented (e.g., administration of analgesics to control pain) or implementing measures to avoid its occurrence on future occasions (acclimation and habituation for fear). Additionally, this approach makes it possible to compare the welfare consequences of using different models and the effectiveness of strategies for alleviation of adverse states, such as administering an analgesic agent. This makes it possible to refine procedures by reducing their severity without necessarily adversely affecting the scientific objective. Examples of additional strategies to reduce suffering and to improve the science are given in Tables 14.1 and 14.2.

Score sheets (also known as observation sheets or welfare assessment sheets) are lists of relevant observations. A sheet should be drawn up that is specific to each scientific procedure, for each species and sometimes for each strain. Experimental outcomes vary in the same way as the impact on the animal; for example, the predictable side effects of a failed skin graft differ from those for a failed liver or kidney graft. Behavioural characteristics of species, and even strains within a species, can differ considerably.

TABLE 14.1
Some Considerations to be Taken Into Account When Assessing Severity and Animal Suffering

1. ENVIRONMENTAL ASPECTS
   Husbandry (e.g., enrichment):
   - Complexity of environment
   - Interaction with humans and other animals (avoiding single caging or single penning of animals, stable groups, in sight of others
   - Novelty (e.g., diet, enrichment objects)

2. CAN THE SCIENTIFIC PROCEDURES BE IMPROVED
   - Controls (necessity, historical data)
   - Staging of experiments (critical experiments, terminal anaesthetic before recovery)
   - Surgery and anaesthesia (asepsis, modern anaesthetics, analgesics)
   - Euthanasia (humane, skilled persons)
   - Statistics and experimental design (too many/too few, right design—obtain professional advice before starting work)
   - Endpoints (humane and scientific, avoid painful endpoints where possible)
   - Monitoring adverse effects
   - Education, training and competence
   - Question tradition and obtain data to support scientific and humane suppositions
   - Database for animal models

TABLE 14.2
Some Considerations to be Taken Into Account When Assessing an Animal Model for Its Usefulness, Fidelity and Acceptability

1. What is being modelled?
2. Purpose of model (e.g., human disease, drug evaluation)
3. Advantages and disadvantages (scientific and in relevance to human or animal disease or treatment)
4. Clinical trials or evaluation of a medicine, indicating relevance or predictability of the model to its intended purpose (2 above)
5. Species of animal and experimental details (e.g., strain, inbred/outbred, sex, age, weight, health status, acclimation period, habituation, other)
6. Husbandry special or critical requirements (e.g., caging or pen type, animal kept singly or in groups, diet, bedding, isolator or filtration boxes, breeding details, other)
7. Methodological details (e.g., equipment and skills required, dose, timings, where advice can be obtained)
8. Refinement aspects (e.g., humane endpoints, score sheet giving cardinal signs, useful tests and any other information)
9. Scoring of signs and severity grading (severity grading: give average and maximum expected)
10. Replacement (are alternative methods for all or part of the model available, e.g., pre-screening?)
11. Success rate of the model (give mean and range, and methods used to determine success)
12. Ethical commentary
13. References
14. Keywords for literature searches

Note: See also Chapter 7.

14.3.1 Constructing a Score Sheet

During the planning stage, those responsible for the welfare of the animals (the researcher, senior animal caretaker and veterinarian) should determine the probable impact of the procedure on the subjects and the signs these are likely to cause. Predictions should be based on publications, past experience and communication with colleagues. If these are unreliable or evidence is lacking, a critical anthropomorphic approach may be adopted; that is, assessing what the impact of the procedure on a human would be, deducing the signs which would arise, then adjusting these in light of the biology of the experimental model. Some signs may be common to all animals (i.e., a zoomorphic approach), for example, decreased activity, reduced eating or drinking and so on. In addition, the experimental procedure may be anticipated to have specific effects, monitoring of which will lead to a more useful score sheet. When a list of key indicators of possible adverse effects has been drafted, their absence or presence is recorded, for example, present (+) or absent (–) or, if unsure, (+/–). Reduction of clinical signs to this binary scoring method, avoids misinterpretation or subjective evaluation. One benefit of this scheme is that it reduces the likelihood of observer error and when this does occur it is usually at the lower levels of severity where only small changes from normality have occurred.

By convention negative signs are used to indicate normality (i.e., within the normal range), whilst positive signs indicate that the animal is outside the normal range. For example, if an animal develops convulsions, this would be scored a (+); similarly, if food intake falls below normal, this too would also be scored as a (+). When specific clinical states, such as diabetes mellitus are produced experimentally, it is usually possible to incorporate specific clinical chemistry data, such as glucose concentration in urine, thus increasing the precision of assessment, although this information may not help measure the impact of the procedure on the animal's mental state.

Score sheets should be carefully constructed in consultation with all those involved in the care and use of the animals, but may be re-evaluated and modified on the basis of observations carried out during a pilot study* or the first group of animals undergoing a procedure, so as to identify actions that can improve care or establish earlier endpoints and so allow prompter intervention.

14.3.2 Using Score Sheets

Although the completion of score sheets requires effort and resources, it is relatively easy for an experienced person to determine whether each animal is well, and to annotate the sheet "NAD" (Nothing Abnormal Diagnosed) rather than to carry out a full examination.

Score sheets drawn up following the convention outlined above can be quickly evaluated to gain an overall impression of animal well-being: the more plusses, the more an animal has deviated from normality and hence, the likelihood that the impact of the scientific procedure has been greater (Morton and Griffiths 1985). It is also possible to establish overall categories of suffering such as: None, Mild, Moderate and Severe.

The overall impact of an experimental procedure can be assessed from the range of abnormalities as well as their magnitude, particularly where "banding" of quantifiable signs has been used. Table 14.3 shows an example of a score sheet, based on subjective and objective signs during a study involving heterotopic kidney transplant in rats; although this technique is now rarely used (see Chapter 11), the adverse effects are fairly widely known.

Another method of reviewing score sheets is to allocate a score and a weighting to each sign, and to calculate a mean score. Thus, 0 could be allocated for normal body weight, 1 for a loss of 5–10%, 2 for 11–20% and 3 for greater than 20%. Similarly for the eye: normal would score 0, half-closed 1,

---

* A pilot study precedes the main study. In a pilot study, a small number of animals (say 1–3 sequentially) is subjected to the proposed procedure to provide information that is unknown at the time but which is of key importance for the successful design of the main study, such as level of response, inter-individual variation, clinical effects and so on.

TABLE 14.3
Completed Score Sheet for Heterotopic Kidney Transplant

RAT No.	HN1		ISSUE No:	234			
DATE OF OPERATION: 20 June at 11.00 hour							
PRE-OPERATION WEIGHT: 250G							
DATE		20	20	21	21	22	22	22
DAY								
TIME		13.30	17.30	8.00	4.00	8.00	11.00	14.00
FROM A DISTANCE								
Inactive		—			—	–/+	+	+
Inactive? Try red light response	^				—	+	+	
Isolated		—			—	+	+	+
Hunched posture		+			+/–	–/+	+	+
Staring coat		—			+	+	+	+
Rate of breathing		54			60	64	70	40
Type of breathing[a]						R	R	L
ON HANDLING								
Not inquisitive and alert		—			—	—	—	+
Eating/Jelly Mash? amount eaten[b]		—		50%	—	?	?	?
Not drinking		—			—	?	?	?
Bodyweight (g)			254		260	250	221	215	205
% change from start		—	6%	0%	–12%	–14%	–18%	
Body temperature (°C)		35	36.5	37	38	38	36.5	35.5
Crusty red eyes/nose		—			—	—	+	+
Excessive wetness on lower body		—			—	—	—	+
Sunken eyes	^	—			—	—	—	—
Dehydration	^	—			—	—	+	+
Coat/wet soiled		—			—	—	—	+
Pale eyes and ears		—			—	—	—	—
Blue extremities	^	—			—	—	—	+
Stitches OK?/Date removed		—			—	—	—	
Swelling of graft[c]	^				—	—	—	–/+
NAD			✓	✓	—	—	—	—
Dosing								
OTHER		Diarrhoea^					+	+
SIGNATURE:								

Special husbandry requirements:
Animals should be put on a cage liner with tissue paper and a small piece of VetBed
Scoring details:
[a] Type of breathing: R = rapid, S = shallow, L = laboured, N = normal
[b] Eaten/Jelly mash: amount? Record as 0/25/50/75/100%
[c] Swelling score: 0 = normal, 4+ = rejection and large swelling
✓ means checked and nothing found. The animals have been looked at and all was well.
Humane endpoints and actions:
1. Weight loss of 15% or more inform the investigator, veterinarian and technician I/C
2. Pre-moribund state (indicating a failing graft)
3. Any major clinical sign recurs after 24 hours (marked ^, less than 35°C)
Scientific measures:
Take 1 ml of blood and urine, if possible, place at 4°C
Place transplanted kidney into 10 ml formal saline

and fully closed 2. If the cornea is ulcerated or there is a discharge, the score might be 3. In this way a total score is obtained that indicates the extent of deviation from normal and can be used to initiate action, for example, to seek veterinary advice or treatment. Tables 14.4A and B show how scoring is applied to a study of the effects of endotoxin and how findings from several studies are evaluated.

The sheet should also provide information about how observations and qualitative clinical signs (such as lameness, diarrhoea or quality of respiration) should be recorded in order to help ensure consistency of scoring between observers. In addition, explicit instructions should be included about any special requirements for husbandry and care, and implementation of humane endpoints (see below) or points at which the researcher or veterinarian should be asked for advice. Finally, if an animal has to be killed for humane reasons there should be instructions about special actions to be taken, for example withdrawal of a blood sample or tissue harvesting, so that information relevant to the study, including possible evidence of unanticipated adverse effects, is gathered.

14.4 LEVELS OF SUFFERING THAT SHOULD NOT BE EXCEEDED

The use of score sheets provides a method for determining when prompt action is needed to ensure that unnecessary suffering is not allowed to occur. As explained earlier, the repertoire of possible clinical signs is large and each must be considered in the context of the animal species, the experimental protocol and the objectives of the study. However, there are several clinical signs that are frequently seen when welfare is seriously compromised and that should alert those responsible for the animals to the need for further action.

Inability to eat and/or drink. This might result from the animal feeling unwell—for example associated with general disease, but might have a more specific cause, such as obstruction of the upper digestive tract or abnormalities of the teeth.

Rapid or continuous weight loss. It can be difficult to accurately assess whether an animal is losing weight if it is growing, or body mass is increasing for other reasons, for example growth of a tumour. In the former case, comparisons should be made with age-matched controls but in other cases body condition can be assessed from the fleshiness of the animal in the pelvic area (so called body condition assessment; Ullman-Cullere and Foltz 1999). As a guideline, a weight loss of over 20% during a period of a few days or a chronic weight loss of more than 15% is the maximum that should be tolerated, unless there are strong reasons for discounting such losses—for example during studies into anti-obesity therapies or involving diabetes mellitus.

Dehydration due to decreased fluid intake or excessive fluid loss. The degree of dehydration can be determined using the turgor test, which involves gently pinching and twisting the skin then releasing it and observing how long it takes for the skin fold to flatten.

Generalised decrease in grooming behaviour. Decreased grooming, as evidenced by a rough and dirty hair coat, red tears (chromodacryorrhea) and nasal orifices, and soiled perianal and perioral regions. These may represent a physiological side effect of ageing, but when seen in young and adult animals such signs might point to suffering and indicate a need for attention.

Neurological disorders such as convulsions, circling, tremors, reduced reflexes and so on that might be the result of an intoxication or an infectious disease process.

Apathy/lethargy that results from exhaustion (e.g., after a surgical procedure or as a result of general malaise).

Conditions indicating severe pain, distress or suffering. The most appropriate action depends on what is causing these signs and the context of the experiment. For instance, in an experiment studying rheumatoid arthritis, pain and distress may be an intrinsic part of the study. However, in other cases, pain relief might not interfere with the investigation or it may be decided to kill the animal humanely.

Severe or continuing respiratory distress (e.g., laboured breathing or cyanosis) are strong indicators for application of humane endpoints; if these features are anticipated to occur, advice should be sought from the animal welfare officer/veterinarian and the ethical review body should specifically consider the protocol.

TABLE 14.4A
Mean Scores Used in Mouse Endotoxin Experiment

SIGN	SCORE
Appearance	
Hunched	1
Pilo-erection	1
Closed eyes	1
Behaviour	
Decreased activity	1
Clinical	
Diarrhoea	1
Haematuria	1
Each °C drop	1

Source: Townsend, P. and Morton, D. B., *Practical Assessment of Adverse Effects and Its Use in Determining Humane End-Points*, paper presented at Proceedings of the 1993 5th FELASA Symposium, Welfare & Science, FELASA, Brighton, UK, 1994.

TABLE 14.4B
Mean Scores for Mice Given Endotoxin and Various Treatments

	Mean scores	N
Controls		
No endotoxin	1.53[a] +/− 0.108	8
50 μg endotoxin only	9.86 +/− 0.57	8
Prophylaxis against endotoxin with:		
Non-immune serum	9.24 +/− 0.36	4
Anti-endotoxin antibody	7.22 +/− 0.27	4
Pre-incubation of endotoxin with:		
Normal serum	10.49 +/− 1.20	5
Anti-endotoxin antibody	9.11 +/− 0.59	5
Post-endotoxin treatment with:		
Normal serum	8.90 +/− 0.64	6
Anti-endotoxin antibody	6.89 +/− 0.79	6

N = number of animals.
Source: Townsend, P., and Morton, D. B., *Practical Assessment of Adverse Effects and Its Use in Determining Humane End-Points*, paper presented at Proceedings of the 1993 5th FELASA Symposium, Welfare & Science, FELASA, Brighton, UK, 1994.
[a] Significant differences at $p < 0.05$ when control untreated group was compared with all other groups.

14.5 THE DEVELOPMENT AND APPLICATION OF HUMANE ENDPOINTS

14.5.1 THE Ws OF HUMANE ENDPOINTS

Application of humane endpoints should be included in policies for refining experimental procedures, especially those anticipated to involve severe suffering. The decision to use a humane endpoint is a balanced one: application too early, when scientific endpoints have not been met, might frustrate the experiment; on the other hand, too late an application allows avoidable pain and suffering for the animals and is generally unacceptable. This section further refines the definition of humane endpoints and addresses three "W" questions: What are humane endpoints? Why use them? and When should they be used?

14.5.1.1 What is a Humane Endpoint?

A humane endpoint constitutes a refinement strategy designed to minimise pain, suffering or distress experienced by animals during an experiment. Examples of when they should be used include:

> *'In experiments involving animals, any actual or potential pain, distress, or discomfort should be minimised or alleviated by choosing the earliest endpoint that is compatible with the scientific objectives of the research. Selection of this endpoint by the investigator should involve consultation with the laboratory animal veterinarian and the animal care committee.' (Bhasin et al. 1999)*
>
> *'... The earliest indicator in an animal experiment of severe pain, severe distress, suffering, or impending death and this indicator can then be used to kill the animal.' (OECD 2000)*
>
> *'... The limits placed on the amount of pain and distress any laboratory animal will be allowed to experience within the context of the scientific endpoints to be met.' (Wallace 2000)*

These three recommendations differ in important ways. According to the CCAC, application of humane endpoints includes the taking of measures to alleviate pain and distress, while the OECD definition clearly proposes killing an animal that experiences severe pain or severe distress. Both the OECD and CCAC definitions use the term "earliest indicator" (CCAC 1998; OECD 2000), which might include both pre-clinical and early pathophysiological parameters, while Wallace's definition is based on the assumption that an animal will experience pain and distress. Both Wallace and CCAC balance the limit of pain and distress against scientific endpoints. Moreover, Wallace infers that there may be a limit to the ethical acceptability of the scientific endpoint, whereas the OECD definition separates the humane endpoint from the context of the experiment.

A modification that incorporates the provisions of the above recommendations would be:

> *'... The earliest indicator in an animal experiment of (potential) pain and/or distress that, within its scientific context and moral acceptability, can be used to avoid or limit adverse effects by taking actions such as humane killing, terminating the study or alleviating the pain and distress.'*

14.5.1.2 Why Apply Humane Endpoints?

There are several moral/social, legislative and scientific reasons to make use of humane endpoints.

14.5.1.2.1 Moral and Social Reasons

Continuing an experiment beyond justifiable limits of suffering is morally unacceptable. As discussed above, these limits can be defined only in the context of a particular experiment. For example, severe pain associated with research into arthritis might be acceptable for a certain period of time whereas pain in an animal resulting from dental malformation that is not associated with any anticipated scientific or medical benefit requires it to be killed immediately. Many Europeans are critical about the use of laboratory animals for scientific purposes, especially if it involves pain and/or distress (see Chapter 1; also Aldhous, Coghlan, and Copley 1999). Consequently, application of humane endpoints helps to address societal concerns about limiting pain and distress in animal studies.

14.5.1.2.2 Legislative Reasons

Most national and international laws and regulations on animal experimentation require that laboratory animals should not be subjected to unnecessary pain and suffering. For example, the Article

13 of the revised text of the current European Directive 86/609/EEC* states: "Death as the end-point of a procedure shall be avoided as far as possible and replaced by early and humane end-points. Where death as the end-point is unavoidable, the procedure shall be designed so as to:

(a) result in the deaths of as few animals as possible; and
(b) reduce the duration and intensity of suffering to the animal to the minimum possible and, as far as possible, ensure a painless death" (http://ec.europa.eu/environment/chemicals/lab_animals/home_en.htm).

14.5.1.2.3 Scientific Reasons

Pain and distress might confound animal experimentation by interfering with scientific outcomes such as immunological or hormonal responsiveness, and thereby invalidate the findings. For example, if an animal were to die at a time when staff members were not in attendance (e.g., during the night), autolytic changes may have taken place by the time it is discovered or the carcase may have been cannibalised by cage-mates. Consequently, there is loss of data because histopathology is not possible and key organs/tissues/tissue fluids might have been lost, as a result of which the experiment might need to be repeated.

14.5.1.3 When to Apply a Humane Endpoint

Generally speaking, there are four specific situations in which the application of humane endpoints should be considered:

1. When the scientific objectives have been met and there is no reason to continue the investigation.
2. When unexpected suffering occurs. In this situation, the suffering is not related to the experiment but is unexpected and had not been anticipated at the outset. Examples include: rectal prolapse, bone fractures, bite wounds and so on.
3. When suffering was anticipated at the start of the experiment, but has become more severe than predicted. This could occur when animals have been exposed to a noxious agent that proves to be more toxic than anticipated from previous studies or literature searches, or a mammary tumour develops that is located in a position that impairs mobility of an animal.
4. When pain and/or distress are an inherent part of the experiment. In this situation, pain and/or distress are anticipated at the outset. For example, a challenge experiment involving a virulent micro-organism that can be predicted to result in clinical signs characteristic of the disease with which it is associated.

Each situation requires a specific decision to be made with regard to the application of humane endpoints.

In situations 1 and 2, humane endpoints should normally be applied. Failure to kill or treat an animal can be justified only in exceptional cases, and when authorised also by a veterinarian, animal welfare officer or other competent person who has no direct interest in the experiment. A similar approach should be used in situation 3, although the prediction of suffering might have resulted from critical information being absent or overlooked, such as exaggerated toxicity of a substance in a particular strain of rats. In this situation a small pilot study should be performed before the main study is continued.

---

* The European Parliament and the Council have agreed to a text which is pending final adoption and publication, foreseen in autumn 2010.

In the case of situation 4, it is particularly important that a careful evaluation is carried out of the scientific and other (e.g., social) benefits of the experiment and these are balanced against the potential pain and/or distress of the animals involved. Any experiment in this category should be considered and approved by the ethical evaluation process (see Chapter 5). A pre-condition for studies of this type is that scientific endpoints are carefully established before the investigation starts, and that humane endpoints are chosen that take account of these so that the study is not continued beyond the stage at which the scientific objectives have been met. For example, in an investigation into the efficacy of an anti-cancer drug to prevent metastasis, an animal should be killed in event of tumour metastasis.

14.5.1.4 Setting Humane Endpoints

The objective of a humane endpoint is to limit the level of suffering that an animal experiences. Consequently, the endpoint can be set at any time between the first detectable deviation from normality or onset of a disease process to the first indications of severe suffering. From this point of view, the onset of disease can also be defined as the appearance of events at a molecular level that trigger the cascade of reactions that ultimately lead to death or severe suffering. If it can be demonstrated that a molecular marker reliably predicts death or severe suffering, then evidence of up- or down-regulation of that molecule can be used as a humane endpoint. Figure 14.1 shows how disease typically progresses from molecular triggering to pre-clinical changes, pathophysiological alterations, mild clinical/behavioural signs, severe clinical/behavioural signs and finally death. Of course when severe pain or distress is a result of trauma, the initiating process is different although its progress may be similar.

Clinical/behavioural signs: Humane endpoints are generally set on the basis of clinical and/or behavioural signs, although, as discussed above, many other opportunities are available. Clinical/behavioural signs include all those deviations from normal health status that can be gathered using a score sheet in the manner discussed above.

Early clinical and non-clinical pathophysiological changes: Pathophysiological changes include deviations outside the normal physiological range of variables such as blood pressure, heart rate, body temperature or blood oxygen saturation. In some cases these deviations lead directly or indirectly to clinical signs. For example, an increase of body temperature may be detected by touching hairless parts of the skin, while reduced oxygen saturation of the blood increases the rate of breathing and

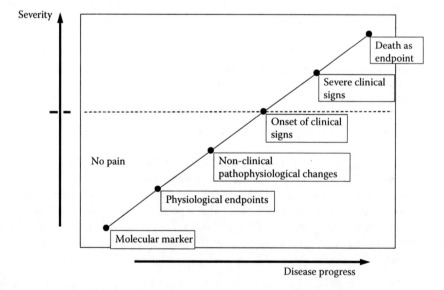

FIGURE 14.1 The application domain of humane endpoints. (Adapted from Richmond, J., *Criteria for Humane Endpoints,* paper presented at Humane Endpoints in Animal Experiments for Biomedical Research—Proceedings of the International Conference, Zeist, The Netherlands, 1999.)

results in blueness of the skin at the extremities. Technological innovations have increased opportunities for early detection of pathophysiological changes. Temperature-sensitive transponders can be used to monitor body temperature (Kort et al. 1998; Hartinger et al. 2003), telemetry allows real-time monitoring of body temperature, blood pressure and heart rate and so on (see Chapter 10; also Kinter and Johnson 1999; Kramer et al. 1999; Kramer 2000). Laparoscopy has been described as a way of detecting early liver metastases (Kobaek-Larsen et al. 2004). Bioluminescence imaging, based on release of photons by luciferase genes (which can be inserted into tumour cells or pathogenic micro-organisms) can be used to monitor internal tumour growth/metastasis and bacteraemia, respectively (see Chapter 12; also e.g., Wang 2003). In addition, clinical chemistry can provide valuable early information about the progress of physiological changes by establishing when hormonal, enzymatic, immunological and haematological characteristics deviate significantly from the normal range. Examples are: acute phase proteins, cytokines and stress hormone levels (prolactin, corticosteroids). Clinical chemistry can be carried out on specimens of blood, urine (Poon and Chu 1999), saliva, faeces, lachrymal fluid and so on. It should be recognised that application of new technologies might have a negative impact on refinement. For example, biophotonic imaging might make it possible to set early endpoints, but requires the animals to be anaesthetised at regular intervals to monitor progress of the disease (Chapter 12: de Boo and Hendriksen 2005).

Pre-clinical changes indicate that a healthy animal is susceptible to a particular disease process, for example antibody concentrations below the level required for protection against a particular disease. Consequently, instead of challenging a vaccinated animal with a lethal dose of virulent micro-organisms, vaccine potency can be determined by measuring the blood concentration of antibodies induced by vaccination (Hendriksen and Steen 2000).

Molecular parameters: Up- or down-regulation of genes might indicate that severe pain or suffering is likely to occur later in the study. The ability to recognise early changes such as these allows a molecular marker be selected as a humane endpoint. New innovative technologies such as transcriptomics and proteomics make it possible to study the effect of test compounds at the molecular level by detecting changes in the genome. For example, chemically induced carcinogenesis might be associated with up- or down-regulation of a cluster of specific genes, and these can be monitored to provide an early indication of tumour induction.

The purpose of the experiment is a major consideration when selecting humane endpoints. For example the use of temperature-sensitive transponders may be very valuable in studies related to infectious disease but they may not be relevant within cancer studies (Hartinger et al. 2003).

Effective implementation of humane endpoints requires that several conditions are met; these include establishing who is responsible for what actions, ensuring that those involved possess appropriate attitudes and expertise, and instituting suitable observation and monitoring procedures, such as the use of score sheets and so on.

14.5.2 Responsibilities

Responsibilities must be clearly defined before the experiment starts; this must include defining actions to be taken if certain personnel are not immediately available, for example during weekends. In this respect, several levels of responsibility can be recognised, for example: the ethical review body, the study director or principal investigator, the animal welfare officer and/or veterinarian, the animal care technician and the pathologist. The responsibility of each is summarised briefly below:

- *The ethical review body* sets standards of animal welfare within institutions and assesses the social/scientific need for experiments in relation to their impact on animals (see Chapter 5). For example, the ethical review body might define severity limits beyond which an experiment is unacceptable, and consequently should be provided with all available information, including retrospective data relating to investigations where the severity level of the experiment differed from what had been anticipated (Mench 1999).

- *The study director or principal investigator* is primarily responsible for conduct of the experiment. He or she defines the study hypothesis, experimental design and scientific endpoints and is responsible for ensuring incorporation of the Three Rs, including the establishment of humane endpoints. In this respect, a full appreciation of critical steps in the protocol is important—for example are severe clinical signs to be expected and if so, at which stage of the study? Often, a pilot study involving a small group of animals is performed. Also in case of unforeseen outcomes, the study director/principal investigator is responsible for deciding what action should be taken.
- *The animal welfare officer/veterinarian* operates at different levels: as an advisor to the study director or principal investigator when the experiment is being planned; as an independent advisor to the ethical review body; also during the course of the experiment, for example by monitoring the health and welfare of the animals. If suffering exceeds that foreseen in the planning phase (e.g., as a result of unplanned injury or when severity is higher than anticipated), this person should have authority, after consultation, to overrule decisions of the study director/principal investigator (Medina 2004).
- *The laboratory animal technician, or caretaker* is responsible for day-to-day care of the animals. He or she should be competent not only in tasks involved with this, but also in monitoring the animals' well-being and health status. If animals become unwell or appear to be suffering, the animal caretaker may need to take decisions about treatment or humane killing based on information contained in the protocol. Decisions have to be communicated to the study director/principal investigator. When unanticipated adverse effects occur, the technician or caretaker should consult the study director/principal investigator and the animal welfare officer or veterinarian as appropriate.
- *The pathologist*: animals that are humanely killed or die during the course of an experiment should be subject to autopsy. The report of the pathologist can provide valuable information about severity that can aid the interpretation of clinical signs and facilitate preparation of better guidance for subsequent studies (Fentener van Vlissingen et al. 1999).

14.5.2.1 Attitude and Expertise

Two key concepts in the application of humane endpoints are attitude and expertise. Attitude is underpinned by the phrase "only those who are prepared to see will see". For the study director or principal investigator it refers to willingness to take full account of suffering arising from the experiment. As discussed earlier, this need is based on ethical, legal and scientific considerations. Approaches may vary from passive (doing what is well-known) to pro-active, which involves searching for earlier and/or innovative endpoints. In this respect, the role of the ethical review body and the animal welfare officer/veterinarian is primarily to interact with the study director/principal investigator and the animal technicians, engaging in questioning, discussion, feed-back and dialogue. Animal technicians/caretakers and pathologists play an essential role in observing, monitoring and communicating observations relevant to the animals' well-being. Taken together, the activities of these people and the management of the institute or department should lead to establishment of a culture of care.

Attitude is ineffective in the absence of expertise. Correct application of humane endpoints requires knowledge and understanding of the behaviour and physiology of normal, healthy animals as a starting point for recognising deviations from these. Training underpins this—study directors and principal investigators should have an appreciation of the basic biology of animals they are working with. Recommendations for training laboratory animal scientists (category C) have been established by FELASA and are mandatory in several European countries (FELASA 1995; Hau 1999; see also Chapter 16). The recommendations identify experimental design and humane endpoints as compulsory topics, including familiarity with normal behaviour/physiology, clinical signs and humane killing. The curriculum for training animal caretakers (category A), animal technicians (category B) and laboratory animal specialists (category D) should include in-depth information about these topics (FELASA 1995; Nevalainen et al. 1999; Nevalainen et al. 2000). Additional training material, such as a recently developed website on humane endpoints (http://www.humane-endpoints.info/) provides advice about recognising critical clinical signs.

14.5.2.2 Observation and Monitoring

Observation should be an ongoing process, with new information being integrated with that from preceding observations (Clark, Rager, and Calpin 1997); this is an efficient way of improving the way animal well-being is assessed. The score sheet is an indispensable aid for monitoring welfare, general health and clinical signs that are particularly relevant for the research model (Lloyd and Wolfensohn 1999; Morton 2000). Animals must be observed at least once each day, including weekends. The frequency of observation should be increased at critical time points in the experiment, for example when physiological changes are anticipated, adverse toxic effects of a chemical are expected to be manifest or the incubation period of an infectious disease agent comes to an end.

14.6 WHY ARE HUMANE ENDPOINTS NOT ALWAYS USED, EVEN WHEN THEY CAN BE APPLIED?

It has been shown earlier that the use of score sheets makes it possible to define levels of suffering at which action is needed to provide relief for the animal. Some clinical signs may themselves act as indicators that welfare is being seriously compromised, and in such circumstances, the application of humane endpoints is usually appropriate. However, a survey of vaccine production facilities found that humane endpoints are not always used in the case of test protocols conducted in accordance with regulatory protocols, even when included in the pharmacopoeia's requirements and required by law (e.g., Article 7.4 of Council Directive 86/609/EEC). Why are humane endpoints, which have been accepted, not adopted? There may be several reasons.

Economics: The application of humane endpoints requires more frequent and more detailed monitoring of the animals, which is time consuming and consequently more costly. Furthermore, it is necessary to apply resources to train animal technicians and scientists to recognise key clinical signs.

Subjectivity: Death is an indisputable endpoint, whereas assessment of clinical signs is subjective, and consequently open to challenge. Figure 14.2 demonstrates that this concern can be overcome by training staff to recognise and interpret clinical signs when assessing the potency of pertussis vaccine. However, Figure 14.2 also reveals that training is not always successful; some personnel appear to lack a 'clinical eye,' or the ability to recognise key clinical signs despite training.

Lack of harmonisation: When tests are performed in accordance with regulatory requirements, the use of humane endpoints in place of traditional endpoints such as lethality has to be accepted by regulatory bodies. As long as different regulatory bodies fail to harmonise their requirements, non-humane endpoints will continue to be used even though suitable humane endpoints are available. An example of successful harmonisation from the area of vaccine quality control, is the requirement in Europe and the United States that protocols for potency testing of rabies vaccine shall explicitly specify the use of humane endpoints. Although the wording in the European Pharmacopoeia and the USDA is not identical, vaccine manufacturers can easily comply with both regulations. There are other areas where similar strategies could be followed and, in the interest of animal welfare, industry and regulators, it is essential that progress should be made in this area.

14.7 WHY IS VALIDATION NEEDED?

In the context of humane endpoints, validation means demonstrating the relevance and reliability of the humane endpoint concerned as a replacement for an existing less humane endpoint. Relevance

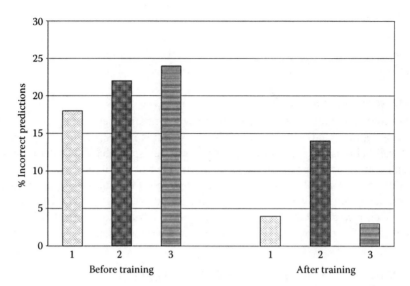

FIGURE 14.2 Percentage of incorrect humane endpoint predictions of three animal technicians (1, 2 and 3) in *Bordetella pertussis* potency testing before and after training on the spot. Before training means that technicians were only told what clinical signs were to be used as humane endpoints instead of a lethal outcome after intracerebral infection with virulent *B. pertussis*.

refers to the need for the humane endpoint to consistently and accurately predict the currently accepted scientific endpoint. For example, if lethality is used as a the scientific endpoint during an investigation (for example when testing potency of vaccines), the clinical signs or pathophysiological parameters used to replace it must accurately predict lethality. Incorrect predictions might result in the animal being killed even though it would have survived the observation period. Reliability refers to the robustness and reproducibility of the humane endpoint. Wherever possible, endpoints should be independent of the laboratory or technicians responsible for the study.

Generally, quantifiable endpoints such as those based on measurement of heart rate or body temperature (e.g., using temperature-sensitive transponders) are more reliable than other endpoints such as clinical signs although these can be useful, even though not quantifiable (e.g., blood in the urine, diarrhoea). Consequently, as mentioned earlier, staff should be trained to collect data relating to these more subjective clinical signs. It is also important to assess the sensitivity of the humane endpoint chosen, to ensure that it coincides with the earliest reliable prediction of death or serious illness.

14.8 CONCLUSIONS

About 15% of all animal experiments performed in The Netherlands induce severe pain and distress in the animals involved. For both ethical and for scientific reasons there is a strong need to reduce this percentage. Application of humane endpoints is a powerful refinement strategy and if correctly done, makes it possible to minimise the amount of pain and distress that animals experience during experiments and this addresses a major animal welfare concern. The traditional approach to using humane endpoints is to avoid the occurrence of severe clinical signs by predicting their imminence from milder clinical signs. Innovative technologies such as genomics and proteomics have now become available that might offer new opportunities for moving endpoints to the pre-clinical stages of disease processes. The attitude of all those involved in animal experimentation is fundamental to successful implementation. However, on its own this is not enough and these approaches must be accompanied by proper training and acceptance of well-defined responsibilities. These, together with commitment by management, should provide a platform for phasing out severe pain and distress in animal experiments.

14.9 QUESTIONS UNRESOLVED

- More authoritative texts on the basic ethology of the common laboratory species would raise awareness of normal behaviour and facilitate the use of clinical observation for setting humane endpoints.
- There is a need for better harmonisation of regulatory requirements to promote the uptake of humane endpoints. In addition, regulatory bodies should support initiatives for the development of earlier humane endpoints, preferably before serious illness occurs.
- The ability to establish and validate earlier humane endpoints would be promoted by research into the use of new technologies to better understand the processes that lead up to sensations of suffering or pain.

REFERENCES

Aldhous, P., A. Coghlan, and J. Copley. 1999. Animal experiments—Where do you draw the line? Let the people speak. *New Scientist* (1971) 162:26–31.

AVMA. 1987. AVMA colloquium on recognition and alleviation of animal pain and distress. May 15–17, 1987, Chicago, IL. Proceedings. *Journal of the American Veterinary Medical Association* 191:1184–298.

Bhasin, J., R. Latt, E. Macallum, K. McCutcheon, E. Olfert, D. Rainnie, and M. Schunk. 1999. *Canadian Council on Animal Care guidelines on choosing an appropriate endpoint in experiments using animals for research, teaching and testing.* Canada: CCAC.

BVAAWF. 1986. *The recognition of pain, distress and discomfort in small laboratory mammals and its assessment.* London: BVA, 2nd BVA Animal Welfare Foundation Symposium.

Clark, J. D., D. R. Rager, and J. P. Calpin. 1997. Animal well-being. IV. Specific assessment criteria. *Laboratory Animal Science* 47:586–97.

CCAC. 1998. Guidelines on: choosing an appropriate endpoint in experiments using animals for research and testing. Ottawa, Ontario, Canada, 30.

de Boo, J., and C. Hendriksen. 2005. Reduction strategies in animals research: A review of scientific approaches at the intra-experimental, supra-experimental and extra-experimental levels. *Atla-Alternatives to Laboratory Animals* 33:369–79.

Diesch, T. J., D. J. Mellor, C. B. Johnson, and R. G. Lentle. 2008. Responsiveness to painful stimuli in anaesthetised newborn and young animals of varying neurological maturity (wallaby joeys, rat pups and lambs). *AATEX* 14:549–52.

Duncan, I. J. H., and V. Molony. 1986. *Assessing pain in farm animals.* Luxembourg: Publications of the Commission of the European Communities.

EFSA. 2009. EFSA scientific opinion of the panel on animal health and welfare on a request from European commission on general approach to fish welfare and to the concept of sentience in fish. *The EFSA Journal* 954:1–26.

European Commission. 1986. *European directive 86/609/EC on protection for animals used for experimental and other scientific purposes.* Available from http://ec.europa.eu/environment/chemicals/lab_animals/home_en.htm

FELASA. 1994. Pain and distress in laboratory rodents and lagomorphs. Report of the Federation of European Laboratory Animal Science Associations (FELASA) working group on pain and distress accepted by the FELASA board of management. November 1992. *Laboratory Animals* 28:97–112.

FELASA. 1995. FELASA recommendations on the education and training of persons working with laboratory animals: Categories A and C. Reports of the Federation of European Laboratory Animal Science Associations working group on education accepted by the FELASA board of management. *Laboratory Animals* 29:121–31.

Fentener van Vlissingen, J. M., M. H. M. Kuijpers, E. C. M. van Oostrum, R. B. Beems, and J. E. van Dijk. 1999. *Retrospective evaluation of clinical signs, pathology and related discomfort in chronic studies.* Paper presented at Humane Endpoints in Animal Experiments for Biomedical Research—Proceedings of the International Conference, Zeist, The Netherlands.

Fitzgerald, M. 1994. The neurobiology of foetal and neonatal pain. In *The textbook of pain*, ed. P. Wall, 46–62, 3rd ed. Edinburgh: Churchill Livingstone.

Garner, J. P. 2005. Stereotypies and other abnormal repetitive behaviors: Potential impact on validity, reliability, and replicability of scientific outcomes. *ILAR Journal* 46:106–17.

Hartinger, J., D. Kulbs, P. Volkers, and K. Cussler. 2003. Suitability of temperature-sensitive transponders to measure body temperature during animal experiments required for regulatory tests. *ALTEX: Alternativen Zu Tierexperimenten* 20:65–70.

Hau, J. 1999. *Humane endpoints and the importance of training.* Paper presented at Humane Endpoints in Animal Experiments for Biomedical Research—Proceedings of the International Conference, Zeist, The Netherlands.

Hendriksen, C. F., and B. Steen. 2000. Refinement of vaccine potency testing with the use of humane endpoints. *ILAR Journal* 41:105–13.

ILAR/NRC. 1992. *Committee on pain and distress in laboratory animals: Recognition and alleviation of pain and distress in laboratory animals.* Washington, DC: National Academy Press.

Kinter, L., and D. K. Johnson. 1999. *Remote monitoring of experimental endpoints in animals using radiotelemetry and bioimpedance technologies.* Paper presented at Humane Endpoints in Animal Experiments for Biomedical Research—Proceedings of the International Conference, Zeist, The Netherlands.

Kobaek-Larsen, M., L. Rud, F. Oestergaard Soerensen, and J. Ritskes-Hoitinga. 2004. Laparoscopy of rats with experimental liver metastases: A method to assess new humane endpoints. *Laboratory Animals* 38:162–68.

Kort, W. J., J. M. Hekking-Weijma, M. T. TenKate, V. Sorm, and R. VanStrik. 1998. A microchip implant system as a method to determine body temperature of terminally ill rats and mice. *Laboratory Animals* 32:260–69.

Kramer, K. 2000. Applications and evaluation of radio-telemetry in small laboratory animals. PhD Thesis. Utrecht, The Netherlands: University of Utrecht.

Kramer, K., J. Grimbergen, B. Brockway, B. Brockway, and H. P. Voss. 1999. Circadian respiratory rate rhythms in freely moving small laboratory animals using radiotelemetry. *Laboratory Animals* 28:38–41.

Lloyd, M. H., and S. E. Wolfensohn. 1999. *Practical use of distress scoring systems in the application of humane endpoints.* Paper presented at Humane Endpoints in Animal Experiments for Biomedical Research—Proceedings of the International Conference, Zeist, The Netherlands.

Medina, L. V. 2004. How to balance humane endpoints, scientific data collection and appropriate veterinary interventions in animal studies. *Contemporary Topics in Laboratory Animal Science* 43:56, 58, 60–62.

Melzack, R., and P. Wall. 1982. *The challenge of pain.* Harmsworth, UK: Penguin books.

Mench, J. 1999. *Defining endpoints: The role of animal care committee.* Paper presented at Humane Endpoints in Animal Experiments for Biomedical Research—Proceedings of the International Conference, Zeist, The Netherlands.

Morton, D. 1992. A fair press for animals. *New Scientist* (1971) 134:28–30.

Morton, D. B. 1998. The importance of non-statistical design in refining animal experimentation. ANZCCART facts sheet. *ANZCCART News* 11:1–12.

Morton, D. B. 2000. A systematic approach for establishing humane endpoints. *ILAR Journal* 41:80–86.

Morton, D. B., G. M. Burghardt, and J. A. Smith. 1990. Animals, science, and ethics—Section III. Critical anthropomorphism, animal suffering, and the ecological context. *The Hastings Center Report* 20:S13–S19.

Morton, D. B., and P. H. Griffiths. 1985. Guidelines on the recognition of pain, distress and discomfort in experimental animals and a hypothesis for assessment. *The Veterinary Record* 116:431–36.

Morton, D. B., and P. Townsend. 1995. Dealing with adverse effects and suffering during animal research. In *Laboratory animals—An introduction for experimenters*, ed. A. A. Tuffery, 215–31. Chichester, UK: John Wiley & Sons Ltd.

Nevalainen, T., E. Berge, P. Gallix, B. Jilge, E. Melloni, P. Thomann, B. Waynforth, and L. F. van Zutphen. 1999. FELASA guidelines for education of specialists in laboratory animal science (category D). Report of the Federation of Laboratory Animal Science Associations working group on education of specialists (category D) accepted by the FELASA board of management. *Laboratory Animals* 33:1–15.

Nevalainen, T., I. Dontas, A. Forslid, B. R. Howard, V. Klusa, H. P. Kasermann, E. Melloni, et al. 2000. FELASA recommendations for the education and training of persons carrying out animal experiments (category B). Report of the Federation of European Laboratory Animal Science Associations working group on education of persons carrying out animal experiments (category B) accepted by the FELASA board of management. *Laboratory Animals* 34:229–35.

OECD. 2000. *Guidance document of the recognition, assessment and use of clinical signs as humane endpoints for experimental animals used in safety evaluation.* Paris: OECD.

Poole, T. 1997. Identifying the behavioural needs of zoo mammals and providing appropriate captive environments. *RATEL* 24:200–211.

Poon, R., and I. Chu. 1999. *Urinary biomarkers as humane endpoints in toxicological research.* Paper presented at Humane Endpoints in Animal Experiments for Biomedical Research—Proceedings of the International Conference, Zeist, The Netherlands.

Reese, E. P. 1991. The role of husbandry in promoting the welfare of animals. In *Animals in biomedical research*, eds. C. F. M. Hendriksen and H. B. W. M. Köeter, 155–92. Amsterdam: Elsevier.

Richmond, J. 1999. *Criteria for humane endpoints.* Paper presented at Humane Endpoints in Animal Experiments for Biomedical Research—Proceedings of the International Conference, Zeist, The Netherlands.

Smith, J. A., and K. Boyd, eds. 1991. *Lives in the balance.* Oxford: Oxford University Press.

Townsend, P. 1993. Control of pain and distress in small laboratory animals. *Animal Technology* 44:215–26.

Townsend, P., and D. B. Morton. 1994. *Practical assessment of adverse effects and its use in determining humane end-points.* Paper presented at Proceedings of the 1993 5th FELASA Symposium: Welfare & Science. Brighton, UK: FELASA.

Ullman-Cullere, M. H., and C. J. Foltz. 1999. Body condition scoring: A rapid and accurate method for assessing health status in mice. *Laboratory Animal Science* 49:319–23.

Wallace, J. 2000. Humane endpoints and cancer research. *ILAR Journal* 41:87–93.

Wang, L. V. 2003. Ultrasound-mediated biophotonic imaging: A review of acousto-optical tomography and photo-acoustic tomography. *Disease Markers* 19:123–38.

Wiepkema, P. R., and J. M. Koolhaas. 1993. Stress and animal welfare. *Animal Welfare* 2: 195–218.

Zimmermann, M. 1983. International association for the study of pain committee for research and ethical issues: Ethical guidelines for investigations of experimental pain in conscious animals. *Pain* 16:109–10.

15 Euthanasia

Luis Antunes, Portugal

CONTENTS

Objectives...355
Key Factors 356
15.1 Introduction 356
15.2 Avoiding the Need for Unnecessary Euthanasia 356
15.3 Euthanasia Methods...357
    15.3.1 Physical Methods 357
        15.3.1.1 Cervical Dislocation 358
        15.3.1.2 Cerebral Concussion 358
        15.3.1.3 Decapitation 358
        15.3.1.4 Microwave Irradiation 359
        15.3.1.5 Captive Bolt Shooting 359
    15.3.2 Chemical Methods 358
        15.3.2.1 Inhalational Agents...359
        15.3.2.2 Injectable Anaesthetic Agents 362
    15.3.3 Methods Applicable Only to Unconscious Animals................................362
    15.3.4 Euthanasia of the Foetus and Newborn Animal 363
15.4 Practical Considerations 363
    15.4.1 Training..363
    15.4.2 Facilities for Euthanasia 364
    15.4.3 Handling...364
    15.4.4 Equipment 364
    15.4.5 Choice of Method..364
    15.4.6 Study Considerations and Alternatives 365
    15.4.7 Animals/Tissue Sharing...366
    15.4.8 Confirmation of Death 366
    15.4.9 Disposal of Carcasses 366
    15.4.10 Proposals for the Amended European Directive 366
15.5 Conclusion 367
15.6 Questions Unresolved 367
Acknowledgements...367
References...367

OBJECTIVES

Euthanasia, or humane killing may be necessary at the end of an experiment, to alleviate suffering or to dispose of animals that are surplus to requirements. It is an ethically contentious subject that places moral responsibilities on all those breeding, caring for and using laboratory animals.

There is no ideal way of ending an animal's life, and the method adopted is a compromise between the requirements of the experiment, the facilities and expertise available, and minimising distress to the animal. Methods based on narcosis may rely on the administration of anaesthetic agents, or induction of severe hypercapnoea or anoxia. Several physical methods are also available, most of which rely on delivering instantaneous trauma to the brain. All of these methods should be followed by measures to assure the irreversibility of the process. This chapter reviews the relative merits of these different methods and stresses the importance of providing appropriate facilities and training of staff involved in carrying it out.

KEY FACTORS

- All those engaged in laboratory animal science must recognise the ethical sensitivity and close regulatory environment in which euthanasia takes place.
- Staff carrying out euthanasia should be trained, competent, experienced and have access to well-maintained facilities and equipment.
- The method chosen should take account of welfare, research needs, safety and aesthetical and ethical issues, and should assure rapid, stress-free loss of consciousness (LOC) without subsequent recovery.
- Unnecessary distress should be avoided by using the home cage and established social groupings wherever possible, careful handling, skilful manipulation and avoiding alarm communication between animals.
- Establishments should develop a broad strategy to minimise the number of animals euthanised, for example by encouraging good experimental design, efficient breeding policies and germ-line banking.

15.1 INTRODUCTION

The word euthanasia is derived from the Greek terms "eu" meaning good and "thanatos" meaning death. A "good death" is one that is accompanied by minimal pain and distress. Euthanasia of terminally ill human patients or as an ultimate sanction in penal systems remains a controversial issue and raises conflicting ethical opinions, the interpretation and resolution of which differs between individuals and cultures. In applying the term to animals, it is necessary also to consider moral issues relating to the killing of animals that may be clinically healthy, and also economical and commercial factors. From an ethical viewpoint, it can be argued that a death that does not have the purpose of ending suffering can never be considered a "good death" and the term euthanasia is currently often replaced by more precise terms such as humane death or humane killing. Throughout this review the term euthanasia will be used to describe killing by means that minimise adverse experiences on the part of the animal, whatever the reason for carrying it out.

Laboratory animals may be euthanised for various reasons: for example when an experiment is finished and there is no justification for continuing to house them in laboratory conditions or because tissues must be removed for analysis; when adverse effects such as pain, distress or suffering can no longer be justified and when animals are unwanted stock because of overbreeding or have unsuitable biological characteristics, or they are no longer suitable for breeding (Close et al. 1996). Euthanasia is particularly difficult to justify ethically when the animals are a result of overbreeding, or are of unwanted sex, genotype or phenotype. Euthanasia is much easier to justify when animals are experiencing pain or suffering, which cannot be relieved or when a pre-determined humane end point has been reached.

15.2 AVOIDING THE NEED FOR UNNECESSARY EUTHANASIA

Wherever possible, measures should be taken to avoid overproduction by carefully planning and monitoring breeding programmes to ensure that these match research needs, with respect to number, uniformity and health (APC 2004). The situation is complicated somewhat by the aggressive nature of some strains of male mice and rabbits, which makes it difficult to house them socially.

Because of this, many scientists prefer to work with female animals; this issue is discussed further in Chapter 4. It is important too that the quality (e.g., demeanor and health status) of the animals is sufficiently high to enable experimental procedures to be conducted using minimal numbers, thereby maximising the efficiency of experimental programmes (EFSA 2005; see also Chapter 6). In practise, wastage is usually minimised by purchasing animals from a commercial breeder rather than establishing in-house breeding programmes. In the case of specific genetic strains such as most genetically altered lines, this may not be possible. In such cases, if it is necessary to maintain a specific strain of animals for future use, consideration should be given to cryopreservation of embryos or gametes rather than maintaining small breeding colonies over many generations simply to ensure that the genetic line is maintained. To some extent, the practicalities of animal supply mean that deaths in such cases are inevitable, but all measures possible should be taken to minimise the number involved (see Chapters 8 and 9). Good experimental design is also an important contributor to minimising the number of animals required for experiments and therefore the number that ultimately needs to be killed; this is discussed at length in Chapter 6.

15.3 EUTHANASIA METHODS

Euthanasia techniques should induce rapid loss of consciousness (LOC) unless instant death can be effectively achieved, for example by an appropriate physical method. In the former case, euthanasia should ensue without distress or pain. Techniques such as neuromuscular blockage or electrocution are only acceptable when applied to animals in which unconsciousness has already been established, because they have the potential to cause pain and distress. Euthanasia should only be carried out by personnel who are well-trained and experienced in the techniques being used. The selection of euthanasia method should be based upon welfare, effectiveness, distress, pain imposed, speed of inducing unconsciousness, ease of performance, safety for operator, aesthetical and ethical implications within the context of the research needs.

It is important that euthanasia is carried out in an area free from disturbances, where all the facilities requirements are available, both for carrying it out and ensuring that it is maintained in a clean and efficient condition, and that it should not be carried out in the presence of other, conscious animals that may be sensitive to alarm calls, release of pheromones, sudden noises or unusual movements by animals being killed. Facilities must also be readily available for making measurements or the collection of tissue samples where necessary, and for the safe and effective disposal of carcases.

The methods most commonly adopted for adult rodents and rabbits may be divided into physical, chemical and methods used only after LOC. A comparison of different methods of euthanasia applicable to rodents and rabbits is provided in Table 15.1.

15.3.1 Physical Methods

Physical methods should provoke immediate and assured LOC by causing physical trauma to the brain; this may cause less distress to the animal but requires expertise and careful handling and restraint of the animal (Close et al. 1997). These are often the methods of choice although in all cases death should be confirmed by exsanguination (e.g., by opening major blood vessels) or destruction of the brain. The expertise of the operator has a great impact on the humaneness of these techniques and some find them distasteful; such persons may become anxious or hesitant and should not carry out them. It is also important to ensure that the methods are not carried out in a hurried way and that the person performing them out does not become fatigued and consequently less efficient.

Whilst physical methods provide a rapid and, when skillfully carried out, humane death, their use is limited when large numbers of animals must be killed because the techniques must be performed by skilled operators for whom the procedure is often time consuming, demands concentration and is sometimes perceived by personnel as unpleasant.

TABLE 15.1
Comparison of Different Methods of Euthanasia Applicable to Rodents and Rabbits

Methods for Animals Other Than Foetal and Embryonic Forms	Animals for Which the Technique is Appropriate/Special Notes
Cervical dislocation of the neck	Mice and rats under 150 g
	Should be followed by immediate exsanguination, or immediate destruction of the brain, or death to be ensured by another method
Concussion of the brain by striking the cranium	Rats
	Should be followed by immediate exsanguination, or immediate destruction of the brain, or death to be ensured by another method
Decapitation	Mice, rats under 200 g and rabbits under 1 Kg
Microwave irradiation	Rodents only, specialised equipment required
Captive bolt	Rabbits, to be followed by immediate exsanguination, or immediate destruction of the brain, or death to be ensured by another method
Anaesthetics overdose using a route and anaesthetic agents appropriate for the size and species of animal	Mice, rats and rabbits
	Prior sedation of the animal may be recommended, specially in rabbits
Exposure to carbon dioxide gas	Rodents only
	To be used in a rising concentration only
Carbon monoxide	Danger to operator
Inert gases (nitrogen, argon)	To be followed by immediate exsanguination, or immediate destruction of the brain, or death to be ensured by another method

15.3.1.1 Cervical Dislocation

This method is commonly used in small rodents; the objective is to cause extensive damage to the brain stem resulting in immediate LOC and death. In mice and rats the thumb and index finger are placed on either side of the neck at the base of the skull, or a rod is pressed at the base of the skull and the other hand grasps the pelvis or hind limbs and the base of the tail; a sharp, quick pull is applied that separates the cervical vertebrae from the skull (AVMA 2007). Rabbits are usually suspended briefly by the hind-limbs and a sharp blow is delivered to the back of the cranium. Rodents or rabbits weighing over 150 g shoud be sedated before this procedure is carried out (Close et al. 1997; AVMA 2007). Death should be confirmed at once.

15.3.1.2 Cerebral Concussion

Cerebral concussion is a method of stunning that provokes an instantaneous but potentially reversible LOC without dislocating the spine from the base of the skull and thereby minimising gross anatomical changes in the brain stem (EFSA 2005). Concussion is achieved by striking the back of the head, thereby rendering the animal unconscious. This is acceptable for animals weighing less than 1 kg and death must always be carefully confirmed.

15.3.1.3 Decapitation

Decapitation is also used as a physical euthanasia method. The head is separated from the body usually using a mechanical device with a sharp and clean blade such as a guillotine; this provides a means to obtain fluids without chemical contamination and anatomically undamaged brain tissue.

This technique provides a method of recovering tissues and body fluids that are chemically uncontaminated (AVMA 2007) and physical disruption of the structure of the central nervous system is minimised. It is particularly important that the equipment used for decapitation is clean and maintained in good working order and that the blade is kept sharp.

15.3.1.4 Microwave Irradiation

Microwave irradiation is also used as a euthanasia method for rats and mice (AVMA 2007). Domestic microwave equipment is not suitable—the technique needs specialised equipment and was first used to enable sampling of brain metabolites in their *in vivo* state whilst maintaining the anatomic integrity of the brain. The equipment delivers microwave irradiation directly to the head of the animal (EFSA 2005); personnel using this equipment must be suitably trained to ensure correct positioning of the beam and application of radiation of suitable power.

15.3.1.5 Captive Bolt Shooting

The captive bolt pistol has been used for euthanasia of large laboratory rabbits. It relies on the instantaneous induction of extensive trauma and concussion to the brainstem and cerebral hemispheres. Captive bolt pistols are powered by explosive charges or compressed air; only devices designed specifically for use on rabbits should be used and personnel must be well-trained and able to ensure correct positioning of the weapon.

15.3.2 CHEMICAL METHODS

Chemical methods are well-suited for the euthanasia of larger numbers of animals. This group of methods includes use of compounds delivered by *inhalation* such as carbon dioxide (CO_2), carbon monoxide (CO), inert gases and volatile anaesthetics; also *injectable agents* such as intravenous anaesthetics.

15.3.2.1 Inhalational Agents

15.3.2.1.1 Volatile Anaesthetic Agents

When inhalational substances are used, rodents have to be placed into an anaesthetic chamber to which the anaesthetic agent is introduced, usually with oxygen. Older equipment consisted of a chamber within which volatile anaesthetic agents were released from a receptacle containing gauze soaked with the agent being used; in the case of such an apparatus, it is extremely important to prevent the animal coming into direct contact with the anaesthetic agent (which is irritant), but the conditions in the chamber are not easily controlled and except for very occasional use, such devices are best avoided. For best practise, a calibrated vaporiser should be used to deliver a predetermined concentration of the anaesthetic into the chamber at a controlled rate. During induction with volatile anaesthetic agents, air or oxygen should be provided (Close et al. 1997) to ensure that hypoxia does not develop. Waste anaesthetic gas should be scavenged in order to protect the health of the operator. These agents should not be delivered by face mask because this is often distressing for the animal and increases the risk of human exposure (EFSA 2005).

To exert their effect, anaesthetic agents must exceed a certain alveolar concentration, which can take some time, depending on the rate of flow, concentration of agent and the size of the chamber. Particular care is needed to ensure that animals do not become distressed between initiating induction and before consciousness is lost—they may become excited or struggle and show anxiety because of issues related to handling and separation from their peers, and some anaesthetic vapours may be mildly irritating (AVMA 2007). To reduce stress, animals placed together in chambers should be of the same species and wherever possible, kept in their social groups. Chambers should not be overloaded and need to be kept clean to minimise odours that might distress animals subsequently placed in them (AVMA 2007).

Inhalational anaesthetic agents such as halothane, isoflurane, enflurane, desflurane and sevoflurane have been suggested as more humane agents than CO_2, although there is no consensus on whether distress during induction with these agents is less than with CO_2. Inhalational anaesthesia is also used to induce unconsciousness before the application of methods that would otherwise compromise their welfare. A wide range of anaesthetic agents is available and a more in-depth review of their pharmacological characteristics can be found elsewhere (Brunson 2008).

Halothane induces anaesthesia smoothly and rapidly and is an effective inhalant anaesthetic for euthanasia, although because it is no longer used for human anaesthesia it is more difficult to find it on the market. In comparison with halothane, isoflurane undergoes much less hepatic metabolism and is therefore more suitable for studies of metabolic processes. The agent is available in many laboratories; it is less soluble than halothane, and would be expected to induce anaesthesia more rapidly. However, it has a slightly pungent odour and animals often hold their breath, delaying the onset of LOC and this may lead to stress. It may be necessary to use larger quantities of isoflurane to achieve euthanasia, compared with halothane. Although isoflurane is an acceptable agent for euthanasia, halothane is preferred. Sevoflurane is less soluble than halothane and does not appear to have an objectionable odour. It is less potent than isoflurane or halothane and has a lower vapour pressure. Anaesthetic concentrations can be achieved and maintained easily and rapidly. Desflurane has become available only relatively recently and is currently the least soluble potent inhalant anaesthetic available; however, the vapour is quite pungent, which may prolong induction. Desflurane is so volatile that it could displace oxygen and induce hypoxemia during induction if supplemental oxygen is not provided. Enflurane is less soluble in blood than halothane, but, because of its lower vapour pressure and lower potency, the rate of induction is similar to that of halothane. Enflurane is not readily available in some countries. At deep anaesthetic planes, animals may show seizures, which may be disturbing to personnel (AVMA 2007).

15.3.2.1.2 Carbon Dioxide

Carbon dioxide (CO_2) is used to euthanise groups of rodents using specially designed chambers. This gas induces widespread tissue acidosis, including within the central nervous system, and inhibition of neurones leads to LOC, insensibility and finally death (EFSA 2005). It is recommended that cylinders of compressed carbon dioxide (industrial grades should not be used because of impurities they contain) should be used in conjunction with calibrated flow-meters to allow precise regulation of the inflow to the chamber. The gas flow should be constant and a rate of 15–35% of the chamber volume minute^{-1} is commonly used, although the rate of flow may be increased once consciousness is lost to shorten the time until death. The American Veterinary Medical Association (AVMA 2007) recommends a minimum flow rate of 20% volume minute^{-1}, however several studies using objective behavioural measurements suggest that this figure is an absolute maximum (Young 2006; Niel, Stewart, and Weary 2008). The gas flow should be maintained for at least 1 minute after cessation of respiratory movements and apparent clinical death (AVMA 2007).

Major concerns have been raised over the use of CO_2 and there is ongoing debate about whether use of pre-filled or gradually filled chambers is preferable (Artwohl et al. 2005). Breathing of high concentrations of CO_2 appears to be both aversive and painful. Euthanasia with pre-filled chambers is certainly rapid but is bad practice because there is a high possibility of animals experiencing potentially severe pain and distress, if only for a relatively short-period. It is recommended that animals should be placed into a chamber that contains room air and that this is then gradually replaced by CO_2 (Artwohl et al. 2005). For example, the UK Home Office (Scientific Procedures Act, 1986) and the current draft of the new EC directive (European Commission 2008) specifies that euthanasia of rodents with CO_2 should be performed only with a gradually rising concentration of CO_2.

Recent studies by Niel, Stewart, and Weary (2008) have shown that rats find gradual-fill CO_2 exposure to be aversive, even at relatively low CO_2 concentrations (ca. 15%). These studies suggest that even during gradual-fill procedures, mechanisms other than pain may cause distress. Thus, animals placed in a chamber with a rising concentration of CO_2 may show aversive behaviour and may experience dyspnoea and "air hunger". This situation is known to be very distressing in humans (Banzett and Moosavi 2001) and supports the need to identify alternative methods of euthanasia. One approach would be to replace CO_2 by an anaesthetic vapour as discussed earlier. Because the time taken for these agents to cause death may be prolonged, they could be used to induce LOC and replaced with CO_2 until death supervenes. However, a move away from the use of CO_2 for euthanasia faces two obstacles: practicality and economics (Marris 2006). Anaesthesia-based techniques are slower and more expensive in terms of anaesthetic and equipment. Arguments in favour of the continued use of carbon dioxide as a sole agent are that it is a practical and effective agent, which minimises the need for animal handling in comparison with techniques that require greater technical expertise such as cervical dislocation, and that consequently there are fewer possibilities of errors being made during the killing process. Moreover, for some authors the benefit to be achieved by using an anaesthetic agent for induction of anaesthesia before killing by CO_2 is still not established and it is also sometimes argued that the question about whether CO_2 induces stress or distress has not been convincingly answered (Sharp, Azar, and Lawson 2006).

15.3.2.1.3 Carbon Monoxide

Carbon monoxide (CO) binds irreversibly with haemoglobin to form carboxyhaemoglobin thereby blocking uptake of oxygen by erythrocytes and resulting in tissue hypoxemia, leading finally to unconsciousness and death (AVMA 2007). During unconsciousness, muscular convulsions and spasms may occur as a result of the stimulatory effect of CO on the motor centres of the brain; local cerebral bleeding may result in paralysis (EFSA 2005). Carbon monoxide induces LOC without pain; there is minimal discernible discomfort and hypoxemia induced by CO is insidious, so that the animal appears to be unaware (AVMA 2007). Rodents should be put in a container pre-filled with at least 6% CO by volume. At concentrations in air greater than 10% CO is highly explosive and it is toxic to personnel, so it should be used with appropriate gas scavenging. Commercial sources of CO are preferable to local generation of CO by other means, because it is not contaminated with other gasses and it avoids problems associated with adjusting the concentration, cooling of the gas and equipment maintenance; personnel using CO must be well-trained in its effective and safe use and must understand its hazards and limitations (AVMA 2007).

15.3.2.1.4 Inert Gases

Nitrogen and argon are neither flammable nor explosive; they are inert, colourless and odourless gases. For euthanasia, a container is usually pre-filled with a minimum of 98% by volume of nitrogen or argon and after transferring animals into the chamber death is induced by severe, acute hypoxemia. Nitrogen has a similar density to air, and specialised equipment is needed for its administration; argon is denser than air and is contained easily. The oxygen content of the argon gas must be kept below 2% by volume to hasten death by anoxia (EFSA 2005). There is no consensus concerning the welfare aspects of the use of argon or nitrogen and further studies are needed. Early studies had suggested that although benign in pigs (Raj 1999), argon is aversive to rodents (Niel and Weary 2007; Makowska et al. 2008). The AVMA Guidelines on Euthanasia, indicate that these techniques are conditionally acceptable only if the concentration of oxygen is rapidly reduced to below 2%, and the animals are previously heavily sedated or anaesthetised (AVMA 2007).

In conclusion, there is insufficient evidence to indicate whether these or alternative gaseous agents or gas mixtures that cause death by hypoxia or anoxia, provide a humane way for killing rats or mice. The use of volatile anaesthetic agents may be an appropriate alternative to CO_2 but the vapour of some of these is aversive. They can be used either as the sole euthanasia agent, accepting that, time to death may be very prolonged (sometimes in excess of 20 minutes) and require large quantities of the agent that renders their use in this way impractical and very expensive. Alternatively, for practical and often technical research reasons, the animal may be killed after anaesthesia has been induced by switching to CO_2, administration of a parenteral injection of pentobarbitone or the use of a physical method (Hawkins et al. 2006).

15.3.2.2 Injectable Anaesthetic Agents

Injectable anaesthetic agents may be used for euthanasia when employed at concentrations much higher than those used for anaesthesia (i.e., by administration of an overdose) via the intraperitoneal (i.p.), intravenous (i.v.), intramuscular (i.m.) or subcutaneous (s.c.) routes. Most commonly the i.p. route is used in mice and rats and the i.v. route in rabbits. The dose should be sufficient to induce respiratory arrest followed by cardiac arrest. The i.v. route is the most rapid and reliable, although it may be difficult to perform in rats and mice without causing them distress; the i.p. route is the easiest to use; the rate of absorption from both i.m. and s.c. routes is slower and more variable so that LOC takes too long—they are not acceptable as euthanasia routes. Intracardiac (i.c.) injections are very painful and unacceptable in conscious animals. Usually, injection of twice the anaesthetic dose of the agent provokes respiratory arrest; four times the anaesthetic dose produces cardiac arrest in ventilated animals, while in non-ventilated animals three times the dose is satisfactory (Close et al. 1997); there is little confirmation of these doses in the literature. If sedatives are required prior to overdosing by an injected or inhaled agent, they may be administered by s.c. or i.m. routes.

Barbiturates are the most commonly used and accepted agent for euthanasia. They depress the central nervous system causing progressive LOC and leading to respiratory and cardiac arrest.

Sodium pentobarbitone is the most commonly used injectable agent for euthanasia; preparations specifically designed for euthanasia are available commercially and can be administered intravenously or intraperitoneally. However, pentobarbitone administered via the i.p. route may cause irritation of the serosa because of its alkalinity. To avoid this, the drug can be diluted with sterile saline, or the solution can be buffered to about neutrality or combined with a local anaesthetic agent (EFSA 2005).

T-61 is another injectable agent used for euthanasia; it is a non-barbiturate and non-narcotic. T-61 is a combination of three drugs—tetracaine hydrochloride, embutramide and mebenzonium iodide—that provide local anaesthetic, hypnotic and paralytic actions, respectively. This agent should only be administered by slow intravenous injection because it is otherwise painful (Close et al. 1997; AVMA 2007). Barbiturates are preferable to T-61 since the neuromuscular blocking effect may provoke respiratory muscle paralysis before LOC, thus causing substantial distress (Hellebrekers et al. 1990). Therefore, in some countries T-61 is not allowed without premedication with an anaesthetic, and is generally banned for rodent euthanasia in others.

15.3.3 METHODS APPLICABLE ONLY TO UNCONSCIOUS ANIMALS

There are several methods that should be used only in animals already rendered unconscious. Two of these will be considered here: rapid freezing and exsanguination.

Rapid freezing is useful for biochemical analysis of tissues because it minimises enzyme activity. This technique is acceptable only under specific circumstances when justified by experimental requirements and subject to careful ethical evaluation; it is applicable to very small animals such as embryonic rodents or rabbits or to neonatal rodents; specialised equipment and trained personnel are required (Close et al. 1997). After induction of anaesthesia or rendering insensible, the animal is completely immersed in liquid nitrogen; it takes 10–90 seconds to freeze the deep structures of the brain.

Killing by exsanguination also requires unconsciousness to be induced beforehand, because of the stress caused by hypovolaemia and by the pain of cutting deeper blood vessels. It is particularly important to use a different room to kill the animals so as to avoid other rodents smelling the blood.

TABLE 15.2
Methods of Euthanasia Appropriate for Foetal and Embryonic Forms of Rodents and Rabbits

Methods for Foetal and Embryonic Forms	Animals for Which Appropriate
Overdose of anaesthetics	Mice, rats and rabbits
Cooling of foetuses followed by immersion in cold tissue fixative	Mice, rats and rabbits under 4 g
Decapitation	Mice and rats

Exsanguination is also used to confirm death after the application of other procedures such as concussion and cervical dislocation (Close et al. 1997; AVMA 2007).

15.3.4 EUTHANASIA OF THE FOETUS AND NEWBORN ANIMAL

Two critical factors must be considered when choosing the euthanasia method for foetal or newborn animals. They are resistant to hypoxia and they metabolise drugs more slowly than older animals. In view of this, it is recommended that two methods acceptable for the species should be combined and death confirmed by careful observation of clinical signs.

Methods of euthanasia acceptable for neonates are injection of chemical anaesthetics (e.g., pentobarbital), cervical dislocation or decapitation (Klaunberg et al. 2004); these are summarised in Table 15.2. In foetal or neonatal rodents decapitation should be performed using a sharp blade or scissors. Reflex withdrawal responses to noxious stimulation is observed in rodent embryos starting in late gestation, for example, on embryonic day 17 in the rat foetus (Narayanan, Fox, and Hamburger 1971) and the method of killing should take account of this. Immersion in liquid nitrogen may be used only if preceded by anaesthesia, as discussed above. The use of hypothermia for euthanasia is no longer considered appropriate (AVMA 2007).

In the case of neonatal rodents, Pritchett and others (2005) have confirmed the very extended period between exposure to an atmosphere of CO_2 and the time of death as compared with adults, a consequence in part of their resistance to hypoxia (Singer 1999; AVMA 2007). In rats, the time to death decreased by 3 minutes for every 1 day increase in age between days 0 and 10 (Pritchett-Corning 2009). By 10 days of age all died after exposure for 5 minutes. There were also pronounced differences between inbred and outbred strains (Pritchett et al. 2005). Methods appropriate for euthanasia of neonatal animals with CO_2 are therefore substantially different from those employed for adults.

15.4 PRACTICAL CONSIDERATIONS

15.4.1 TRAINING

The killing of animals by inadequately trained or unskilled people can cause significant, unnecessary pain, or distress. The competence of persons carrying out this operation is paramount. Only authorised and trained persons, experienced with the equipment employed and using a humane method that is considered appropriate to the species and the scientific purpose, should carry out euthanasia. In most instances this will involve maintaining a register of persons competent in specific techniques and making arrangements to review periodically the continuing competence of those registered.

Researchers should be provided with training to enable them to perform euthanasia, although if they do not feel confident using the chosen technique, they should be encouraged to arrange in advance for staff in the animal facility to provide further instruction or to carry the procedure out on their behalf. During weekends and holidays, special arrangements may be necessary.

15.4.2 Facilities for Euthanasia

Animals should be euthanised in a clean quiet environment, in the absence of other animals. If an animal is frightened, it may make stress-related behavioural responses (e.g., vocalisation, urination, and defecation) that may cause fear and distress to other animals. Where possible, animals should be killed within their home cage or as soon as possible after they are removed from it. When euthanising social groups of animals it is advantageous to kill them together in the home cage, which causes much less stress than removing them to an unfamiliar location (Wolfensohn and Lloyd 2003). McIntyre and others (2007) have described a number of innovations, including a method of home cage euthanasia and the use of gas chambers with fill rates or gas mixtures that are tailored to minimise stress. In the event that euthanasia is required in an emergency, for example if an animal has been involved in a fight and is badly wounded, it may be better to kill it immediately, rather than to move it away to another room. Records should be kept of all animals euthanised.

15.4.3 Handling

Animals must be handled carefully and competently without causing them stress. Euthanasia should be carried out in designated places, suitably equipped and maintained, and never in the presence of other, conscious animals. Animals are often calmer if they are already familiar with the personnel accustomed to handling them and there may be considerable advantage in animal care staff becoming actively involved in performing euthanasia on behalf of researchers, who may be less familiar to the animals and perhaps have less experience with the various techniques.

15.4.4 Equipment

Several methods of euthanasia require equipment of varying complexity. This may vary from disposable needles and syringes to gaseous anaesthesia equipment incorporating vaporisers, scavenging equipment, and induction chambers. Regardless of its complexity, the availability of the equipment and its effective functioning should be verified beforehand. The apparatus should be checked for cleanliness and cleaned again between animals or batches to prevent transmission of odours and heightening animals' apprehension. Rodents make extensive use of pheromones in urine and faeces for communication (Close et al. 1996) and it is good practise to line the floor of the induction chambers or bench workplace with disposable paper, and to change this each time after use.

15.4.5 Choice of Method

One key consideration when selecting the euthanasia method is the method's ability to cause unconsciousness and death without evoking apprehension, distress, fear, or pain in the animals. For example, although published studies draw conflicting interpretations from the behavioural responses to CO_2 exposure, it is very probable that the formation of carbonic acid on mucous membranes, hypoxia, and hypercapnoea leading to hyperventilation are likely contributors to stress in animals killed in chambers pre-filled with CO_2. One recommended procedure is to place animals into a chamber that contains room air and then to gradually introduce CO_2 (Artwohl et al. 2005). Conversely, although decapitation is a rapid process that quickly induces LOC, the handling and restraint required may be distressful to rats together with exposure to novel equipment and surroundings (Stutler et al. 2007). When the protocol allows, use of anxiolytic agents or induction of light general anaesthesia prior to decapitation will reduce the associated stress.

15.4.6 Study Considerations and Alternatives

It is essential that experiments are planned and performed in a way that ensures the validity of the data produced. If the method used for euthanasia interferes with the ultimate goals of the research study and makes the data unusable, then the lives of the animals have been wasted. Researchers should always perform a literature review and critically evaluate the techniques used by others before carefully assessing the possible adverse effects of the various options available. There may be special circumstances in which options not listed in this document might be considered acceptable. Whatever method is chosen, it must be carefully considered and justified by the investigator, the local ethical evaluation process, and the national competent authority to assure the best outcome for the animals as well as the study.

For example, gross structural or histopathological changes arising from the method used for euthanasia may impact on the usability of samples collected post-mortem; for example, cervical dislocation may damage the brain or the tissues in the cervical region and most methods affect the lungs to some degree, although often this is simply mild congestion of the alveolar capillaries. Intraperitoneal injection of sodium pentobarbitone causes splenic enlargement, focal vascular congestion beneath the intestinal serosa and subcapsular necrosis of the liver and kidneys. Cervical dislocation and decapitation are suitable methods for examination of the abdominal viscera and to obtain biochemically unaltered tissues and cells. The use of agents such as halothane or isoflurane may have a greater impact on hepatic metabolism and are therefore unsuitable for metabolic studies.

The impact of the method of euthanasia on the research outcome is illustrated by a study performed to evaluate the effects of seven euthanasia methods on sperm motility in mature rats (Stutler et al. 2007). It was found that euthanasia by CO_2 inhalation met the pre-defined criteria, with the added benefits of easy administration, personnel safety and low cost. Other gaseous agents were acceptable providing they were used only to rapidly induce anaesthesia and were immediately followed by a physical method of euthanasia, such as decapitation or exsanguination. This combined approach reduced the duration of exposure to the agent, time to clinical death and likelihood of adverse effects on sperm motility parameters. The authors concluded that in a research setting, euthanasia procedures may directly impact on study outcome measures.

Additional information may be found in reports from several working groups that provide recommendations for euthanasia of experimental animals in Europe (Close et al. 1996, 1997; EFSA 2005) and in the United States (Artwohl et al. 2005; AVMA 2007). However, ultimately it is the researcher's responsibility to select the most appropriate euthanasia method. Briefly, the following general considerations should be taken into account when selecting the method (Wolfensohn and Lloyd 2003):

- Ability to induce LOC and death without causing pain, distress, anxiety or apprehension
- Compatibility with scientific requirements, purposes and subsequent evaluations, examination or use of tissue
- It should be simple and easy to carry with few possibilities for errors
- LOC and death should occur as quickly as possible
- Reliability and irreversibility
- Safety of personnel carrying it out
- Emotional effect on observers or operators
- Drugs and equipment should be readily available and have minimal potential for human abuse.
- Compatibility with species, age and health status
- Ability to maintain equipment in proper working order
- The method should be economically acceptable
- The facilities for carcase disposable need to be considered

15.4.7 Animals/Tissue Sharing

Animal tissue and organs are used for the development of *in vitro* methods. Where practicable, tissues of animals being killed should be made available to researchers and academics as a way of reducing the number of animals used. A central database at each institution, listing the researchers and projects involved, should simplify the coordination of arrangements for collecting tissues at the moment of death, helping reduce the number of animals used and costs. Such practises help avoid unnecessary duplication of research and foster the spirit of the current proposals for the new European Directive, which promotes application of the principle of reduction. Member states may wish to consider establishing local, regional or national programmes for sharing the organs and tissue of animals killed for other purposes.

15.4.8 Confirmation of Death

After performing euthanasia by any method and before disposing the carcases it is essential to confirm death to ensure that the animal will not recover. The process of killing may be complemented with one of the following methods:

- Confirmation of permanent cessation of heartbeat and circulation
- To ensure destruction of the brain absence of corneal or palpebral reflexes and glazing of eyes
- Dislocation of the neck
- Exsanguination
- Confirming the onset of rigor mortis
- Instantaneous destruction of the body in a macerator

15.4.9 Disposal of Carcases

After an animal has been euthanised and its death confirmed, it is the responsibility of the individual performing the procedure to ensure proper disposal of the carcase, for example by transferring it to the designated cold room or freezer. Animal carcasses are classified as clinical waste and as such they should be disposed of in a safe way and in accordance with national legislation and local regulations. In Europe, these should follow regulation (EC) No 1774/2002 of the European Parliament and of the Council of the European Union of 3 October 2002 laying down health rules concerning animal by-products not intended for human consumption (European Commission 2002). Usually carcases are placed in clinical waste bags that may then be incinerated on site or collected and disposed of by an approved contractor. Where there is doubt, the advice of the safety and health administration office should be sought.

Special care is necessary with biohazardous carcases, which have been exposed to hazardous chemicals, infected with organisms pathogenic to humans, inoculated with human derived tissues, fluids, or cell lines. Where carcases are known to be radioactive, special disposal procedures may be in place and radiation safety personnel should be contacted. Always contact your safety personnel, as different institutions may have different procedures.

15.4.10 Proposals for the Amended European Directive

The current draft of the new European directive requires member states to ensure that animals are killed in an authorised establishment, by an authorised person, and with a minimum of pain, suffering, and distress and using the appropriate humane method of killing for the species as set out in Annexes proposed for the Directive, although provision is made in case of field studies for animals to be killed outside an authorised establishment. Competent local authorities may grant exemptions from this on the basis of scientific justification that the purpose of the procedure cannot be achieved by the use of a standard method of humane killing. Provision is also made for an animal to be killed in emergency circumstances for welfare reasons and individual member states are empowered to determine the emergency circumstances.

15.5 CONCLUSION

There remains a great deal of research to be done on the topic of laboratory animal euthanasia. There is a lack of consensus between individual opinions regarding the best euthanasia techniques, this may reflect a wide range of experiences between users of the techniques and high variability related with subjective concepts (e.g., distress, pain, level of expertise).

15.6 QUESTIONS UNRESOLVED

The ethical awareness of staff responsible for supplying and using animals should be raised to ensure that numbers of animals killed are minimised. Examples are the avoidance of overbreeding and greater uptake of gene-line archiving, which avoids the need to maintain small breeding colonies whose purpose is only to perpetuate a strain of animals.

There is an urgent need to gain a clearer understanding of the welfare implications of different methods of euthanasia and in particular to develop protocols for carrying them out, which are robust and reliable.

The systematic collection of data on the biochemical and structural changes induced by different euthanising methods would provide a more rational basis for ensuring the scientific validity of the method chosen.

Currently pentobarbitone sodium is the most commonly used injectable euthanising agent; it has several drawbacks including tissue irritancy and induction of histopathological and biochemical changes that may confound scientific measurements. The availability of a different agent without these drawbacks would be welcomed.

All of the current inhalational methods have short-comings and it is essential to continue the search for novel methods that better achieve a "good death" for laboratory animals. For example, there is no consensus concerning the welfare aspects of the use of argon or nitrogen, and studies to establish these needs to be performed.

The range of olfactory, auditory, and visual cues that animals use when distressed needs to be better understood so that the degree of isolation appropriate to areas where euthanasia is carried out can be more precisely estimated.

ACKNOWLEDGEMENTS

This work was supported by research grant POCTI/CVT/59056/2004 from Fundação para a Ciência e Tecnologia (FCT), co-funded by FEDER, Lisbon-Portugal and European COST Action B24 Laboratory Animal Science and Welfare.

REFERENCES

APC. 2004. *Report of the animal procedures committee for 2003*. House of Commons, London: The Stationery Office, HC 1017.

Artwohl, J., P. Brown, B. Corning, and S. Stein. 2005. *Public statements: Report of the ACLAM task force on rodent euthanasia*. ACLAM.

AVMA. 2007. *AVMA guidelines on euthanasia (formerly report of the AVMA panel on euthanasia)*. AVMA, Available from http://www.avma.org/issues/animal_welfare/euthanasia.pdf

Banzett, R., and S. Moosavi. 2001. Dyspnea and pain: Similarities and contrasts between two very unpleasant sensations. *American Pain Society Bulletin* 11:6–8.

Brunson, D. 2008. Pharmacology of inhalation anesthetics. In *Anesthesia and analgesia in laboratory animals*, eds. R. Fish, P. Danneman, M. Brown, and A. Karas, 83–95, 2nd ed. Amsterdam: Elsevier.

Close, B., K. Banister, V. Baumans, E. M. Bernoth, N. Bromage, J. Bunyan, W. Erhardt, et al. 1996. Recommendations for euthanasia of experimental animals: Part 1. DGXI of the European commission. *Laboratory Animals* 30:293–316.

Close, B., K. Banister, V. Baumans, E. M. Bernoth, N. Bromage, J. Bunyan, W. Erhardt, et al. 1997. Recommendations for euthanasia of experimental animals: Part 2. DGXI of the European commission. *Laboratory Animals* 31:1–32.

EFSA. 2005. Aspects of the biology and welfare of animals used for experimental and other scientific purposes. *Annex to the EFSA Journal* 202:1–136.

European Commission. 2002. *Laying down health rules concerning animal by-products not intended for human consumption in union*, eds. EPAOTCE EC. Available from http://europa.eu/legislation_summaries/food_safety/specific_themes/f81001_en.htm

European Commission. 2008. *Proposal for a directive of the European Parliament and of the council on the protection of animals used for scientific purposes.* Available from http://www.europarl.europa.eu/sides/getDoc.do?pubRef=-//EP//NONSGML+TA+P6-TA-2009-0343+0+DOC+PDF+V0//EN

Hawkins, P., P. Playle, H. Golledge, M. Leach, R. Banzett, A. Coenen, J. Cooper, et al. 2006. Newcastle consensus meeting on carbon dioxide euthanasia of laboratory animals. [Cited September 30, 2009]. Available from http://www.nc3rs.org.uk/downloaddoc.asp?id=416&page=292&skin=0

Hellebrekers, L. J., V. Baumans, A. P. M. G. Bertens, and W. Hartman. 1990. On the use of T61 for euthanasia of domestic and laboratory-animals—An Ethical evaluation. *Laboratory Animals* 24:200–204.

Home Office. 1986. *Animals (scientific procedures) Act.* Trans. UK Home Office.

Klaunberg, B. A., J. O'Malley, T. Clark, and J. A. Davis. 2004. Euthanasia of mouse fetuses and neonates. *Contemporary Topics in Laboratory Animal Science* 43:29–34.

Makowska, I. J., L. Niel, R. D. Kirkden, and D. M. Weary. 2008. Rats show aversion to argon-induced hypoxia. *Applied Animal Behaviour Science* 114:572–81.

Marris, E. 2006. Bioethics: An easy way out? *Nature* 441:570–71.

McIntyre, A. R., R. A. Drummond, E. R. Riedel, and N. S. Lipman. 2007. Automated mouse euthanasia in an individually ventilated caging system: System development and assessment. *Journal of the American Association for Laboratory Animal Science* 46:65–73.

Narayanan, C. H., M. W. Fox, and V. Hamburger. 1971. Prenatal development of spontaneous and evoked activity in the rat *(Rattus norvegicus albinus)*. *Behaviour* 40:100–134.

Niel, L., and D. M. Weary. 2007. Rats avoid exposure to carbon dioxide and argon. *Applied Animal Behaviour Science* 107:100–109.

Niel, L., S. A. Stewart, and D. A. Weary. 2008. Effect of flow rate on aversion to gradual-fill carbon dioxide exposure in rats. *Applied Animal Behaviour Science* 109:77–84.

Pritchett, K., D. Corrow, J. Stockwell, and A. Smith. 2005. Euthanasia of neonatal mice with carbon dioxide. *Comparative Medicine* 55:275–81.

Pritchett-Corning, K. R. 2009. Euthanasia of neonatal rats with carbon dioxide. *Journal of the American Association for Laboratory Animal Science* 48:23–27.

Raj, A. B. M. 1999. Behaviour of pigs exposed to mixtures of gases and the time required to stun and kill them: Welfare implications. *Veterinary Record* 144:165–68.

Sharp, J., T. Azar, and D. Lawson. 2006. Comparison of carbon dioxide, argon, and nitrogen for inducing unconsciousness or euthanasia of rats. *Journal of the American Association for Laboratory Animal Science* 45:21–25.

Singer, D. 1999. Neonatal tolerance to hypoxia: A comparative-physiological approach. *Comparative Biochemistry and Physiology. Part A, Molecular & Integrative Physiology* 123:221–34.

Stutler, S. A., E. W. Johnson, K. R. Still, D. J. Schaeffer, R. A. Hess, and D. P. Arfsten. 2007. Effect of method of euthanasia on sperm motility of mature Sprague-Dawley rats. *Journal of the American Association for Laboratory Animal Science* 46:13–20.

Wolfensohn, S., and M. Lloyd. 2003. *Handbook of laboratory animal management and welfare*, 3rd ed. Oxford: Blackwell Publishing.

Young, A. 2006. Halothane induction results in different behaviours compared with carbon dioxide mixed with oxygen when used as a rat euthanasia agent. *Animal Technology and Welfare* 5:49–59.

16 Education, Training, and Competence

Bryan Howard, United Kingdom; Katey Howard, Spain; and Peter Sandøe, Denmark

CONTENTS

Objectives 369
Key Factors 370
16.1 Background 370
    16.1.1 Category A Persons 371
    16.1.2 Category B Persons 372
    16.1.3 Category C Persons 372
    16.1.4 Category D Persons 372
16.2 Teaching Methodology and Learning Styles 372
    16.2.1 The Teacher 373
    16.2.2 Teaching Strategies 374
        16.2.2.1 Didactic Teaching 375
        16.2.2.2 Group Learning 376
        16.2.2.3 Individual Learning 377
        16.2.2.4 The Development of Practical Skills 378
        16.2.2.5 Teaching Ethics 380
        16.2.2.6 Feedback 381
16.3 Lifelong Learning or Continuing Professional Development (CPD) 382
    16.3.1 Monitoring CPD 383
16.4 Assessment of Competence 384
16.5 Course Evaluation and Learning Outcomes 385
16.6 Oversight of Education and Training 386
    16.6.1 Managing Education 387
    16.6.2 Quality Assurance and Transferability 388
16.7 Conclusions 388
Questions Unresolved 389
References 389

OBJECTIVES

Effective education and training in laboratory animal science underpins consistent, high quality care and use and must address not only the technical expertise of individuals, but also the institutional and societal context of their activities. Learning outcomes should be capable of being benchmarked to assure attainment of satisfactory standards and to facilitate the movement of staff between establishments both nationally and internationally.

To achieve these objectives, education and training must focus on developing the full range of competencies appropriate to the individual's role. Entrants may be required to fulfil certain pre-requisites in terms of experience or education and training, and at the conclusion of the programme, an independent assessment must be made of the individual's learning attainment—or competence; not only must shortcomings be corrected, but measures introduced for continuous updating of knowledge and skills. These measures should be formalised and reviewed through continuous professional development, to ensure skills remain matched to current needs.

This chapter reviews the core competencies expected of those who work with laboratory animals, and explores the factors that promote the development of a learning environment within establishments where laboratory animals are kept and used. The differences in individual learning styles are shown to influence the preferred teaching methodologies and advice is offered on establishing, monitoring and evaluating the effectiveness of teaching as a means of assuring high standards of animal welfare whilst maximising the quality of science. Although there is an extensive literature on teaching methodologies, recent developments have tended to focus on the politics of education and exciting new developments in distance and IT learning. The former is strongly influenced, even dominated by the legislative climate in Europe, whereas the latter remains largely unexploited at the time of writing (December 2009). Consequently, the views expressed and developed in this chapter are drawn from the authors' experience in delivering and evaluating training programmes, including service by one on the FELASA Education and Training Board.

KEY FACTORS

- All those engaged in laboratory animal science must recognise that they are working in an ethically sensitive and closely regulated environment.
- Delivery of education and training does not of itself ensure the development of competence; the trainee must also be committed and express a genuine desire to assure animal well-being and the conduct of sound science.
- Although care staff and scientists may believe they have improved their skills this does not mean that they will implement what they have learned. Training should be embedded into practise by continual supervision, coaching and development.
- Training programmes may be arranged in-house or contracted to outside deliverers. In either case, responsibility for the outcomes and quality of training remains with the host institution which must monitor the development of competence amongst its employees. This takes considerable effort and requires an effective oversight mechanism.
- Successful training requires enthusiastic, well-motivated teachers who establish a professional relationship with their students.

16.1 BACKGROUND

The knowledge and skills required of scientists, care staff and others who are engaged with the use of animals in scientific investigations are key determinants of their ability, not only to meet the expectations of society in safeguarding animals through in-depth application of the Three Rs, but also to ensure high quality and integrity of their scientific findings. Investigations that involve the use of living animals not only receive close public scrutiny, but require complex planning and implementation strategies because of inherent biological variability between animals, the need for careful model selection, and use and rational interpretation of results.

Scientists must be familiar with the range of experimental approaches available that use non-sentient models; the value of training in reducing the use of animals has been reviewed by Van der Valk and Van Zutphen (2004). Scientists must also possess an intimate knowledge of the biology of the species to which the findings will be applied (usually the human), and where animals are required, that of the model used. Where no suitable replacement exists, they must design studies that use the smallest number of animals of the lowest neurophysiological sensitivity and in the least invasive way. An example of how a training initiative can be used to set humane endpoints and thereby minimise the severity of a protocol is provided by Hau (1999) and discussed further in Chapter 14. Similarly, those who care for laboratory animals must be acquainted with the needs of those species and capable of ensuring their well-being whilst supporting the objectives of the investigation. Furthermore all those involved in animal research must understand the ethical discussion surrounding the use of animals in research.

Article 7 of the European Directive (European Commission 1986), currently under revision, requires that "Experiments shall be performed solely by competent authorised persons, or under the direct responsibility of such a person, or if the experimental or other scientific project concerned is authorised in accordance with the provisions of national legislation". In addition, the European Science Foundation (ESF) has urged that "investigators and other personnel involved in the design and performance of animal-based experiments should be adequately educated and trained" (European Science Foundation 2001). The ESF called on member organisations to promote accredited courses on laboratory animal science, including information on animal alternatives, welfare, and ethics.

The competencies appropriate for those involved with laboratory animals were addressed by the Council of Europe (1993) in a resolution on education and training, which was adopted by the Multilateral Consultation of the Parties to the Convention ETS No 123 on 3 December 1993. In brief, this resolution recognised the four categories of competency proposed by FELASA, which were elaborated and published 2 years later (FELASA 1995):

Category A: Persons taking care of animals
Category B: Persons carrying out animal experiments
Category C: Persons responsible for directing animal experiments
Category D: Laboratory animal science specialists

16.1.1 Category A Persons

These are responsible for the everyday care of animals; they are generally known as animal care staff or, when their duties are principally related to routine animal husbandry and care tasks, as animal caretakers. Within this category, four levels of competence are recognised. The first of these—Level 1—is appropriate for persons providing basic laboratory animal care whilst closely supervised by an experienced staff member. They follow established working practises and procedures and verbal or written instructions such as standard operating procedures (SOPs). They are expected to have a general interest in the well-being of animals and understand their needs, know the legal principles applicable to their work and appreciate the ethical responsibility of their role. Amongst other things, their duties may include operating and maintaining cleaning and sterilisation equipment, daily animal care routines including monitoring and recording environmental conditions, cleaning, feeding and watering, and observation and examination of animals for their well-being.

Level 2 persons possess more experience and can work without close supervision. Their duties are likely to involve extensions of those at Level 1, requiring greater knowledge and practical skills, for example within specialised animal facilities such as specified pathogen free and barriered areas or caring for genetically modified animals. They may be responsible for daily routines, breeding programmes, specialist care of animals undergoing procedures, and so on. They may also help with the training of new animal care personnel. By the time Level 3 is reached, supervisory and managerial skills would have been acquired and responsibility may be assumed for more complex activities such as organising and supervising animal care and husbandry routines, co-ordinating resources, managing breeding colonies, overseeing biocontainment programmes, and regular involvement in training and developing staff.

At the highest level (4) care persons are familiar with theoretical and practical aspects of laboratory animal science; most are likely to be senior managers skilled in oversight of animal facilities. Their abilities may overlap with those of FELASA Category D: laboratory animal specialists (FELASA 1995).

16.1.2 Category B Persons

Category B persons are responsible for the conduct of animal experiments. Article 26 of the Council of Europe Convention (1986), requires that persons 'who carry out, take part in or supervise procedures on animals, or take care of animals used in procedures, shall have had appropriate education and training'. There is still considerable diversity in national approaches to the education and training of persons responsible for carrying out experimental procedures, although wider adoption of the FELASA scheme would create greater uniformity and thereby facilitate the interchange of personnel through mutually recognised criteria. The competencies identified by FELASA include awareness of European and national legislation and of the ethical framework within which experiments are conducted, an understanding of local regulations and procedures, skill in animal handling and the techniques to be used, an ability to recognise pain and discomfort in the relevant species, and to take appropriate action to minimise them (Nevalainen et al. 2000).

16.1.3 Category C Persons

These are persons who direct experiments involving animals; they should possess a university degree or equivalent in a biomedical discipline and have completed a course of basic training equivalent to 80 hours. Additional training may be needed in subjects such as surgery and other specialised techniques. Because of their role in experimental design, Category C persons must be able to identify alternatives to animals to obtain the scientific results needed and where none are available, to design a study that uses as few animals as possible whilst maintaining scientific validity, and minimising any suffering or distress imposed on those animals. In addition to being familiar with the legal framework within which they operate and the ethical concerns that have led to development of that framework, they must understand the basic requirements and care of animals, the importance of maintaining them in good health, and hazards to the health of personnel working in the animal facility (FELASA 1995).

16.1.4 Category D Persons

These are specialists in laboratory animal science, and may be veterinarians, nutritionists, facility design specialists, microbiologists, pathologists and so on. Usually a specialist is a graduate who has acquired additional training and experience relevant to laboratory animal science and has an appreciation of all key topics relating to the care and use of animals for scientific purposes, in addition to any specialist knowledge they possess (Nevalainen et al. 1999).

16.2 TEACHING METHODOLOGY AND LEARNING STYLES

If learning is considered to be the accumulation of knowledge, whereas education is the process by which that is achieved, then teaching is the promotion and facilitation of learning. The purpose of teaching is to help the learner to acquire, understand and interpret information in an effective way. The term "training" describes the gaining of skills, understanding and attitudes that underpin competence. The European Directive 86/609/EEC (European Commission 1986) and its successor Directive (in draft form at the time of writing), use the word "competence" to imply both understanding and skill—in other words it involves both education and training. It is a statement about a person's ability to carry out his/her duties in a proficient manner, whilst assuring the well-being of animals and the integrity of scientific findings. From this perspective, teaching involves instruction and practise aimed at improving performance and equipping the learner to respond appropriately to expected and unanticipated situations. In this chapter, the term "student" applies to any learner within the broad field of laboratory animal science—varying from animal facility care staff to senior scientists.

An effective trainer has credibility with the learner and is recognised as knowledgeable in the subject area whilst creating an environment that encourages dialogue. Staff, especially if they are new in the role, may feel intimidated and unable to ask questions or express views in the presence of senior persons and particularly line managers and this may impede learning and hence incorporation of theory into practise. Care is needed in selecting the most appropriate trainer and sometimes it may be appropriate to make changes if the trainee's line management alters.

At the outset, students should be given an overview of the whole course, including the content of individual classes and arrangements for assessment. Different people learn in different ways, and Felder and Silverman (1988) distinguish between active and reflective learning styles—people who find it easier to assimilate knowledge by carrying out an activity, or by reflecting on the process, respectively. The concept has been further developed, for example by Honey and Mumford (2000) who describe four principal approaches to learning:

- Activists who learn from current, immediate experiences
- Reflectors who learn by thinking about issues and considering different approaches
- Theorists who base knowledge on principles, theories, models and systems
- Pragmatists who identify new ideas and learn by experimenting

In selecting the best teaching strategy, it must be appreciated that an approach suitable for teaching one student may be less appropriate for another. There is value in establishing the preferences of students at the beginning of the programme, although it is unlikely that a single approach will be found that benefits all in the group. Learning is influenced not only by variables arising from the learner but also the subject matter, teacher and learning environment. Some people learn better when interacting with the teacher and colleagues and for these, group teaching might be preferred. Others find interaction difficult, and use the classroom largely to gather information, which they assimilate subsequently on their own. Some prefer to learn by reading and listening, others through performing mental or practical tasks. A well-designed course will comprise a range of different teaching strategies, so that one aspect at least will appeal to each student and so ensure all can learn in a manner appropriate to themselves. Even diligent, activist students may find it difficult to maintain interest in formal lecture presentations for more than an hour (Bligh 2000).

Students should be allowed to work at their own pace, to develop an awareness of how their learning is progressing, and be provided with opportunities to seek help or advice when needed. Some learners struggle with large quantities of written text and can be helped by supplementing this with visual tools such as computer-based learning (CBL), video, pictures and simulations. Two broader considerations apply to the selection of teaching strategies: what the students must learn and the number of students present in the group. As most students will be working in parallel to their other duties, it is important also to offer flexibility in the scheduling of study time, so that it does not impact negatively on their work.

16.2.1 The Teacher

The teacher plays a critical role in the learning experience. The teacher's role is to promote development of professional competence and the other attributes required for the professional role of the individual, not just the knowledge required to pass examinations. To achieve this, a stimulating learning environment must be established that encourages the development of enthusiastic self-motivated learners with flexible but predominantly deep learning styles, who are self-critical, and reflective practitioners. Lampert (2002) has shown that effective teachers share a number of common behaviours, including direct and sympathetic engagement with students, excellent subject knowledge, purposeful teaching, attention to feedback and adherence to promises. In addition teachers of laboratory animal science must demonstrate an awareness of the ethical sensitivities of the subject, and instil this in the students themselves. The importance of effective educational engagement with students is discussed at length by Biggs and Tang (2007). Those new to teaching should adopt a relatively simple approach with which they are comfortable and that enables them to feel in control of the learning process; this may be modelled on someone whose performance the teacher admires. Asking an experienced teacher to sit in on some of their sessions can provide invaluable feedback. As teachers gain confidence and first-hand experience of how students learn, they can extend the range of teaching strategies and monitor the effects on students, learning.

Having recognised the importance of the teacher's credibility in effective training, his or her motivation is also important. An enthusiastic teacher is more likely to motivate students than one who is disinterested or is preoccupied with other matters. To these considerations must be added practicalities relating to the level of the course, teaching environment, group size and the teacher's own preferences and abilities. Particularly when training is delivered to those at an advanced stage of their career (e.g., FELASA Categories C and D), account must be taken of students' differing background experiences. Although younger and less experienced learners also have differing backgrounds, the content of laboratory animal science courses is often rather technical, and relatively formal teaching strategies—such as lectures and self-learning packages—are frequently used.

Many who deliver training in laboratory animal science are not professionally trained teachers, but rather specialists in their own discipline, and they may need support in developing strategies for sharing their knowledge with colleagues. This may involve providing them with basic skills in training/teaching. In larger organisations, or where trainers can come together to discuss issues (as for example the Universities' Training Group in UK) there is benefit in creating a database of materials that can be used to support less experienced trainers and provide consistency in the quality and content of training programmes. Additionally, creation of a "community of teachers", involving different programmes and organisations, can ensure the maintenance of key skills and the dissemination of information across the wider professional community.

16.2.2 Teaching Strategies

The attention span of many learners is less than 30 minutes and for some may be much shorter (Mellis 2008), so spending extended periods of time on single activities may not necessarily prove beneficial. Generally a mixture of training styles is more appropriate than blocking together lectures, practical classes, or seminars. In this section only a few of the more commonly used methods are considered, but no one method is suitable in all circumstances. A distinction is sometimes drawn between formally taught learning and the gaining of knowledge and skills informally through experience in the workplace and private study. Often the latter lacks direction and may be subject to influence by poor role models or inadequate instructional experiences; not only may informal training be inefficient, but because such private learning is not easily assessed or benchmarked, it is not always recognised.

Despite this, skills development outside formal training can be a useful adjunct to improving performance, and involves techniques such as using posters to raise awareness, team meetings and job shadowing (where an employee observes a fully competent colleague completing their work activities). Another effective training format is a relatively informal lunch or evening meeting, lasting no more than 1 hour, at which refresher training is delivered, a specific topic addressed, or staff are updated about work in progress or discuss a "hot-topic". The timing of these sessions needs to be carefully planned to avoid interrupting critical workplace activities, and periods when employees are likely to be less attentive (for example, directly after lunch).

16.2.2.1 Didactic Teaching

Lectures are the traditional means of presenting factual information and are particularly suited to imparting knowledge and comprehension to large groups of learners. The presentation should be broken up into smallish segments, lasting ideally no more than 45–60 minutes, so that students can use the breaks to assimilate what they have learned and to prepare themselves for what is to follow. Rose (2003) recommends that after each 30 minutes there should be a break lasting 5 minutes consisting of a contrasting activity. The teacher should monitor the attention of the group to decide when a break is required. Where a great deal of theoretical material has to be covered, it is useful to arrange that this is delivered by more than one teacher.

Lectures need very careful planning; generally they should open with an introduction to the content, then present the detail and finish with a summary of key points. Whatever method of delivery is adopted, students should be encouraged to network and discuss issues between themselves, an activity that not only reinforces learning but also enables those who are having difficulty with particular topics to address these informally with peers; for example, discussion about an ethical issue or some aspect on animal care.

Views on handouts are divided. Some teachers prefer to distribute lecture notes in advance (preferably not immediately before the lecture, because students may read them rather than listening to the lecture), others consider them a distraction and distribute them at the end of the lecture (although students will not know how much detail or information they contain), and some may not distribute notes at all, in which case the students should be made aware of this beforehand. Handouts should be designed to support students who have attended and concentrated on the lecture and may serve as a framework for note-taking or following the material as it is delivered, otherwise a good textbook would be as valuable. When handouts are provided in the form of web-based materials, usually on an intranet and password protected, it is important that students have ample opportunity not only to log-on to the relevant material, but also to print it out. If it is found that a high proportion of students are printing out lecture notes, it is worth examining the cost of such downloads in comparison with more traditionally printed materials.

It is important to start and finish lectures on time, allowing opportunities for questions at the end (some students may prefer to approach a lecturer in private after the open session). Material should be presented in an un-rushed manner, clearly and easily heard and understood by everyone in the class; a microphone should be used if necessary.

Video presentations are also suitable for large classes, and can either introduce practical skills, such as handling or the conduct of technical procedures, or used to support other teaching strategies to provide variety and realism. There is however a danger that the teacher may become subsidiary to the images on the screen, so care is needed to ensure that the teacher and not the video controls the class and that students remain active. One approach is to preview the presentation and identify issues requiring emphasis so that the presentation can be paused and the issue highlighted or the presentation played at slow speed to clarify aspects of a technique. Another strategy is to prepare a questionnaire for the students, which prompts them to take notes as they are viewing and provides a permanent record of the contents.

Demonstrations provide a valuable adjunct to the use of videos as a way of helping to develop practical skills. Their purpose is to help students recognise and appreciate the stages or steps involved in the skill. Some tasks cannot be performed slowly (for example picking up a ferret or administering an anaesthetic agent) because of adverse consequences for animal welfare or the proficiency of the procedure. In such cases the demonstration can be carried out at normal speed, and then repeated whilst describing the sequence of events. One or more students can then be asked to repeat the demonstration and describe what they are doing whilst other students watch. This gives the teacher an opportunity to refer again to the key issues and any mistakes should be immediately pointed out, but without overt criticism; involvement of students helps identify difficulties they have in completing the task.

Facility visits can be of particular benefit to those preparing to work as caretakers or animal care technicians (FELASA 1995); valuable information can be obtained by working for several days or weeks at different establishments. Similarly, those training to become research technicians (Nevalainen et al. 2000) and others may observe or work for extended periods of time with experienced colleagues in laboratories where relevant techniques are performed. The range of tasks that can be carried out may be restricted if prior authorisation by the relevant National Authority is required (for example in UK or Switzerland) although students may be able to assist with non-contact tasks. This type of activity may be considered as a greatly extended practical class, and should be supported by clear learning objectives and regular assessment and feedback. It is also essential that the facility within which the students work is selected on the basis not only of the knowledge and inter-personal skills of personnel who work there, but the relevance of the equipment used and the nature of work being conducted.

Organised tours of animal holding facilities or laboratory areas where animals are kept or investigations are performed, may be of value to personnel at all levels. Such visits are generally of short duration (typically less than one day) and focus on specific aspects of animal care and use, so the quality of learning should be maximised by presenting the desired learning outcomes and sufficient background information beforehand. Biosecurity measures must be closely followed. Students may be asked to compile information about specific aspects of the visit, and to present these subsequently to other members of the class; alternatively, general aspects of what has been learned may be discussed at the end of the visit.

16.2.2.2 Group Learning

Whilst the lecture is an efficient means of conveying information to large groups of students, it is not always apparent whether students have learned key messages. To some extent, this can be assessed by posing key questions to the class during or at the end of the lecture, either addressing a student by name or the entire class. Such questioning enables class members to assess their own knowledge and provides the teacher with an opportunity to summarise key topics.

Another disadvantage of lectures is the challenge that they present to supporting different individuals' learning styles. Students may be more actively involved in learning by techniques such as brainstorming, which encourages them to contribute to identifying or solving problems. Responses are collated into categories, which are considered in turn. This process is best suited to large groups and need last no longer than 5 minutes; it can be used to provide a break from formally presented material or to introduce new perspectives or topics.

Larger classes can be subdivided into smaller groups for interactive sessions involving discussion of actual or hypothetical topics or scenarios, to explore the issues involved and the merits of different viewpoints and solutions. This is known as small group teaching, and is a useful way of enabling participants to network, share opinions and experiences, and explore attitudes; it is particularly valuable where class members have a wide range of different backgrounds, which may be the case when teaching at Category C level. Students find it easier to engage with doubts and uncertainties within peer groups and to develop more positive attitudes towards the subjects they are studying. Similar dialogue can be initiated within the context of a lecture by asking students to discuss with others sitting nearby, an issue that has just been presented. Care should be taken in selecting topics to ensure that they are unambiguous and contribute to the learning objectives; it is a good plan to ensure that the topic for discussion is seen by everyone—for example, by displaying it on a screen, whiteboard or blackboard, or by circulating the question in printed format, in which case additional (background) information can be included. Ideally, groups should comprise no more than six or eight persons, and participants should sit so they can hear, see, and interact with each other. Results of the discussion may be reported back to the whole class or sometimes just the teacher—this and the time available should be made clear at the outset. Adequate time must be allowed if feedback is to the whole class.

Discussion provides a mechanism to help students understand different perspectives of problems, or to explore issues and draw conclusions. It is a useful way of exploring ethical dilemmas and of generating change in attitudes and opinions, but it needs even more careful preparation than a lecture. Not only is it important to identify the material to be learned and to determine the learning outcomes, but learning pathways must be identified and pointers placed along the way to encourage students to find their own way through the material. Sometimes the teacher participates directly, in which case he or she should interact with the groups not as an authority figure, but as an equal participant or to facilitate learning by ensuring that everybody is encouraged to contribute. Where several groups are discussing issues simultaneously, the teacher should move between groups, but contribute only to confirm the appropriateness of discussion or to open up new ways of thinking. Although the process of learning is passed to or shared with the students, the teacher remains responsible for its direction, and must ensure that the class remains focused on the topic, guided by clear objectives, and in accordance with agreed ground rules.

This type of interactive teaching is often the preferred mode of teaching smaller groups of students—no more than eight or 10 and preferably only three or four; it allows a mix of different inputs and encourages everyone to participate. Lectures may even be replaced by directed reading, supplemented by group discussions, seminars, or tutorials.

Seminars also involve interactive learning groups within which, one or more students are required to research a specified topic and to present their findings to other group members, followed by open questioning and discussion. Students develop skills in identifying and accessing information, presenting an argument, discussing its implications and defending their interpretation in face of questions from colleagues. Case studies similarly involve small groups of students researching and discussing a real or simulated problem, which is the focus for learning. Often case studies are used to compare strategies for dealing with particular scenarios or to analyse technical or ethical issues, for example, a comparison of different analgesic regimes for postoperative care following laparotomy in rabbits.

Tutorials involve the teacher setting students a topic to research, guiding them through this and subsequently discussing and evaluating the quality of the learning. They may be used to support work on assignments, to evaluate problems that students encounter in their learning, or to review progress on the course as a whole. They are typically one-to-one sessions (groups of two or even three can sometimes be accommodated) and the objective is for the teacher to listen to students, to form an opinion about what they know and what they do not know, and to discuss and agree with them strategies for correcting any shortcomings.

All these forms of small-group teaching take advantage of the fact that students often learn most effectively from their peers, and that as long as the educational content of the course is clearly identified and the progress of learning is monitored, the educational experience is enhanced. To achieve these advantages, it is essential that the teacher prepares lessons carefully and ensures that the rooms in which teaching occurs are appropriately laid out. Sometimes it is advantageous to carefully select the composition of groups in order to ensure an appropriate mix of skills and abilities; for example, groups may consist of persons working in similar (or differing) scientific disciplines or differing technical and scientific expertise.

Assignments are rather different insofar as students are normally expected to work alone. The student is given a topic to analyse in-depth and supplied with prepared materials, access to a library, or guidance on sources of information. Assignments can also be used to teach practical skills, for example imaging techniques or aspects of animal care. A PhD in the United Kingdom can be considered as individual learning, often commencing with small group teaching (the taught element) before the assignment commences. Again, it is very important that appropriate ground rules are laid down including the breadth and depth of the topic, the timescale and the method of reporting—written or verbal; usually the progress of assignments is monitored by periodic tutorials. A project that involves a group of students working together on different aspects of one problem is a hybrid of individual and small-group teaching.

16.2.2.3 Individual Learning

It is sometimes appropriate for students to work on their own, for example within very small organisations where access to more formal training is difficult, where staff cannot be freed to attend classes, or occasions when no suitable course is available but individuals need interim training (e.g., to accommodate a change in job role). Perhaps surprisingly, the amount of effort and time needed to manage this type of learning is as much or greater than that needed for other teaching methods, because learners require individual assistance and monitoring.

Coaching and mentoring are effective methods of consolidating and contextualising learning. Managers of animal facilities and senior members of scientific teams may benefit from these processes to help them assimilate and utilise feedback, and to ensure that learning is structured and objective.

Mentoring techniques have developed significantly over recent years, and in particular reverse mentoring (when the mentee provides structured feedback to the mentor) can help mentors develop their own skills by reflecting on how effective they have been in communicating and establishing progress.

Home study is another valuable way of promoting learning, but it must be properly recognised and appropriately monitored, for example by regular assessment of assignments or within a class situation; alternatively, small groups or individuals may be tutored in a more informal environment. Online learning provides an alternative strategy for individual study and makes it possible to provide information with which nobody within the establishment has expertise (e.g., presentation of a new technology) and allows experts who are uncomfortable with face-to-face training, to contribute to development as tutors and mentors. Distance and computer-based learning have the potential to support and enhance acquisition of knowledge and to some extent skills; packages have been devised with all FELASA Categories in mind, including some for Category D persons (for a review of some of those related to veterinarians see Colby, Turner and Vasbinder 2007) as well as some additional applications such as supporting persons involved with ethical review processes. These approaches do not require access to structured learning environments and are useful when no suitable classes are available locally, or if leave of absence is not an option (although students must be allowed adequate time to study and arrangements must be made to support and guide learning, e.g., by tutorials). However it is particularly important to monitor progress and to ensure that the content is appropriate to the learning outcomes desired.

Webinars, notice boards, chat rooms, discussion groups, and blogs can create peer learning opportunities in a similar way to group teaching, although to be used to best effect a means needs to be found of ensuring that the breadth, direction and content of learning meets the needs of the organisation and individual, and to provide the learner with a means of monitoring progress.

16.2.2.4 The Development of Practical Skills

The purpose of training in laboratory animal science is to enable practitioners to enhance their performance, so the development of practical skills is a key element of that training, although often it receives less attention than knowledge and understanding. The objective of practical training is to transform theoretical knowledge into the technical, procedural skills of a competent animal care or research professional.

Practical classes provide an opportunity for students to develop skills in a supervised environment and to link theory with practise. Where handling of living animals is concerned, some theory must be taught before a practical session commences; this should include relevant aspects of ethology and recognition of distress, discomfort, or pain. Wherever possible, preparation should be carried out beforehand by use of videos and inanimate training aids and should make students aware of their ethical and legal responsibilities and relevant health and safety matters. Arrangements must be made to ensure that there are sufficient skilled tutors available to ensure that all groups are properly supervised and that animal well-being is not compromised; a tutor:student ratio of 1:4 or 1:5 is generally satisfactory. In many countries, living animals cannot be used to acquire technical skills that have the potential to cause pain or suffering, and even where such use is permitted, ethical review should be carried out beforehand to ensure that the learning objectives are worthwhile and that the Three Rs have been fully implemented. The learning outcomes should be explicit, a requirement facilitated by providing students with a worksheet that identifies the tasks to be carried out and an assessment schedule detailing learning outcomes. In any case, only the smallest number of animals necessary to develop the required skills should be used, whilst ensuring that no animal is caused unnecessary distress. At the start of the class, the teacher should check that all equipment is available and working correctly. Rooms in which manual skills are to be developed should be free from distractions and if necessary, registered with the regulatory authorities. Students should be encouraged to work at their own pace and where necessary, time made available for additional practical training. The teacher should ensure that there are plenty of opportunities to ask questions and for providing feedback and correcting faults. A rigorous and sufficiently frequent schedule for assessment ensures that the progress of learning is monitored by both the student and his or her supervisor.

To date there are few published guidelines detailing the practical skills expected of either animal care staff or of scientists, although the UK's Institute of Animal Technology has set out a detailed syllabus for education and training to the lower levels of FELASA Category A competencies (IAT 2006). In the case of Category A persons, most activities to be learned are neither invasive nor distressing and the time available for developing competence is very considerable—often many months or even years. During this period, a desire for professionalism, based on independent learning and enquiry should be fostered. Students may acquire basic practical skills during formal training, alongside study of theoretical and practical components by arrangements such as block- or day-release. Scientists (Category B or C persons) cannot easily develop all the practical skills they require during the short period of formal training, which is necessarily brief because they usually occupy positions that make other demands on their time. Consequently a strong argument can be made for them to develop competence whilst supervised by senior colleagues, after they complete their basic scientific training. A general account of approaches to developing competence in technical skills is provided by Rothwell and Benkowsk (2008).

Scientists may work in many different environments, ranging from academic establishments to pharmaceutical companies, contract testing houses and so on, but in all cases there must be support by enthusiastic, experienced senior scientists. Nonetheless formal training is the foundation for all future learning and practise as a professional, and it must provide a framework of theoretical knowledge on which practical aspects of animal experimentation can be developed in a safe, ethical, and systematic way and also ensure that the learner realises when to seek help and from whom. Several commercial and academic organisations offer structured training (such as AALAS and the American College of Laboratory Animal Medicine).

Not only must researchers develop a range of basic technical skills (e.g., the administration of substances and withdrawal of samples), but also more specialised techniques appropriate to their planned work. They must also develop communication skills necessary for working with the range of other professionals involved in the study and an understanding of their own responsibilities, roles, and duties. Newly qualified categories B and C animal researchers should be offered ample opportunities to acquire the necessary skills under careful supervision. Some commercial establishments use senior scientists to deliver structured primary and secondary support to new scientific workers. Within academic establishments, the administrative structure is rather looser and skills training is often less formalised. Sometimes relatively junior scientists deliver bench-training to new recruits and it has been argued that they may find it easier to relate to new trainees; they should be allocated time for this. It is during this early period of training that the relationship with more experienced scientists becomes a factor that determines many of the attitudes that will be eventually developed. This argues for the establishment of a more structured environment to ensure that training in practical skills is conducted thoroughly and efficiently and with a high priority of ensuring high standards of animal welfare. It also means that scientists in academic institutions must drive the learning process themselves to take maximum benefit from the opportunities around them. Log books or web-based databases may be used to ensure development is properly recorded against a list of essential competences, thereby enabling progress to be tracked.

16.2.2.5 Teaching Ethics

Underpinning training in laboratory animal science is the need to instil in potential laboratory animal carers and users, an appreciation of the ethical issues surrounding their activities. This is different from merely acquiring information about the subject. It involves the ability to stand in the shoes, so to speak, not only of the animals but also of people holding ethical views that differ from the learner. Furthermore, it involves the ability to engage in ethical discussion. Consequently, there is a need for teaching methods that provoke and challenge students rather than simply telling them how things are, and that encourage students to engage with one another and with teachers in discussions encompassing a wide range of differing viewpoints.

Ethics relating to laboratory animal use is a subject that is taught best not by presenting facts, but rather by encouraging assessment, analysis, and fresh appraisal of personal attitudes and an appreciation of diverse viewpoints about this very complex subject area. It can be seen as combining three elements:

1. Identification and reflection on different ethical views regarding man's interaction with and exploitation of animals for different purposes. Here, laboratory animal science should be seen and discussed as one form of animal use among others. The ethical views considered could include contractarianism, utilitarianism, and the animal rights view (Sandøe and Christiansen 2008).
2. Presentation of ethical guidelines regarding the use of animals for research. Here there is a well-established consensus that it is only morally acceptable to use animals for research provided the following two things are in place: (i) the research must have a relevant purpose, addressing issues of agreed importance, such as new ways of preventing, curing or alleviating serious human diseases, and with a realistic prospect of achieving a satisfactory answer; and (ii) there must be no other way of achieving the same results that is less harmful to the animals, and every effort must be made to ensure that as few animals as possible are involved, and that these experience no more discomfort or suffering than is strictly required by the experiment. These two provisions include the Three Rs principle and legislation and guidelines derived from this.
3. A critical discussion about the application of the ethical guidelines. During the course of this, it will become clear that the principles presented under Element 2 are rarely straightforward and are sometimes internally contradictory, and that they must be interpreted in the light of a general ethical outlook. The discussion frequently leads back to Element 1 (Sandøe and Christiansen 2008).

The topic clearly does not lend itself to didactic teaching or to individual learning, and is usually best approached by interactive group work, which may take the form of discussion groups, seminars, or tutorials. If they are to critically evaluate and develop their own ethical thinking, laboratory animal carers and users need to appreciate different ethical viewpoints and especially the rationale on which they are based. This includes awareness of other frameworks for thinking about ethical issues and an appreciation of how perspectives might differ—including within project teams (e.g., between animal care staff, researchers, and managers of specific studies), within institutions (e.g., within an institutional ethical review process), and/or with a wider public. An understanding of these issues helps to ensure that ethical decision making is responsive to a wide range of factors and interests. For example, people may have different views about what counts as a good animal life (Lassen, Sandøe, and Forkman 2006; Sandøe and Christiansen 2008). Also when it comes to the use of biotechnology (e.g., in the form of genetic modifications or cloning) there is a wide diversity of views (Lassen, Gjerris, and Sandøe 2006), and whilst these are not always based on underlying scientific principles, they may be strongly held nonetheless.

An understanding and respect for these different views can enhance personal, institutional and help develop a wider confidence in judgements that are made (Smith and Boyd 1991) and ultimately, this might lead towards consensus on a body of common ethical standards. Knowledge about 'ethical theories,' such as utilitarianism and animal rights views, may be of value in helping researchers to analyse how a particular programme of research may be viewed from different ethical perspectives (Olsson et al. 2002).

Underlying the views presented here is a pluralist approach to ethics teaching. Such teaching should encourage students to develop an appreciation of the strengths of different perspectives and why people have been drawn to these views. Adopting a pluralist approach does not mean that ethics teachers do not have views of their own. However, it is founded on a strong conviction about the best way to promote learning, understanding and application of ethical principles and the right way to handle public controversies. Ethics teaching should not take the form of moral lecturing. Rather, the aim of teaching should be to give students state of the art tools with which they can interact with the subject and an appreciation that at present leading scholars in the field of ethical theory disagree strongly about many, if not most, matters!

An effective way of presenting ethics to students on an introductory course is to describe the underlying principles of competing theories, showing that each has certain strengths, but making it obvious at the same time that they cannot all be correct because they lead to conflicting conclusions.

One clear advantage of this approach is that it obliges students to become engaged in ethical reflection. They are not simply presented with facts to learn. They are challenged to make up their own mind on matters that require resolution but where there are no correct answers to set this before them. No answer can be completely right or completely wrong so, the students have no easy way, which they might find if they were asked merely to choose one theory or another. Each view has its own strengths and weaknesses.

In contrast to this completely open way of thinking, when it comes to the ethical guidelines regarding the use of animals for research students must be given much more guidance. They must be told that certain principles and policies are not up for discussion. At this point ethics teaching may go hand in hand with explanations of current rules and regulations concerning the use of animals for research. Of course, there is interesting ethical ground to explore about why free thinking is inappropriate when dealing with societal viewpoints that drive government and other policy makers, and if there is time to pursue this line of enquiry, consideration of social cohesion and the nature of democracy, but that is a much longer lesson.

Teaching of ethics calls for a combination of different teaching strategies. Some parts, such as the background and rationale of ethical principles and theories and presentation of guidelines based on the Three Rs principle, can be presented through lectures. For other topics it is necessary to engage students more actively, as for example in structured tutorials, seminars, or discussion groups. One approach is to engage students in case studies about how best to apply the Three Rs principle to examples which are illustrative of particular research protocols; in such cases it is wise to discuss this first with those engaged in the investigation to ensure that they can contribute to the dialogue if they wish and provide reassurance that selection of their work is not an implicit criticism of it. Cases can either be identified in advance or can be proposed by participants themselves. One useful way of deploying individual learning techniques to prompt students to reflect on the principles underlying various forms of animal use is the internet-based educational tool Animal Ethics Dilemma (http://ae.imcode.com//; Hanlon et al. 2006).

16.2.2.6 Feedback

The quality of feedback provided to students can have a huge impact on the learning process; it should be:

1. Delivered fairly in a non-confrontational and factual way.
2. Offered frequently—this may be difficult during short courses and requires careful planning and close monitoring of learning.

3. Given immediately as the opportunity arises, when the details of what the learner has written, said, or done are still clear in his/her mind.
4. Discriminating—establishing the difference between poor, acceptable, and exceptional performance.

Both care staff and scientists must be able to evaluate their own effectiveness in the workplace and feedback provides teachers with an opportunity to encourage students to do this. The process may be introduced initially by asking groups of students to generate and discuss appropriate criteria for evaluating and assessing their own competence (Ashcroft and Foreman-Peck 1994). The use of structured forms can help training organisers and senior staff to present their feedback in a constructive way and address key issues of animal care. Moreover, teachers, particularly those at an early stage in their career, may find that feedback provides a means of developing an objective, self-critical approach to their own performance.

16.3 LIFELONG LEARNING OR CONTINUING PROFESSIONAL DEVELOPMENT (CPD)

Despite the increasing duration of formal education in the natural sciences, the knowledge and expertise acquired during this period are insufficient on their own, for a professional career spanning several decades. Additionally, the twenty-first century has brought an accelerated pace of innovation and technological development, so it is more important than ever that all professionals and particularly those working in ethically sensitive areas are committed to maintaining their knowledge and skills. In other words, individuals must sustain the development of their skills and knowledge throughout their working lives and must be supported by their employers in this process.

Continuing professional development (CPD) is a process of systematic life-long learning directed at maintaining and improving knowledge, skills and competence. The European Commission paper on adult learning "It's Always a Good Time To Learn" (European Commission 2007), argues that life-long learning has a key role to play in developing and maintaining competence and promoting citizenship.

Life-long learning is essential not only for researchers, but also for those who care for laboratory animals, oversee their husbandry and use, or advise on or monitor activities in biomedical facilities. Laboratory animal science progresses by constant technological innovation and there is a need throughout the course of an investigation, continuously to re-assess implementation of the Three Rs. To some extent these matters are also the remit of the ethical review process. However, CPD should not be restricted to purely theoretical aspects—laboratory animal science incorporates manual skills and practical expertise that can be lost without continuing performance and reassessment of skills.

To be effective, CPD should constitute an extension of the formal education and training of all staff. In an effective establishment all people involved in the care and use of animals contribute to a multi-professional, team-based process that is sometimes described as a "community of care". Because of the differing educational backgrounds of people involved in this community, it is important to establish agreed and mutually recognised CPD expectations that meet the priorities of the organisation and safeguard animal welfare. Individuals must be enabled to expand and fulfil their potential by taking control of their own learning, responding to new thinking and new technology, and thereby increase their efficiency and effectiveness.

There are many approaches to developing and delivering a formal CPD programme. Learners may study pre-defined topics presented formally by lectures; attend short courses; participate in conferences, seminars, or workshops; undertake assigned reading; study online learning packages; act as a coach or mentor for others, and so on. A quite different approach is to give learners freedom to choose those topics that they study and the means by which this is done, including attendance of scientific meetings, conferences, distance or e-learning, home study, or correspondence courses and

Education, Training, and Competence

so on. There is considerable benefit in developing training plans in consultation with staff, to ensure that their own interests and needs are taken into account. Where periodic assessments of workplace performance are conducted—such as performance appraisals or periodic staff reviews—this occasion can be used to establish needs and to monitor progress of CPD. Internal training programmes may be used by organisations to promote a more dynamic employee base, acquainted with the priorities of the organisation and able to react swiftly and appropriately to a rapidly changing commercial climate. CPD may be used to raise awareness of ethical, technical and scientific developments and may include work-related seminars or technical presentations open to scientists, animal care staff and others.

16.3.1 Monitoring CPD

Whatever means is used for the delivery of CPD, a mechanism must be found to monitor progress so that individuals can identify and measure their CPD, both for their own development and in order to demonstrate it to others. This enables the student to remain compliant with changing regulations and legislation. In only a few countries are mandatory schemes of CPD in place. For example in Switzerland the Ordinance on Education and Training of Specialized Staff for Animal Experiments of 12 October 1998 (summarised at http://www.bvet.admin.ch/suchen/index.html?lang=en&keywords=SR+455&go_search=Suchen&site_mode=intern&nsb_mode=no&search_mode=AND#volltextsuche) requires that persons engaged in animal experimentation shall be appropriately trained (to FELASA Category B or C level) and in addition must attend at least 4 days CPD during each four-year period, at officially recognised courses. Although it is likely that the new Directive, which is proposed to replace Directive 86/609/EEC (European Commission 1986), will require Member States to publish requirements for delivering and maintaining education and skills, there is no intention to introduce mandatory CPD at this stage.

Despite the lack of legal enforcement, CPD fulfils an important role in assuring continuing competence. Recognition must be given for practise-based as well as academic learning, and the process must ensure that less formal learning scenarios such as web-based learning and directed reading or private time developing bench-top skills are incorporated into the credit framework. Monitoring systems established by some professions simply stipulate a number of hours to be completed each year as a prerequisite for recertification. Although easily assessed and quickly completed by the learner, these methods record only the time spent learning rather than what has been learned. Credit-based approaches to assessing CPD are generally based on completion of modules that deliver clearly specified learning outcomes, and although this is a considerable improvement over simply monitoring time expended, it is important that the specification of outcomes incorporates competencies. A more meaningful approach to monitoring may be the completion of personal portfolios in which learners reflect on their own CPD and evaluate their achievements. These can take the form of web-based personal development records, summaries of which can be incorporated into online CVs and personal websites.

The validation of CPD is especially important for academics, biotechnologists and others who need to demonstrate their education, training and competence in another country; currently this can be very difficult. In addition to developing methods to deliver and record CPD outcomes, a method must also be found to provide reassurance that quality criteria are applied impartially. FELASA has established a working group that has made a number of proposals in this regard (http://www.felasa.eu/index.php). These include recommendations that everyone working with animals should be required to maintain state-of-the-art knowledge and skills, assured by provision of opportunities for undertaking CPD in a way that accords to the size and nature of the establishment, the skills needed, and facilities available. CPD should commence as soon as an individual starts working with animals and continue throughout his or her working career, according to a development plan agreed with and periodically reviewed by the line manager, head of department, QA department, ethical review body, and so on as appropriate. Progress can be monitored by accumulating credits over a range of different practical- and knowledge-based activities, with a requirement to achieve a minimum number of credits during a time interval determined by their category. FELASA has indicated its readiness to encourage, oversee, and review provision of suitable CPD activities within and between different member states. It would be the responsibility of the individual to establish the need for his or her CPD (objectives), to maintain any records required, and to communicate with member associations for endorsement of activities that are not listed.

16.4 ASSESSMENT OF COMPETENCE

The effective assessment of competence has two prerequisites:

a. Establishment, before the programme begins, of unambiguous learning outcomes that must, collectively, constitute competence;
b. Ensuring that all those involved (students, teachers, and supporting technical staff) are clear about what constitutes satisfactory performance and how it is measured.

Students should be aware what topics they need to develop for themselves.

In traditional, didactic courses, examinations often provide the only opportunity for the teacher to determine whether students have achieved the learning objectives; this process is known as "summative assessment" and its principal role is to establish what has been learned. Such assessments might be used, for example, to establish the knowledge of students studying at the lower levels of Category A as a way of monitoring progress. Examples of summative assessments are multiple-choice questions, written short answers, performance of technical activities under closely specified conditions or identifying materials presented for a short time, commonly known as "spot tests". Longer written essays that are awarded a mark and reported back with little or no analytical critique are also considered summative.

Of these methods, the most frequently used is probably multiple choice questioning (MCQ), designed to establish whether the student can recall key information relating to core topics. This type of examination has been called objective—the same student would get the same outcome, regardless of who marked the paper. The setting of MCQ questions requires considerable expertise to avoid ambiguity (Holsgrove and Elzubeir 1998), but assessments are amenable to rapid marking and the examination can be presented and evaluated using computer software. This makes the method attractive where large numbers of students are involved or where the examination is taken and marked at several different centres, because no interpretation of student response is required. Moreover, software can perform rapid and comprehensive analysis to show how many students sitting the examination answered (or omitted to answer) each option of each question and to correlate individual student's selections for different questions, thereby establishing whether particular subject areas or particular types of question have presented difficulties. The findings can be used to modify the wording of questions or to make changes to teaching practises. However, it is much more difficult to ensure that MCQ questions test understanding or the ability to interpret facts (Morrison and Free 2001). For assessing these outcomes, the methods most frequently used range from requiring students to complete short or long answer written questions under formal examination conditions (they may be seen or unseen), oral interrogations, or the preparation of written assignments, projects, or portfolios. The consistent marking of these is not as straightforward as for MCQs but the questions are generally easier to prepare. Spot tests are most relevant to assessment of observational and interpretive skills, but must be carefully designed to avoid ambiguities and where a specific observation is required (e.g., down a microscope) the image should be checked periodically to ensure that it remains constant.

Practical competencies may be assessed by direct observation (e.g., of handling animals or the conduct of simple experimental procedures), computer simulation (e.g., using interactive online software) or laboratory work such as preparation and microscopic examination of samples of blood or epithelial smears or chemical analysis of tissue fluid. An objective scoring sheet should be prepared beforehand to ensure that all key aspects of the task are assessed and that marking is consistent. Assessment of attitudes towards experimental procedures, laboratory animals and the Three Rs is much more difficult to achieve, and most frequently relies on interviews, structured short questions "such as complete the sentence ....." or long answer questions such as essays. None of these is very satisfactory and it may be that feedback from supervisors provides a better view of workplace attitudes than more structured methods. A useful summary of the characteristics of these and other assessment strategies is provided by Bull (1993).

It is good practise to examine retrospectively the assessment performance of all students and to identify topics where performance has been generally poor or good. A view can then be taken about whether changes to the teaching or assessment process are appropriate.

Most persons studying for Categories C and D, and many may have already completed a university degree course or have undertaken extensive training within other contexts. In this situation, learning is often non-linear and a more complex process of feedback and assessment is generally used, known as "formative or educative assessment". Formative assessments are designed to enhance the quality of learning, by requiring assimilated information to be applied in a contextual way and provide the teacher/instructor with an opportunity to feedback the effectiveness of learning. Assessments of this sort may incorporate exercises, questions and/or scenarios that possess a real-life context within the subject area to be assessed. In this type of assessment, the teacher typically asks learners to reflect on how their knowledge can be used to resolve a realistic problem. Generally there is no right or wrong solution and a range of possibilities can be explored, many of which will be equally valid; the teacher responds to the student's assessment by providing feedback and assessing the effectiveness with which learned material has been deployed. The assessment used may take a variety of forms, including writing an essay discussing a contentious issue, preparing a portfolio of evidence on a given topic or performing and assessing a complex technical task (Ashcroft and Foreman-Peck 1994).

In practise, it is best to use a combination of summative and formative assessment to determine competence. Regardless of method, teachers should always explain clearly the task, the criteria that will be used in assessment and the standards expected.

16.5 COURSE EVALUATION AND LEARNING OUTCOMES

The quality of training is a function of the delivery of material and its assimilation by the learner. Course evaluation by students (and, where appropriate by their supervisors) is just as important a measure of the quality of training as the outcome of examinations, and provides a mechanism for ensuring that the desired learning outcomes are not only achieved, but achieved efficiently, see for example, Carlsson and others (2001). The fact that employees believe they have improved their skills through a training intervention does not mean that they will implement these in their daily role, and students' perceptions of the effectiveness of training may differ from the view of their supervisors. Kirkpatrick and Kirkpatrick (2006) have developed an evaluation model comprising four levels:

- The reaction of students—what they thought and felt about the training (assessed through course evaluation forms, discussion etc.)
- Learning—the resulting increase in knowledge or capability (assessed through examinations)
- Behaviour—improved behaviour, capability and implementation/application (assessed through supervisor feedback)
- Results—the effects on the facility, business or environment resulting from the trainee's performance. This is the most difficult level of evaluation to implement and is usually assessed by reference to the success of projects and interventions that enhance the student's environment.

When a course has been recently established or has been substantially modified, specific questions can be asked that, for example, may probe certain aspects of training such as the allocation of time, the use of audio–visual aids, perceptions of the practical classes, or whether reference materials were suitable. In the case of a course that has been running for some time, broader information may be sought about the effectiveness of course delivery. Questions may also be included about the performance of individual teachers or the presentation of particular topics. Evaluation forms should be concise and easy to complete, taking no longer than 5 or at most 10 minutes. Sometimes, responses are guided by supplying a series of questions and either asking each item to be graded on a scale (for example 1–4, or 1–5) or students may simply be asked for general comment, for example, to identify three aspects of the course they found most helpful and three unhelpful. Unstructured course evaluation encourages students to think more widely about course delivery, but may fail to provide information about specific issues that might otherwise not be mentioned. Alternatively, evaluation can be solicited verbally either at the end of, or during the programme on an informal basis, although anonymity is clearly not possible.

An alternative approach, particularly appropriate within larger organisations, is to use web-based questionnaires, which can be arranged to preserve student anonymity and may also facilitate comments by teachers and summarising of responses.

Whatever method is used to gather evaluations of training, it is essential that the information received is not biased. Bias can arise if only a small subset of students respond (a return rate of 90% or more is desirable), or evaluation is requested during or immediately after completion of the course, because recent classes are freshest in the mind and most likely to be recalled. Students may be reluctant to provide balanced evaluation if there is a perception that this may influence the outcome of the training (e.g., examination results) and many course organisers prefer to gather evaluations anonymously so that respondents cannot be identified, although this has the disadvantage that it is difficult to clarify or explore concerns that may be expressed. A good compromise is to seek comments shortly after the end of the course. One opportunity for this may be immediately after the assessment, or when the results are announced. Students can then talk more freely and can include the assessment itself in their evaluation.

There is no point is collecting the views of participants if nothing happens to the findings. For example they may be reviewed by the teaching committee, course presenters, ethical review process, or student focus groups. In addition there must be genuine consideration of opportunities to modify the syllabus, timetable or teaching methodology and students must be made aware that issues they have raised receive serious consideration, even if change is not implemented. One approach is to supply students with summaries of student evaluations, together with notes of actions taken, so that they can participate in further discussion if they so wish. From the point of view of the course organiser, evaluation provides assurance that the best possible quality of training has been delivered and that students are satisfied with their learning.

16.6 OVERSIGHT OF EDUCATION AND TRAINING

At establishments where laboratory animals are kept and used, ultimate responsibility for the provision of education and training rests with senior management, who must be aware of national, regional and local requirements and the range and depth of competencies required to meet these. This information may be used to establish profiles for each type of function—for example based on the FELASA Categories. A training plan should be developed that assures accountability and transparency. For example it should:

1. Develop a profile of job competencies required, on the basis of current and anticipated needs
2. Establish the training needs of each staff member with advice of senior staff
3. Prepare training proposals in consultation with staff, identifying an overall development plan
4. Deliver training

5. Formally evaluate outcomes (e.g., by testing or performance monitoring)
6. Incorporate ongoing coaching and mentoring to identify further areas for development and to reinforce new behaviours

The training plan should identify clear targets; for example, avoiding infringements, progressively implementing new Three Rs initiatives, breeding animals of a defined quality, minimising over-breeding, attaining a defined quality and quantity of scientific output, and so on. Such objectives must be informed by the business or academic objectives of the organisation itself. Additional issues may arise where there is a need to harmonise cultural differences, for example when staff have been recruited from an establishment with different operating objectives, or where persons are moving from a facility with different legislative and ethical expectations.

Care should be taken to ensure that the training needs of animal care staff are not overlooked or under-estimated; they are just as important, if not more so, than those of FELASA Category B or C personnel. An efficient working environment requires all those engaged in the care and use of laboratory animals to be competent and aware of their dual role in minimising the harms inflicted on animals and assuring high quality science.

Even if it were possible to recruit only skilled and knowledgeable staff (e.g., by use of an agency), the need remains for management to maintain those skills and to prepare for future developments by regular updating of knowledge and skills. All technological establishments require mechanisms for promoting and monitoring education and training and providing a supportive, open environment and practical opportunities for development. In return, staff working with laboratory animals should understand their responsibility for their own learning and development. In achieving this, supervisors must focus on quality learning outcomes and effective returns on investment. Key to these responsibilities are:

- focus on the learner;
- innovative approaches to learning;
- effective determination of training needs;
- efficient administration and allocation of resources;
- oversight of the quality of delivery;
- strong, evidence-based monitoring and evaluation;
- acquaintance with the workplace context, including learners' peers and professional bodies;
- ensuring quality learning outcomes and effective returns on investment.

16.6.1 Managing Education

Educational management and leadership, even in academia, tends to be dominated by the expectations of immediate supervisors and attendees themselves, rather than the needs of the establishment, regulatory authorities and the wider public. This can be addressed by establishing a group responsibility for training; for example, a committee or less formal body, which oversees, delivers, and monitors education and training. For convenience this is referred to here as an education committee; it should convene (and it may do this electronically) sufficiently frequently to be able to oversee the content and effectiveness of training and report periodically to a more senior level of management, for example, the ethical review body or a senior member of the institution. This committee should periodically review the content of the curriculum, monitor indicators of learning such as performance at assessment and in the laboratory, review evaluation from students and their supervisors and make recommendations for maximising the effectiveness of training. It must keep up-to-date not only with current teaching methodologies, but also with regulatory requirements and anticipated technical demands on staff. In addition to key stakeholders (managers, trainers, supervisors, and at least one "student" member), the committee may include a range of other expertise, for example quality assurance or business development staff, staff development personnel and in academic establishments a representative of the faculty teaching board.

16.6.2 Quality Assurance and Transferability

Because training is expensive in terms of commitment, time and resources, programmes should always be monitored to ensure that delivery is effective, at an appropriate level and time, and within budget; the supervisor's role is key in this. Monitoring is particularly important when training is delivered by an "external" provider who may require feedback on the effectiveness of the programme delivered. Wherever possible the assurance of quality should involve a mechanism quite independent of the training process, so as to avoid conflicts of interest. At its simplest this may involve scrutinising the results of assessment, but a more thorough approach is to consider also the quality of work before and after training and changes in productivity using metrics developed at the outset of the training programme. This issue is considered further by Foshay and Tinkey (2007).

There is a need within Europe to address issues of mutual recognition and acceptance of qualifications; this has prompted the European Commission to introduce a qualifications framework (2007) to facilitate movement and recognise the lifelong learning of European citizens. At an undergraduate level, the Bologna Declaration signed in June 1999, established a system that made it easy to compare academic grades and included a diploma supplement, designed to improve international transparency and to facilitate wide recognition of qualifications. There is pressure for lifelong learning to be treated in a similar way, with development of reference levels, a basis for certification, and other measures, including a credit transfer system.

Within universities, Category B, and sometimes Category C training may be incorporated into Master or Doctor-level programmes and occasionally into the undergraduate syllabus; sometimes these opportunities may be available also to those working for commercial companies. Wherever possible in such cases, the educational programme should be incorporated into the relevant quality assurance process. FELASA has developed an independent scheme for the assurance of quality in the delivery of education and training that enables organisations to describe their course as "FELASA Accredited" (Nevalainen et al. 2002). Organisations wishing to apply for accreditation are required to have run at least two similar courses beforehand. They then submit key indicators of course quality, including details of the curriculum, course size and frequency, oversight mechanisms, learning materials, assessment outcomes, results of course evaluation by students and the credentials of teaching staff. The board overseeing the accreditation scheme reviews the application and may award or decline accreditation or may ask for changes to be made. Accredited courses are visited at least once during the period of accreditation (5 years in the case of B and C Category courses, 10 years in the case of A and D Categories) and annual reports must be submitted for review by the accreditation board.

16.7 CONCLUSIONS

An effective training programme requires identification of training needs and the development of personal development plans relevant to the FELASA category within which the individual will be working. There is a wide range of techniques for delivering training, and the most appropriate of these depend on the topic being presented, the individual's preferred learning style, and the facilities available. In general, interactive learning is an effective way of developing understanding and encouraging students to develop interpersonal skills that are relevant to all persons working in laboratory animal science. Learning outcomes should always be defined up-front, and assessed by methods well suited to objectively determining the desired learning outcomes and competence of the student.

One major objective of teaching and training is encouragement of professionalism, including a desire to maintain and develop new knowledge and skills throughout life, to provide support for others, and take pride in working to the highest welfare and scientific standards. Whilst some topics lend themselves to relatively formal teaching methodologies—for example understanding the law or experimental design—others, such as technical aspects of work with genetically modified animals, non-invasive imaging or radiotelemetry are less amenable. In such cases more in-depth specialist training may be preferable, such as workshops, training schools, or specific modules, rather than relying solely on material included in the core instructional programme. Recognition of in-depth learning may take the form of certificates of attendance or attainment, inclusion on a register of competencies, or more formal acknowledgment in the workplace.

QUESTIONS UNRESOLVED

- There is a need to ensure that the development of practical and other workplace skills takes place within a supportive and professional environment. In particular, guidance is needed on the training and expectations of supervisors and the monitoring of learning.
- Reliable strategies are needed for assessing the "ethical attitudes" of learners.
- There are informal mechanisms by which teachers and trainers can communicate with each other and discuss issues relating to training and development. However these are generally few, and a means by which trainers can network with others to share good practise and seek help when necessary would be highly advantageous.
- Continuing professional development is fundamental to staying up-to-date and to career development. A means is needed to ensure wide recognition of achievements in this area.
- The FELASA Accreditation Scheme is still relatively little utilised and an increase in the number of accredited facilities within Europe would promote development of more uniform levels of competency and raise the academic status of persons involved in laboratory animal science.

REFERENCES

Ashcroft, K., and L. Foreman-Peck. 1994. *Managing teaching and learning in further and higher education.* London: Routledge Falmer.
Biggs, J., and C. Tang. 2007. *Teaching for quality learning at university,* 3rd ed. Berkshire, England: Open University Press, McGraw-Hill Education.
Bligh, D. A. 2000. *What's the use of lectures?* San Francisco: Jossey-Bass Publishers.
Bull, J. 1993. *Using technology to assess student learning.* Sheffield: CVCP, University of Sheffield.
Carlsson, H. E., J. Hagelin, A. U. Hoglund, and J. Hau. 2001. Undergraduate and postgraduate students' responses to mandatory courses (FELASA category C) in laboratory animal science. *Laboratory Animals* 35:188–93.
Colby, L. A., P. V. Turner, and M. A. Vasbinder. 2007. Training strategies for laboratory animal veterinarians: Challenges and opportunities. *ILAR Journal* 48:143–55.
Council of Europe. 1986. *European convention for the protection of vertebrate animals used for experimental and other scientific purposes, publications and documents division, Strasbourg, 1986.* Strasbourg: Trans. Council of Europe, Strasbourg.
Council of Europe. 1993. *Multilateral consultation of the parties to the council of Europe convention ETS no 123 on 3 December 1993.* Available from http://eurlex.europa.eu/smartapi/cgi/sga_doc?smartapi!celexplus!prod!DocNumber&lg=en&type_doc=Directive&an_doc=2003&nu_doc=65, Accessed 2 July 2010
European Commission. 1986. Council directive 86/609/EEC of 24 November 1986 on the approximation of laws, regulations and administrative provisions of the member states regarding the protection of animals used for experimental and other scientific purposes. *Official Journal of the European Union* L358:1–29.
European Commission. 2007. *Communication from the commission to the council, the European parliament, the European economic and social committee and the committee of the regions action plan on adult learning: It is always a good time to learn,* Vol. COM(2007). Brussels.

European Science Foundation. 2001. *Policy briefing 15 (August 2001-second edition): Use of animals in research.* Available from http://www.esf.org/activities/science-policy/science-policy-activities-1998-2001.html, Accessed 2 July 2010

FELASA. 1995. FELASA recommendations on the education and training of persons working with laboratory animals: Categories A and C. Reports of the Federation of European Laboratory Animal Science Associations working group on education accepted by the FELASA board of management. *Laboratory Animals* 29:121–31.

Felder, R. M., and L. K. Silverman. 1988. Learning and teaching styles in engineering education. *Engineering Education* 78:674–81.

Foshay, W. R., and P. T. Tinkey. 2007. Evaluating the effectiveness of training strategies: Performance goals and testing. *ILAR Journal* 48:156–62.

Hanlon, A., T. Dich, T. Hansen, H. Loor, P. Sandøe, and A. Algers. 2006. Animal ethics dilemma—An interactive learning tool. Available from http://ae.imcode.com

Hau, J. 1999. *Humane endpoints and the importance of training.* Paper presented at Humane Endpoints in Animal Experiments for Biomedical Research—Proceedings of the International Conference, Zeist, The Netherlands.

Holsgrove, G., and M. Elzubeir. 1998. Imprecise terms in UK medical multiple-choice questions: What examiners think they mean. *Medical Education* 32:343–50.

Honey, P., and A. Mumford. 2000. *The learning styles helper's guide.* Maidenhead: Peter Honey Publications Ltd.

IAT. 2006. *Institute of animal technology syllabus.* Summertown, Oxford, UK.

Kirkpatrick, D. L., and J. D. Kirkpatrick. 2006. *Evaluating training programs: The four levels,* 3rd ed. San Francisco: Berrett-Koehler.

Lampert, M. 2002. Appreciating the complexity of teaching and learning in school. *Journal of the Learning Sciences* 2:365–69.

Lassen, J., M. Gjerris, and P. Sandøe. 2006. After Dolly—Ethical limits to the use of biotechnology on farm animals. *Theriogenology* 65:992–1004.

Lassen, J., P. Sandøe, and B. Forkman. 2006. Happy pigs are dirty—Conflicting perspectives on animal welfare. *Livestock Science* 103:221–30.

Mellis, C. M. 2008. Optimizing training: What clinicians have to offer and how to deliver it. *Paediatric Respiratory Reviews* 9:105–12; quiz 112–13.

Morrison, S., and K. W. Free. 2001. Writing multiple-choice test items that promote and measure critical thinking. *The Journal of Nursing Education* 40:17–24.

Nevalainen, T., E. Berge, P. Gallix, B. Jilge, E. Melloni, P. Thomann, B. Waynforth, and L. F. van Zutphen. 1999. FELASA guidelines for education of specialists in laboratory animal science (category D). Report of the Federation of Laboratory Animal Science Associations working group on education of specialists (category D) accepted by the FELASA board of management. *Laboratory Animals* 33:1–15.

Nevalainen, T., H. J. Blom, A. Guaitani, P. Hardy, B. R. Howard, and P. Vergara. 2002. FELASA recommendations for the accreditation of laboratory animal science education and training. *Laboratory Animals* 36:373–77.

Nevalainen, T., I. Dontas, A. Forslid, B. R. Howard, V. Klusa, H. P. Kasermann, E. Melloni, et al. 2000. FELASA recommendations for the education and training of persons carrying out animal experiments (category B). Report of the Federation of European Laboratory Animal Science Associations working group on education of persons carrying out animal experiments (category B) accepted by the FELASA board of management. *Laboratory Animals* 34:229–35.

Olsson, A. S., P. Robinson, K. Pritchett, and P. Sandøe. 2002. Animal research ethics. Handbook of laboratory animal science, second edition: Essential principles and practices. 2nd ed. In Vol. I, 13–31. Boca Raton, FL: CRC Press.

Rose, C. Accelerated learning. 2003 Available from http://www.andreileca78.ro/wp-content/uploads/2010/01/accelerated_learning_colin_rose.pdf, Accessed 2 July 2010

Rothwell, W. J., and J. A. Benkowsk. 2008. *Building effective technical training: How to develop hard skills within organizations.* New York: Pfeiffer & Company.

Sandøe, P., and S. B. Christiansen. 2008. *Ethics of animal use.* Oxford: Blackwell.

Smith, J. A., and K. Boyd, eds. 1991. *Lives in the balance: The ethics of using animals in biomedical research.* Oxford: Oxford University Press.

Van der Valk, J., and B. F. M. Van Zutphen. 2004. Reduction through education: The insight of a trainer. *Alternatives to Laboratory Animals* 32:1–4.

17 Animal Experimentation and Open Communication

Ann-Christine Eklöf, Sweden; Anne-Grethe Berg, Norway; and Jon Richmond, United Kingdom

CONTENTS

Objectives 392
Key Factors 392
17.1 Background 392
17.2 Analysis 394
    17.2.1 Why is a Communications Strategy Required? 394
    17.2.2 What is the Strategy Intended to Achieve? 395
    17.2.3 Who is Involved, and for What Purpose? 395
        17.2.3.1 Who has Information? 395
        17.2.3.2 Who needs Information? 396
        17.2.3.3 Who is Responsible for Providing Information? 396
        17.2.3.4 Who can Best Communicate the Information and Who has the Resource To Do So? 397
    17.2.4 What Information is to be sought and Provided; What is the Strategy Intended to Achieve? 397
    17.2.5 How Will Information, Opinions and Outputs be Communicated? 398
    17.2.6 How Organisations Involved in Animal Research Publicise/Communicate this to Stakeholders: an Overview 399
        17.2.6.1 Regulators: Those Who Regulate Animal Research and Those Who Require Animal Use and Data 399
        17.2.6.2 Public Sector Funding Bodies 400
        17.2.6.3 Private Sector Funding and Private Sector Establishments Undertaking Animal Research 400
        17.2.6.4 Public Sector Establishments and Scientists Undertaking Animal Research 401
        17.2.6.5 Professional Bodies 401
        17.2.6.6 Lobbyists and Pressure Groups 401
        17.2.6.7 The Media 402
17.3 Conclusions 402
17.4 Questions Unresolved 403
References 403

OBJECTIVES

It is essential to inform the public, researchers, decision makers and policy makers about checks and balances that are applied to the experimental use of animals. This includes justifying the need to use animals for experimental and other scientific purposes, explaining the ethical evaluation of research proposals and setting out the principles applied and outcomes obtained in the harm-benefit analysis.

This chapter provides general insights into what constitutes an effective communications strategy, and contemplates the approach currently taken by some stakeholder groups. It presents an introduction to good practice when planning and executing communications strategies related to the use of animals in science and has been prepared with the European regulatory framework (European Commission 1986) in mind. No attempt is made to examine the various agendas that different stakeholder groups might have, rather to explore the strengths and weaknesses of the tools they use. You should not regard these tools as "sector specific", rather you should consider them all as options when developing your own communications strategies.

The starting point to planning any strategy to inform and influence current thinking, and future policy and practise are a sound understanding of how to communicate effectively. Properly planned and well-executed communication is particularly important in this area where conflicting information is presented by different interest groups and mistrust may polarise opinion, and colour stakeholder perceptions and interactions.

This text does not express an opinion about what point of view is to be communicated; it does not assume that any single ethical framework or point of view is either timeless or correct nor does it provide a preferred framework for ethical evaluation or harm–benefit assessment or set out the arguments needed to defend the use of animals for scientific purposes. It offers general principles and insights that can be used in many contexts and examines how effective communications strategies can be designed, conducted and evaluated. No best practice blueprint is offered, rather the general principles of good practice and an evaluation of the main communications tools.

The advice offered here is of necessity generic, and should be critically evaluated and adapted to the needs of different objectives, messages and audiences. However, whilst money and manpower are sometimes considered to be the most important resources when developing and running a communications strategy, nothing positive or lasting will be achieved without understanding and applying the basic principles set out in this chapter.

KEY FACTORS

The key factors to be researched and defined before contemporary communications strategies are launched can be listed as (ESRC 2008):

- Why is a communications strategy needed
- What is the current situation
- What are the objectives and desired outcomes
- Who are the target audiences
- What are the essential messages
- What tools and activities can be used
- What resources are required
- What are the timescales
- How and when will the impact of the strategy be evaluated, and the strategy updated or phased out

17.1 BACKGROUND

Effective communications strategies are carefully planned and properly resourced and effective communication is a two-way process—communicating with and receiving feedback from others

(Fairclough 2003). Intentions and desired outcomes and how they will be evaluated need to be clearly defined before launching a communications strategy. Communication must establish an open stakeholder dialogue, facilitating both clarifications and exchanges of views, rather than simply providing for the one-way transmission of information, denying recipients the opportunity to consider and comment before decisions are taken or change effected.

It has long been established that the main elements underlying effective communication are:

- the medium through which the communication takes place;
- the subject matter and message; and
- the person to whom information is addressed.

It is the resulting impact on the understanding and behaviour of the last of these that is the only meaningful measure of success (Rhys Roberts 1924; Lasswell 1948).

In the context of the use of animals for scientific purposes, one of the common general communications objectives is to increase the shared knowledge necessary to have a proper and informed understanding both of the ethically sustainable and the scientifically valid use of animals in research. Current ethical norms and the resulting regulatory frameworks have determined that if scientific validity and utility cannot be assumed or demonstrated, then there is no defensible or sustainable ethical basis for the use of animals in procedures that have the potential to cause them pain, suffering, distress, or lasting harm for experimental or other scientific purposes (see Chapter 5). Only if these pre-conditions are met, can a sound communication strategy be developed (Russell and Burch 1959).

Never assume that the public and opinion makers automatically accept either that there is a need for animal-based research, or that they necessarily believe animal-based research remains scientifically valid. Defending scientific validity and utility by reference to the past is seldom a priority, and argument alone will never win over those who take a different view of contemporary animal research.

Animal research has been important in the past—but debate and decision making must take place in relation to its value at present and in the future. Times change: the focus is on the present and the future of both fundamental and applied research. For those seeking to explain the continued need for animal-based research and testing, recapitulating the historical importance of animal research is not enough, indeed if the objective is to manage the present and prepare for the future; arguments rooted in the past will be seen by some as over-defensive and counterproductive.

It can also be counterproductive to emphasise only the most socially acceptable classes of animal use. For example using information and material relating solely to the development of new health care opportunities may be seen as a tacit acknowledgement that other classes of animal use are less important or dispensable.

It is also of little importance to debate the relative merits of fundamental versus applied research—the boundaries are not rigid; individual experiments, projects and programmes seldom, if ever, indisputably determine priorities in product development, regulatory decisions, or health care outcomes; it is basic, fundamental research that provides the new insights that are subsequently exploited in product development and regulatory affairs. That is not to say that these issues do not have to be acknowledged, but they should not play a central role in a communications strategy aimed at better understanding the present or planning for the future. The most effective factors in communications strategies share three common features; they:

- play a central role in making authoritative information and expertise available to raise awareness; improve understanding; and inform debate, policy development, practice and outcomes;
- provide and involve stakeholders within a structured and transparent framework for raising awareness, providing a forum for discussion and consideration of relevant

activities and issues in a way that lets them feel they are part of the decision making process; and
- evaluate and publicise the outputs, and their impact, as progress is made and change is effected.

Although openness and transparency are essential for successful communication, in this field where there can be mistrust and even conflict between different classes of stakeholders, issues such as personal security, the illegal activities of animal rights activists and commercial confidentiality must also be considered when deciding what and when to communicate. Whilst this does not exclude responsible stakeholders from the process, it does mean that, for example, those who consider themselves at risk of attack from extremists may be reluctant to play a highly visible part in the process or to share detailed information of their research or their most significant doubts and concerns.

High-quality statistical and other data are not available for all aspects of the use of animals in science. As a result case studies can be particularly powerful and are an important means of displaying, informing and determining current good practise and standards and illustrating the need for, and potential scope and limitations of, options for change. Ideally case studies should be real and meaningful examples, rather than fictional or abstract artificial constructs devised solely to set out a particular point of view of a theoretical situation (Robert and Yin 2008).

A useful and well-maintained resource of material for real-life case studies is the UK Home Office website (http://tna.europarchive.org/20100413151426/http://scienceandresearch.homeoffice.gov.uk/animal-research/publications-and-reference/001-abstracts/index.html), which publishes abstracts of animal research currently being undertaken in the United Kingdom.

17.2 ANALYSIS

Little can be achieved with a one-off round of communication. The intention must be to develop and maintain a sustainable communications strategy. Internal and external communication tools are required; that is, there is a need to consider communications and information flow both within and between stakeholder groups.

It is sometimes considered that government and industry can deploy large amounts of resource to communications. The truth is that, faced with many other pressures and priorities, in practise they commonly deploy much less resource on any given topic than single-issue interest groups. An effective communications strategy must ensure that the right information is provided to, considered by, and acted on by the right people at the right time. To be effective it must also make provision for feedback on the needs and responses of the stakeholders.

From the outset there is a need for clarity with respect to:

- *why* a communication strategy is required;
- *what* is the strategy intended to achieve;
- *who* is involved, and for what purpose;
- *what* information is to be sought and provided;
- *where* and by what means communication should be effected;
- *when* key activities should take place;
- *how* information, opinions, and outputs will be evaluated and communicated; and
- *how and when* progress, and success or failure will be judged.

17.2.1 WHY IS A COMMUNICATIONS STRATEGY REQUIRED?

An effective communications strategy must:

- inform and educate;
- correct misinformation and misunderstandings;

- when necessary, change behaviour and performance with respect to animal welfare and humane experimental technique;
- influence opinion formers;
- promote feedback and informed debate; and
- by changing attitudes and behaviours, shape policy, encourage best practice, and promote progress.

17.2.2 What is the Strategy Intended to Achieve?

Different strategies will have different objectives and preferred outcomes. The unifying objectives in the context of this chapter are likely to be the desire to:

- create or display a "culture of care" at establishments where animals are bred and used;
- equip operational and managerial staff and others to demonstrably provide this culture;
- support good animal welfare and good science; and
- promote public and political understanding and debate of, and confidence in, the responsible and justifiable use of animals for experimental and other scientific purposes and the societal benefits this can bring.

Only when the specific objectives and preferred outcomes are defined can the "how" and the "why" be understood and elaborated.

17.2.3 Who is Involved, and for What Purpose?

Only when the specific objectives are defined, and the relevant outcome measures agreed, will stakeholders be in a position to judge if they are competent and willing to participate. When planning the strategy identify the following classes of person; those:

- with the knowledge;
- who need information;
- with the responsibility and resources to evaluate information and ensure effective communication takes place; and
- with the credibility, skills, and resources to communicate effectively.

17.2.3.1 Who has Information?

High-quality information—facts, figures, analysis and opinion—of known and robust provenance is required to devise and deliver any evidence-based, effective and meaningful communications strategy. To participate effectively, stakeholders must be able to identify, locate, share and evaluate objective material. The aim is to identify and work with information that can be presented and accepted as authoritative by all parties. To properly inform the desired understanding, debate, decision-making activities and outcomes, all reasonable efforts must be made to identify all those who may have useful information, views or contacts.

An appropriate balance has to be struck between involvement with individuals, local/national organisations, and inter- and trans-national bodies. Although the larger bodies have larger constituencies, of necessity they tend to offer a corporate, consensual view rather than articulating the much broader range of views held by, and diverse interests of, their individual members.

It should not automatically be assumed that those charged with providing information will themselves have access to (or will have accessed) all of the relevant information, can always provide authoritative and objective information and comment, or are best qualified to communicate effectively. Assess the quality of all contributions and judge sources, no matter how apparently impartial or prestigious, on the basis of the quality of the input provided rather then their general reputation.

Over-reliance is sometimes placed on facts and figures and information in the peer-reviewed literature; remember that publication bias (the views and agenda of the funders, authors, editors and journals, and the difficulty in publishing some unexciting findings and non-conforming views) may need to be taken into account.

A robust communications strategy must also be able to identify, evaluate, and display reliable unpublished, informed, third-party opinions and views. The communication strategy must clearly set out the measures taken to find such information, to establish and maintain quality standards, and to correct misinformation.

The World Wide Web can be a good resource; but it can also be a major source of prejudice and misinformation (Allen et al. 1999). In assessing the validity of web-based material, caution is needed as to its provenance, the author's underlying agenda, and its currency and quality.

17.2.3.2 Who needs Information?

Those who seek to provide information must themselves be very well informed. In many situations expert stakeholders, trained primarily to communicate with other experts, are requested or need to communicate with opinion formers outside their own peer groups. Outputs prepared with this in mind must be easily understood by those who are not experts—some of whom may have little pre-existing knowledge of the subject matter or specialist vocabulary. Clear non-technical material for the public and others is a particularly important part of any strategy to inform, challenge, establish and reflect the range of beliefs, societal values, priorities and concerns that underpin informed public opinion and public policy.

Even when the underlying messages are the same, the different needs of different audiences will often require that information is packaged and presented in a range of ways to best meet their particular needs, preferences and expectations. If you choose not to present your readers with what they prefer or need, do not assume either that they will have understood, or even that they will have made the effort to understand.

The nature of the audience determines what information is required and how it is:

- packaged;
- presented and made accessible; and
- how effectiveness is evaluated in terms of outcomes.

The development of public policy is often informed and determined by non-technical factors, including perceived public preferences. Would-be opinion formers neglect this at their peril. Those planning high-quality communications strategies know that both experts and non-experts have legitimate and important concerns, opinions and insights—and that the public, and in many cases the decision makers, are not experts. Proper consideration of the needs of non-experts is essential to understand and inform public and political understanding and debate of, and decisions affecting, issues relating to animals in science.

17.2.3.3 Who is Responsible for Providing Information?

Begin by considering who has a legal duty or a moral obligation to provide information. This will include those who:

- determine and implement public and institutional policy;
- fund or undertake animal research; and
- lobby or seek to influence, shape, or change public opinion or policy about animal research.

All too often those who act in these capacities have adopted a fixed position and see standard practise as primarily being to explain and justify their established views, policies and actions without encouraging engagement, dialogue or change.

Knowledge of the interests and agenda of those providing information is essential for any critical evaluation of the objectivity of the information and analysis supplied by those with a direct interest and a particular established point of view. As a matter of routine, this should invite questions as to what should and does motivate (and resource) even those seemingly non-partisan groups claiming to provide objective and balanced information for its own sake: a hidden agenda is more dangerous than no agenda.

Also, be aware that material with a high emotional content, seeking to win over hearts rather than minds, often conceals a lack of objective evidence and sound arguments. Nevertheless, as animal welfare is an emotive issue for many people, good communications strategies must make appropriate use of evidence, language and images to appropriately engage both hearts and minds. Those who do not critically evaluate information provided in this light:

- may not appreciate the need for such critical evaluation;
- will not be fully aware of the dangers of taking a position having only considered one point of view; and
- have formed their opinions based primarily on the impact of material with a high emotional and low factual content.

Sources of information include the scientific community itself, and the funding bodies and the regulatory authorities who take decisions on the basis of animal test data. Their information is generally factually accurate, and is increasingly targeted at a wider range of audiences including the general public. Communications strategies and outputs must be tailored to provide the information needed by the target audience to inform influence and determine, rather than simply to describe and defend, established policy and practice.

17.2.3.4 Who can Best Communicate the Information and Who has the Resource To Do So?

In determining who is best placed to be the public face of the communications strategy, never lose sight of the fact that all information and means of communication must resonate both with the objectives of the strategy and the needs of the audience (Rhys Roberts 1924). Take account of the preferences, expectations, and needs of those who must be provided with material for consideration and action when determining both the skills and resources required to communicate effectively. A communications strategy that does not take account of these basic elements will fail.

Remember that those who are most expert or authoritative are not necessarily the best communicators. Remember too that not all audiences believe that those who are most expert or authoritative are the most credible or trustworthy. Even in highly technical areas, celebrity rather than a technical background will resonate with some audiences.

17.2.4 WHAT INFORMATION IS TO BE SOUGHT AND PROVIDED; WHAT IS THE STRATEGY INTENDED TO ACHIEVE?

Take time to define what has to be communicated and why, and how your effectiveness will be evaluated. As individuals our attitudes and opinions are not always "evidence based"; and our behaviours are not always rational. The same can be true of organisations and their policies and practises.

In the real world amongst the general public there is a broad spectrum of sincerely and firmly held and very different views with respect to animals in science. You may dispute the validity of some of these views—but never underestimate the strength with which they are held.

This breadth of understanding and opinion is not based upon a "deficiency model"—generally summarised as the belief that if we all had access to the same factual information we would all automatically from then on hold the same views. In the context of animals in science, even if everyone had access to the same authoritative information no single and agreed view would prevail.

Consider, having defined your specific objectives, whether your communications strategy should next explore when, why, and how non-evidence-based views are formed, and how they can be changed. Many single issue pressure groups do this to better understand and exploit the power of emotional content, including human interest stories, in getting their key messages across. An emotive case study, involving real people and animals, can have more impact on the public and decision makers than a thoroughly researched and definitive comprehensive technical report based on impersonal facts and figures.

Thought next has to be given to ensure the accuracy, balance and currency of the information selected for use, and how informed societal values, concerns and needs can be gauged and incorporated.

The interests and agenda of those who commission or compile information are key considerations: perhaps, less so when there is a statutory requirement to do so; more so when it is intended to support a particular policy or point of view; and particularly so when there is no immediately visible agenda, and no clear indication of, why such an apparently altruistic resource has been provided. Wherever possible place reliance on first-hand data.

In practice there is seldom access to complete original datasets (including public opinion polls) and a full description of how it is compiled (Crespsi 1980). Even when this is possible, making sense of the facts and figures is generally impossible without additional information being available to understand the scope and limitations of the dataset, details of how it was complied, its context, and therefore the inferences that can be drawn.

Complete reliance can seldom be placed on second- and third-hand descriptions of datasets or media coverage. Such sources:

- are not authoritative;
- information has generally been selectively gathered and summarised or paraphrased;
- important qualifications with respect to the quality and utility of the information discarded; and
- some form of editorial view is generally imposed on the material selected for publication.

Reliance is best based on the material published by those who generated the data, mindful of what is known of the purposes for which it was complied and presented.

17.2.5 How Will Information, Opinions and Outputs be Communicated?

Whilst there will always be a place for the written word and peer-reviewed publications, increasing use is being made of inter-active, web-based tools that allow the use of both text and images, and allow the readership to interact with the authors. The web is a particularly dynamic environment and the material displayed, and the addresses of the sites displaying it, are frequently subjected to change without notice.

Organisations planning to use this medium as part of a communications strategy need to take advice not only on how material is to be presented on the web, but how the website is to be designed to ensure it is easily found by the common search engines.

Nevertheless, written material is still the norm (albeit increasingly often using electronic media) and, notwithstanding the benefits of web-based and other web-enabled solutions, means should always be sought to ensure key outputs are recorded and commented on in the peer reviewed literature and media.

Other forms of publications, meetings and peer-pressure can also have a role to play in a communications strategy. For a contemporary communications strategy a pro-active and flexible approach is recommended; keeping information and messages as user-friendly, consistent and simple as possible across the different communications media; and producing material that everyone can understand, have confidence in, and can discuss, promote, evaluate, debate and defend.

There will be some material and audiences where selective distribution will be best (although processes and planning must always be perceived as inclusive and transparent rather than selective and exclusive), and others where the wider the circulation the better. Existing networks, forums and contacts are often the most efficient and effective means of eliciting and distributing information and fostering debate—at least in the early stages. Key considerations are what information is available and, more importantly, how widely material will be circulated, and how feedback will be obtained.

Presentations, meetings, and discussion groups allow for a dialogue with targeted stakeholders, generally with small selective specialist groups, but their use can create difficulties in verifying and recording any new information and opinions elicited. Whenever possible material used for presentations should also be archived and made available by other means. All presentations given in support of a particular communications strategy should be logged centrally and the feedback systematically captured, recorded and evaluated.

Professional advice should be taken to determine how to use media coverage to best effect. It needs to be remembered that media coverage is not always accurate, and you will have no editorial control over what is actually published or how it is presented. Specialist science correspondents are more likely to be familiar with the subject matter and best able to offer objective coverage than general news reporters.

17.2.6 How Organisations Involved in Animal Research Publicise/Communicate this to Stakeholders: an Overview

A clear understanding of how different stakeholder subsets prefer to communicate, and the timelines to which they work can assist with the planning of communications strategies. The Economic & Social Research Council offers a range of web-based resources providing practical advice and templates for doing this; and additional guidance for making the best use of the media (ESRC 2009). The Research Councils UK: "Dialogue with the Public: Practical Guidelines", available from the Research Councils UK website (RCUK 2009), is another particularly useful resource.

17.2.6.1 Regulators: Those Who Regulate Animal Research and Those Who Require Animal Use and Data

European Union political and scientific institutions share a number of simple, key, and often repeated messages and values:

- animal welfare is important;
- the use of animals for experimental or other scientific purposes is necessary (particularly in the context of human and environmental safety and improvements in health care provision for humans and animals), but must incorporate the Three Rs (Russell and Burch 1959);
- steps must be taken to develop, validate, and adopt alternative methods; and
- ethical review promotes good science and good welfare.

These tend to be communicated primarily through the EU institutions' official websites.

Key documents include a published animal welfare strategy and action plan: technical reports (including advice and opinions) and other expert documents; press releases and newsletters; parliamentary questions and debates; and formal consultations and impact assessments. Although the original sources of the information are often paper documents, in reality they are generally accessed and circulated in electronic format. Disappointingly, a critical evaluation of these top-level EU documents would have to concede that, other than showing a shared common position, these outputs should not be adopted as role models.

Official EU websites are not particularly user-friendly, and tracking down contemporary relevant documents is not easy. At times the documents themselves fail to communicate key information to relevant stakeholders in a timely manner. Although media coverage and stakeholder mailings are used by these bodies, the former cannot always be guaranteed to be accurate or balanced, or to contain the caveats and qualifications required to properly understand the issues; and the latter, as they are cascaded through others to wider and wider numbers of stakeholders, often have additional third-party editorial comments attached.

National regulators also maintain informative websites that are increasingly providing accessible information for the general public as well as specialist information advice to those that they regulate.

17.2.6.2 Public Sector Funding Bodies

The key messages of these bodies tend to include that the animal research funded is important, necessary and ultimately in the public interest, placing animal research in the context of other funded work aimed at producing tangible societal benefits and stressing that funding streams are intended to meet carefully defined strategic scientific needs and priorities.

Corporate publications are generally accessed through the funding bodies' websites: and, on completion, funded work is published in peer reviewed scientific literature. Their websites commonly include:

- relevant areas of science and research priorities;
- information on the processes and criteria applied to select proposals for funding;
- summary details of what is being funded;
- assurances that legal requirements and appropriate ethical norms are applied, including statements of ethical principles and good practice; and
- assertions that scientific validity and animal welfare are both important funding considerations.

Almost universally they stress that the Three Rs (Russell and Burch 1959) are applied, and that the funded research produces new science and methodological progress that is reflected in progress with the Three Rs and societal benefits.

Although funding bodies are sometimes thought of as being essentially reactive—for example publicising new scientific developments—they can also provide important proactive contributions through highly professional, more strategic communications strategies though media coverage, reports, lectures and meetings.

17.2.6.3 Private Sector Funding and Private Sector Establishments Undertaking Animal Research

These organisations often lead with their strategic and economic importance, whilst stressing they are socially responsible and committed to high ethical and animal welfare standards. Many increasingly stress the linkage between high-welfare standards and good science. They acknowledge that animal use is still required, and generally state that they support and fund the Three Rs (Russell and Burch 1959), are highly regulated, and operate to high standards.

Many have a good story to tell. The pharmaceuticals and cosmetics sectors in particular have contributed significantly to the development and validation of alternative testing methods and strategies. High profile charitable or *pro bono* activities, including funding Three Rs initiatives, and part-funding of public sector work, will be prominently displayed. They understand the need for effective communications, and communications strategies are often led by trade associations rather than individual companies. Although the focus sometimes seems to be on quasi-advertising and public relations, they make good use of information packs and other educational material for more specialist audiences. Direct lobbying activities are also undertaken. The most relevant material can be accessed through corporate websites.

Although different strategies, materials and methods are used to maintain investor, political and public confidence, they are all very professional and well resourced. Nevertheless their public image at times is volatile, and often dependent on the tone and content of recent media coverage.

17.2.6.4 Public Sector Establishments and Scientists Undertaking Animal Research

Key messages tend to include that continuing animal use is essential, important and in the public interest, and that it is undertaken to high standards and is highly regulated. They tend to be particularly good at engaging at local level. In addition, high-quality institutional web-based material is often supplemented by individual scientist's home-pages, summaries of on-going research and ethics committee outputs. Information is generally prepared and presented as middle-of-the-road and non-contentious. Direct reference to animal use sometimes has little visibility.

One problem is regularly seen when new public sector scientific findings are newsworthy. The source scientific publication may be balanced, but in the hands of the popular media it is often summarised and paraphrased in a way that tends to simplify and sensationalise, removing any caveats and qualifications the authors thought necessary to include in the source publication. This risks raising unrealistic expectations in patient groups, and provides ammunition for critics to claim that the original authors have deliberately over-interpreted or misrepresented their findings.

Another pitfall can arise from misuse of the analysis and discussion sections in peer reviewed publications reporting the original findings. These sections of scientific papers are generally candid about the scope and limitations of the reported findings, and the nature of any significant gaps in our knowledge. A number of animal protection campaigns have recycled these doubts and uncertainties, provided by the authors in the original publications, as original critiques without acknowledging that they are not only well known to the researchers, but were included in the original publication in order that the informed reader could take account of them.

17.2.6.5 Professional Bodies

Professional bodies rightly stress their high standards and professionalism, through events, policy statements, discussion forums, guidelines, codes of practice, newsletters and journals. Key publications and position statements are generally available on their websites.

Even the longest established and most traditional and conservative organisations increasingly prefer to use electronic formats and web-based information systems; catering primarily for their own members and interest groups, with less resource targeted at the public and policy makers unless there is a cause to be fought.

17.2.6.6 Lobbyists and Pressure Groups

These are campaigning organisations: that is why they exist, and that is what they do. Key messages are determined by, and reflect, their constitution, membership and agenda. Anti-vivisection groups tend to focus on criticising the scientific validity of animal use; and claims of unnecessary animal suffering, animal exploitation and other welfare and ethical issues including misuse of public resources and a failure to make best provision for human safety. Rebutting such claims is not always straightforward, as at times there can be a lack of clarity whether their arguments are based on scientific claims that animal research is poor science because animals are too dissimilar to man; or whether they believe the case to answer is that, morally, animals should not be used because of their close similarities to man.

Pro-animal use groups tend to focus on unmet needs and potential clinical benefits; often playing down animal use for other scientific purposes, and playing up case studies, at times using emotive and human interest content. Sometimes, in reaction to a specific criticism, the response will relate to the general importance of the field of study, rather than addressing the specific concern or criticism.

These groups typically use the full range of tools available for communication and awareness raising, often involving their members in communications activities such as local action, media coverage including letters to editors, reports, videos, flyers, petitions, mass mailings and political lobbying. Target audience may include the public, opinion formers and policy makers, but seldom the scientists in the case of the animal protectionists, and seldom the animal protectionists in the case of the pro-science groups.

Their websites are well-resourced, professional-looking and display well-presented campaign material, and provide contact details for those from the media wanting more information. They will generally invite those who share their views to express support by joining or providing a financial contribution. Lobbyists and pressure groups understand the need to work with the media and provide ready-made briefing to journalists and others. Many have celebrity patrons and spokespersons in addition to high profile web-presences.

Communication, engagement and short-term attention and wins, rather than constructive dialogue, is sometimes perceived as a preferred tactic. Impact rather than accuracy may seem at times to colour material produced and views expressed.

They are always focused, often on a single issue, with the central messages being crystal clear, even though the evidence and analysis may not always withstand rigorous evaluation. Issues are often presented as black or white, right or wrong. Emotive content can be high, and sometimes issues are portrayed as having victims and villains—with animals, and those who have suffered adverse effects as the result of medical treatments, presented as victims—and regulators and scientists as villains.

17.2.6.7 The Media

For many people the popular media is their primary source of information on animal welfare and scientific issues (MORI 2002). The key messages are context specific and liable to change over time both in terms of what is newsworthy and how it is reported. News stories and features appearing in traditional broadcast and print media are now increasingly followed up by Internet BLOGs allowing viewers and readers to comment, though currently what these feedback sites lack in balance they also tend to lack in impact.

Bad news and conflict are generally more newsworthy than good news, and the majority of stories on the biological sciences are bad news stories (Paulos 1996), with many stories and features uncritically based upon information provided directly to the media "ready to print" by third parties. There are often only limited opportunities for corrections or right to reply.

As news articles tend to be produced to very short deadlines and to tight word limits by non-experts, they are often and perhaps inevitably simplified and abbreviated. Accuracy and balance can suffer as a result. Features, and the more reflective coverage provided in the Sunday papers, tend to be better researched and to include a point of view rather than just reporting the facts.

Often animal research is not mentioned in the good news stories describing medical progress, but it is often blamed for medical failures in bad news stories even though many *in vitro* tests will also have informed decision making.

17.3 CONCLUSIONS

It does not matter how good a story you have to tell—without a carefully planned communications strategy no one will ever know, a badly planned or poorly executed communications strategy will damage your cause, and a better presented alternative view, even if it has less merit, will invariably be better received.

Similarly, it does not matter how unbalanced the agenda of others might be—those who adopt the best communications strategies will be the talking point, making it difficult to correct any beliefs that are formed on the basis of misinformation that has already been mistaken for fact.

It is important to look behind any communications strategy and dataset at who owns it and what interests they represent, to determine how to evaluate the messages it contains; particular care is needed when the information resource has been provided by an apparently disinterested party. When providing advice about communication in this area, remember that it should be undertaken in three phases:

- first, to raise target audience awareness of, and interest in, the drivers and processes that inform practise and progress with respect to humane animal research;
- then, to evaluate, plan, coordinate and publicise the exchange of information using a range of relevant initiatives to change attitudes, behaviours and outcomes; and
- finally, to effectively and objectively determine and describe the resulting impact and progress.

17.4 QUESTIONS UNRESOLVED

This chapter has not addressed three crucial components of any effective communications strategy:

- what is the baseline of knowledge and understanding before the communications strategy begins;
- how is the strategy being received; and
- what, ultimately was its impact?

All of these need to be addressed at the outset, and require time and money to address well.

Base-lining offers the opportunity to evaluate not just current attitudes, but how they are formed, and how they can be challenged and changed. Seeking feedback throughout the communications programme provides the opportunity for measuring outcomes and continuously improving the communications strategy. And evaluating outcomes is required to know that is has worked.

REFERENCES

Allen, E. S., J. M. Burke, M. E. Welch, and L. H. Rieseberg. 1999. How reliable is science information on the web? *Nature* 402:722.
Crespsi, I. 1980. Polls as journalism. *Public Opinion Quarterly* 44:462–76.
ESRC. 2008. *Working with the media—A best practice guide.* London: ESRC. Available from http://www.esrcsocietytoday.ac.uk/ESRCInfoCentre/Images/Media_Booklet_tcm6-26393.pdf
ESRC. 2009. Economic and social research centre, home page. [Cited December 21, 2009]. Available from http://www.esrc.ac.uk/ESRCInfoCentre/about/CI/CP/index.aspx
European Commission. 1986. *European directive 86/609/EC on protection of animals used for experimental and other scientific purposes.*
Fairclough, N. 2003. *Analysing discourse; Textual analysis for social research.* See Chapter 4. London: Routledge.
Lasswell, H. D. 1948. The structure and function of communication. In *Society, the communication of ideas,* ed. L. Bryson, 37. New York: Institute for Religious and Social Studies, Jewish Theological Seminary of America.
MORI. 2002. *Science and the media: Research study conducted for the science media centre,* 120. London: Mori.
Paulos, J. A. 1996. *A mathematician reads the newspaper,* 224. London: Penguin Books.
RCUK. 2009. Research councils UK, home page. [Cited December 21, 2009]. Available from http://www.rcuk.ac.uk/default.htm
Rhys Roberts, W. 1924. *Rhetorica, the works of aristotetle,* ed. W. D. Ross, 1358, Vol. XI. London: Oxford University Press.
Robert, K., and R. K. Yin. 2008. *Case study research: Design and Methods. Volume 5 of applied social research methods,* 240. Los Angeles, CA: Sage Publications Inc.
Russell, W. M. S., and R. L. Burch. 1959. *The principles of humane experimental technique.* Potters Bar, England: Special edition, Universities Federation for Animal Welfare.

Index

A

Acoustic environment
    in animal rooms, 46–48
    auditory sensitivity, 45
    physiology and behaviour, sound effects, 46
    vocalisation, 45–46
Additive transgenesis, 182–183
    mediated by ES cells, 187
    nuclear transfer (cloning), 187
    use of transposons, 187–188
    vector mediated, 185–187
Ad libitum feeding, 52
Administration methods, anaesthesia
    endotracheal intubation
        airway maintenance and protection, 318
        drugs, 318
    induction chambers
        plexiglas, 317–319
    mask delivery
        sedation, 318
        sedatives, 318
    open jar, 318
    vaporiser
        delivery systems, 317
        types, 317
Adult rats, 78
Agnotoxenic animals, 165
Alarm system, 22
Albino rats, hearing range, 45
Ammonia impact on animals, 42
Anaesthesia
    agents
        inhalation, 315–318
        injection, 318–322
    artificial ventilation, 306
    death, 306
    Doppler/ultrasound investigations, 306
    fluid balance
        overhydration and oedema, catheters, 306
        sterile physiological, 306
    genetic differences, 306
    invasive procedures
        catheter, 306
        oedema, 306
    monitoring during
        body temperature maintenance, 324–325
        cardiovascular function, 324
        depth, 323–324
        fluid and electrolyte balance maintenance, 325
        respiratory function, 324
    NMBAS, 322–323
    physiological functions, 305
    post-anaesthetic care
        fluid and nutritional support, 326
        warmth and comfort, 325
    pre-anaesthetic consideration
        acclimatisation, 314
        anticholinergic agents, 315
        drug dose, 315
        hypoglycaemia., 315
        physiological changes, 314
        sedative/analgesic agents, 315
        use, large species, 315
    respiration sensors and ECG, 305
    use, 314
    ventilation, assisted
        artificial, 323
        imaging, 323
        IPPV, 323
        tubes and intubation techniques, 323
    volatile, 306
Analgesia
    delivery methods, 329
    doses and drugs
        local anaesthetics agents, 328
        NSAIDS, 328
        opioids, 327–328
    drugs, pain relief, 326
    pain evaluation
        behaviour, 326–327
        physiological parameters, 326
    small laboratory animals, use, 328
    surgery, 326
    use, 314
Analysis, communication strategy
    achievement, 395
    internal and external tools, 394
    involvement persons and purpose
        information, 395–396
    requirement, 394–395
Analysis of variance (ANOVA), 132, 145–146
Animal models
    categories, 164
    clinical/field studies, 154
    components, 153
    criteria of good, 170
    defined, 153
        objective of study, 154
    discrimination, 153
    efficacy and, 153
    factors, selection of, 152–153
    for human diseases, 155
    investigation, 154
    for investigations of animal diseases, 163
    preclinical studies, 154
    selection, 152
        requirements and constraints, 170
        and source, 169–171
    specification for, 153
    steps in, 152

Animals
  based research, 120
  experiments, 108
  micro-flora, 166–167
  of non-laboratory species, 109–110
  as sample, 116
  use in
    ethical acceptability, 2
    European Directive 86/609/EEC on protection, 2
  use in scientific testing, 1
Animal suffering
  dystress
    eustress, 335
    occurrence, 335
  fear
    and anxiety, 336
    occurrence, 336
  fish, amphibians and reptiles, 334
  lasting harm, 336
  mental distress
    behavioural patterns, 336
    description, 336
    and physical/physiological, 336
  pain
    description, 335
    intensity and duration, 335
    protective function, 335
    thresholds, 335
  physiological and behavioural measures, 335
  severity, 334–335
Animal welfare
  animal motivating for needs, 82
    building behaviour, 83
    of captive animals, 86
    careful handling, 82
    chronic thwarting, 86
    contrafreeloading, 86
    dominance relationships, 84
    foraging, 85–86
    hiding, 83
    hormonal status, 86
    locomotory activity, 86
    motor activity, 86
    nesting, 83
    species-specific needs, 86
  assessment, 92
  barbering, 80
  and defensive strategies, 81
  depression, 80
  deprivation of food and, 77
  discomfort and, 78–79
  emotions and, 81
  environmental challenges, 77
  fasting for longer period, 78
  foraging activities and, 81
  freedom from
    discomfort, 77–79
    to express normal behaviour, 77, 81–82
    from fear and distress, 77
    Five Freedoms approach, 77
    hunger and thirst, 77–78
    pain/injury, 77, 79–81
  frustration, 80
  grooming behaviour, 78
  health problems, 78
  helplessness and, 80
  and inappropriate housing, 80
  maintaining full health, 77
  normal behaviour and, 81
  and physical injury, 79
  and potentially dangerous situation, 81
  reproductive condition, 78
  scientific results, 92
  and socially stressful environment, 80
  sub-clinical infections, 80
  territorial defence and, 79
  traditional approach, 76
Antibody-free animals, 167
Aseptic hysterectomy, 167–168
Atropine, 315
Auditory environment, 88

B

Background noise, 212
Bacterial artificial chromosomes (BACs), 184
Barriers; see also Laboratory animals
  bioexclusion, 23
  corridors, dual/single, 14
  experimentally infected animals, 24
  health monitoring, 23–24
  immunocompetent animals, 24
  immunocompromised animals, 24
Bedding materials, 54–56, 88
Behavioural abnormalities, 89
Behavioural syndromes, 80
Biocontainment
  health and safety
    groups, 24–25
  housing systems, 163
Bioexclusion, 23; see also Barriers
Bioluminescence imaging, 347
Biophotonic imaging, 347
Biosecurity, engineering control and
    maintenance, 57
  hazardous biological agents, 58
Bloodoxygen-level dependent (BOLD) signal, 292
Blood sampling
  accuracy and precision in mice
    ALT values, 269
    BASO/lobularity channel, 269
    glucose and insulin, 269–270
    methods use, 270
    RBC and platelet counts, 269
    WBC, 269
  accuracy and precision in rats
    haematological evaluation, 273
    haematological parameters, 272–273
    retro-orbital puncture technique, 273
    Sprague-Dawley rats, 272–273
    WBC counts, 273
  approaches to refinement in rats
    ACTH and corticosterone in, 271
    anaesthesia in, 270–271
    corticosterone concentration, 272
    isoflurane, 270
    radiotelemetry use, 270
    retro-orbital puncture, 270–272

Index

refinement aspects in mice, 266
   anaesthesia, 268
   corticosterone concentration in, 268
   plasma catecholamine, study, 268–269
   retro-orbital bleeding, 268
   tail-clip technique, 268
urine and faeces
   qualitative urinalysis, 273
   uncontaminated rodent urine samples, collection, 274
BMS, *see* Building management system (BMS)
BOLD signal, *see* Bloodoxygen-level dependent (BOLD) signal
Bordetella pertussis potency testing, 350
Brain vascularisation, 89
Breeding
  genotyping
   DNA markers, 209
   techniques, 209
  phenotyping
   protocols, 213–214
   strategy, 209–212
  records, 208–209
   of individual animals, 208
   lineage records, 208–209
  schemes, 208
   congenic breeding, 206–207
   harem mating, 207
   pair mating, 207
   trio mating, 207
Building management system (BMS)
  facility and equipment, maintenance programmes, 20
  room sensors, 20
  temperature and RH values, 20
Buprenorphine
  before K/M anaesthesia, 328
  postoperative pain, 328

C

Caenorhabditis elegans as animal model, 169
Cages, 31
  equipment, 88
  food within, 89
  furnishings, 88–89
  stereotypies, 92
  stocking density, 80
  structuring of, 88–89
Captive bolt shooting, 359
Carbon dioxide impact on animals, 42
Cerebral concussion, 358
Cervical dislocation, 358
CFD software programmes, *see* Computational fluid dynamics (CFD) software programmes
Chow diets, 172
Circadian rhythm of temperature (CRT), 238
Coisogenic strains, 158
Cold storage room, 22
Colony management programmes (CMPs), 208–209
Colony termination, 169
Combined models, 162–163
Commensal/resident micro-flora, 166
Committees, ethical evaluation
  animal technician, 122
  chairperson, 121
  ethicists, 122
  expert on replacements, 122
  lawyers, 122
  lay members, 122
  representatives of
   animal welfare organisations, 122
   patients organisations, 123
  scientists, 121
  social responsibilities of, 123–124
  specialists in animal welfare, 122
  statisticians, 122
  veterinarian, 121
Communication and animal experimentation, 22
  analysis
   and achievement, 395, 397–398
   information and opinions, 398–399
   internal and external tools, 394
   and outputs, 398–399
   research publicise organisation, 399–402
   stakeholders, 395–397, 399–402
   strategy required, 394–395
  case studies, 394
  elements, 393
  features, 393–394
  issues, 394
  objectives, 393
  two-way process, 392–393
Competence assessment
  animal care staff
   responsibilities, 387
  Article 7 of European Directive, 371
  assessment, 384–385
  Categories C and D, study, 385
  Category A persons
   duties, 371
   level 1, 371
   level 2, 371
   level 3, 371
   level 4, 371
  Category B persons
   Article 26 of Council of Europe Convention, 372
   FELASA scheme, 372
  Category C persons, 372
  Category D persons, 372
  course evaluation and learning outcome
   bias, 386
   evaluation forms, 386
   model, levels, 385
   questions, 386
   by students, 385
   Web-based questionnaires, 386
  European Science Foundation (ESF), 371
  formative/educative assessment, 385
  lifelong learning/CPD, 382–383
  managing
   education committee, 387–388
   group responsibility, 387
  multiple choice questioning (MCQ), 384
  practical, 384–385
  prerequisites, 384
  quality assurance and transferability
   Category B and Category C, 388
   FELASA, 388
   monitoring, 388

scientists, 370–371
summative assessment, 384
teaching methodology and learning
    style, 372–373
  didactic, 375–376
  ethics, 380–381
  feedback, 381–382
  group, 376–377
  practical skills, 378–379
  strategies, 374
  teacher, 373–374
training plan
  accountability and transparency, 386–387
  targets, 387
Complex burrows, 83
Computational fluid dynamics (CFD) software programmes, 16
Computed tomography (CT)
  and conventional X-radiography, 293
  X-ray transmission, 293
Conditional mutagenesis, 192–194
Congenic strains, 158
Consomic strains, 159
Construct design, 195
  of gene targeting constructs, 196–197
  of transgenic constructs, 195–196
Continuing professional development (CPD)
  community of care, 382
  description, 382
  internal training programmes, 383
  life-long learning, 382
  monitoring
    competence, 383
    credit-based approaches, 383
    FELASA, 383–384
    mandatory schemes, 383
    validation, 383
  programme development and delivery, 382–383
Corticotropin-releasing factor (CRF), 234
CPD, *see* Continuing professional development (CPD)
Cre-loxP construct, 193
Cre-recombinase, 192
CRF, *see* Corticotropin-releasing factor (CRF)
CRT, *see* Circadian rhythm of temperature (CRT)
Cryoprotective agents, 217, 220

D

Data record form, 214
Decapitation, 358–359
Dependent variables of experiment, 134–135
Design development of experiment
  considerations in, 132
  improvements, 132
    three Rs, 133
  independent and dependent variables, 134–135
  principles
    ExpU identifying, 134
    purpose of, 133
    types, 133–134
  requirements
    absence of bias, 135–136
    amenable to statistical analysis, 139–140
    power of, 136–138

range of applicability, 139
    simplicity, 139
Designing components in laboratory animals, 12
  animal holding rooms
    barrier/containment, 15
    electrical supply system, 15
    floor materials, 14
    HVAC system, 15
    light sources in, 49
    size of, 14
    types, 14
  architectural finishes and materials, 19
  barriers and elements, 13
  flow chart, 13
  flow cycles for, 14
  procedure rooms
    furniture and equipment, 16
    laboratories, 16
    necropsy room, 17
    surgical facilities, 16–17
  reception areas
    elevators, 14
    loading docks, 14
    non-barrier laboratories, 14
    offices, 14
    public corridors, 14
    supply and storage rooms, 14
    UV sterilising port, 14
  teaching and training rooms
    FELASA recommendations, 18
  transgenic core facilities
    biosafety risk assessment criteria, 17–18
    breeding programme, 17
    components, 18
    cryogenic storage, 18
    health and barrier standards, 17
  ventilation system
    cost of, 15
    health and safety norms, 15
    software programmes, 16
    type of, 15
Didactic teaching
  animal holding facilities/laboratory areas tours, 376
  demonstrations, 375
  facility visits, 376
  lectures
    handouts, lecture notes, 375
    planning, 375
    presentation, 375
    video presentations, 375
Digital image capture techniques, 293
Disposable cages, 35–36
Distribution systems
  drains, 19
  watering systems, 19–20
DNA markers, SNPs and SSLPs, 207
Domestication, 81
Dosing
  administration routes
    characteristics, 260
    dermal application, 265–266
    inhalation, 265
    intradermal (ID), 266
    intragastric gavage (IG), 261–263

Index

intramuscular (IM) injections, 264
intravenous (IV), 265
oral, 259–261
physiological effects, 259
subcutaneous (SC), 263
blood sampling
approaches, 270–272
methods in, 267
in mice, 269–270
in rats, 272–273
refinement aspects, 266, 268–269
urine and faeces, 273–274
Drugs
development process, 107
discovery, *in vitro* studies, 107
Dual-energy X-ray absorptiometry (DEXA)
bone mineral content (BMC), fat and lean tissues, 294
bone mineral density (BMD)
BMC and area, 295–296
measurements, *ex vivo*, 293
osteoporosis, 293
fan-beam, 294
rat whole body scanning, 294, 296
region of interest (ROI), 294
trabecular bone, 294

E

Education
animal care staff
responsibilities, 387
Article 7 of European Directive, 371
assessment, 384–385
Category A persons
duties, 371
level 1, 371
level 2, 371
level 3, 371
level 4, 371
Category B persons
Article 26 of Council of Europe Convention, 372
FELASA scheme, 372
Category C persons, 372
Category D persons, 372
course evaluation and learning outcome
bias, 386
evaluation forms, 386
model levels, 385
questions, 386
by students, 385
Web-based questionnaires, 386
European Science Foundation (ESF), 371
lifelong learning/CPD, 382–383
managing
education committee, 387–388
group responsibility, 387
quality assurance and transferability
Category B and Category C, 388
FELASA, 388
monitoring, 388
scientists, 370–371
teaching methodology and learning style, 372–373
didactic, 375–376
ethics, 380–381
feedback, 381–382
group, 376–377
individual, 377–378
practical skills, 378–379
strategies, 374
teacher, 373–374
training plan
accountability and transparency, 386–387
targets, 387
Embryo transfer procedures, 168, 219–220
EMPReSS protocols, 210, 214
Environment
changes, assessment of impact, 91
conditions, 171
constraints and animal welfare, 81
definition and control
acclimatising and quarantine period, 173–174
diet and nutrition, 171–172
environmental conditions, 171
transport conditions, 172–173
exploration of, 85
monitoring systems, 22
Environment refinement
and animal environment, 87
auditory, 88
components, 87
enriched conditions, 87
enrichment, 86
housing, 87
of physical, 87
cage structure and furnishings, 88–89
feeding, 89
physical exercise and exploration, 89–90
programmes, 91
of social, 90–91
stimulation, 92
validation of
behavioural and physiological parameters, 91
impact on scientific results, 92
welfare assessment, 92
Ethical evaluation of animal experiments
checklist
aspects of, 102, 106
composition and dynamics, 106
description and purpose of study, 104
experts and competent persons, 106
information about researcher and institute, 105
reduction, 105
refinement, 104–105
replacement, 104
retrospective information from similar studies, 105
collaborative action COST B 24, 102
committee composition
animal technicians, 122
chairperson, 120
ethicists, 122
experts on animal welfare, 122
experts on replacements to use of animals, 122
lawyers, 122
lay persons, 122
representatives of animal welfare organisations, 122
representatives of patient organisations, 123
scientists, 121

statisticians, 122
veterinarians, 121
communication with public, 120
conflict of interest, 118–119
considering animals with special needs, 108–109
documents, 125–126
ethical review body, 104
evaluation
  benefit, denying license, 107
  experiment/protocol level, 124
  lifetime experience, 125
  previous studies, 118
  procedure level, 124–125
  project level, 124
  proposals submitted for, 103
  of replacement methods, 107–108
genetically modified animals, 109–110
HEPs, responsibility and authority, 110–111
reduction, standardisation and importance, 117
repeated use, 117–118
scientific investigation, 103
scientific validity, 119
severity of procedures, 111
social responsibility of committee, 123–124
study design, 111
Ethical review bodies, 107, 116–119, 125
Europe
  animals use in scientific testing
    poll conducted, 1
  European Directive
    ethical evaluation of animal investigations, 102
  European Mouse Disease Clinic, 213–214
  European rabbits, olfactory stimuli role in, 44
  European Science Foundation (ESF), 371
  European Union Framework Programmes, 119
  European Union Mouse Research
    for Public Health and Industrial
      Applications, 213
EuroPhenome resource, 210
Euthanasia, 81
  animals/tissue sharing
    in vitro methods, 366
  carcase, disposal
    animal, classification, 366
    biohazardous, 366
  death confirmation
    methods, killing process, 366
  equipment
    gaseous, 364
    needles and syringes, 364
  European directive proposals, 366
  facilities
    environment, 364
    home cage method, 364
  handling
    animals, 364
  ill human patients, 356
  laboratory animals, 356
  method choice
    anxiolytic agents/light induction, 364
    CO_2, 364
  methods
    chemical, 359–362
    foetus and newborn animal, 363
    physical, 357–359
    unconscious animals, 362–363
  rodents and rabbits, 358
  study considerations and alternatives
    data validity, 365
    method selection, 365
    sperm motility, rats, 365
    structural/histopathological
      changes, 365
  training
    humane method, 363
    researchers, 363–364
  unnecessary, avoiding
    animal quality, 357
    breeding programmes, 356
    experimental design, 357
    genetic strains, 357
    male and female animals, 356–357
Exercise/behavioural enrichment, 89
Experimental induction
  advantages, 162
  categories, 162
  drawbacks, 162
Experimentally infected animals, 24
Experimental unit (ExpU), 134

F

F1 hybrids, 158, 160
Facility oversight in laboratory animals
  animal house director
    duties of, 11
    qualifications, 10–11
  qualifications of
    persons taking care, 12
    veterinarian responsible for, 11–12
  supervisor of technical staff, 12
Facility planning in laboratory animals
  capacity, 10
  defining activities
    containment, 9
    education and training, 10
    in-house breeding, 10
    regulatory control, 9
    sources of animals, 10
    species, 9
    transgenic animals, 10
    types of research activities, 9
  design team
    professional team, 9
    representatives of client, 9
  flow chart, 13
  strategic limits and questions
    flexibility and allowance, 10
Factorial design, 143–144
FCA, *see* Freund's complete adjuvant (FCA)
Federation of International Mouse Resources (FIMRe), 217
FELASA Education and Training Board, 370
FELASA working group, 102
FIA, *see* Freund's incomplete adjuvant (FIA)
Filter-top cage, 33–34
FIMRe, *see* Federation of International Mouse Resources
  (FIMRe)
Fire alarm systems, 22

Five Freedoms to laboratory animals, 77–82
Foetal/newborn animals, euthanasia method
  CO_2, 363
  decapitation, 363
  factors, 363
  liquid nitrogen, 363
  rodents and rabbits, 363
Follicle stimulating hormone (FSH), 218
Food; *see also* Macro-and micro-environment
  diets
    batch-analysis reports, 52
    commercial, 51
    experimental results and, 52
    formulations, 51
    supply and treatments, 51–52
Food restriction in laboratory
  ad libitum feeding, 241–245
  CD rat pups, weights, 244
  feeding station, 245
  group size calculations for, 246
  labour-intensive method, 244
  male B6C3F1 mice, response, 245
  practical aspects
    diet board, 247
    food pellets, 247
    restricted feeding, 246–247
  rabbits, 247
    ad libitum feeding, 248
  restricted feeding (RF), effect of, 242
  rodents mice
    control, 244–245
    health, 241–243
    power calculations, 243–244
    reduced mortality rate, 240–241
    sensitivity, 245–246
    uniformity, 243–244
  Sprague-Dawley rats
    body weights of, 243
    survival rates, 240
  survival rates, 241
Formal experimental designs, 139
  completely randomised design, 140
  factorial design, 143–144
  Latin square design, 142
  randomised block design, 140–141
  repeated measures/cross-over, 141
  sequential designs, 143
  split-plot designs, 142
Freedom of information (FOI), 120
Freezing method, 217
Freund's complete adjuvant (FCA), 275
Freund's incomplete adjuvant (FIA), 275
Friedman test, 146
Functional magnetic resonance imaging (fMRI)
  BOLD signal, 292
  brain/spinal cord, hemodynamic changes, 292
  pathological condition, brain function, 292

G

Gene of interest (GOI), 183
Gene targeting, 188–189
  gene transfer in ES cells, 188–189
  nuclear transfer, 190

Genetically engineered model organism, 181
  maintaining databases, 181
Genetically modified (GM), 180
  animal model, 180
    factors, 180
  animals, 109, 179, 210
    breeding performance of, 208
    housing and transport of, 215
    husbandry and care of, 215
    phenotyping, 212
  line, 207
    production of, 215
  models, 215
  mouse models, 180
Genetically modified organisms
    (GMOs), 180
Genetic drift, 216
Gene transfer techniques, 183–184
Gene trapping, 195
Genotyping techniques, 199–200
  DNA markers, 209
German mouse clinic (GMC), 214
GLP, *see* Good laboratory practise (GLP)
Glycopyrrolate, 315
GMC, *see* German mouse clinic (GMC)
GM strains cryopreservation, 206, 217, 221–222
  embryo and recovery
    embryo transfer, 219–220
    quick freezing method, 219
    slow/equilibrium method, 218
    vitrification method, 219
  spermatozoa and recovery, 220
    ICSI, 221
    oocyte and ovarian tissue
      cryopreservation, 221–222
    in vitro fertilisation (IVF), 221
Gnotoxenic animals, 164
Good laboratory practise (GLP), 9
GraphPadPRISM, 147
Grass cubes, 89

H

Handling of animals
  BN rats, 232
  cage complexity, 230–231
  communication
    stress response in, 230
  comparison of
    species differences, 233–234
  environment, familiarity
    HPA, 230
  frequent handling, 230
  habituation
    ACTH, 233
    nape of neck, 233
    prolactin secretion, 233
    scruffing, 233
  methods
    comparisons, 231–232
    experience, 231
    habituation, 233
    person, 231
    scruffing, 233

stocks/strains
  temperament and behaviour of, 232
  telemetry, use, 232
Hazardous biological agents, 58
Health
  additional SPF definitions, 166
  antibody-free animals, 167
  aseptic hysterectomy, 167–168
  aseptic hysterotomy/caesarean section, 168
  biocontainment and health monitoring, 168–169
  bioexclusion, 168
  care programme
    preventive veterinary medicine for recently arrived animals, 57
    veterinary surveillance, 57
  carriers, 167
  colony termination and recycling policy, 169
  definitions, 163
  embryo transfer, 168
  genetic issues, 168
  micro-flora, 166–167
  monitoring
    FELASA recommendations, 23
  pathogen free health standards, 165
  re-derivation techniques, 167
  standards
    gnotoxenic animals, 164
    holoxenic/conventional animals, 164
Healthy carrier animals, 167
Heating, ventilation and air conditioning (HVAC), 15
HEPs, see Humane endpoints (HEPs)
High-frequency ultrasound waves, imaging techniques
  Doppler ultrasonography
    blood flow, image, 300
    intravascular ultrasound (IVUS), 300
    pulsed wave, 300
  ultrasonography
    broadband ultrasound attenuation (BUA) and speed of sound (SOS), 300
    contact gel, 300
    echo display modes, 299
    fluid, 300
    high-frequency sound waves, 299
    image contrast, 300
High intra-cage ventilation, 79
Holoxenic/conventional animals, 164
Homologous models, 159
House director
  duties of, 11
  principal responsibilities, 11
  qualifications, 10–11
Housing conditions, 91
Human–animal interaction, 82
Human chorionic gonadotropin (hCG), 217
Human diseases models, 159
  combined models, 162–163
  experimental induction, 162
    categories of, 162
  inbred strains, used as disease models, 159
  monogenic models, 160–161
  polygenic models, 159–160
Humane endpoints (HEPs)
  conflict, 110
  decision-making and setting, 110
  Directive 86/609, 111
  early recognition of, 110
  ETS 123 guidelines, 111
  euthanise animal, 111
  harm-benefit evaluation, 111
  inspection and clinical evaluation, 110
  objective of, 110
  planning and application, 110–111
  score sheets, 110
  severity classification, 111
  time point and conditions, 110
  use, 110
  U.S. regulations/legislation, 111
Humane endpoints (HEP) suffering
  application, 343–344
    domain, 346
    legislative reasons, 344–345
    moral and social reasons, 344
    OECD and CCAC, 344
    scientific reasons, 345
    situations, 345–346
  attitude
    description, 348
    ineffective, expertise, 348–349
  observation and monitoring, 349
  reasons for
    Bordetella pertussis potency testing, 350
    death, 349
    economic, 349
    harmonisation lack, 349
  responsibilities
    actions, defining, 347
    animal welfare officer/veterinarian, 348
    ethical review body, 347–348
    laboratory animal technician/caretaker, 348
    pathologist, 348
    study director/principal investigator, 348
  setting
    clinical/behavioural signs, 346
    early clinical and non-clinical pathophysiological changes, 346–347
    molecular parameters, 347
    objective, 346
    pre-clinical changes, 347
  vaccine production, 349
  validation
    relevance, 349–350
    reliability, 350
    sensitivity, 350
HVAC, see Heating, ventilation and air conditioning (HVAC)
Hypnorm, 320
Hypothermia
  blood pressure and cardiac output fall, 325
  prevention, 325

I

ICSI, see Intracytoplasmic sperm injection (ICSI)
IETS, see International Embryo Transfer Society (IETS)
Illumination; see also Macro-and micro-environment
  physiological processes, 50
  vision and housing, 49–51
  visual perception, 48–49

Index

Imaging techniques
  animal
    detecting light, 300–301
    high-frequency ultrasound waves, 299–300
    infrared/near infrared radiation (NIR), 301–302
    ionising radiation, 292–299
    nuclear magnetic resonance, 288–292
  reduction and refinement
    and animal imaging, 302–303
    animal welfare and scientific quality, setting targets, 305–307
    MRI, 303
    quality testing and facility planning, 307–308
    severity classifications, 303–304
Immunisation
  administration routes
    characteristics, 260
    dermal application, 265–266
    inhalation, 265
    intradermal (ID), 266
    intragastric gavage (IG), 261–263
    intramuscular (IM) injections, 264
    intravenous (IV), 265
    oral, 259–261
    physiological effects, 259
    subcutaneous (SC), 263
  blood sampling
    accuracy and precision in mice, 269–270
    accuracy and precision in rats, 272–273
    methods in mice and rats, 267
    refinement aspects in mice, 266, 268–269
    refinement in rats, approaches, 270–272
    urine and faeces, 273–274
  for PABS
    adjuvant, 275–276
    animal, 276–277
    antigens, 274–275
    protocol, 277–279
    rabbits, 274
Immunocompetence, 90
  immunocompetent animals, 24
Immunocompromised animals, 24
  research models, 164
Implantable telemetry
  reduction possibilities, 238
  refinement possibilities
    anaesthesia, reversal, 237
    CRT, 238
    ketoprofen and carprofen in, 237
    MAP, 237–238
    oral ibuprofen, 237
    peri-and post-operative care, 237–238
    transmitter size, 236–237
  scientific integrity
    observer effect, 238–239
    telemetry, accuracy and precision, 239–240
Inbred models, 156
  characteristics, 157
  chromosomal deletion, 157
  derived models, 158–159
  genetic nomenclature, 157
  retinal degeneration, 157
  strains and sub-strains, 157
  used as disease models, 159
Inbred strains, 157
Independent variables of experiment, 134–135
Individually ventilated cages (IVCs), 23, 34–35, 39–40
  housing systems, 79, 215
Inducible transgenesis, 190–191
  tet-off system, 191
  tet-on inducible transgenesis, 191
Industrial applications
  management
    genome resource banking management, 222
    health monitoring, 215–216
    housing and transport, 215
    husbandry, animal care and welfare, 215
    preservation and recovery, 216–222
  mouse clinic concept, 214
  MuTrack system, 214
  PhenoSITE, 214
Information, communication
  authoritative and objective, 395–396
  communication and resource, 397
  facts and figures, 396
  identification and working, 395
  needs
    audience nature, 396
    public policy, 396
    stakeholders, 396
  opinions and outputs
    material and audiences, 399
    presentations, meetings, and discussion groups, 399
    professional advice, 399
    publications, meetings and peer-pressure, 399
    web-based tools, 398
    written material, 398
  persons responsible for providing
    interests and agenda, 397
    legal duty/moral obligation, 396
    sources, 397
  planning and
    deficiency model, 398
    evidence based, 397
    objectives, 398
    sources, 398
  third-party opinions and views, 396
  World Wide Web, 396
Infrared/near infrared radiation (NIR) imaging
  remote thermography and
    image acquisition, 302
    near infrared cameras, 302
    skin temperature, 302
    and MRI/PET, 302
Inhalational agents, euthanasia
  carbon dioxide
    anaesthesia-based techniques, 361
    dyspnoea and air hunger, 361
    flow rate, 360
    gradual-fill procedures, 360–361
    pre-fill chambers, 360
    tissue acidosis, 360
  carbon monoxide
    commercial sources, 361
    during unconsciousness, 361

inert gases
   and carbon dioxide, 362
   nitrogen and argon, 361
   volatile anaesthetic agents, 361–362
volatile anaesthetic
   alveolar concentration, 359
   calibrated vaporiser, 359
   desflurane, 360
   enflurane, 360
   halothane, 360
   isoflurane, 360
Inhalation anaesthesia
   administration methods, anaesthesia, *see* Administration methods
   advantage, 315
   benefits
      induction delivery and, 315
      oxygen carrier gas, hypoxia, 315
   injectable agents, 315, 317
   operator safety, 318
   physiological parameters, 317
   potency, volatile anaesthetic, 317
   volatile agents (*see* Volatile agents, anaesthesia)
Injection anaesthesia
   administration methods
      dose and route, 321
      intramuscular, 322
      intraperitoneal, 322
      intravenous (IV) injection, 321–322
      subcutaneous, 322
   barbiturates
      long-acting (*see* Thiobutabarbital)
      medium long-acting, 320
      ultra-short acting (*see* Thiopental)
   fentanyl/fluanisone/midazolam
      rabbits, 320
      rodents and rabbits, 320
   injectable anaesthetic agents, euthanasia
      barbiturates, 362
      dose, 362
      intracardiac, 362
      sodium pentobarbitone, 362
      T-61, 362
   ketamine/acepromazine, 319
   ketamine/medetomidine
      atipamezole, 319
      IM injection, rabbit, 319
      invasive procedures, 319
      mice and guinea pigs, 319
   ketamine/xylazine
      guinea pigs, 319
      sedative acepromazine, 319
      side effects, 319
   local anaesthetics
      advantages, 321
      benefit, 320
      overdoes effect, 321
   propofol
      description, 320
      hepatic metabolism, 320
Intellectual property (IP) rights, 120
Interfering agents, 167
Intermittent positive pressure ventilation (IPPV), 305
International Embryo Transfer Society (IETS), 217

InterPhenome project, 210
Intracytoplasmic sperm injection (ICSI), 221
Intradermal (ID) administration, 266
Intragastric gavage (IG)
   administration, 261
      flexible catheters, 263
      refine strategies, 262–263
      stress test, 262
      systolic and diastolic blood pressure, changes, 262
Intramuscular (IM) injections
   administration, 264
Intravascular ultrasound (IVUS)
   arterial atherosclerotic plaques/vascular stenosis, 300
   catheterisation, 300
   disadvantage, 300
Intravenous (IV) administration, 265
Invasive sampling methods, 209
In vitro fertilization (IVF), 221
   protocol, 217
Ionising radiation, imaging techniques; *see also* Computed tomography (CT)
   detection, sensitivity, specificity and accuracy, 299
   DEXA (*see* Dual-energy X-ray absorptiometry (DEXA))
   fluorodeoxyglucose (FDG), 298
   half-life definition, 298
   μCT (*see* Micro-computed tomography (μCT))
   medical diagnostics, 298
   neutrons and protons, unstable condition, 297–298
   nuclear medicine studies, 298–299
   positron and electron annihilate and gamma rays emmision, 299
   pQCT (*see* Peripheral quantitative computed tomography (pQCT))
   X-radiography, 293
Isolation defined, 13
Isolators, 23, 40–41
Isomorphic models, 159
"It's Always a Good Time To Learn," 382
IVCs, *see* Individually ventilated cages (IVCs)
IVF protocol, *see In vitro* fertilization (IVF)

J

Japanese mouse mutagenesis programme, 214

K

Ketamine
   description, 318
   side effects, 318
Kruskal–Wallis test, 146

L

Laboratory animals
   barriers
      bioexclusion, 23
      experimentally infected animals, 24
      health monitoring, 23–24
      immunocompetent animals, 24
      immunocompromised animals, 24
   biocontainment
      health and safety, 24–25

Index

communication systems, 22–23
culture of care
  animal suffering, 4
  authorisation, 4
  regulatory and legal requirements, 4
  scientific and welfare integrity, 4
designing, components in, 12
  animal holding rooms, 14–15
  animals, scientists and equipment, flow cycles for, 14
  architectural finishes and materials, 19
  barriers and elements, 13
  procedure rooms, 16–17
  reception areas, 14
  teaching and training rooms, 18–19
  transgenic core facilities, 17–18
  ventilation system, 15–16
distribution systems
  drains, 19
  watering systems, 19–20
ethical acceptability, 2
facilities, 86
  animal house director, 10–11
  qualifications of persons taking care, 12
  qualifications of veterinarian responsible for, 11–12
  supervisor of technical staff, 12
facility planning
  capacity, 10
  defining activities, 9–10
  design team, 9
  strategic limits and questions, 10
Five Freedoms for, 77–82
functional areas
  cold room, 22
  security, 22
  storage, 21
  washing area (WA), 21
  waste disposal, 22
housing, standardisation, 76
human and veterinary therapies, 2
immunological status of, 109
management systems, building and monitoring
  maintenance programmes, 20, 80
quarantine
  rules, 25
replacement of
  strategies for, 3
  using computer-generated models, 2
technical advances
  anaesthetic and analgesic techniques, 3
  biocontainment conditions, 3–4
  gamete analysis and manipulation, 4
  genetically altered, 4
  implantation techniques, 4
  monitoring and dosing techniques, 3–4
  physiology and anatomy, 3
  reproductive manipulation techniques, 4
  somatic and stem cell cloning, 4
  surgically prepared, 4
Lactated Ringer's solution, 325
Laparoscopy, 347
Latin square design, 142

Light detection based imaging techniques
  bioluminescence imaging
    limitation, 301
    luciferase gene, 301
    signal-to-noise ratio, 301
    and MRI/PET, 301
  fluorescence optical imaging
    genes coding, *Aequorea victoria*, 300–301
    laser light, 300–301
    markers, 301
    use, 301
    and MRI/PET, 301
Light intensity
  endocrine functions, 50
  and health of laboratory animals, 49
Local anaesthetic agents
  bupivacaine, 328
  lidocaine, 328
  toxicity, 328
Locomotor activity, 89
  in rodents, 90
Loss of consciousness (LOC), 357

M

Macro-and micro-environment
  acoustic environment
    in animal rooms, 46–48
    auditory sensitivity, 45
    physiology and behaviour, sound effects, 46
    vocalisation, 45–46
  bedding material, 54
    care staff and, 55–56
    effect on experimental results, 55
  food
    commercial diets, 51
    experimental results and, 52
    supply and treatments of diets, 51–52
  illumination
    vision and housing, 49–51
    visual perception, 48–49
  odour and housing, 44
  RH and gases, 41–42
  temperatures for
    behavioural thermoregulatory responses, 43
    C57BL mice testing, 43
    REM sleep time, 42
  water supply
    dispensing, 52
    water treatment, 53–54
Magnetic resonance imaging (MRI)
  anatomical abnormalities, 292
  for cardiac function, 292
  field strengths, 291
  internal organs and tumours
    grey and white matter, brain, 291–292
    pathological processes, brain, 292
  magnetic objects, 292
  proton spin, 288, 290–291
  radio frequency (RF) pulse, 290–291
  receiver coil systems, 291
  recovery/relax, proton, 290

in renal studies, 292
signal processing, 290
tissue relaxation times, 290
time, 290
tumour size and growth
determination, 292
Magnetic resonance spectroscopy (MRS), 292
Mammalian phenotype (MP) ontology, 210
Mean arterial blood pressure (MAP), 237
Metabolic cages, 35, 79
Mice
auditory sensitivity, 45
barbering in, 80
laboratory, 83
minimum enclosure dimensions and
space allowances, 32
olfaction in, 44
ultrasonic vocalisations, 45
Micro-computed tomography (μCT)
neoplasms, abnormal cells, 297
osseointegration quality, 297
skeletal research, 297
and X-ray techniques, 297
Microwave irradiation, 359
MINITAB software, 147
MMB, see Mutant mice behaviour
network (MMB)
Monitoring, anaesthesia
body temperature maintenance
heat loss, 324–325
hypothermia (see Hypothermia)
cardiovascular function
arterial blood pressure, 324
ECG, 324
heart rate, 324
depth
level, 324
light/medium/deep, 323–324
fluid and electrolyte balance
haemorrhage, 325
lactated Ringer's solution, 325
respiratory function
depression, 324
oxygenation, haemoglobin, 324
Monogenic models
Banbury Conference Guidelines, 161
$Cftr^{tm1Hgu}$ mutation, 161
Fec^{m1Pas} mutation, 161
$Lepr^{db}$ mutation, 161
ways in genetic variation, 160
Mouse clinic concept, 214
Mouse Genome Informatics (MGI), 180
Mouse Phenome Database (MPD), 210
Mouse phenome project, 213
Mouse phenotype database integration consortium
(MPDIC), 210
MPD, see Mouse Phenome Database (MPD)
MPDIC, see Mouse phenotype database integration
consortium (MPDIC)
MRI, see Magnetic resonance imaging (MRI)
Mutant mice behaviour network (MMB), 212
MuTrack system, 214
Mycoplasma pulmonis as animal
model, 168

N

National Institute of Health's neuromutagenesis
programme, 214
National security, 120
Natural behaviour
of laboratory rodents and rabbits, 81
Near infra-red spectroscopy (NIRS)
advantages and disadvantage, 302
equipment, 302
property based on, 301–302
regional perfusion, 302
tissue content calculation, 302
tissue oxygenation, 301
Necropsy room, 17; see also Procedure rooms
Nervous system, functional modifications, 86
Nesting material for rodents, 88
N-Ethyl-N-nitrosourea (ENU) mutagenesis, 189, 194
project, 210
Neurogenesis, 90
Neuromuscular blocking agents (NMBAS)
anaesthetic regimen, 322–323
animal paralyse and nociceptive
stimulation, 322
respiratory function, 322
use
laboratory animals, 323
paralysis, 322
Noise and animal facilities, 46–48
Non-animal methods in biological research, 108
Non-steroidal anti-inflammatory agents (NSAIDS)
behaviour and recovery, after surgery, 328
carprofen, 328
moderate post-operative pain, 328
Nuclear magnetic resonance, imaging techniques
fMRI (see Functional magnetic resonance imaging
(fMRI))
MRI (see Magnetic resonance imaging (MRI))
MRS (see Magnetic resonance
spectroscopy (MRS))
Nuclear transfer (NT), 187

O

Olfaction, 44
Open cage
type IL, IIL and IVS, 33
types I–IV, 32–33
Opioids
buprenorphine (see Buprenorphine)
ceiling effect, 327
side effects, 327
use, 327
Ordinance on Education and Training of Specialized
Staff for Animal Experiments of 12
October 1998, 383
Osteoporosis, 79
Outbred animals; see also Physiological
models
genotypic and phenotypic variation, 156
population and historical data, 156
from randomly mating population, 155
scientific issues, 156
Oxygen toxicity, 90

Index

P

Pain evaluation, analgesia
  behaviour
    assessment techniques, laboratory animals, 327
    body weight and food and water intake, 327
    NSAID, 327
    opioids, 327
    spontaneous, 326–327
  physiological parameters
    corticosterone measurements, 326
    plasma cortisol/cortisone, 326
    sympatho-adrenal and hypothalamic–pituitary–adrenal systems, 326

Pathogenic agents, 166–167
P1-derived artificial chromosomes (PACs), 184
Pen housing, 36, 79
Peripheral quantitative computed tomography (pQCT)
  anesthetised rat positioned for, 296
  cortical and trabecular bone analysis
    morphometry measurements, 296
    scout view, 295, 297
    strength strain index (SSI) and strength, 295–296
  gantry and resolution, 295
  translation–rotation principle, 295

Persons taking care of laboratory animals, 12
PhenoSITE, 214
Phenotyping breeding
  aim of, 211
  protocols, 213–214
    mouse phenome project, 213
  recording of, 208
  strategy
    animals to test, 212
    hierarchical approach, 211
    selection of tests, 211–212
  test, 212, 215

Physical activity, 90
Physical methods, euthanasia
  brain trauma, LOC, 357
  captive bolt shooting pistols, 359
  cerebral concussion, 358
  cervical dislocation
    mice and rats, 358
    objective, 358
    rabbits, 358
  decapitation, 358–359
  microwave irradiation
    brain metabolites, *in vivo*, 359
    domestic, 359
  whilst, 357

Physiological models, 154
  F1 hybrids
    features of, 158
  inbred models, 156–157
    derived models, 158–159
  outbred animals, 155–156

Physiological stress markers, 92
PMSG, *see* Pregnant mare serum gonadotropin (PMSG)

Polyclonal antibodies (PABS)
  adjuvant
    categories of, 276
    FIA and FCA, 275
    water-in-oil (W/O) emulsions, use, 275
  animal
    age, 277
    sex, 277
    species, 276–277
    strain-stock, 277
  antigens
    protein-carrier molecule, 274
    purity, 275
  immunisation protocol
    booster immunisation, 278–279
    dose of antigen, 278
    guidelines for, 279
    injection route, 277–278
    volume of injection, 278
  rabbits, 274

Polygenic models generated by mono-or bi-directional selection, 159–160
Positron emission tomography (PET)
  energy positron emitting tracers, 297
  isotopes, 298
  micro, 298
  PET/CT modality, 299
  PET–MRI solutions, 299

Post-anaesthetic care
  fluid and nutritional support
    hypoglycemia, 326
    ileus, rabbits, 326
  warmth and comfort
    rodents, 325
    sawdust/wood shaving bedding, 325

Pregnant mare serum gonadotropin (PMSG), 218
Preservation
  general aspects, 216–217
  GM strains, cryopreservation of, 217–222
  public and private repositories, 217
  and recovery, 216–222

Preventive veterinary medicine, 57
Primary enclosures
  cages and housing, 31
    mice, 32
  complex housing systems, 36
  disposable cages, 35–36
  filter-top cage, 33–34
  IVCs, 34–35
  metabolic cages, 35
  open cage
    type IL, IIL or IVS, 33
    types I–IV, 32–33
  pens, 36

Procedure rooms; *see also* Designing components in laboratory animals
  furniture and equipment, 16
  laboratories, 16
  necropsy room, 16–17
  surgical facilities, 16–17

Pronuclear microinjection, 183, 185
Pseudopregnancy, 219
Public concerns
  for animals use in scientific testing, 1–2

Q

Quality testing and planning, imaging
    equipment testing, living animals, 307
    genetically modified animals, functional genomics
        models, 307–308
        non-invasive imaging techniques, 307
    multi-modal, software packages, 307
    PET and SPEC, 307
    reduction, refinement and legitimising, animal
        experiment, 308
    shaving procedure, 307
Quarantine, 25
Quick freezing method, 219

R

Rabbits
    in cages, 78
    chinning activity, 44
    food restriction in laboratory, 247
        ad libitum feeding, 248
    hearing threshold, 45
    housing in small cages, 78
    minimum enclosure dimensions and space
        allowances, 33
    natural behaviour, 81
    Oryctolagus cuniculus
        visual acuity in, 48
    pen for, 89
Rack system, 36–37; *see also* Secondary enclosures
Randomised block design, 140–141
Rats
    auditory threshold, 45
    defensive behaviours, 84
    dominant male, 84
    environmental constraints, 84
    exploratory behaviour, 85
    housing system for, 79
    infanticidal attacks, 84
    minimum enclosure dimensions and space
        allowances, 32
    nocturnal feeders, 85
    parturition, 84
    social encounters, 84
    social isolation, 84
    socially cohesive behaviours, 84
    social organisational behaviour, 84
    vocalisation studies, 45
R Commander package, 147
Recycling policy, 169
Red blood cell (WBC), 269
Re-derivation techniques, 167
    re-derive colony, 168–169
        genetic issues, 168
Refinements of social environment, 90–91
Regulatory authorities, 108
Relative humidity (RH)
    breeding performance, 42
    role in animal thermoregulation, 41
Research animal legislation, 110
Research publicise/organisations
    establishments, public sector
        problem, 401

    lobbyists and pressure groups
        anti-vivisection, 401
        pro-animal use, 401–402
        websites, 402
    media, 401
    private sector funding and establishment
        pharmaceuticals and cosmetics, 400–401
        public image, 401
    professional bodies, 401
    public sector funding bodies
        three Rs and societal benefits, 400
        websites, 400
    regulators, animal use and data
        documents, 399–400
        EU institutions' official
            websites, 399
        national, 400
Restraint of animals
    AMP concentrations in, 234
    CRF, 234
    F344 rats, 234
    handling, comparison, 235–236
    HPA axis, 234
    methods, comparison
        habituation to, 234
        SHR rats, 234
        species differences, 233–234
    in mice and rats studies, 235
    WKY and SHR rats, 234
Reverse genetics, 181
Risk assessment in animal welfare, 82
Robot systems, 21
Rodent mice
    control, 244–245
    health, 241–243
    power calculations, 243–244
    reduced mortality rate, 240–241
    sensitivity, 245–246
    uniformity, 243–244
Rodent rats
    housing in small cages, 78
    natural behaviour, 81
    neutral thermal zones of, 43
    visual acuity in, 48
Rodent Refinement Working Party, 81
Rodent strains, nomenclature of, 181

S

Sampling
    administration routes
        characteristics, 260
        dermal application, 265–266
        inhalation, 265
        intradermal (ID), 266
        intragastric gavage (IG), 261–263
        intramuscular (IM) injections, 264
        intravenous (IV), 265
        oral, 259–261
        physiological effects, 259
        subcutaneous (SC), 263
    blood
        approaches, 270–272
        methods in, 267

Index

in mice, 269–270
in rats, 272–273
refinement, 266, 268–269
urine and faeces, 273–274
SAS software, 147
Scientific experimentation, hierarchical structure, 124
  experiment level, 124
  lifetime experience, 125
  procedure level, 124–125
  project level, 124
Scientific papers surveys, 132
Score sheet, suffering assessment
  constructing
    benefit, 340
    negative and positive signs, 340
    planning stage, 340
    signs, animals, 340
  description, 338
  mouse endotoxin experiment, 343
  use
    heterotopic kidney transplant, 340–341
    nothing abnormal diagnosed (NAD), 340
    observations and clinical signs recording, 342
    sign weighting, 340, 342
Secondary enclosures
  isolators, 40–41
  IVC systems, 39–40
  open rack, 36, 38
  special housing, 41
  temperatures for, 42–44
  ventilated cabinet
    HVAC facilities, 38
Sentient animals, for scientific research, 102
Sequential designs, 143
Severity classification imaging
  anaesthesia, 303
  animal health and procedure
    invasive, 304
    no invasive, 304
  for evaluation process, 111–115
  non healthy animals
    anaesthetic risk, 304
    brain tumours, 304
    invasive procedure, 304
SHIRPA test, 211
Silver-based photographic emulsions, 293
Simple sequence length polymorphisms (SSLPs), 207
Single nucleotide polymorphisms (SNPs), 207
Single photon emission computed tomography (SPECT)
  γ emitting isotopes, 297
  and SPECT, 298–299
Slow/equilibrium methodology, 217
Social behaviours, 83–85
  and animal welfare, 81
Social housing, 78, 90
Social responsibilities of ethical committee, 123–124
Solid floor cages, 33
Southern blotting, 209
Species selection, 108
  genetic characteristics, 108
  housing and husbandry standards, 109
  microbiological status, 109
  sex and age, 109
Specific and opportunistic pathogen free standards, 164
Specified pathogen-free (SPF) animals, 3, 53, 163
Speed congenics process, 207
SPF exclusion list, to incorporate opportunistic agents, 166
Split-plot designs, 142
S-PLUS software, 147
Spontaneous hypertensive rat (SHR), 160, 232, 234
SPSS software, 147
SSLPs, Simple sequence length polymorphisms (SSLPs)
Stakeholders communication
  establishments, public sector
    problem, 401
  lobbyists and pressure groups
    anti-vivisection, 401
    pro-animal use, 401–402
    websites, 402
  media, 401
  private sector funding and establishment
    pharmaceuticals and cosmetics, 400–401
    public image, 401
  professional bodies, 401
  public sector funding bodies
    three Rs and societal benefits, 400
    websites, 400
  regulators, animal use and data
    documents, 399–400
    EU institutions' official websites, 399
    national, 400
Standard animal cages, 79
Statistical analysis
  discrete data, 146
  examining data, 145
  interpreting and reporting results, 146–147
  non-parametric tests, 146
  parametric statistical analysis, 145–146
  statistical software, 147
Stereotypic behaviours, 92
Sterilisation systems, 163
Storage area, 21
Strain designations, 217
Stress-evoked behavioural response, 80
Subcutaneous (SC) administration, 263
Suffering minimisation
  adverse states recognising
    appearance, posture and behaviour, 337–338
    assessments, 338
    behavioural observation, 338
    biology and biography, experiment, 336
    husbandry and living conditions, 337
    mammals, 336
    physiological and behavioural changes, 337
    rodent and humans, 337
  animal
    dystress, 335
    fear, 336
    fish, amphibians and reptiles, 334
    lasting harm, 336
    mental distress, 336
    pain, 335
    physiological and behavioural measures, 335
    severity, 334–335

assessment
  advantage, 338
  environmental aspects and scientific
      procedures, 339
  score sheets, 338, 340–342
  usefulness, fidelity and acceptability in
      models, 339
HEPs, application and development, 343–344
  attitude and expertise, 348–349
  death, 349
  economics, 349
  harmonisation, 349
  legislative reasons, 344–345
  moral and social reasons, 344
  observation and monitoring, 349
  OECD and CCAC, 344
  responsibilities, 347–348
  scientific reasons, 345
  setting, 346–347
  situation, 345–346
  use, 343–344
  validation, 349–350
levels, not be exceeded
  apathy/lethargy, 342
  conditions, pain/distress, 342
  dehydration, 342
  eat/drink, 342
  grooming behaviour, 342
  neurological disorders, 342
  respiratory distress, 342
  weight loss, 342
Sunset and sunrise duration
  biological impact on animals, 50

T

Teaching methodology and learning style
  competence, 372
  didactic
    lectures (see Didactic teaching)
  feedback
    impact, 381–382
    staff and scientists, 382
  group learning
    advantages, 377
    assignments, 377
    Category C level, 376
    discussion, students, 376–377
    lecture disadvantages, 376
    results, 376
    seminars, 377
    small group teaching, 376
    tutorials, 377
  individual learning, 377
    coaching and mentoring, 378
    distance and computer-based, 378
    home study, 378
    management, effort and time, 378
    online, 378
    peer learning, 378
  influence, 373
  practical skills
    academic establishments, 379
    Category A persons, 379
    communication skills, 379
    guidelines, 379
    living animals, theory, 378
    objective, 378
    rooms and teachers, 379
    scientists, 379
    three Rs, 378–379
  principal approaches, 373
  strategies, 373
    awareness, team meetings and job
        shadowing, 374
    training styles, 374
  students, 373
  teacher
    behaviours, common, 373–374
    laboratory animal science, 374
    role, 373
    students, 374
    training, 374
  teaching ethics
    advantage, 381
    interactive group work, 380
    laboratory animal use, 380
    pluralist approach, 381
    students, 381
    three Rs principle, 381
    utilitarianism and animal rights, 381
  training
    description, 372
    trainer, 373
  visual tools, 373
Telemetry
  accuracy and precision
    EMG signals, 240
    recording of pressure, 239–240
    SBP and DBP with, 239
  implantable
    refinement possibilities, 236–238
    scientific integrity, 238–240
    use in handling of animals, 232
Tennessee mouse genome consortium, 214
Thermoneutral zone (TNZ), 42
Thermoregulation, 83
Thiobutabarbital, 320
Thiopental, 320
Transgenesis, 181
  identification, 199–200
  production efficiency, 198
Transgenic core facilities; see also Designing
        components in laboratory animals
  biosafety risk assessment criteria, 17–18
  breeding programme, 17
  components, 18
  cryogenic storage, 18
Transit micro-flora, 166
Transport of animals
  conditions, 172–173
  type of, 56–57
t-test, 146

U

UK Animal Procedures Committee (2003), 107
UK Farm Animal Welfare Council, 77

Index

UK Institute of Animal Technology, 379
Ultrasonic vocalisations, 45
Unconscious animals, euthanasia method
　exsanguinations, 362–363
　rapid freezing, 362

V

Ventilated cabinet, 38
Ventilation rate, 79
Ventilation system, 15–16
Vertebrate animals, protection and, 78
Veterinary care in laboratory animal units, 11
　diagnosis and medical treatment, 12
　qualifications, 12
Veterinary surveillance, 57
Vitrification method, 219
Vocalisation, 45
Volatile agents, anaesthesia
　halothane and desflurane, 317
　isoflurane
　　induction and recovery, 317
　　MAC, 317
　sevoflurane
　　induction and recovery, 317
　　MAC, 317

W

Washing area (WA)
　robot systems, 21
Waste disposal programme, 22
Water supply; *see also* Macro-and micro-environment
　dispensing, 52
　water treatment
　　acidification, 53
　　chlorination, 53
　　demineralisation, 53
　　desalination, 53
　　experimental results, effect on, 53–54
　　heat sterilisation, 53
　　ozone treatment, 53
　　sterile filtration, 53
　　ultraviolet disinfection, 53
Well-designed experiments, 133
Whisker trimming, 80
White blood cell (WBC), 269, 273
Wild rabbits, 84
Wild rats, 84
Wild-type (WT) female animal, 206
Wire-grid cages, 33
World Organisation for Animal Health, 217

X

X-radiography
　principle, 293
　tissue densities, 293

Y

Yeast artificial chromosomes (YACs), 184

Z

Zoonotic agents, 167